MULTIBAY FRAMES

ADOLF KLEINLOGEL
and *ARTHUR HASELBACH*

MULTIBAY FRAMES

Ready-to-use formulas for single- and two story continuous frames comprising any number of bays, elastically restrained at the supports, and with or without sidesway

90 frame shapes with 192 general and 103 special loading conditions, two worked numerical examples, and 450 illustrations in all

Translated from *Mehrfeldrahmen*, the 7th revised and considerably enlarged edition of KLEINLOGEL'S *Mehrstielige Rahmen*
by **C. V. AMERONGEN**

FREDERICK UNGAR PUBLISHING CO.
NEW YORK

The original German edition appeared under the title
MEHRFELDRAHMEN
published by Wilhelm Ernst & Sohn, Berlin

Copyright © 1963 by Frederick Ungar Publishing Co., Inc.

Printed in the United States of America

Library of Congress Catalog Card No. 63-10863

Preface

The main section of the seventh edition of Volume I contains formulas for 50 Frame Shapes with laterally restrained joints (no sidesway) and for 40 Frame Shapes with laterally unrestrained joints (with sidesway). These 90 different Frame Shapes in all — as compared with only 40 in earlier editions — are subdivided into 70 single-story and 20 two-story frames having from one to eight bays (see the key diagrams in the table of contents). The formulas for the frames with three and more bays are so constructed as to be readily extendable to similar multibay frames having any number of bays. In addition, all supports are treated as being "elastically restrained," i.e., capable of some degree of elastic rotation. If the particular coefficient ϵ associated with each external support of a frame member is equated to zero, this limiting value will correspond to a "pinned" (i.e., freely rotating) bearing. Similarly, the value $\epsilon = \frac{1}{2}$ corresponds to the other extreme case of "complete restraint" (i.e., rigid fixity). Appropriate intermediate values of ϵ can be estimated or calculated or may, alternatively, be available as predetermined data of the problem.

In the earlier editions the formulas were given as independent self-contained expressions. In the present edition, however, this practice has been abandoned, since it yields unwieldy results for frames which are statically indeterminate to the fourth or a higher degree. Instead, the method of solution by means of so-called reduction sequences — already introduced in the seventh edition of the book entitled *Durchlaufträger* ("Continuous Beams") — has been systematically extended to laterally restrained continuous frames. Whereas in *Durchlaufträger* this method of solution was confined to the application of the so-called Method of Forces (here referred to as Method I), it has in the present volume been made to include the application of the alternative method called the Deformation Method (Method II). The user can decide for himself whether to use Method I or Method II in any particular case. In general, with Method I the computation of the coefficients is somewhat more time-consuming; but, on the other hand, the series of end moments of the beam spans is directly obtained. With Method II the computation of the coefficients is shorter, but is associated with the drawback that the moments are only indirectly obtained.

Apart from the theoretical merits or demerits of the well-known conventional assumptions commonly adopted in engineering structural analysis, all the formulas contained in the present volume are mathematically "exact." In addition, the rigorously prescribed procedure of solution by means of reduction sequences is only to a relatively slight extent sensitive to numerical errors, so that such errors, should they occur, will rapidly die out. Also, forward and backward reduction automatically provide a mutual check. On the whole, the computation procedures indicated for each Frame Shape are much more agreeable to work with than were the earlier "self-contained" formulas. For example, whereas the coefficients in previous editions of this book were often inconvenient and obscure, each set of coefficients now has its own easily remembered and directly appreciated significance in the statical and elastic sense.

The method of solution by means of reduction sequences adopted in this volume is, however, admissible only for sets of trinomial elastic equations — such as can be established for laterally restrained "open" systems of members (e.g., for Frames Shapes 1 to 50), both by Method I and Method II. Direct elastic equations for laterally unrestrained systems (e.g., the Frame Shapes designated in this book by the subscript "v") have, as a rule, more than three terms. In order, nevertheless, to retain the advantages associated with the use of trinomial equations, in the case of laterally unrestrained frames the "beam" (i.e., the continuous horizontal member connecting the columns) is, as a first stage of the calculation procedure, assumed to be "laterally restrained" in the sense that horizontal movement (sidesway) is prevented. The "restraining force" (unbalanced horizontal shear) associated with this condition is then calculated. Next, a horizontal concentrated load is applied to the beam (now assumed to be "laterally unrestrained"), and the displacement effect of this load is taken into account by means of an additional elastic equation. Superposition of the two stages of the calculation procedure — namely, the laterally restrained and the laterally unrestrained condition — finally yields the results for the actual laterally unrestrained frame.

For the analytical treatment of multibay frames, the "reduction procedure on a conventional basis," as employed here, is probably the most suitable method of calculation. This assertion is based upon some twenty years' practical experience in the use of this method and upon the numerous comparative computations which have been performed in connection therewith by the authors of this book. This reduction procedure is in itself merely the rational solution of a set comprising any number of trinomial equations by means of the well-known Gaussian algorithm (see, for example, the standard works of Müller-Breslau, Pasternak, Domke, Beyer and others). According to this method, the solution can be performed in the form of easy-to-remember reduction sequences and recursion formulas. The reduction factors appear as moment carry-over factors in Method I and as rotation carry-over factors in Method II. The "definitive solution scheme for the multibay frame," which is really very simply and quickly determined by the outlined procedure and which is characterized by the so-called "conjugate matrix of the influence coefficients," will then not only be valid for any external loading applied to the system, but also for any internal loading such as temperature variations and displacements of supports. In addition, the definitive solution scheme also serves as the starting point for taking account of the displacements of joints in the treatment of laterally unrestrained multibay frames having the same basic shape. In the present volume, however, the definitive solution scheme for Method I has, for practical purposes, been split up into three Main Loading Conditions, namely, arbitrary vertical loading on the beam, arbitrary horizontal loading on the columns, and loading of the joints by external rotational moments. On the other hand, with Method II such a subdivision of the loading into different types does not yield any simplification.

The so-called "moment distribution methods" of structural analysis (Professor Hardy Cross's method, iteration methods, etc.), on which a considerable number of publications have appeared in the literature, undoubtedly offer great advantages when dealing with systems embodying a number of closed frames. On the other hand, for multibay rigid frames of the kind with which the present volume is concerned, the superiority of such moment distribution methods is more apparent than real. If only one loading condition had to be considered — permanent load, say — a single iteration procedure would probably be simpler and shorter to perform. But if two or more loading conditions have to be considered, as is usually the case in actual practice (e.g., loads on alternate bays), then the advantage is definitely on the side of the computation rules given in this book with a view to obtaining universally applicable solution schemes in the form of "general loading conditions."

In the Introduction are given all the important data relating to the arrangement and subdivision of the book, symbols and definitions, sign conventions, and design postulates. In addition to the pairs of "load terms" $\mathfrak{L}$ and $\mathfrak{R}$ (actual load terms), $\mathfrak{S}_r$ and $\mathfrak{S}_l$ (statical moments of the span load resultants) and $\mathfrak{M}_l$ and $\mathfrak{M}_r$ (fixed-end moments), two further pairs of load terms — referred to as "restraining moments" — have been introduced, namely, $\mathfrak{E}_l$ and $\mathfrak{E}_r$ for elastically restrained members without sidesway, and $\mathfrak{F}_l$ and $\mathfrak{F}_r$ for elastically restrained members with sidesway.

All the individual formulas, formula sequences and formula matrices given in the main section of the book, as well as the computation schemes for the laterally unrestrained frames (with subscript "v" denoting sidesway), are so clear that they do not require any further clarification. Nevertheless, the authors have considered it useful to add an Appendix giving some explanatory notes and derivations. More particularly, it appeared advisable to elucidate the concepts of "elastic restraint," "flexibility" and "stiffness" from first principles. In addition, the many possible applications of the formulas contained in the main section are examined and discussed — e.g., with reference to oblique members, oblique loads, temperature variations, and displacements of joints. Also, the Appendix deals at some length with the problem of systematically constructing and establishing influence line equations. Finally, two rigorously worked numerical examples are given which illustrate the different procedures of treatment of a problem in accordance with Method I and Method II.

TABLE OF CONTENTS

INTRODUCTION
General information and definitions

	page
1. Arrangement and subdivision of this book	XVII
2. Symbols and definitions	XX
3. Sign conventions	XXIX
4. Design postulates	XXXI

MAIN SECTION
Formulas for 90 "Frame Shapes" in all [1]

Note: The formulas for all "Frame Shapes" provided with an asterisk (*) can readily be extended to similar continuous frames with any number of bays.

*	Frame Shape 1 — Page 1 to 8
*	Frame Shape 1v — Page 9 and 10
	Frame Shape 2 — Page 11 to 15
	Frame Shape 2v — Page 15
*	Frame Shape 3 — Page 16 to 23
*	Frame Shape 3v — Page 24 and 25
	Frame Shape 4 — Page 26
	Frame Shape 4v — Page 26
*	Frame Shape 5 — Page 27 to 41

[1] The meaning of the key diagrams in the Table of Contents is explained on page XVI.

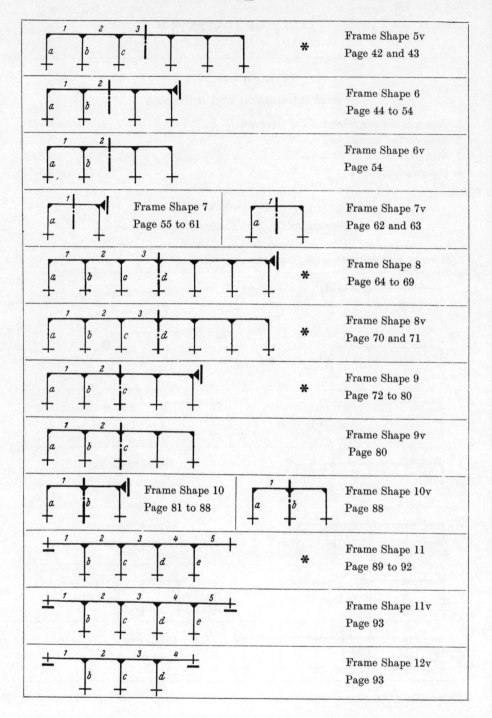

— XI —

(frame with points 1,2,3,4 and supports b,c,d)	*	Frame Shape 12 Page 94 to 98
(frame with points 1,2,3 and supports b,c)	*	Frame Shape 13 Page 99 to 103
(frame with points 1,2,3 and supports b,c)		Frame Shape 13v Page 104
(frame with points 1,2 and support b) — Frame Shape 14, Page 104 to 107	(frame with points 1,2 and support b)	Frame Shape 14v Page 107 to 111
(frame with points 1,2,3,4 and supports b,c,d,e)	*	Frame Shape 15 Page 112 to 118
(frame with points 1,2,3,4 and supports b,c,d,e)		Frame Shape 15v Page 118
(frame with points 1,2,3,4 and supports b,c,d)	*	Frame Shape 16 Page 119 to 124
(frame with points 1,2,3,4 and supports b,c,d)		Frame Shape 16v Page 124
(frame with points 1,2,3 and supports b,c,d)	*	Frame Shape 17 Page 125 to 134
(frame with points 1,2,3 and supports b,c,d)		Frame Shape 17v Page 134
(frame with points 1,2,3 and supports b,c)	*	Frame Shape 18 Page 135 to 144
(frame with points 1,2,3 and supports b,c)		Frame Shape 18v Page 144

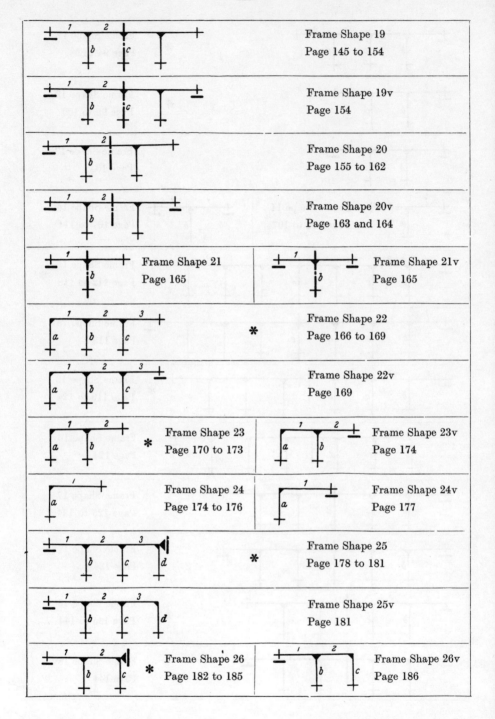

— XIII —

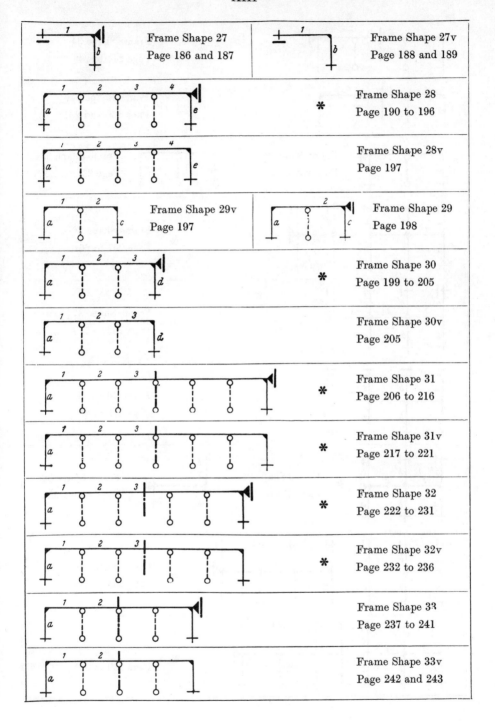

— XIV —

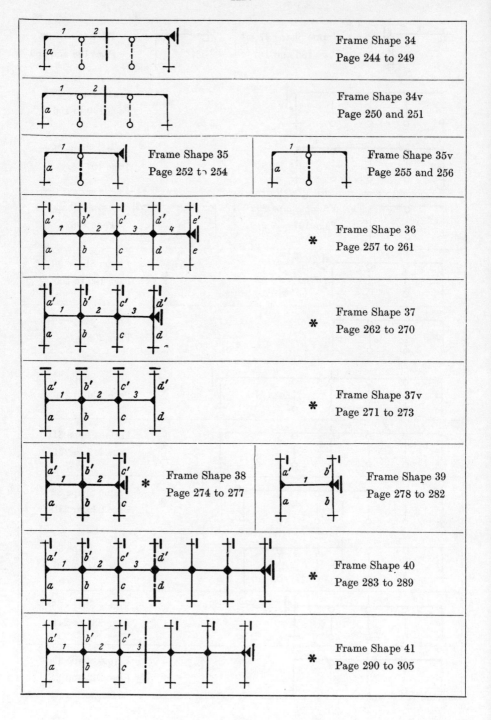

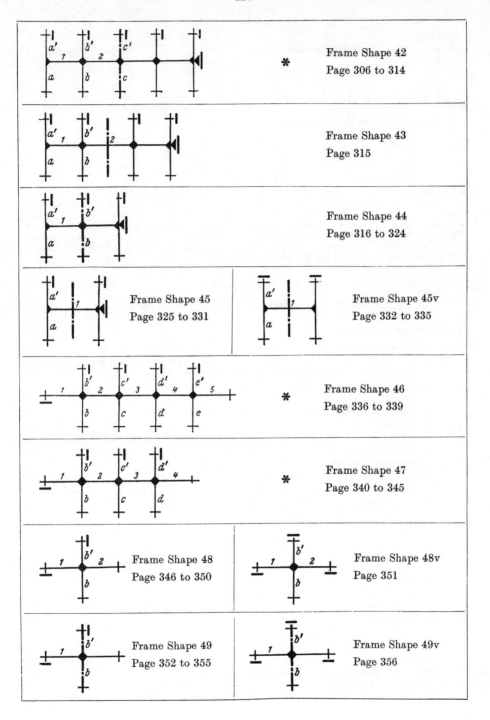

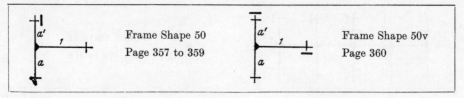

Frame Shape 50
Page 357 to 359

Frame Shape 50v
Page 360

Explanation of the key diagrams:

a) Rigid corner of frame – b) Immovable, elastically restrained beam or column support (comprising the limiting cases of "complete restraint" and "pinned bearing") – c) and d) Elastically restrained support, as before, but permitting horizontal and vertical movement respectively – e) Elastically restrained column top, capable of horizontal (and, in general, also vertical) movement – f) Vertical sliding bearing – g) Rocker column (has the same function as a horizontally sliding bearing) – h) Symmetry axis of the frame – i) Numbers denoting the columns and the bays; also used to indicate members of different length and different moment of inertia.

APPENDIX
Explanatory notes, derivations and numerical examples

	page
A. General considerations	361
B. Straight members and joints of frames	363
1. Frame member, with constant moment of inertia, subjected to arbitrary load	363
a) The general slope deflection equation and special cases	363
b) Elastically restrained end conditions, the spring constant and its practical substitute – the moment carry-over factor	369
c) Frame member with elastic end restraints and side-sway	373
d) Flexibility and rigidity	375
e) Flexibility and rigidity in the case of a member with one end free to undergo lateral displacement	381
2. Joints in rigid frames	385
a) The concept of "joint stiffness"	385
b) The concept of "joint flexibility"	389
c) The laterally restrained four-member joint	390
C. The laterally restrained multibay rigid framework or continuous beam on elastically restrained supports	410
1. Analysis by the Method of Forces (Method I)	410
2. Analysis by the Deformation Method (Method II)	417
3. Validity of the formulas for frames with oblique members	422
4. Internal loads	424
5. Influence lines and influence line equations	426
6. Numerical example 1	442
D. The laterally unrestrained multibay framework	453
1. Analysis by Method I and Method II	453
2. Validity restrictions for oblique members	458
3. Internal loads	459
4. Influence lines and influence line equations	459
5. Numerical example 2	461

Introduction

General Information and Definitions

1. Arrangement and Subdivision of this Book

The present compilation *Multi-Bay Frames*, Volume I, which is a translation of the German *Mehrfeldrahmen*, is a self-contained book in so far as the ready-to-use formulas it presents are concerned. Apart from applied moments and horizontal point loads, however, the loads to which the formulas relate are arbitrary loads acting on the members of the frames. Hence it will also be necessary to refer to the auxiliary book entitled *Belastungsglieder*.[1] This auxiliary book contains all the necessary formulas, as well as tables of numerical data, for 135 special Loading Conditions. Furthermore, in the present volume the formulas presented for the individual cases only go as far as giving all the final end moments of the members, whereas the formulas for determining horizontal reactions at supports have been grouped together and relate to families of analogous types of frames. Formulas for determining vertical reactions have—with few exceptions—been omitted altogether.

The main part of this book contains, in the first place, formulas for 50 'laterally restrained frame shapes' (i.e., types of frames in which 'side-sway' —lateral displacement of joints and supports—is prevented) comprising 35 single-story and 15 two-story frames. In general, each 'laterally restrained frame shape' is accompanied by its 'laterally unrestrained' counterpart, i.e., a structure of the same basic shape but capable of 'side-sway' in the sense that the continuous horizontal member or, where applicable, the upper story of the structure can undergo lateral displacement under the action of the loads. The frame shapes of the last-mentioned kind are referred to by the same number but have, in addition, the subscript 'v'. In dealing with these laterally unrestrained frames it has been considered sufficient merely to give five characteristic computation schemes. Only in the case of 15 single-column and symmetrical two-column laterally unrestrained frame shapes have complete sets of formulas been given.

Generally speaking, each 'frame shape' constitutes a self-contained chapter. At the same time, however, cross-references linking related or similar frame shapes are given in the form of notes in the text and footnotes.

---

[1] *Belastungsglieder*. Statische und elastische Werte für den einfachen und eingespannten Balken als Element von Stabwerken. By Prof. Dr.-Ing. A. Kleinlogel and Bauing. A. Haselbach. 8th completely revised and considerably enlarged edition. Wilhelm Ernst & Sohn, Berlin, 1956. A translation entitled "Beam Formulas" has been published by Frederick Ungar Publishing Co. New York.

The object is to urge the reader to make a closer study of the systematic construction of the formulas, so as to enable him to write down for himself all the necessary formulas, equations, etc., for types of frame not directly given in the book.

The formulas presented here are based on two different methods of structural analysis, and in each case the formulas derived by both methods are given. These methods are referred to as the Method of Forces (Method I) and the Deformation Method (Method II). The former is based on the Maxwell-Mohr energy equation (principle of virtual displacements), whereas the starting point for the latter method is provided by the somewhat modified slope deflection equations of Guldan[1] in conjunction with Ostenfeld's method.

The 50 frame shapes dealt with in this book can be divided into the following seven groups (apart from the general main division into single-story and two-story frames):

Group 1 — Frame Shapes 1–10: with initial column and end column
Group 2 — Frame Shapes 11–21: with initial beam and end beam
Group 3 — Frame Shapes 22–27: with initial column and end beam, or vice versa
Group 4 — Frame Shapes 28–35: with rocker type intermediate columns
Group 5 — Frame Shapes 36–45: with initial column line and end column line
Group 6 — Frame Shapes 46–49: with initial beam and end beam
Group 7 — Frame Shape 50: with column line and one beam span

With the exception of Groups 3 and 7, which comprise basically asymmetrical shapes, each group in turn contains three sub-groups, namely: asymmetrical shapes, symmetrical shapes with odd numbers of bays, and symmetrical shapes with even numbers of bays. The subdivision of the symmetrical shapes into those with odd and those with even numbers of bays (or spans) is of importance in connection with the use of the 'load transposition method' (Belastungs-Umordnungs-Verfahren).

As a rule, in each sub-group at least one frame shape is treated both by Method I and Method II, while the other frame shapes are treated either by one or the other of these methods. Owing to the characteristic construction of the formulas pertaining to each group or sub-group, nearly all the sets of formulas actually given in the book can readily be rewritten to suit any other frame belonging to the same group but having more or fewer bays. The latter alternative is indicated by dotted lines in the case of some frame shapes.

For reasons of expediency the 'definitive solution scheme for general and arbitrary frame loading' has, for treatment by Method I, as a rule been split

---

[1] *Rahmentragwerke und Durchlaufträger.* By Prof. Dr.-Ing. habil. R. Guldan. Springer-Verlag, Vienna, 1952.

up into three main loading conditions, namely: 'Loading Condition 1: all the beam spans carrying arbitrary vertical loading', 'Loading Condition 2: all the columns carrying arbitrary horizontal loading', and 'Loading Condition 3: external rotational moments applied at all the joints of the horizontal member.' On the other hand, for treatment by Method II a subdivision of this kind would not produce any simplification of the procedure; hence in this case the 'definitive solution scheme' is designated as 'Loading Condition 4: overall loading condition (superposition of Loading Conditions 1, 2 and 3)'. For laterally unrestrained multi-bay rigid frames there is, in addition, the characteristic case constituted by 'Loading Condition 5: horizontal concentrated load W acting at level of horizontal member'.

In the case of symmetrical frames there are, in addition to Loading Conditions 1, 2, 3 and 4—which may be arbitrarily asymmetrical in character—two further loading conditions that must be taken into consideration, namely (in relation to the axis of symmetry of the framework): 'symmetrical load arrangements' (Loading Conditions 1a, 2a, 3a and 4a) and 'antisymmetrical load arrangements' (Loading Conditions 1b, 2b, 3b and 4b). In the case of symmetrical laterally unrestrained frames, Loading Condition 5 is by nature always antisymmetrical. In the diagrams representing the symmetrical Loading Conditions 1a, 2a, 3a and 4a the bearing preventing horizontal displacement of the horizontal member (i.e., the 'restraining bearing' which prevents side-sway) has intentionally been omitted, in order to indicate that these cases are also applicable to the corresponding laterally unrestrained frame shapes (designated by the subscript 'v').

In principle, any symmetrical frame shape can also be analysed with the aid of the formulas given for the asymmetrical frames of the same general type. In the symmetrical case, however, the reduction processes in both directions will be the same, and the relevant main matrix of the influence coefficients will become bisymmetrical. This method of calculation gives the desired result almost as quickly as the 'load transposition method' and is always clear and convenient. However, if the loading cases to be dealt with are all symmetrical or antisymmetrical, it will obviously be simpler to use the given formulas directly applicable to these cases.

The user of this book is, of course, at liberty to treat symmetrical frames by means of the 'load transposition method'. It can here be assumed that the reader is familiar with this method (for example, see also the present authors' book *Durchlaufträger*, Volume II, Section N, Par. 2, pages 381–92). In using the method, special attention must be paid to the load transformations of the centre span in frames with odd numbers of bays (spans), or of the centre column in the case of frames with even numbers of bays (spans). It should furthermore be noted that with the 'load transposition method' the calculation should be confined only to the left-hand half of the frame; all the load terms are always referred to this half of the structure. The affinity between the coefficients for asymmetry, on the one hand, and the coefficients of symmetry and antisymmetry, on the other hand, is given due emphasis, at the relevant place, for each symmetrical frame shape dealt with in this book.

2. Symbols and Definitions

a) Preliminary note

The following is a summary of all the signs, symbols and concepts that occur in this book. Despite the apparent profusion of the signs and symbols employed, and of the—in some cases new—concepts presented, they are quite easy to remember, thanks to the logical system adopted. It has not always been possible to avoid using one and the same letter to denote two different things. This is unlikely to cause confusion, however, because in such cases the letter is used in entirely different contexts and represents entirely different quantities and concepts.

For the purpose of indicating the unit of a quantity, forces are expressed in pounds (lb) and lengths in feet (ft).

b) Ends of members, lengths of members, and subscripts

Each of the fifty 'frame shapes' is preceded by two key diagrams. The first of these represents the statical system of the structure, together with the relevant dimensions and symbols. These have the following significance:

$A, B, C \ldots N \ldots$ Points of support (bearings) and joints of the continuous horizontal member (or, more briefly, 'the beam') of a multi-bay framework.

$a, b, c \ldots n \ldots$ Feet (or bases) of the columns of a single-story framework, or of the lower columns of a two-story framework.

$a', b', c' \ldots n' \ldots$ Heads (or tops) of the upper columns of a two-story framework.

$(1), (2), (3) \ldots$ Numbers denoting the individual spans of the continuous horizontal member.

(v) Symbol denoting any particular beam span[1] or, alternatively, any member of the framework.

$(a), (b), (c) \ldots (n) \ldots$ Symbols denoting the columns of a single-story framework, or the lower columns of a two-story framework.

$(a'), (b'), (c') \ldots (n') \ldots$ Symbols denoting the upper columns of a two-story framework.

---

[1] (Translator's note). By 'beam span' is meant the portion of 'the beam' (i.e., the continuous horizontal member) spanning between two successive columns.

Note: The parentheses ( ) enclosing the symbols serve to give prominence to them and are used in diagrams only.

1, 2, 3, 4 ... i ... k ... Sequence numbers of the ends of the beam spans (required only in connection with Method I as the sequence numbers of the statically redundant moments X).

Note: In the diagrams these sequence numbers i, k are marked by ‖ placed close beside the beam joints concerned. Sequence numbers not directly required are enclosed in square brackets [ ] in the diagrams.

All sets of numbers or symbols run from left to right. Span 1 is always characterised by the points of support A and B and by the sequence numbers 1 and 2; span 2 is always characterised by the points of support B and C and by the sequence numbers 3 and 4; etc. Furthermore each 'column line' (i.e., each unit comprising an upper and a lower column in the case of a two-story framework) is characterised by a–A–a', b–B–b', etc., denoting respectively: column foot, beam joint, column head.

The symbols N (referring to joints), v (referring to spans) and n and n' (referring to columns) are appended to other symbols as subscripts relating to the joint or member in question. [For convenience, the prime ('), where it occurs, is applied to the main symbol, although it belongs, strictly speaking, to the subscript of that symbol]. In the case of double subscripts, the subscript referring to the joint always precedes the other, and both subscripts are placed at the same level.

All the other symbols relating to the upper columns of two-story frameworks are, in addition, provided with a prime ('). As a rule, a prime is also applied to symbols relating to formulas and expressions established by proceeding from right to left, whereas those established by proceeding from left to right are not provided with a prime. Furthermore the prime (') and the double prime(") are employed as designations for symmetrical and antisymmetrical loading conditions in connection with the 'load transposition method'.

The 'sequence numbers i, k' are applied as single or double subscripts to all symbols relating to the auxiliary moments X in Method I.

In cases where it is necessary to distinguish a symbol relating to a laterally unrestrained framework from the same symbol relating to a laterally restrained framework, the first-mentioned symbol is provided with the superscript v.

The following symbols should furthermore be noted:

l_v Lengths of beam spans.

h_n, h'_n Height of lower and upper column respectively in the case of a two-story framework (alternatively, in the case of a single-story framework, h_n is the column height).

L, R	Left-hand and right-hand end of a frame member when considered individually as a separate unit. The subscripts l and r similarly denote 'left' and 'right' (see also the auxiliary book *Belastungsglieder*).
s_ν	Length of any particular frame member (also used as a general symbol in place of l_ν and h_ν).
I', I	Left-hand and right-hand 'fixed point' of a member; subscript: the number or symbol denoting the member concerned.
i', i	Distance of 'fixed point' as measured from the nearest end point of the member.
E_n, E'_n	Outermost 'fixed point' of, respectively, the lower and upper elastically restrained column of a two-story framework.
e_n, e'_n	The distance of the above-mentioned 'fixed point' from the column end n and n' respectively.

Note: In the case of elastically restrained end spans of the continuous horizontal member the symbols E and e are provided with the subscript l' and v respectively.

c) Forces and moments

For an explanation of the symbols for external loads and their positions on the 'single-span beam as an element of framed structures', and of the symbols for the shear forces and bending moments occurring in such single-span beams, the reader is referred to the auxiliary book *Belastungsglieder*.

The second key diagram accompanying each of the fifty 'frame shapes' shows the conventions adopted as defining the positive direction of action of all external loads and all support reactions The following symbols are employed:

$\boldsymbol{S}_\nu$	Vertical resultant load of the beam span ν of a multi-bay framework or, alternatively, of any particular member of a framework.[1]
$\boldsymbol{W}_n, \boldsymbol{W}'_n$	Horizontal resultant load on lower column n and on upper column n' respectively.[1]
$\boldsymbol{D}_N$	Given external rotational moment (applied moment) acting at the beam joint N.

Note: The above symbols $\boldsymbol{S}$, $\boldsymbol{W}$ and $\boldsymbol{D}$ are always printed in boldfaced type.

W or P	Horizontal concentrated load, acting at the level of the beam, in the case of a laterally unrestrained framework.
V_n	Vertical reaction at support (support reaction) at base of column N.

---

[1] The letter Q, which was used to denote the resultant load in the 2nd–6th editions of *Mehrstielige Rahmen*, is now used only to denote shear force.

Note: In the case of end supports of the continuous horizontal member in a framework having no end columns the subscript n is replaced by the relevant subscript N.

H_n, H_n' Horizontal support reaction (horizontal shear) at the base of the lower column and at the top of the upper column, respectively.

H Horizontal 'restraining force' acting on the beam (continuous horizontal member) in the case of a laterally restrained framework.

$Q_l, Q_r; T$ General symbols denoting shear forces on members (see also *Belastungsglieder*); the number or symbol denoting the member concerned is added as a further subscript.

M_N^0 Support moment of the column line n–N–n' considered individually (auxiliary moment).

$X_1, X_2 \ldots X_i$ End moments of beam spans, constituting auxiliary unknowns (in Method I).

$Y_A, Y_B \ldots$ Angles of rotation of joints, expressed dimensionally as moments, constituting auxiliary unknowns (in Method II).

$M_{A1}, M_{B1}, M_{B2} \ldots$ Final end moments of beam spans.

$M_{Aa}, M_{Bb} \ldots M_{Nn} \ldots$ Final end moments of lower columns at beam joints.

$M_a, M_b \ldots M_n \ldots$ Final moments at lower column bases.

$M_{Aa}', M_{Bb}' \ldots M_{Nn}' \ldots$ Final end moments of upper columns at beam joints.

$M_a', M_b' \ldots M_n' \ldots$ Final moments at upper column tops.

x, y, m Specific moments (unit moments); same subscripts as for X, Y, M.

d) Load terms

All 'load terms' are printed in boldface type. This applies both to load terms in a more restricted sense and to those symbols which, in a more general sense, are likewise designated as load terms (such as the symbols in boldface type indicated in the foregoing section). See also the auxiliary book *Belastungsglieder*. The following symbols are defined:

𝔖$_r$, 𝔖$_l$ Statical moment of the load resultant S or W, referred to the right-hand and left-hand end of the member respectively.[1]

𝔏, ℜ Load terms in respect of flexibility (these are the 'primary' load terms: the tangent angles of a simply supported beam, expressed dimensionally as moments); subscripts: v, n, n'.[2]

L, R Load terms multiplied by the flexibility coefficient k of the same span, i.e., 'k-fold load terms'; subscripts (also for the coefficients k): as above.[2]

---

[1] The symbols 𝔐$_r$ and 𝔐$_l$ used in the 2nd–6th editions of *Mehrstielige Rahmen* to denote these statical moments will, from now on, be employed (but in boldface type) as symbols for the restraining moments associated with full end-fixity of a member. (See also footnote 2 in *Belastungsglieder*, page 4.)

[2] The symbols 𝔖 for (𝔏 + ℜ), 𝔇 for (𝔏 − ℜ), S for (**L + R**) and D for (**L − R**), as employed in the 2nd–6th editions of *Mehrstielige Rahmen*, have been discarded.

— XXIV —

Hence we may write in general:

$$L_\nu = \mathfrak{L}_\nu \cdot k_\nu \quad \text{and} \quad R_\nu = \mathfrak{R}_\nu \cdot k_\nu.$$ (a)

$\mathfrak{M}_l, \mathfrak{M}_r$ — End moments of the member fully fixed at both ends (load terms with respect to stiffness) (fixed-end moment); second subscript: v (in the case of columns fully fixed at the ends: also n and n').

$\mathfrak{M}'_r, \mathfrak{M}'_l$ — End moment of the member fully fixed at the right-hand end or at the left-hand end respectively, the other end of the member in each case being assumed to have a pin-joint, or hinge, permitting rotational freedom; second subscript v (in the case of columns with pin-jointed ends: also n and n').

$\mathfrak{E}_r, \mathfrak{E}_l$ — End moment of the member fully fixed at the right-hand end or at the left-hand end respectively, the other end of the member in each case being assumed to have a given degree of elastic rotational restraint; second subscript: v, v', n, n'.

$\mathfrak{F}_{rn}$ — End moment at the laterally unrestrained top of the lower column n, which is assumed to have a given degree of elastic restraint at the base.

$\mathfrak{F}'_{ln}$ — End moment at the base of the upper column n', whose top is assumed to be laterally unrestrained and to have a given degree of elastic restraint.

$\mathfrak{B}$ — Composite column load term or joint load term in Method I; subscripts: n, n', N.

$\mathfrak{K}_N$ — Composite joint load term in Method II (balancing moment).

Note: All the load terms in this section d) have the dimension of a bending moment (ft-lb). The pairs of letters $\mathfrak{L}, \mathfrak{R}$ and L, R themselves, as well as the pairs of subscripts l, r symbolically denote 'left' and 'right' with reference to the member, which should always be considered from the side marked by the broken line in the diagram (i.e., beam spans should be considered from below, and columns from the right-hand side).

e) Material constant, constants for the members, and distortion factors

E — Modulus of elasticity of the framework material (lb/ft^2).
$J_\nu; J_n, J'_n$ — Moments of inertia of the cross-sections of the members, in the plane of bending (ft^4).

J_k — Comparative moment of inertia: constant for the entire framework, but of arbitrarily selected value (ft^4).

s_k or l_k — Comparative length, corresponding to the above (ft).

k, K — Flexibility coefficient and stiffness coefficient of a frame member; subscripts: v, n, n':

$$k = \frac{J_k}{J_\nu} \cdot \frac{s_\nu}{s_k}; \qquad K = \frac{J_\nu}{J_k} \cdot \frac{s_k}{s_\nu} = \frac{1}{k}. \qquad (b)$$

Distortion factor for a member or for the system (ft-lb):

$$\mathfrak{T}_\nu = \frac{6EJ_\nu}{s_\nu} \quad \text{and} \quad \mathfrak{T}_k = \frac{6EJ_k}{s_k}. \qquad (c)$$

Note: Instead of choosing arbitrary values for J_k and s_k individually, an arbitrary suitable numerical value may be chosen for the ratio J_k/s_k or, alternatively, for s_k/J_k (dimensions: ft³ and 1/ft³ respectively).

Interrelations between k, K and $\mathfrak{T}$:

$$\mathfrak{T}_k = \mathfrak{T}_\nu \cdot k_\nu; \qquad \mathfrak{T}_\nu = \mathfrak{T}_k \cdot K_\nu. \qquad (d)$$

h_k Arbitrary comparative column height for laterally unrestrained framework (ft).

$\gamma_n = h_k/h_n$ Comparative values for columns (dimensionless).

f) Spring constants, carry-over factors, flexibilities and stiffnesses

f_L, f_R Spring constant for elastically rotating 'abutment' at left-hand and right-hand end of a member, respectively (1/lb. ft.)

ι_ν, ι_ν' Moment carry-over factors of the beam span v, for carry-over from left to right and from right to left, respectively (dimensionless).

$\varepsilon_n, \varepsilon_n'$ Given external moment carry-over factor (determined by estimate, measurement or some other means) for the lower column elastically restrained at the base and for the upper column elastically restrained at the top, respectively.

$\varepsilon_1', \varepsilon_\nu$ External carry-over factors, as above, for the first beam span, elastically restrained at the left-hand support, and for the last beam span, elastically restrained at the right-hand support.

Important note: The factors ε serve as a convenient substitute for the unwieldy spring constants. Each of these factors ε has a value between 0 and $\frac{1}{2}$. The limiting value 0 corresponds to a pin-jointed end (i.e., full rotational freedom), while the limiting value $\frac{1}{2}$ corresponds to full end-fixity (i.e., complete rigid restraint). The mean value $\varepsilon = \frac{1}{4} = 0.25$ is commonly accepted as corresponding to 'semi-rigid restraint' in the sense of an end connection half-way between the two above-mentioned limiting cases. (The respective values of the spring constant corresponding to these two limits are ∞ and 0. Further information is given in the Appendix).

$$\varepsilon = \frac{1}{2 + \mathfrak{T} \cdot f} \quad \text{or} \quad f = \frac{1 - 2\varepsilon}{\mathfrak{T} \cdot \varepsilon}. \qquad (e)$$

σ, σ' Transfer factors for the beam moments at the joint, operating from left to right and from right to left respectively; subscript: n or N.

b_v, b_v' Left-hand and right-hand flexibility, respectively, of the beam span v or of any particular member of a framework.

b_n, b_n' Flexibility of the lower and of the upper column, respectively, referred to the beam joint N. The following relationships exist:

$$\boxed{b_n = k_n(2 - \varepsilon_n); \qquad b_n' = k_n'(2 - \varepsilon_n').} \tag{f}$$

s_N Flexibility of the column line n–N–n' of a two-story multi-bay framework, referred to the beam joint N.

η_n, η_n' Distribution factors for the columns of a two-story column line.

m_n, m_n' Free flexibility of the lower and upper column respectively—referred to the beam joint N—in the case of a laterally unrestrained frame. The following relationships exist:

$$\boxed{m_n = k_n(1/\varepsilon_n + 4); \qquad m_n' = k_n'(4 + 1/\varepsilon_n').} \tag{g}$$

S_v, S_v' Left-hand and right-hand stiffness, respectively, of the beam span v or of any particular member of a framework.

S_n, S_n' Stiffness of the lower and of the upper column, respectively, referred to the beam joint N. The following general relationships exist:

$$\boxed{S = 1/b; \qquad S' = 1/b'.} \tag{h}$$

$S_N, (S_N^v)$ Stiffness of the set of members intersecting at the joint N in the case of a single-story or two-story framework. (The superscript v is added if the structure is laterally unrestrained, i.e., can undergo side-sway).

$\mu, (\eta)$ Moment distribution factors at a joint; subscripts: the numbers or symbols denoting the beam spans and columns intersecting at that joint. (If the framework is laterally unrestrained, the symbol η is used).

F_n, F_n' Free stiffness of the lower and upper column respectively—referred to the beam joint N—in the case of a laterally unrestrained frame. The following relationships exist:

$$\boxed{F_n = 1/m_n \quad \text{and} \quad F_n' = 1/m_n'.} \tag{i}$$

j_v, j_v' Rotation carry-over factors for the beam span v, operating from left to right and from right to left, respectively (in Method II).

General interrelations between the moment carry-over factors and rotation carry-over factors for the same span and same direction:

$$j = \frac{1-2\iota}{2-\iota} \quad \text{or} \quad \iota = \frac{1-2j}{2-j}. \tag{k}$$

Note: All the flexibilities b, s and m have been multiplied by the factor $\mathfrak{T}_k$, and the stiffnesses S and F have correspondingly been multiplied by $1/\mathfrak{T}_k$. (Where necessary, the true values of these flexibilities and stiffnesses could be denoted by adding the superscript w).

When using the "load transposition method" the symbols l and g or j are provided with the additional subscript s in the case of symmetrical loading, and the additional subscript t in the case of anti-symmetrical loading.

g) Auxiliary coefficients and influence coefficients

Symbols used in connection with Method I:

O_i — Diagonal coefficients or support coefficients (coefficients of the diagonal terms of the elasticity equations).

r_i, v_i — Auxiliary coefficients in connection with backward reduction and forward reduction, respectively.

n_{ii} — Principal influence coefficients (diagonal values of the conjugate matrix: enclosed in heavy parentheses in the square computation scheme).

$n_{ik} = n_{ki}$ — All the other influence coefficients of the conjugate matrix.

s_{iv} — Composite influence coefficients for symmetrical loading of the beam spans.

w_{in}, w_{iN} — Composite influence coefficients for column loading and column line loading respectively.

Symbols used in connection with Method II:

K_N — Joint coefficients (coefficients of the diagonal terms of the conditional equations).

r_n, v_n — Auxiliary coefficients in the reduction sequences.

u_{nn} — Principal influence coefficients (diagonal values).

$u_{ik} = u_{ki}$ — All the other influence coefficients.

The influence coefficients n_{ik} or u_{ik} can always be arranged in the form of a square matrix, symmetrical with respect to its principal diagonal. In the double subscript ik the first subscript denotes the row, and the second subscript denotes the column of the matrix.

The composite influence coefficients occur only in matrix columns. The subscripts have the same meaning as indicated above, except that now the second subscript (indicating the number of the matrix column) belongs to the other number sequences (v, n, N).

If the 'load transposition method' is applied to symmetrical multi-bay frameworks, the values r_i, n_{ii}, n_{ik}, s_{in}, w_{in}, w_{iN} or r_n, u_{nn}, u_{ik} (as also the auxiliary unknowns X_i or Y_N) are additionally provided with a prime (') in the case of symmetrical load arrangement, and with a double prime (") in the case of antisymmetrical load arrangement. As for the coefficients O_i, v_i or K_N, v_n, only the terminal value requires a prime or double prime.

h) Rotations, deflections and influence lines

τ_l, τ_r	Left-hand and right-hand tangent angle of a member; second subscript: symbol or number denoting the member concerned.
φ	Angle of rotation of a joint; subscript: letter denoting the joint concerned; if necessary, a second subscript may be added, consisting of the symbol or number denoting a particular member.
ψ	Angle of rotation of a member; subscript: symbol or number denoting the member concerned.
δ	Displacement of the end of a member at right angles to the axis of the member.
y	Ordinates of deflection curves; subscripts: symbol or number denoting the member concerned.
ω	Müller-Breslau's omega functions.
η	Ordinates of influence lines; subscripts: v, n; A, N.
e, m	Coefficients of the influence line equations for member end moments (end moments of beam spans and columns) and for beam span moments.
$\vartheta; \alpha, \beta, \gamma$	Coefficients of the influence line equations for shear forces and support reactions.
f	Coefficients of the influence line equations for the restraining forces at the continuous horizontal member.
μ, ζ, τ	Auxiliary coefficients (ratios).

i) Further comments

The representation of the different types of support or bearing conditions in the diagrams calls for little comment. Elastic end restraint is represented in the same manner as full end-fixity, except that the oblique hatching has been omitted. Freedom of horizontal or vertical movement of a support is indicated by the addition of a thick horizontal or vertical line (symbolizing a sliding bearing). The explanatory notes in the Table of Contents should also be consulted.

Signs and symbols other than those defined above—particularly in the Appendix—are explained where they occur.

3. Sign Conventions

a) General principle

All computations should be performed algebraically, that is to say, each quantity should always be used with its proper algebraic sign. The sign of each type of quantity is fully defined by the rules given below. In this connection the designation of one of the sides of a member by means of a broken line (------) plays an important part.[1] In all the multi-bay frameworks considered in this book the horizontal beams are always provided with such a broken line on the underside, and all the columns of the framework are provided with such a broken line on the right-hand side (also in the case of symmetrical frames).

b) External forces and moments

Vertical loads on horizontal members (beams) of frameworks are positive if they act in the downward direction. *Horizontal forces* on vertical columns of frameworks are positive if they act from left to right. In general, *loads of arbitrary direction* acting on arbitrarily sloping frame members are positive if they would produce tensile stresses on the side of the member marked by the broken line if the member in question were a simply-supported beam.

In all the loading diagrams the load is assumed to be positive and is represented as such. When dealing with a load acting in the direction opposite to that of the loading envisaged in this book, such load should be introduced with a negative sign into the calculations. In the case of symmetrical frameworks with antisymmetrical arrangement of the entire loading on the horizontal member, the loads on the right-hand half of the frame have, of course, from the outset been introduced as negative loads. The same applies to symmetrical frameworks with symmetrical arrangement of the entire loading acting on the columns.

Vertical support reactions are positive if they act in the upward direction. *Horizontal support reactions* are positive if they act from right to left. In general, *support reactions* are positive if they correspond to the directions designated as positive in the general loading diagram for a 'Frame Shape'.

If the calculated value of any particular support reaction is found to be negative, this means that the reaction acts in the direction opposite to the positive direction as defined above. This similarly applies to every one of the statical and other quantities mentioned below.

External *applied moments* (rotational moments, cantilever moments, joint fixing moments and balancing moments) are positive if they have a clockwise direction of rotation.

---

[1] For further information on this point see *Belastungsglieder*, page 1 and 2, and Fig. 1.

The sign conventions for *restraining moments* (including fixed-end moments of frame members) and in general for *support moments* are the same as those for bending moments (see section c).

c) Internal forces and moments

The determining factor with regard to the sign convention for the so-called 'stress resultants' (forces and moments equivalent in their effect to the internal stresses) is their effect upon the portion of the member in which they occur.

Bending moments are positive if they produce tension on the side of the member marked by the broken line. The algebraic sign of a bending moment is not, therefore, directly associated with its direction of rotation—unless the relationship is defined as follows: a bending moment is positive if it acts in a clockwise direction upon the left-hand end of a beam and if it acts in a counter-clockwise direction upon the right-hand end of a beam. The so-called 'member end moments' (end moments of beam spans and columns) are, of course, also to be treated as bending moments.[1]

Bending moment diagrams are always plotted on the side of a member at which the relevant bending moments produce tension. Positive bending moment diagrams are therefore always plotted on the side marked by the broken line, negative bending moment diagrams on the side of the member not provided with the broken line.

A *shear force* is positive if it is directed upward on the left-hand side of a section, and downward on the right-hand side of a section. It makes no difference from which side the member is considered. The algebraic sign of the shear force is not related to that of the bending moment and is therefore independent of the position of the broken line.

Longitudinal forces (axial forces in members) are positive if they produce compression in the member.[2]

d) Rotations, deflections and influence lines

Tangent angles are positive if they open towards the side of the member marked by the broken line. In other words: a tangent angle is positive if its direction of opening corresponds to the direction of rotation of a positive end moment.

The *angle of rotation of a joint* or of a *member* is positive if it opens in the clockwise direction.

Deflections (ordinates of the deflection curve) and *end displacements* of members are positive if they are directed towards the side of the member

---

[1] With regard to the algebraic signs of bending moments, see also *Belastungsglieder*, page 7 and 8.

[2] A recent trend in structural analysis is to define a tensile axial force as positive. If the reader prefers to adopt this convention, it will merely be necessary to add a negative sign to the forces N in Fig. 376, 390 and 391.

marked by the broken line. A *relative displacement of the two ends of a member* is positive if it is associated with a positive angle of rotation of the member.

Influence line ordinates are positive when plotted upwards from the axis of a horizontal member, or to the left from the axis of a column, or, in general, when plotted to the side of the member not marked by the dotted line.

4. Design Postulates

(a) The derivation of the ready-to-use formulas contained in this book is based primarily on the following assumptions.

In the case of the fifty 'laterally restrained' principal frame shapes, all the supports and joints are assumed to be incapable of undergoing displacement in any direction. However, in order not to have to reckon with subsidiary effects produced by axial forces, not all the supports of the structure should be conceived as being subject to complete linear restraint. Thus, in the case of the continuous horizontal member only one such 'restraining bearing' is required, which is always represented as being located at the right-hand end of that member (in a case where a bearing is present at the left-hand end, it is a horizontal sliding bearing). Of course, the restraining bearing (i.e., the bearing or external support which prevents side-sway) may actually be located at any other point of the horizontal member without entailing any change in the relevant formulas. The position of the restraining bearing affects only the distribution and algebraic sign of the axial forces in the horizontal member. (In this connection the reader is referred to 'Considerations on the influence and the distribution of axial forces in continuous beams' in the present authors' book *Durchlaufträger*, Volume II, Section N, Par. 3, pages 393–6). Similarly, for each two-story column line it will suffice to assume one bearing producing complete linear restraint at the foot of the lower column, while the bearing at the top of the upper column is to be conceived as a vertical sliding bearing. Despite the assumption of these horizontal and vertical sliding bearings, the 'non-displaceability' of all the supports and joints of the framework is assured on geometrical grounds.

In the 'laterally unrestrained'[1] frame shapes (indicated by the subscript 'v') the restraining bearing at the right-hand end of the horizontal member is in any case omitted. In two-story frameworks, the tops of the upper columns may be kept restrained in their original positions or, alternatively, they may undergo side-sway (lateral displacement) just like the continuous horizontal member.

All the single-member bearings of every frame shape are assumed to constitute end connections presenting elastic rotational restraint. The 'degree of fixity' (or 'degree of restraint') of the elastically rotatable abutment is to be regarded as given, i.e., it is known (estimated, measured or calculated by some

---

[1] The term 'laterally unrestrained' in the present translation is merely a conventional term denoting the absence of an external 'restraining bearing' or similar device preventing side-sway.

means). This is expressed in the external moment carry-over factor ε of the member concerned. The limiting cases of full end-fixity (ε = 0.5) and full rotational freedom (ε = 0) are directly provided for by the formulas given in this book. (For further information see the Appendix).

(b) In arriving at the formulas for the statically indeterminate quantities (member end moments) and for the geometrically indeterminate quantities (rotations of joints) only the effect of the bending moments has been taken into account, as is indeed usual in structural analysis. Practical experience has shown that it is permissible to neglect the effect of the axial forces and shear forces, inasmuch as these forces contribute very little to the elastic deformations of the structure. If the effect of these forces were taken into account, the amount of computational labour involved would be quite disproportionate to the slightly greater accuracy achieved. It must nevertheless be pointed out that neglecting the effects of axial forces and shear forces must not be adopted as a rule that can be blindly applied to all structures. Thus, in the case of short deep beams and similar members the axial and shear forces will play a significant part.

(c) In all the frameworks considered in this book, it has been assumed that, generally speaking, each individual member has a different moment of inertia. At the same time, however, the moment of inertia is assumed to be constant over the entire length of each member. This is expressed in the flexibility coefficients k (in Method I) and in the stiffness coefficients K (in Method II), each member having one such coefficient associated with it.

(d) The modulus of elasticity E (Young's modulus) of the material is assumed to be of constant magnitude for all the members of a framework. Hence E cancels out and therefore does not occur in any of the formulas for statical quantities due to external loading.

(e) The diagrams representing deflections and influence lines in the Appendix are highly distorted representations. In reality the angles due to deformation of the structure are so small that the following simplifications are valid for them (e.g., for τ):

$$\sin \tau \approx \text{arc } \tau \approx \tan \tau \quad \text{and} \cos \tau \approx 1. \tag{l}$$

Consequently, all the trigonometrical relationships are simplified, in particular the most commonly employed one:

$$\tan (\tau_1 \pm \tau_2) \approx \tan \tau_1 \pm \tan \tau_2 \approx \tau_1 \pm \tau_2. \tag{m}$$

Further information on the subject is given in the present authors' *Durchlaufträger*, Volume I, page 51 *et seq.*[1]

---

[1] *Durchlaufträger. Ausführliche theoretische Entwicklungen, gebrauchsfertige Formeln und Zahlentafeln.* By Prof. Dr.-Ing. A. Kleinlogel and Bauing. A. Haselbach. 7th completely revised and considerably enlarged edition. Vol. I: Derivations and ready-to-use formulas for continuous beams with spans of arbitrary stiffness. Vol. II: Derivations and numerical tables for continuous beams with spans with particular stiffness properties, and comprehensive supplement. Wilhelm Ernst & Sohn, Berlin, 1949 and 1952.

Frame Shape 1

Single-story three-bay frame with laterally restrained horizontal member and four elastically restrained columns

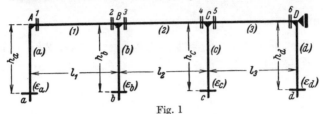

Fig. 1
Frame shape, dimensions and symbols

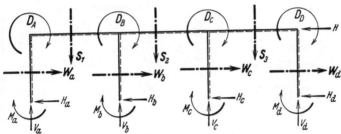

Fig. 2
Definition of positive direction for all external loads on joints and members, for all column reactions, and for the lateral restraining force

---

Important

All the formula sequences and formula matrices given for Frame Shape 1 can readily be extended to suit frames of similar type having any odd number of bays.*

Note: The *collective subscripts* have the following meaning:

ν = 1, 2, 3   denoting the bays (or beam spans)
N = A, B, C, D denoting the joints
n = a, b, c, d denoting the columns

I. Treatment by the Method of Forces

Coefficients

Flexibility coefficients for all individual members (J_k and l_k are arbitrary):

$$k_\nu = \frac{J_k}{J_\nu} \cdot \frac{l_\nu}{l_k} \; ; \quad k_n = \frac{J_k}{J_n} \cdot \frac{h_n}{l_k} \, . \tag{1}$$

---

* The dotted lines in the sets of formulas define the scope of these formulas as applicable to single-bay frames. See also Frame Shape 2, page 11.

Flexibilities of the elastically restrained columns:

$$b_n = k_n(2 - \varepsilon_n); \text{(the values } \varepsilon_n \text{ are assumed to be known).} \quad (2)$$

Diagonal coefficients:

$$\left. \begin{array}{lll} O_1 = b_a + 2\,k_1 & O_3 = b_b + 2\,k_2 & O_5 = b_c + 2\,k_3 \\ O_2 = 2\,k_1 + b_b & O_4 = 2\,k_2 + b_c & O_6 = 2\,k_3 + b_d. \end{array} \right\} \quad (3)$$

Moment carry-over factors and auxiliary coefficients:

$$\left. \begin{array}{ll} \text{backward reduction:} & \text{forward reduction:} \\[4pt] \iota_3 = \dfrac{k_3}{r_6} \quad r_6 = O_6 & \iota'_1 = \dfrac{k_1}{v_1} \quad v_1 = O_1 \\[6pt] \sigma_c = \dfrac{b_c}{r_5} \quad r_5 = O_5 - k_3 \cdot \iota_3 & \sigma'_b = \dfrac{b_b}{v_2} \quad v_2 = O_2 - k_1 \cdot \iota'_1 \\[6pt] \iota_2 = \dfrac{k_2}{r_4} \quad r_4 = O_4 - b_c \cdot \sigma_c & \iota'_2 = \dfrac{k_2}{v_3} \quad v_3 = O_3 - b_b \cdot \sigma'_b \\[6pt] \sigma_b = \dfrac{b_b}{r_3} \quad r_3 = O_3 - k_2 \cdot \iota_2 & \sigma'_c = \dfrac{b_c}{v_4} \quad v_4 = O_4 - k_2 \cdot \iota'_2 \\[6pt] \iota_1 = \dfrac{k_1}{r_2} \quad r_2 = O_2 - b_b \cdot \sigma_b & \iota'_3 = \dfrac{k_3}{v_5} \quad v_5 = O_5 - b_c \cdot \sigma'_c \\[6pt] \quad\quad r_1 = O_1 - k_1 \cdot \iota_1. & \quad\quad v_6 = O_6 - k_3 \cdot \iota'_3. \end{array} \right\} \quad \begin{array}{l}(4) \\ \text{and} \\ (5)\end{array}$$

Principal influence coefficients by recursion:

$$\left. \begin{array}{ll} \text{forward:} & \text{backward:} \\[4pt] n_{11} = \dfrac{1}{r_1} & n_{66} = \dfrac{1}{v_6} \\[6pt] n_{22} = \dfrac{1}{r_2} + n_{11} \cdot \iota_1^2 & n_{55} = \dfrac{1}{v_5} + n_{66} \cdot \iota'^2_3 \\[6pt] n_{33} = \dfrac{1}{r_3} + n_{22} \cdot \sigma_b^2 & n_{44} = \dfrac{1}{v_4} + n_{55} \cdot \sigma'^2_c \\[6pt] n_{44} = \dfrac{1}{r_4} + n_{33} \cdot \iota_2^2 & n_{33} = \dfrac{1}{v_3} + n_{44} \cdot \iota'^2_2 \\[6pt] n_{55} = \dfrac{1}{r_5} + n_{44} \cdot \sigma_c^2 & n_{22} = \dfrac{1}{v_2} + n_{33} \cdot \sigma'^2_b \\[6pt] n_{66} = \dfrac{1}{r_6} + n_{55} \cdot \iota_3^2. & n_{11} = \dfrac{1}{v_1} + n_{22} \cdot \iota'^2_1. \end{array} \right\} \quad \begin{array}{l}(6) \\ \text{and} \\ (7)\end{array}$$

Principal influence coefficients direct:

$$n_{11} = \frac{1}{r_1} \qquad n_{22} = \frac{1}{r_2 + v_2 - O_2} \qquad n_{33} = \frac{1}{r_3 + v_3 - O_3}$$
$$n_{44} = \frac{1}{r_4 + v_4 - O_4} \qquad n_{55} = \frac{1}{r_5 + v_5 - O_5} \qquad n_{66} = \frac{1}{v_6}. \qquad (8)$$

Computation scheme for the symmetrical influence coefficient matrix:

$$\begin{array}{c|cccccc|c}
\cdot\, \iota_1 & (n_{11}) \downarrow & n_{12} \uparrow & n_{13} \uparrow & n_{14} \uparrow & n_{15} \uparrow & n_{16} \uparrow & \cdot\, \iota_1' \\
\cdot\, \sigma_b & n_{21} \downarrow & (n_{22}) \downarrow & n_{23} \uparrow & n_{24} \uparrow & n_{25} \uparrow & n_{26} \uparrow & \cdot\, \sigma_b' \\
\cdot\, \iota_2 & n_{31} \downarrow & n_{32} \downarrow & (n_{33}) \downarrow & n_{34} \uparrow & n_{35} \uparrow & n_{36} \uparrow & \cdot\, \iota_2' \\
\cdot\, \sigma_c & n_{41} \downarrow & n_{42} \downarrow & n_{43} \downarrow & (n_{44}) \downarrow & n_{45} \uparrow & n_{46} \uparrow & \cdot\, \sigma_c' \\
\cdot\, \iota_3 & n_{51} \downarrow & n_{52} \downarrow & n_{53} \downarrow & n_{54} \downarrow & (n_{55}) \downarrow & n_{56} \uparrow & \cdot\, \iota_3' \\
 & n_{61} & n_{62} & n_{63} & n_{64} & n_{65} & (n_{66}) &
\end{array} \qquad (9)$$

Composite influence coefficients:

$$\begin{aligned}
s_{11} &= +n_{11} - n_{12} & s_{12} &= +n_{13} - n_{14} & s_{13} &= +n_{15} - n_{16} \\
s_{21} &= -n_{21} + n_{22} & s_{22} &= +n_{23} - n_{24} & s_{23} &= +n_{25} - n_{26} \\
s_{31} &= -n_{31} + n_{32} & s_{32} &= +n_{33} - n_{34} & s_{33} &= +n_{35} - n_{36} \\
s_{41} &= -n_{41} + n_{42} & s_{42} &= -n_{43} + n_{44} & s_{43} &= +n_{45} - n_{46} \\
s_{51} &= -n_{51} + n_{52} & s_{52} &= -n_{53} + n_{54} & s_{53} &= +n_{55} - n_{56} \\
s_{61} &= -n_{61} + n_{62} & s_{62} &= -n_{63} + n_{64} & s_{63} &= -n_{65} + n_{66};
\end{aligned} \qquad (10)$$

$$\begin{aligned}
w_{1b} &= +n_{12} - n_{13} & w_{1c} &= +n_{14} - n_{15} \\
w_{2b} &= +n_{22} - n_{23} & w_{2c} &= +n_{24} - n_{25} \\
w_{3b} &= -n_{32} + n_{33} & w_{3c} &= +n_{34} - n_{35} \\
w_{4b} &= -n_{42} + n_{43} & w_{4c} &= +n_{44} - n_{45} \\
w_{5b} &= -n_{52} + n_{53} & w_{5c} &= -n_{54} + n_{55} \\
w_{6b} &= -n_{62} + n_{63} & w_{6c} &= -n_{64} + n_{65}.
\end{aligned} \qquad (11)$$

Loading Condition 1

All the beam spans carrying arbitrary vertical loading

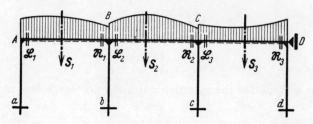

Fig. 3
Beam loading diagram with load terms $\mathfrak{L}$ and $\mathfrak{R}$.

Beam moments $(L_\nu = \mathfrak{L}_\nu k_\nu$ and $R_\nu = \mathfrak{R}_\nu k_\nu)$:

$$\left. \begin{array}{l|cccccc}
 & L_1 & R_1 & L_2 & R_2 & L_3 & R_3 \\
\hline
X_1 = M_{A1} = & -n_{11} & +n_{12} & +n_{13} & -n_{14} & -n_{15} & +n_{16} \\
X_2 = M_{B1} = & +n_{21} & -n_{22} & -n_{23} & +n_{24} & +n_{25} & -n_{26} \\
\hline
X_3 = M_{B2} = & +n_{31} & -n_{32} & -n_{33} & +n_{34} & +n_{35} & -n_{36} \\
X_4 = M_{C2} = & -n_{41} & +n_{42} & +n_{43} & -n_{44} & -n_{45} & +n_{46} \\
X_5 = M_{C3} = & -n_{51} & +n_{52} & +n_{53} & -n_{54} & -n_{55} & +n_{56} \\
X_6 = M_{D3} = & +n_{61} & -n_{62} & -n_{63} & +n_{64} & +n_{65} & -n_{66}
\end{array} \right\} \quad (12)$$

Column moments:

at top:

$$M_{Aa} = +X_1 \quad M_{Bb} = -X_2 + X_3 \quad M_{Cc} = -X_4 + X_5 \quad M_{Dd} = -X_6; \quad (13)$$

at base: $\qquad M_n = -M_{Nn} \cdot \varepsilon_n. \qquad (14)$

Special case: All the beam spans symmetrically loaded $(R_\nu = L_\nu)$:

$$\left. \begin{aligned}
X_1 = M_{A1} &= -L_1 \cdot s_{11} + L_2 \cdot s_{12} - L_3 \cdot s_{13} \\
X_2 = M_{B1} &= -L_1 \cdot s_{21} - L_2 \cdot s_{22} + L_3 \cdot s_{23} \\
X_3 = M_{B2} &= -L_1 \cdot s_{31} - L_2 \cdot s_{32} + L_3 \cdot s_{33} \\
X_4 = M_{C2} &= +L_1 \cdot s_{41} - L_2 \cdot s_{42} - L_3 \cdot s_{43} \\
X_5 = M_{C3} &= +L_1 \cdot s_{51} - L_2 \cdot s_{52} - L_3 \cdot s_{53} \\
X_6 = M_{D3} &= -L_1 \cdot s_{61} + L_2 \cdot s_{62} - L_3 \cdot s_{63}
\end{aligned} \right\} \quad (15)$$

Loading Conditions 2 and 3

All the columns carrying arbitrary horizontal loading (2); external rotational moments applied at all the beam joints (3)

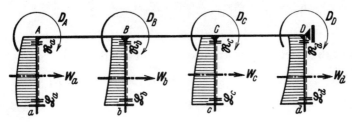

Fig. 4
Column loading diagram with load terms $\mathfrak{L}$ and $\mathfrak{R}$ for the columns and external rotational moments D.

Joint load terms ($L_n = \mathfrak{L}_n k_n$ and $R_n = \mathfrak{R}_n k_n$):

$$\mathfrak{B}_N = D_N \cdot b_N + L_n \cdot \varepsilon_n - R_n. \tag{16}$$

Beam moments*:

	$\mathfrak{B}_A$	$\mathfrak{B}_B$	$\mathfrak{B}_C$	$\mathfrak{B}_D$
$X_1 = M_{A1} =$	$+n_{11}$	$+w_{1b}$	$-w_{1c}$	$+n_{16}$
$X_2 = M_{B1} =$	$-n_{21}$	$-w_{2b}$	$+w_{2c}$	$-n_{26}$
$X_3 = M_{B2} =$	$-n_{31}$	$+w_{3b}$	$+w_{3c}$	$-n_{36}$
$X_4 = M_{C2} =$	$+n_{41}$	$-w_{4b}$	$-w_{4c}$	$+n_{46}$
$X_5 = M_{C3} =$	$+n_{51}$	$-w_{5b}$	$+w_{5c}$	$+n_{56}$
$X_6 = M_{D3} =$	$-n_{61}$	$+w_{6b}$	$-w_{6c}$	$-n_{66}$

(17)

Column moments:

at base:
$$\begin{cases} M_{Aa} = -D_A + X_1 \\ M_{Bb} = -D_B - X_2 + X_3 \\ M_{Cc} = -D_C - X_4 + X_5 \\ M_{Dd} = -D_D - X_6; \end{cases} \tag{18}$$

at top: $\quad M_n = -(\mathfrak{L}_n + M_{Nn})\,\varepsilon_n.$ (19)

---

* In the case of a single-bay frame the composite symbols w in the column under $\mathfrak{B}_B$ must be replaced by the second column of symbols n (i.e., n_{12} and n_{22} take the place of w_{1b} and w_{2b}.)

II. Treatment by the Deformation Method
Coefficients

Stiffness coefficients for all individual members:

$$K_\nu = \frac{J_\nu}{J_k} \cdot \frac{l_k}{l_\nu} = \frac{1}{k_\nu} \; ; \qquad K_n = \frac{J_n}{J_k} \cdot \frac{l_k}{h_n} = \frac{1}{k_n} . \qquad (20)$$

Stiffness of the elastically restrained columns:

$$S_n = \frac{K_n}{2 - \varepsilon_n} = \frac{1}{b_n} \; ; \qquad (21)$$

(the values ε_n are assumed to be known).

Joint coefficients:

$$\begin{aligned} K_A &= 3S_a + 2K_1 \\ K_B &= 2K_1 + 3S_b + 2K_2 \\ K_C &= 2K_2 + 3S_c + 2K_3 \\ K_D &= 2K_3 + 3S_d . \end{aligned} \qquad (22)$$

Rotation carry-over factors and auxiliary coefficients:

backward reduction:

$$\begin{aligned} & & r_d &= K_D \\ j_3 &= \frac{K_3}{r_d} \\ & & r_c &= K_C - K_3 \cdot j_3 \\ j_2 &= \frac{K_2}{r_c} \\ & & r_b &= K_B - K_2 \cdot j_2 \\ j_1 &= \frac{K_1}{r_b} \\ & & r_a &= K_A - K_1 \cdot j_1 . \end{aligned}$$

forward reduction:

$$\begin{aligned} & & v_a &= K_A \\ j_1' &= \frac{K_1}{v_a} \\ & & v_b &= K_B - K_1 \cdot j_1' \\ j_2' &= \frac{K_2}{v_b} \\ & & v_c &= K_C - K_2 \cdot j_2' \\ j_3' &= \frac{K_3}{v_c} \\ & & v_d &= K_D - K_3 \cdot j_3' . \end{aligned} \quad \begin{array}{c}(23) \\ \text{and} \\ (24)\end{array}$$

Principal influence coefficients by recursion:

forward:

$$\begin{aligned} u_{aa} &= 1/r_a \\ u_{bb} &= 1/r_b + u_{aa} \cdot j_1^2 \\ u_{cc} &= 1/r_c + u_{bb} \cdot j_2^2 \\ u_{dd} &= 1/r_d + u_{cc} \cdot j_3^2 . \end{aligned}$$

backward:

$$\begin{aligned} u_{dd} &= 1/v_d \\ u_{cc} &= 1/v_c + u_{dd} \cdot j_3'^2 \\ u_{bb} &= 1/v_b + u_{cc} \cdot j_2'^2 \\ u_{aa} &= 1/v_a + u_{bb} \cdot j_1'^2 \end{aligned} \quad \begin{array}{c}(25) \\ \text{and} \\ (26)\end{array}$$

Principal influence coefficients direct:

$$u_{aa} = \frac{1}{r_a} \qquad u_{bb} = \frac{1}{r_b + v_b - K_B} \qquad u_{cc} = \frac{1}{r_c + v_c - K_C} \qquad u_{dd} = \frac{1}{v_d} . \qquad (27)$$

Computation scheme of the symmetrical influence coefficient matrix:

$$\begin{array}{c} \cdot j_1 \\ \cdot j_2 \\ \cdot j_3 \end{array} \begin{vmatrix} (u_{aa}) & u_{ab} & u_{ac} & u_{ad} \\ \downarrow & \uparrow & \uparrow & \uparrow \\ u_{ba} & (u_{bb}) & u_{bc} & u_{bd} \\ \downarrow & \downarrow & \uparrow & \uparrow \\ u_{ca} & u_{cb} & (u_{cc}) & u_{cd} \\ \downarrow & \downarrow & \downarrow & \uparrow \\ u_{da} & u_{db} & u_{dc} & (u_{dd}) \end{vmatrix} \begin{array}{c} \cdot j'_1 \\ \cdot j'_2 \\ \cdot j'_3 \end{array} \qquad (28)$$

Loading Condition 4

Overall loading condition (superposition of Loading Conditions, 1, 2 and 3)

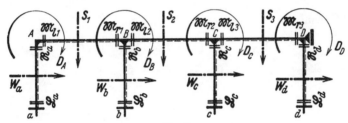

Fig. 5

Loading diagram with fixed end moments $\mathfrak{M}_l$ and $\mathfrak{M}_r$ for the beam spans, load terms $\mathfrak{L}$ and $\mathfrak{R}$ for the columns, and external rotational moments $\boldsymbol{D}$

Column restraining moments at the beam joints:

$$\mathfrak{C}_{rn} = -\frac{\mathfrak{R}_n - \mathfrak{L}_n \cdot \varepsilon_n}{2 - \varepsilon_n}. \qquad (29)$$

Joint load terms (balancing moments):

$$\begin{aligned} \mathfrak{K}_A &= & -\mathfrak{M}_{l1} & +\mathfrak{C}_{ra} + \boldsymbol{D}_A \\ \mathfrak{K}_B &= \mathfrak{M}_{r1} & -\mathfrak{M}_{l2} & +\mathfrak{C}_{rb} + \boldsymbol{D}_B \\ \mathfrak{K}_C &= \mathfrak{M}_{r2} & -\mathfrak{M}_{l3} & +\mathfrak{C}_{rc} + \boldsymbol{D}_C \\ \mathfrak{K}_D &= \mathfrak{M}_{r3} & & +\mathfrak{C}_{rd} + \boldsymbol{D}_D. \end{aligned} \qquad (30)$$

Auxiliary moments ($2EJ_k/s_k$-fold joint rotation φ_N):

$$
\begin{array}{c|cccc}
 & \mathfrak{K}_A & \mathfrak{K}_B & \mathfrak{K}_C & \mathfrak{K}_D \\
\hline
Y_A = & +u_{aa} & -u_{ab} & +u_{ac} & -u_{ad} \\
Y_B = & -u_{ba} & +u_{bb} & -u_{bc} & +u_{bd} \\
Y_C = & +u_{ca} & -u_{cb} & +u_{cc} & -u_{cd} \\
Y_D = & -u_{da} & +u_{db} & -u_{dc} & +u_{dd}
\end{array}
\tag{31}
$$

Beam moments:

$$
\begin{aligned}
M_{A1} &= \mathfrak{M}_{l1} + (2Y_A + Y_B)K_1 & M_{B1} &= \mathfrak{M}_{r1} - (Y_A + 2Y_B)K_1 \\
M_{B2} &= \mathfrak{M}_{l2} + (2Y_B + Y_C)K_2 & M_{C2} &= \mathfrak{M}_{r2} - (Y_B + 2Y_C)K_2 \\
M_{C3} &= \mathfrak{M}_{l3} + (2Y_C + Y_D)K_3; & M_{D3} &= \mathfrak{M}_{r3} - (Y_C + 2Y_D)K_3.
\end{aligned}
\tag{32}
$$

Column moments:

$$
\text{at top:} \quad M_{Nn} = \mathfrak{E}_{rn} - Y_N \cdot 3 S_n; \tag{33}
$$

$$
\text{at base:} \quad M_n = -(\mathfrak{L}_n + M_{Nn})\varepsilon_n. \tag{34}
$$

Check relationships:

$$
\begin{aligned}
D_A + M_{Aa} \phantom{+ M_{Bb}} - M_{A1} &= 0 & D_C + M_{Cc} + M_{C2} - M_{C3} &= 0 \\
D_B + M_{Bb} + M_{B1} - M_{B2} &= 0 & D_D + M_{Dd} + M_{D3}&.
\end{aligned}
\tag{35}
$$

Horizontal thrusts, column shears and lateral restraining force

For Loading Conditions 1 and 3:

$$H_n = +Q_n = M_{Nn}(1 + \varepsilon_n)/h_n; \tag{36}$$

$$H = -(H_a + H_b + H_c + H_d). \tag{37}$$

For Loading Conditions 2 and 4:

$$
\left.
\begin{aligned}
H_n &= +Q_{ln} = \mathfrak{S}_{rn}/h_n + T_n & Q_{rn} &= -\mathfrak{S}_{ln}/h_n + T_n, \\
\text{where} \quad T_n &= (M_{Nn} - M_n)/h_n;
\end{aligned}
\right\}
\tag{38}
$$

$$H = -(Q_{ra} + Q_{rb} + Q_{rc} + Q_{rd}). \tag{39}$$

(Collective subscripts: $N = A, B, C, D$ and $n = a, b, c, d$.)

Frame Shape 1v

(Frame Shape 1 with laterally unrestrained horizontal member)

First step: Calculation of all the requisite loading conditions as well as lateral restraining force H according to Frame Shape 1.

Second Step: Determination of the effect of a horizontal concentrated load W, acting at the level of the horizontal member, with the aid of Loading Condition 5 for Frame Shape 1v, as indicated below.

Third step: Determination of the final moments and forces for Frame Shape 1v by superposition of the results of the first step and those of the second step for $W = H$.

Loading Condition 5

Horizontal point load W at level of horizontal member

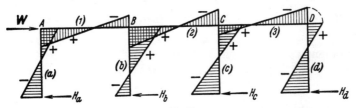

Fig. 6
Loading diagram and diagram of bending moments. (For clarity, only the reactions H_n at the column bases are shown).

Column ratios with arbitrary comparison height h_k: $\quad \gamma_n = \dfrac{h_k}{h_n}.\quad$ (40)

Specific bending moments m (instead of M);
either according to *Method I* from the formulas (17), (18) and (19), i.e., the moments $X_i = m_{Nv}$, m_{Nn} and m_n, for:

$$\mathfrak{B}_N = \gamma_n(1 + \varepsilon_n) \quad \text{and} \quad \mathfrak{L}_n = \gamma_n / k_n; \quad (41)$$

or according to *Method II* from the formulas (31), (32) and (34)—i.e., the moments y_N (instead of Y_N), m_{Nv}, m_{Nn} and m_n—for:

$$\mathfrak{K}_N = \mathfrak{C}_{rn} = S_n \gamma_n (1 + \varepsilon_n) \quad \text{and} \quad \mathfrak{L}_n = \gamma_n K_n. \tag{42}$$

Note: All the other load terms in bold type in the formulas indicated should be omitted.

Specific h_k-fold column shears:

$$p_n = (m_{Nn} - m_n) \gamma_n. \tag{43}$$

Corresponding specific h_k-fold applied load:

$$p = p_a + p_b + p_c + p_d. \tag{44}$$

Conversion moment for a given load W:

$$U = \frac{W h_k}{p}. \tag{45}$$

Final moments and forces for the given point load W

Beam moments:

$$\begin{aligned} M_{A1} &= U \cdot m_{A1} & M_{B2} &= U \cdot m_{B2} & M_{C3} &= U \cdot m_{C3} \\ M_{B1} &= U \cdot m_{B1} & M_{C2} &= U \cdot m_{C2} & M_{D3} &= U \cdot m_{D3}; \end{aligned} \tag{46}$$

Column moments:

$$M_{Nn} = U \cdot m_{Nn}; \quad M_n = U \cdot m_n; \tag{47}$$

Column shears and horizontal shears:

$$Q_n = H_n = \frac{U \cdot p_n}{h_k} = W \cdot \frac{p_n}{p}. \tag{48}$$

(**Collective subscripts:** $n = a, b, c, d$ and $N = A, B, C, D$).

Supplementary note: The computation procedure given here for the "influence of a horizontal concentrated load W acting at the level of the laterally unrestrained horizontal member" is fundamentally applicable also to all the other single-story frames in this book. It should only be noted that the subscript n in each case relates to all the rigidly connected columns and that the subscript N relates to all the corresponding beam joints.

Frame Shape 2

Single-story single-bay frame with laterally restrained horizontal member and two elastically restrained columns

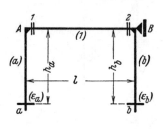
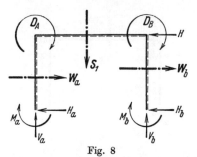

Fig. 7
Frame shape, dimensions and symbols

Fig. 8
Definition of positive direction for all external loads on joints and members, for all column reactions, and for the lateral restraining force

Important

All the formulas required for Frame Shape 2 are already comprised in the sets of formulas that have been given for Frame Shape 1, namely, for the collective subscripts: $\nu = 1$; $N = A, B$; $n = a, b$. All the sets of formulas without collective subscripts are valid up to the dotted lines; conversely, for backward reductions and backward recursions they are valid from those dotted lines onward. On account of the considerable practical importance of the one-bay rigid frame, however, the relevant formulas will now again be given as a self-contained whole.

I. Treatment by the Method of Forces

Coefficients, where $J_k = J_1$ and $l_k = l$, hence $k_1 = 1$:

$$k_a = \frac{J_1}{J_a} \cdot \frac{h_a}{l} \qquad k_b = \frac{J_1}{J_b} \cdot \frac{h_b}{l} \; ; \qquad \begin{aligned} b_a &= k_a(2 - \varepsilon_a) \\ b_b &= k_b(2 - \varepsilon_b) \end{aligned} \Biggr\} \tag{49}$$

$$\iota_1 = \frac{1}{2 + b_b} \qquad b_1 = 2 - \iota_1; \qquad \iota_1' = \frac{1}{b_a + 2} \qquad b_1' = 2 - \iota_1'. \tag{50}$$

$$n_{11} = \frac{1}{b_a + b_1} \qquad n_{22} = \frac{1}{b_1' + b_b} \; ; \qquad n_{12} = n_{21} = n_{11} \cdot \iota_1 = n_{22} \cdot \iota_1'. \tag{51}$$

$$s_{11} = n_{11} - n_{12} = n_{11}(1 - \iota_1) \qquad s_{21} = n_{22} - n_{21} = n_{22}(1 - \iota_1'). \tag{52}$$

Loading Condition 1

Beam span carrying arbitrary vertical loading

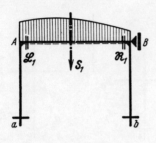

Fig. 9
Beam loading diagram with load terms $\mathfrak{L}$ and $\mathfrak{R}$

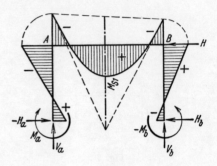

Fig. 10
Diagram of moments and support reactions

Beam moments: ($L_1 = \mathfrak{L}_1$ and $R_1 = \mathfrak{R}_1$, since $k_1 = 1$):

$$\left.\begin{aligned} X_1 = M_{A1} &= -L_1 \cdot n_{11} + R_1 \cdot n_{12} = -\frac{L_1 - R_1 \cdot \iota_1}{b_a + b_1} \\ X_2 = M_{B1} &= +L_1 \cdot n_{21} - R_1 \cdot n_{22} = -\frac{R_1 - L_1 \cdot \iota_1'}{b_1' + b_b} \end{aligned}\right\} \quad (53)$$

Column moments:

at top: $\qquad M_{Aa} = +X_1 \qquad\qquad M_{Bb} = -X_2 \qquad (54)$

at base: $\qquad M_a = -M_{Aa} \cdot \varepsilon_a \qquad M_b = -M_{Bb} \cdot \varepsilon_b . \qquad (55)$

Special case: Span load symmetrical ($R_1 = L_1$):

$$\left.\begin{aligned} X_1 = M_{A1} &= -L_1 \cdot s_{11} = -L_1 \cdot \frac{1 - \iota_1}{b_a + b_1} \\ X_2 = M_{B1} &= -L_1 \cdot s_{21} = -L_1 \cdot \frac{1 - \iota_1'}{b_1' + b_b} \end{aligned}\right\} \quad (56)$$

Loading Conditions 2 and 3

Both columns carrying arbitrary horizontal loading (2); external rotational moments applied at frame corners (3)

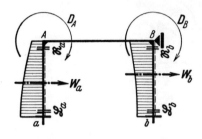

Fig. 11
Column loading diagram with load terms $\mathfrak{L}$ and $\mathfrak{R}$ for the columns and external rotational moments D

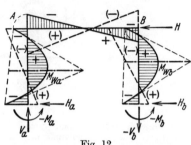

Fig. 12
Moment diagram for Loading Condition 2 (full lines) and for Loading Condition 3 (dotted), and support reactions

Joint load terms ($L_n = \mathfrak{L}_n k_n$ and $R_n = \mathfrak{R}_n k_n$):

$$\mathfrak{B}_A = D_A \cdot b_a + L_a \cdot \varepsilon_a - R_a \qquad \mathfrak{B}_B = D_B \cdot b_b + L_b \cdot \varepsilon_b - R_b . \qquad (57)$$

Beam moments:

$$\left. \begin{array}{l} X_1 = M_{A1} = + \mathfrak{B}_A \cdot n_{11} + \mathfrak{B}_B \cdot n_{12} = + \dfrac{\mathfrak{B}_A + \mathfrak{B}_B \cdot \iota_1}{b_a + b_1} \\[6pt] X_2 = M_{B1} = - \mathfrak{B}_A \cdot n_{21} - \mathfrak{B}_B \cdot n_{22} = - \dfrac{\mathfrak{B}_A \cdot \iota'_1 + \mathfrak{B}_B}{b'_1 + b_b} \cdot \end{array} \right\} \qquad (58)$$

Column moments at top and base:

$$M_{Aa} = - D_A + X_1 \qquad\qquad M_{Bb} = - D_B - X_2 \qquad (59)$$

$$M_a = - (\mathfrak{L}_a + M_{Aa})\,\varepsilon_a; \qquad M_b = - (\mathfrak{L}_b + M_{Bb})\,\varepsilon_b . \qquad (60)$$

II. Treatment by the Deformation Method

Coefficients: where $J_k = J_1$ and $l_k = l$, hence $K_1 = 1$:

$$K_a = \frac{J_a}{J_1} \cdot \frac{l}{h_a} \qquad K_b = \frac{J_b}{J_1} \cdot \frac{l}{h_b}; \qquad S_a = \frac{K_a}{2 - \varepsilon_a} \qquad S_b = \frac{K_b}{2 - \varepsilon_b} . \qquad (61)$$

$$K_A = 3 S_a + 2 \qquad K_B = 2 + 3 S_b; \qquad j_1 = 1/K_B \qquad j'_1 = 1/K_A . \qquad (62)$$

$$u_{aa} = \frac{1}{K_A - j_1} \qquad u_{bb} = \frac{1}{K_B - j'_1}; \qquad u_{ab} = u_{ba} = u_{aa} \cdot j_1 = u_{bb} \cdot j'_1 . \qquad (63)$$

Loading Condition 4

Overall loading condition (superposition of Loading Conditions 1, 2 and 3)

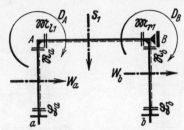

Fig. 13

Loading diagram with fixed-end moments $\mathfrak{M}_l$ and $\mathfrak{M}_r$ for the beam, load terms $\mathfrak{L}$ and $\mathfrak{R}$ for the columns, and external rotational moments D

Column restraining moments at the beam joints:

$$\mathfrak{C}_{ra} = -\frac{\mathfrak{R}_a - \mathfrak{L}_a \cdot \varepsilon_a}{2 - \varepsilon_a} \qquad \mathfrak{C}_{rb} = -\frac{\mathfrak{R}_b - \mathfrak{L}_b \cdot \varepsilon_b}{2 - \varepsilon_b}. \tag{64}$$

Joint load terms:

$$\mathfrak{K}_A = -\mathfrak{M}_{l1} + \mathfrak{C}_{ra} + D_A \qquad \mathfrak{K}_B = +\mathfrak{M}_{r1} + \mathfrak{C}_{rb} + D_B. \tag{65}$$

auxiliary moments:

$$\left.\begin{aligned} Y_A &= +\mathfrak{K}_A \cdot u_{aa} - \mathfrak{K}_B \cdot u_{ab} = \frac{+\mathfrak{K}_A - \mathfrak{K}_B \cdot j_1}{K_A - j_1} \\ Y_B &= -\mathfrak{K}_A \cdot u_{ba} + \mathfrak{K}_B \cdot u_{bb} = \frac{-\mathfrak{K}_A \cdot j_1' + \mathfrak{K}_B}{K_B - j_1'}. \end{aligned}\right\} \tag{66}$$

Beam moments:

$$M_{A1} = \mathfrak{M}_{l1} + (2\,Y_A + Y_B) \qquad M_{B1} = \mathfrak{M}_{r1} - (Y_A + 2\,Y_B). \tag{67}$$

Column Moments at top and base:

$$M_{Aa} = \mathfrak{C}_{ra} - Y_A \cdot 3\,S_a \qquad M_{Bb} = \mathfrak{C}_{rb} - Y_B \cdot 3\,S_b \tag{68}$$

$$M_a = -(\mathfrak{L}_a + M_{Aa})\,\varepsilon_a; \qquad M_b = -(\mathfrak{L}_b + M_{Bb})\,\varepsilon_b. \tag{69}$$

Check relationships:

$$D_A + M_{Aa} - M_{A1} = 0 \qquad D_B + M_{Bb} + M_{B1} = 0. \tag{70}$$

Horizontal thrusts, column shears and lateral restraining force:

For Loading Conditions 1 and 3:

$$H_a = +Q_a = \frac{M_{Aa}(1+\varepsilon_a)}{h_a} \qquad H_b = +Q_b = \frac{M_{Bb}(1+\varepsilon_b)}{h_b} ; \qquad (71)$$

$$H = -H_a - H_b. \qquad (72)$$

For Loading Conditions 2 and 4:

$$\left. \begin{array}{l} H_a = +Q_{la} = +\dfrac{\mathfrak{S}_{ra}}{h_a} + \dfrac{M_{Aa} - M_a}{h_a} \qquad Q_{ra} = -\dfrac{\mathfrak{S}_{la}}{h_a} + \dfrac{M_{Aa} - M_a}{h_a} \\[1em] H_b = +Q_{lb} = +\dfrac{\mathfrak{S}_{rb}}{h_b} + \dfrac{M_{Bb} - M_b}{h_b} \qquad Q_{rb} = -\dfrac{\mathfrak{S}_{lb}}{h_b} + \dfrac{M_{Bb} - M_b}{h_b} ; \end{array} \right\} \quad (73)$$

$$H = -Q_{ra} - Q_{rb}. \qquad (74)$$

Frame Shape 2v

(Frame Shape 2 with laterally unrestrained horizontal member)

Procedure

as for Frame Shape 1v, see page 9/10, with the collective subscripts $\nu = 1$; $N = A, B$; $n = a, b$. The specific moments may be obtained either from formulas (58)—(60) or from formulas (66)—(69). In the case of formulas (44) and (46) only the portions to the left of the dotted boundary line are relevant.

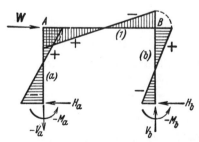

Fig. 14

Loading Condition 5, with diagram of moments and support reactions

Frame Shape 3

Single-story four-bay frame with laterally restrained horizontal member and five elastically restrained columns

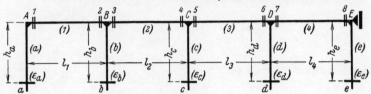

Fig. 15
Frame shape, dimensions and symbols

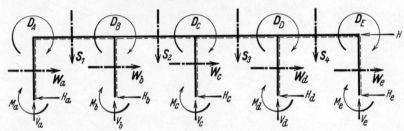

Fig. 16
Definition of positive direction for all external loads on joints and members, for all column reactions, and for the lateral restraining force

---

Important

All the formula sequences and formula matrices given for Frame Shape 3 can readily be extended to suit frames of similar type having any even number of bays.*

Note: The collective subscripts have the following meaning: $\nu = 1, 2, 3, 4$ denoting the bays (or beam spans). $N = A, B, C, D, E$ denoting the joints. $n = a, b, c, d, e$ denoting the columns

I. Treatment by the method of forces:

Coefficients:

Flexibility coefficients for all individual members:

$$k_\nu = \frac{J_k}{J_\nu} \cdot \frac{l_\nu}{l_k} \; ; \qquad k_n = \frac{J_k}{J_n} \cdot \frac{h_n}{l_k}. \tag{75}$$

Flexibilities of the elastically restrained columns:

$$b_n = k_n(2 - \varepsilon_n); \text{(the values } \varepsilon_n \text{ are assumed to be known)}. \tag{76}$$

---

* The dotted lines in the sets of formulas define the scope of these formulas as applicable to two-bay frames.

Diagonal coefficients:

$$O_1 = b_a + 2k_1 \quad O_3 = b_b + 2k_2 \quad O_5 = b_c + 2k_3 \quad O_7 = b_d + 2k_4$$
$$O_2 = 2k_1 + b_b \quad O_4 = 2k_2 + b_c \quad O_6 = 2k_3 + b_d \quad O_8 = 2k_4 + b_e.$$

(77)

Moment carry-over factors and auxiliary coefficients:

backward reduction:

$$r_8 = O_8$$
$$\iota_4 = \frac{k_4}{r_8}$$
$$r_7 = O_7 - k_4 \cdot \iota_4$$
$$\sigma_d = \frac{b_d}{r_7}$$
$$r_6 = O_6 - b_d \cdot \sigma_d$$
$$\iota_3 = \frac{k_3}{r_6}$$
$$r_5 = O_5 - k_3 \cdot \iota_3$$
$$\sigma_c = \frac{b_c}{r_5}$$
$$r_4 = O_4 - b_c \cdot \sigma_c$$
$$\iota_2 = \frac{k_2}{r_4}$$
$$r_3 = O_3 - k_2 \cdot \iota_2$$
$$\sigma_b = \frac{b_b}{r_3}$$
$$r_2 = O_2 - b_b \cdot \sigma_b$$
$$\iota_1 = \frac{k_1}{r_2}$$
$$r_1 = O_1 - k_1 \cdot \iota_1.$$

forward reduction:

$$v_1 = O_1$$
$$\iota'_1 = \frac{k_1}{v_1}$$
$$v_2 = O_2 - k_1 \cdot \iota'_1$$
$$\sigma'_b = \frac{b_b}{v_2}$$
$$v_3 = O_3 - b_b \cdot \sigma'_b$$
$$\iota'_2 = \frac{k_2}{v_3}$$
$$v_4 = O_4 - k_2 \cdot \iota'_2$$
$$\sigma'_c = \frac{b_c}{v_4}$$
$$v_5 = O_5 - b_c \cdot \sigma'_c$$
$$\iota'_3 = \frac{k_3}{v_5}$$
$$v_6 = O_6 - k_3 \cdot \iota'_3$$
$$\sigma'_d = \frac{b_d}{v_6}$$
$$v_7 = O_7 - b_d \cdot \sigma'_d$$
$$\iota'_4 = \frac{k_4}{v_7}$$
$$v_8 = O_8 - k_4 \cdot \iota'_4.$$

(78) and (79)

Principal influence coefficients by recursion:

forward:

$$n_{11} = 1/r_1$$
$$n_{22} = 1/r_2 + n_{11} \cdot \iota_1^2$$
$$n_{33} = 1/r_3 + n_{22} \cdot \sigma_b^2$$
$$n_{44} = 1/r_4 + n_{33} \cdot \iota_2^2$$
$$n_{55} = 1/r_5 + n_{44} \cdot \sigma_c^2$$
$$n_{66} = 1/r_6 + n_{55} \cdot \iota_3^2$$
$$n_{77} = 1/r_7 + n_{66} \cdot \sigma_d^2$$
$$n_{88} = 1/r_8 + n_{77} \cdot \iota_4^2.$$

backward:

$$n_{88} = 1/v_8$$
$$n_{77} = 1/v_7 + n_{88} \cdot \iota'^2_4$$
$$n_{66} = 1/v_6 + n_{77} \cdot \sigma'^2_d$$
$$n_{55} = 1/v_5 + n_{66} \cdot \iota'^2_3$$
$$n_{44} = 1/v_4 + n_{55} \cdot \sigma'^2_c$$
$$n_{33} = 1/v_3 + n_{44} \cdot \iota'^2_2$$
$$n_{22} = 1/v_2 + n_{33} \cdot \sigma'^2_b$$
$$n_{11} = 1/v_1 + n_{22} \cdot \iota'^2_1.$$

(80) and (81)

Principal influence coefficients direct

$$n_{11} = \frac{1}{r_1}; \quad n_{ii} = \frac{1}{r_i + v_i - O_i}, \quad (i = 2, 3, 4, 5, 6, 7); \quad n_{88} = \frac{1}{v_8}. \tag{82}$$

Computation scheme for the symmetrical influence coefficient matrix

$$\begin{array}{c|cccccccc|c}
\cdot \iota_1 & (n_{11}) & n_{12} & n_{13} & n_{14} & n_{15} & n_{16} & n_{17} & n_{18} & \cdot \iota'_1 \\
& \downarrow & \uparrow & \uparrow & \uparrow & \uparrow & \uparrow & \uparrow & \uparrow & \\
\cdot \sigma_b & n_{21} & (n_{22}) & n_{23} & n_{24} & n_{25} & n_{26} & n_{27} & n_{28} & \cdot \sigma'_b \\
& \downarrow & \downarrow & \uparrow & \uparrow & \uparrow & \uparrow & \uparrow & \uparrow & \\
\cdot \iota_2 & n_{31} & n_{32} & (n_{33}) & n_{34} & n_{35} & n_{36} & n_{37} & n_{38} & \cdot \iota'_2 \\
& \downarrow & \downarrow & \downarrow & \uparrow & \uparrow & \uparrow & \uparrow & \uparrow & \\
\cdot \sigma_c & n_{41} & n_{42} & n_{43} & (n_{44}) & n_{45} & n_{46} & n_{47} & n_{48} & \cdot \sigma'_c \\
& \downarrow & \downarrow & \downarrow & \downarrow & \uparrow & \uparrow & \uparrow & \uparrow & \\
\cdot \iota_3 & n_{51} & n_{52} & n_{53} & n_{54} & (n_{55}) & n_{56} & n_{57} & n_{58} & \cdot \iota'_3 \\
& \downarrow & \downarrow & \downarrow & \downarrow & \downarrow & \uparrow & \uparrow & \uparrow & \\
\cdot \sigma_d & n_{61} & n_{62} & n_{63} & n_{64} & n_{65} & (n_{66}) & n_{67} & n_{68} & \cdot \sigma'_d \\
& \downarrow & \downarrow & \downarrow & \downarrow & \downarrow & \downarrow & \uparrow & \uparrow & \\
\cdot \iota_4 & n_{71} & n_{72} & n_{73} & n_{74} & n_{75} & n_{76} & (n_{77}) & n_{78} & \cdot \iota'_4 \\
& \downarrow & \downarrow & \downarrow & \downarrow & \downarrow & \downarrow & \downarrow & \uparrow & \\
& n_{81} & n_{82} & n_{83} & n_{84} & n_{85} & n_{86} & n_{87} & (n_{88}) &
\end{array} \tag{83}$$

Composite influence coefficients:

$$\left.\begin{array}{llll}
s_{11} = n_{11} - n_{12} & s_{12} = n_{13} - n_{14} & s_{13} = n_{15} - n_{16} & s_{14} = n_{17} - n_{18} \\
s_{21} = n_{22} - n_{21} & s_{22} = n_{23} - n_{24} & s_{23} = n_{25} - n_{26} & s_{24} = n_{27} - n_{28} \\
s_{31} = n_{32} - n_{31} & s_{32} = n_{33} - n_{34} & s_{33} = n_{35} - n_{36} & s_{34} = n_{37} - n_{38} \\
s_{41} = n_{42} - n_{41} & s_{42} = n_{44} - n_{43} & s_{43} = n_{45} - n_{46} & s_{44} = n_{47} - n_{48} \\
s_{51} = n_{52} - n_{51} & s_{52} = n_{54} - n_{53} & s_{53} = n_{55} - n_{56} & s_{54} = n_{57} - n_{58} \\
s_{61} = n_{62} - n_{61} & s_{62} = n_{64} - n_{63} & s_{63} = n_{66} - n_{65} & s_{64} = n_{67} - n_{68} \\
s_{71} = n_{72} - n_{71} & s_{72} = n_{74} - n_{73} & s_{73} = n_{76} - n_{75} & s_{74} = n_{77} - n_{78} \\
s_{81} = n_{82} - n_{81} & s_{82} = n_{84} - n_{83} & s_{83} = n_{86} - n_{85} & s_{84} = n_{88} - n_{87};
\end{array}\right\} \tag{84}$$

$$\left.\begin{array}{lll}
w_{1b} = n_{12} - n_{13} & w_{1c} = n_{14} - n_{15} & w_{1d} = n_{16} - n_{17} \\
w_{2b} = n_{22} - n_{23} & w_{2c} = n_{24} - n_{25} & w_{2d} = n_{26} - n_{27} \\
w_{3b} = n_{33} - n_{32} & w_{3c} = n_{34} - n_{35} & w_{3d} = n_{36} - n_{37} \\
w_{4b} = n_{43} - n_{42} & w_{4c} = n_{44} - n_{45} & w_{4d} = n_{46} - n_{47} \\
w_{5b} = n_{53} - n_{52} & w_{5c} = n_{55} - n_{54} & w_{5d} = n_{56} - n_{57} \\
w_{6b} = n_{63} - n_{62} & w_{6c} = n_{65} - n_{64} & w_{6d} = n_{66} - n_{67} \\
w_{7b} = n_{73} - n_{72} & w_{7c} = n_{75} - n_{74} & w_{7d} = n_{77} - n_{76} \\
w_{8b} = n_{83} - n_{82} & w_{8c} = n_{85} - n_{84} & w_{8d} = n_{87} - n_{86}.
\end{array}\right\} \tag{85}$$

Loading Condition 1

All the beam spans carrying arbitrary vertical loading

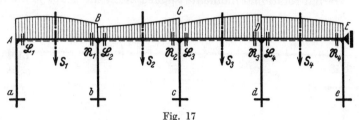

Fig. 17
Beam loading diagram with load terms $\mathfrak{L}$ and $\mathfrak{R}$

Beam moments ($L_\nu = \mathfrak{L}_\nu \, k_\nu$ and $R_\nu = \mathfrak{R}_\nu \, k_\nu$):

(86)	L_1	R_1	L_2	R_2	L_3	R_3	L_4	R_4
$X_1 = M_{A1} =$	$-n_{11}$	$+n_{12}$	$+n_{13}$	$-n_{14}$	$-n_{15}$	$+n_{16}$	$+n_{17}$	$-n_{18}$
$X_2 = M_{B1} =$	$+n_{21}$	$-n_{22}$	$-n_{23}$	$+n_{24}$	$+n_{25}$	$-n_{26}$	$-n_{27}$	$+n_{28}$
$X_3 = M_{B2} =$	$+n_{31}$	$-n_{32}$	$-n_{33}$	$+n_{34}$	$+n_{35}$	$-n_{36}$	$-n_{37}$	$+n_{38}$
$X_4 = M_{C2} =$	$-n_{41}$	$+n_{42}$	$+n_{43}$	$-n_{44}$	$-n_{45}$	$+n_{46}$	$+n_{47}$	$-n_{48}$
$X_5 = M_{C3} =$	$-n_{51}$	$+n_{52}$	$+n_{53}$	$-n_{54}$	$-n_{55}$	$+n_{56}$	$+n_{57}$	$-n_{58}$
$X_6 = M_{D3} =$	$+n_{61}$	$-n_{62}$	$-n_{63}$	$+n_{64}$	$+n_{65}$	$-n_{66}$	$-n_{67}$	$+n_{68}$
$X_7 = M_{D4} =$	$+n_{71}$	$-n_{72}$	$-n_{73}$	$+n_{74}$	$+n_{75}$	$-n_{76}$	$-n_{77}$	$+n_{78}$
$X_8 = M_{E4} =$	$-n_{81}$	$+n_{82}$	$+n_{83}$	$-n_{84}$	$-n_{85}$	$+n_{86}$	$+n_{87}$	$-n_{88}$.

Column moments:

(87) at top:
$$M_{Aa} = +X_1$$
$$M_{Bb} = -X_2 + X_3$$
$$M_{Cc} = -X_4 + X_5$$
$$M_{Dd} = -X_6 + X_7$$
$$M_{Ee} = -X_8;$$

at base:
$$M_n = -M_{Nn} \cdot \varepsilon_n. \qquad (88)$$

Special case: All the beam spans symmetrically loaded: ($R_\nu = L_\nu$):

$$\left.\begin{aligned}
X_1 = M_{A1} &= -L_1 \cdot s_{11} + L_2 \cdot s_{12} - L_3 \cdot s_{13} + L_4 \cdot s_{14} \\
X_2 = M_{B1} &= -L_1 \cdot s_{21} - L_2 \cdot s_{22} + L_3 \cdot s_{23} - L_4 \cdot s_{24} \\
X_3 = M_{B2} &= -L_1 \cdot s_{31} - L_2 \cdot s_{32} + L_3 \cdot s_{33} - L_4 \cdot s_{34} \\
X_4 = M_{C2} &= +L_1 \cdot s_{41} - L_2 \cdot s_{42} - L_3 \cdot s_{43} + L_4 \cdot s_{44} \\
X_5 = M_{C3} &= +L_1 \cdot s_{51} - L_2 \cdot s_{52} - L_3 \cdot s_{53} + L_4 \cdot s_{54} \\
X_6 = M_{D3} &= -L_1 \cdot s_{61} + L_2 \cdot s_{62} - L_3 \cdot s_{63} - L_4 \cdot s_{64} \\
X_7 = M_{D4} &= -L_1 \cdot s_{71} + L_2 \cdot s_{72} - L_3 \cdot s_{73} - L_4 \cdot s_{74} \\
X_8 = M_{E4} &= +L_1 \cdot s_{81} - L_2 \cdot s_{82} + L_3 \cdot s_{83} - L_4 \cdot s_{84}.
\end{aligned}\right\} \quad (89)$$

Loading Conditions 2 and 3
All the columns carrying arbitrary horizontal loading (2); external rotational moments applied at all the beam joints (3)

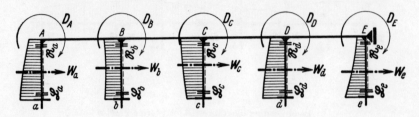

Fig. 18
Column loading diagram with load terms $\mathfrak{L}$ and $\mathfrak{R}$ for the columns and external rotational moments D

Joint load terms: $(L_n = \mathfrak{L}_n k_n$ and $R_n = \mathfrak{R}_n k_n)$:

$$\mathfrak{B}_N = D_N \cdot b_n + L_n \cdot \varepsilon_n - R_n. \tag{90}$$

Beam moments* :

	$\mathfrak{B}_A$	$\mathfrak{B}_B$	$\mathfrak{B}_C$	$\mathfrak{B}_D$	$\mathfrak{B}_E$
$X_1 = M_{A1} =$	$+n_{11}$	$+w_{1b}$	$-w_{1c}$	$+w_{1d}$	$-n_{18}$
$X_2 = M_{B1} =$	$-n_{21}$	$-w_{2b}$	$+w_{2c}$	$-w_{2d}$	$+n_{28}$
$X_3 = M_{B2} =$	$-n_{31}$	$+w_{3b}$	$+w_{3c}$	$-w_{3d}$	$+n_{38}$
$X_4 = M_{C2} =$	$+n_{41}$	$-w_{4b}$	$-w_{4c}$	$+w_{4d}$	$-n_{48}$
$X_5 = M_{C3} =$	$+n_{51}$	$-w_{5b}$	$+w_{5c}$	$+w_{5d}$	$-n_{58}$
$X_6 = M_{D3} =$	$-n_{61}$	$+w_{6b}$	$-w_{6c}$	$-w_{6d}$	$+n_{68}$
$X_7 = M_{D4} =$	$-n_{71}$	$+w_{7b}$	$-w_{7c}$	$+w_{7d}$	$+n_{78}$
$X_8 = M_{E4} =$	$+n_{81}$	$-w_{8b}$	$+w_{8c}$	$-w_{8d}$	$-n_{88}$.

$$\tag{91}$$

Column moments at top:

$$\left. \begin{aligned} M_{Aa} &= -D_A + X_1 \\ M_{Bb} &= -D_B - X_2 + X_3 \\ M_{Cc} &= -D_C - X_4 + X_5 \\ M_{Dd} &= -D_D - X_6 + X_7 \\ M_{Ee} &= -D_E - X_8; \end{aligned} \right\} \tag{92}$$

at elastically restrained base:

$$M_n = -(\mathfrak{L}_n + M_{Nn}) \varepsilon_n. \tag{93}$$

---

* In the case of a 2-bay frame the composite symbols w_{ic} in the column under $\mathfrak{B}_C$ must be replaced by the fourth column of symbols n (i.e., n_{i4} where $i = 1, 2, 3, 4$)

II. Treatment by the deformation method:
Coefficients:

Stiffness coefficients for all individual members:

$$K_\nu = \frac{J_\nu}{J_k} \cdot \frac{l_k}{l_\nu} = \frac{1}{k_\nu}; \qquad K_n = \frac{J_n}{J_k} \cdot \frac{l_k}{h_n} = \frac{1}{k_n}. \tag{94}$$

Stiffnesses of the elastically restrained columns:

$$S_n = \frac{K_n}{2 - \varepsilon_n}; \tag{95}$$

(the values ε_n are assumed to be known).

Joint coefficients:

$$\left.\begin{aligned}
K_A &= 3 S_a + 2 K_1 \\
K_B &= 2 K_1 + 3 S_b + 2 K_2 \\
K_C &= 2 K_2 + 3 S_c + 2 K_3 \\
K_D &= 2 K_3 + 3 S_d + 2 K_4 \\
K_E &= 2 K_4 + 3 S_e .
\end{aligned}\right\} \tag{96}$$

Rotation carry-over factors and auxiliary coefficients:

backward reduction:

$$\begin{aligned}
& & r_e &= K_E \\
j_4 &= \frac{K_4}{r_e} & & \\
& & r_d &= K_D - K_4 \cdot j_4 \\
j_3 &= \frac{K_3}{r_d} & & \\
& & r_c &= K_C - K_3 \cdot j_3 \\
j_2 &= \frac{K_2}{r_c} & & \\
& & r_b &= K_B - K_2 \cdot j_2 \\
j_1 &= \frac{K_1}{r_b} & & \\
& & r_a &= K_A - K_1 \cdot j_1 .
\end{aligned}$$

forward reduction:

$$\left.\begin{aligned}
& & v_a &= K_A \\
j_1' &= \frac{K_1}{v_a} & & \\
& & v_b &= K_B - K_1 \cdot j_1' \\
j_2' &= \frac{K_2}{v_b} & & \\
& & v_c &= K_C - K_2 \cdot j_2' \\
j_3' &= \frac{K_3}{v_c} & & \\
& & v_d &= K_D - K_3 \cdot j_3' \\
j_4' &= \frac{K_4}{v_d} & & \\
& & v_e &= K_E - K_4 \cdot j_4' .
\end{aligned}\right\} \begin{matrix}(97)\\ \text{and} \\ (98)\end{matrix}$$

Principal influence coefficients by recursion:

forward:

$$\begin{aligned}
u_{aa} &= 1/r_a \\
u_{bb} &= 1/r_b + u_{aa} \cdot j_1^2 \\
u_{cc} &= 1/r_c + u_{bb} \cdot j_2^2 \\
u_{dd} &= 1/r_d + u_{cc} \cdot j_3^2 \\
u_{ee} &= 1/r_e + u_{dd} \cdot j_4^2 .
\end{aligned}$$

backward:

$$\left.\begin{aligned}
u_{ee} &= 1/v_e \\
u_{dd} &= 1/v_d + u_{ee} \cdot j_4'^2 \\
u_{cc} &= 1/v_c + u_{dd} \cdot j_3'^2 \\
u_{bb} &= 1/v_b + u_{cc} \cdot j_2'^2 \\
u_{aa} &= 1/v_a + u_{bb} \cdot j_1'^2 .
\end{aligned}\right\} \begin{matrix}(99)\\ \text{and}\\ (100)\end{matrix}$$

Principal influence coefficients direct

$$u_{aa} = \frac{1}{r_a}; \quad u_{nn} = \frac{1}{r_n + v_n - K_N}, \begin{pmatrix} n = b, c, d \\ N = B, C, D \end{pmatrix}; \quad u_{ee} = \frac{1}{v_e}. \tag{101}$$

Computation scheme for the symmetrical influence coefficient matrix

$$\left. \begin{array}{c} \cdot j_1 \\ \cdot j_2 \\ \cdot j_3 \\ \cdot j_4 \end{array} \right| \begin{array}{ccccc} (u_{aa}) & u_{ab} & u_{ac} & u_{ad} & u_{ae} \\ \downarrow & \uparrow & \uparrow & \uparrow & \uparrow \\ u_{ba} & (u_{bb}) & u_{bc} & u_{bd} & u_{be} \\ \downarrow & \downarrow & \uparrow & \uparrow & \uparrow \\ u_{ca} & u_{cb} & (u_{cc}) & u_{cd} & u_{ce} \\ \downarrow & \downarrow & \downarrow & \uparrow & \uparrow \\ u_{da} & u_{db} & u_{dc} & (u_{dd}) & u_{de} \\ \downarrow & \downarrow & \downarrow & \downarrow & \uparrow \\ u_{ea} & u_{eb} & u_{ec} & u_{ed} & (u_{ee}) \end{array} \left| \begin{array}{c} \cdot j_1' \\ \cdot j_2' \\ \cdot j_3' \\ \cdot j_4' \end{array} \right. \quad (102)$$

Loading Condition 4
Overall loading condition (superposition of Loading Conditions 1, 2 and 3)

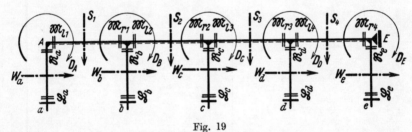

Fig. 19

Loading diagram with fixed-end moments $\mathfrak{M}_l$ and $\mathfrak{M}_r$ for the beam spans, load terms $\mathfrak{L}$ and $\mathfrak{R}$ for the columns, and external rotational moments D

Column restraining moments at the beam joints:

$$\mathfrak{E}_{rn} = - \frac{\mathfrak{R}_n - \mathfrak{L}_n \cdot \varepsilon_n}{2 - \varepsilon_n} . \qquad (103)$$

Joint load terms (balancing moments):

$$\left. \begin{array}{l} \mathfrak{K}_A = \phantom{\mathfrak{M}_{r1}} - \mathfrak{M}_{l1} + \mathfrak{E}_{ra} + D_A \\ \mathfrak{K}_B = \mathfrak{M}_{r1} - \mathfrak{M}_{l2} + \mathfrak{E}_{rb} + D_B \\ \mathfrak{K}_C = \mathfrak{M}_{r2} - \mathfrak{M}_{l3} + \mathfrak{E}_{rc} + D_C \\ \mathfrak{K}_D = \mathfrak{M}_{r3} - \mathfrak{M}_{l4} + \mathfrak{E}_{rd} + D_D \\ \mathfrak{K}_E = \mathfrak{M}_{r4} \phantom{- \mathfrak{M}_{l4}} + \mathfrak{E}_{re} + D_E . \end{array} \right\} \qquad (104)$$

auxiliary moments:

	$\mathfrak{K}_A$	$\mathfrak{K}_B$	$\mathfrak{K}_C$	$\mathfrak{K}_D$	$\mathfrak{K}_E$
$Y_A =$	$+u_{aa}$	$-u_{ab}$	$+u_{ac}$	$-u_{ad}$	$+u_{ae}$
$Y_B =$	$-u_{ba}$	$+u_{bb}$	$-u_{bc}$	$+u_{bd}$	$-u_{be}$
$Y_C =$	$+u_{ca}$	$-u_{cb}$	$+u_{cc}$	$-u_{cd}$	$+u_{ce}$
$Y_D =$	$-u_{da}$	$+u_{db}$	$-u_{dc}$	$+u_{dd}$	$-u_{de}$
$Y_E =$	$+u_{ea}$	$-u_{eb}$	$+u_{ec}$	$-u_{ed}$	$+u_{ee}.$

(105)

Beam moments:

$$\begin{aligned}
M_{A1} &= \mathfrak{M}_{l1} + (2Y_A + Y_B)K_1 & M_{B1} &= \mathfrak{M}_{r1} - (Y_A + 2Y_B)K_1 \\
M_{B2} &= \mathfrak{M}_{l2} + (2Y_B + Y_C)K_2 & M_{C2} &= \mathfrak{M}_{r2} - (Y_B + 2Y_C)K_2 \\
M_{C3} &= \mathfrak{M}_{l3} + (2Y_C + Y_D)K_3 & M_{D3} &= \mathfrak{M}_{r3} - (Y_C + 2Y_D)K_3 \\
M_{D4} &= \mathfrak{M}_{l4} + (2Y_D + Y_E)K_4; & M_{E4} &= \mathfrak{M}_{r4} - (Y_D + 2Y_E)K_4.
\end{aligned}$$

(106)

Column moments:

at top: $\qquad M_{Nn} = \mathfrak{C}_{rn} - Y_N \cdot 3 S_n;$ \hfill (107)

at base: $\qquad M_n = -(\mathfrak{L}_n + M_{Nn})\varepsilon_n.$ \hfill (108)

Check relationships:

$$\begin{aligned}
D_A + M_{Aa} \qquad\quad - M_{A1} &= 0 \\
D_B + M_{Bb} + M_{B1} - M_{B2} &= 0 \\
D_C + M_{Cc} + M_{C2} - M_{C3} &= 0 \\
D_D + M_{Dd} + M_{D3} - M_{D4} &= 0 \\
D_E + M_{Ee} + M_{E4} \qquad\quad &= 0.
\end{aligned}$$

(109)

Horizontal thrusts, column shears and lateral restraining force

Generally as indicated in formulas (36)—(39)
Collective subscripts for (36) and (38):

$$N = A, B, C, D, E \text{ and } n = a, b, c, d, e$$

The expressions for H—(37) and (39)—to include additional terms $-H_e$ and $-Q_{re}$ respectively.

Frame Shape 3v

(Frame Shape 3 with laterally unrestrained horizontal member)

First step: Calculation of all the requisite loading conditions as well as lateral restraining force H according to Frame Shape 3.

Second Step: Determination of the effect of a horizontal concentrated load W, acting at the level of the horizontal member, with the aid of Loading Condition 5 for Frame Shape 3v, as indicated below.

Third step: Determination of the final moments and forces for Frame Shape 3v by superposition of the results of the first step and those of the second step for $W = H$.

Loading Condition 5

Horizontal concentrated load W at level of horizontal member

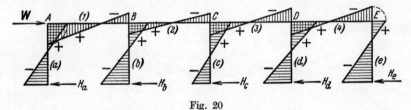

Fig. 20

Loading diagram and anti-symmetrical diagram of bending moments. (For clarity, only the reactions H_n at the column feet are shown).

Column ratios
with arbitrary comparison height h_k $\Big\}$ $\gamma_n = \dfrac{h_k}{h_n}.$ \hfill (110)

Specific bending moments m (instead of M);
either according to *Method I* from the formulas (91), (92) and (93)—i.e., the moments $X_i = m_{Nv}$, m_{Nn} and m_n—for:

$$\mathfrak{B}_N = \gamma_n(1 + \varepsilon_n) \quad \text{and} \quad \mathfrak{L}_n = \gamma_n/k_n; \hfill (111)$$

or according to *Method II* from the formulas (105), (106) (107) and (108)— i.e., the moments y_{NN} (instead of Y_{Nv}), m_{Nn}, m_{Nn} and m_n—for:

$$\mathfrak{K}_N = \mathfrak{E}_{rn} = S_n \gamma_n (1 + \varepsilon_n) \quad \text{and} \quad \mathfrak{L}_n = \gamma_n K_n. \tag{112}$$

Note: All the other load terms in bold type in the formulas indicated should be omitted.

Specific h_k-fold column shears:

$$p_n = (m_{Nn} - m_n) \gamma_n. \tag{113}$$

Corresponding specific h_k-fold applied load

$$p = p_a + p_b + p_c + p_d + p_e. \tag{114}$$

Conversion moment for a given load W:

$$U = \frac{W h_k}{p}. \tag{115}$$

Final moments and forces for the given concentrated load W:

Beam moments:

$$\begin{aligned} M_{A1} &= U \cdot m_{A1} & M_{B2} &= U \cdot m_{B2} & M_{C3} &= U \cdot m_{C3} & M_{D4} &= U \cdot m_{D4} \\ M_{B1} &= U \cdot m_{B1} & M_{C2} &= U \cdot m_{C2} & M_{D3} &= U \cdot m_{D3} & M_{E4} &= U \cdot m_{E4} \end{aligned} \tag{116}$$

Column moments:

$$M_{Nn} = U \cdot m_{Nn}; \quad M_n = U \cdot m_n; \tag{117}$$

Column shears and horizontal thrusts:

$$Q_n = H_n = \frac{U \cdot p_n}{h_k} = W \cdot \frac{p_n}{p}. \tag{118}$$

(Collective subscripts: $n = a, b, c, d, e$ and $N = A, B, C, D, E$.)

Note: The supplementary note on page 10 applies also to the computation procedure for Frame Shape 3v.

Frame Shape 4

Single-story two-bay frame with laterally restrained horizontal member and three elastically restrained columns

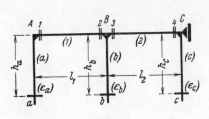

Fig. 21
Frame shape, dimensions and symbols

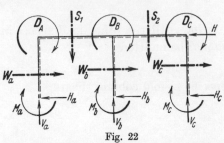

Fig. 22
Definition of positive direction for all external loads on joints and members, for all column reactions, and for the lateral restraining force

---

Important

All the formulas required for Frame Shape 4 are already comprised in the sets of formulas that have been given for *Frame Shape* 3, namely, for the collective subscripts $\nu = 1, 2$; $N = A, B, C$; $n = a, b, c$. All the sets of formulas without collective subscripts are valid up to the dotted lines; conversely, for backward reductions and backward recursions they are valid from those dotted lines onward.

Frame Shape 4v

(Frame Shape 4 with laterally unrestrained horizontal member)

Procedure

exactly as for Frame Shape 3v, with the collective subscripts as for Frame Shape 4, having due regard to the dotted boundary lines in formulas (114) and (116).

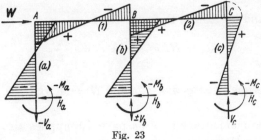

Fig. 23
Loading Condition 5, with diagram of moments and support reactions

Frame Shape 5
Symmetrical single-story five-bay frame with laterally restrained horizontal member and six elastically restrained columns

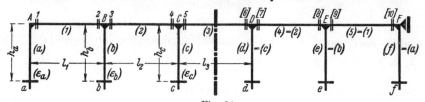

Fig. 24
Frame shape, dimensions and symbols

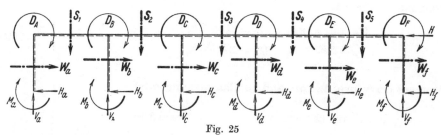

Fig. 25
Definition of positive direction for all external loads on joints and members, for all column reactions, and for the lateral restraining force

All the dimensions and coefficients for the right-hand half of the frame are the same as those for the left-hand half. In the case of the columns the broken line is always placed on the right-hand side of each column, despite the symmetry of the frame.

Note: For Frame Shape 5 the *coefficients* and the *Loading Conditions* 1–4 for *Frame Shape* 1, enlarged by the addition of two bays, may be employed. On account of symmetry, the forward reduction becomes the same as the backward reduction, and the backward recursion becomes the same as the forward recursion, both in Method I and Method II. Furthermore, the *square matrices of influence coefficients* (with 10×10 elements in Method I and with 6×6 elements in Method II) are bisymmetrical, i.e., symmetrical about the principal diagonal and the secondary diagonal. If the *"load transposition method"* is used, however, it is also possible to start from the following symmetrical and anti-symmetrical loading conditions.

Important

All the instructions, formula sequences and formula matrices given for Frame Shape 5 can readily be extended to suit symmetrical frames of similar type having any odd number of bays.

I. Treatment by the Method of Forces

a) Symmetrical load arrangements

Coefficients

Flexibility coefficients of all the members, according to formula (1), for $\nu = 1, 2, 3$ and $n = a, b, c$

Flexibilities of the columns, according to formula (2), for $n = a, b, c$

Diagonal coefficients:

$$\begin{aligned} O_1 &= b_a + 2k_1 & O_3 &= b_b + 2k_2 & O_5' &= b_c + 3k_3 \, . \\ O_2 &= 2k_1 + b_b & O_4 &= 2k_2 + b_c & & \end{aligned} \quad\quad (119)$$

Moment carry-over factors and auxiliary coefficients:

backward reduction: | forward reduction:

$$\begin{aligned} \sigma_{cs} &= \frac{b_c}{r_5'} & r_5' &= O_5' & \iota_1' &= \frac{k_1}{v_1} & v_1 &= O_1 \\ & & r_4' &= O_4 - b_c \cdot \sigma_{cs} & & & v_2 &= O_2 - k_1 \cdot \iota_1' \\ \iota_{2s} &= \frac{k_2}{r_4'} & & & \sigma_b' &= \frac{b_b}{v_2} & & \\ & & r_3' &= O_3 - k_2 \cdot \iota_{2s} & & & v_3 &= O_3 - b_b \cdot \sigma_b' \\ \sigma_{bs} &= \frac{b_b}{r_3'} & & & \iota_2' &= \frac{k_2}{v_3} & & \\ & & r_2' &= O_2 - b_b \cdot \sigma_{bs} & & & v_4 &= O_4 - k_2 \cdot \iota_2' \\ \iota_{1s} &= \frac{k_1}{r_2'} & & & \sigma_c' &= \frac{b_c}{v_4} & & \\ & & r_1' &= O_1 - k_1 \cdot \iota_{1s} \, . & & & v_5' &= O_5' - b_c \cdot \sigma_c' \, . \end{aligned} \quad\quad \begin{matrix}(120)\\ \text{and}\\(121)\end{matrix}$$

Principal influence coefficients by recursion:

forward: | backward:

$$\begin{aligned} n_{11}' &= \frac{1}{r_1'} & n_{55}' &= \frac{1}{v_5'} \\ n_{22}' &= \frac{1}{r_2'} + n_{11}' \cdot \iota_{1s}^2 & n_{44}' &= \frac{1}{v_4} + n_{55}' \cdot \sigma_c'^2 \\ n_{33}' &= \frac{1}{r_3'} + n_{22}' \cdot \sigma_{bs}^2 & n_{33}' &= \frac{1}{v_3} + n_{44}' \cdot \iota_2'^2 \\ n_{44}' &= \frac{1}{r_4'} + n_{33}' \cdot \iota_{2s}^2 & n_{22}' &= \frac{1}{v_2} + n_{33}' \cdot \sigma_b'^2 \\ n_{55}' &= \frac{1}{r_5'} + n_{44}' \cdot \sigma_{cs}^2 \, . & n_{11}' &= \frac{1}{v_1} + n_{22}' \cdot \iota_1'^2 \, . \end{aligned} \quad\quad \begin{matrix}(122)\\ \text{and}\\(123)\end{matrix}$$

Principal influence coefficients direct

$$n'_{11} = \frac{1}{r'_1}; \quad n'_{ii} = \frac{1}{r'_i + v_i - O_i}, \quad (i = 2, 3, 4); \quad n'_{55} = \frac{1}{v'_5}. \qquad (124)$$

Computation scheme for the symmetrical influence coefficient matrix*

$$\begin{array}{c|ccccc|c}
 & (n'_{11}) & n'_{12} & n'_{13} & n'_{14} & n'_{15} & \\
\cdot\, \iota_{1s} & \downarrow & \uparrow & \uparrow & \uparrow & \uparrow & \cdot\, \iota'_1 \\
 & n'_{21} & (n'_{22}) & n'_{23} & n'_{24} & n'_{25} & \\
\cdot\, \sigma_{bs} & \downarrow & \downarrow & \uparrow & \uparrow & \uparrow & \cdot\, \sigma'_b \\
 & n'_{31} & n'_{32} & (n'_{33}) & n'_{34} & n'_{35} & \\
\cdot\, \iota_{2s} & \downarrow & \downarrow & \downarrow & \uparrow & \uparrow & \cdot\, \iota'_2 \\
 & n'_{41} & n'_{42} & n'_{43} & (n'_{44}) & n'_{45} & \\
\cdot\, \sigma_{cs} & \downarrow & \downarrow & \downarrow & \downarrow & \uparrow & \cdot\, \sigma'_c \\
 & n'_{51} & n'_{52} & n'_{53} & n'_{54} & (n'_{55}) & \\
\end{array} \qquad (125)$$

Composite influence coefficients **

$$\begin{aligned}
s'_{11} &= + n'_{11} - n'_{12} = (s_{11} + s_{15}) & s'_{12} &= + n'_{13} - n'_{14} = (s_{12} + s_{14}) \\
s'_{21} &= - n'_{21} + n'_{22} = (s_{21} - s_{25}) & s'_{22} &= + n'_{23} - n'_{24} = (s_{22} + s_{24}) \\
s'_{31} &= - n'_{31} + n'_{32} = (s_{31} - s_{35}) & s'_{32} &= + n'_{33} - n'_{34} = (s_{32} + s_{34}) \\
s'_{41} &= - n'_{41} + n'_{42} = (s_{41} - s_{45}) & s'_{42} &= - n'_{43} + n'_{44} = (s_{42} - s_{44}) \\
s'_{51} &= - n'_{51} + n'_{52} = (s_{51} - s_{55}) & s'_{52} &= - n'_{53} + n'_{54} = (s_{52} - s_{54});
\end{aligned} \qquad (126)$$

$$\begin{aligned}
w'_{1b} &= + n'_{12} - n'_{13} = (w_{1b} + w_{1e}) & w'_{1c} &= + n'_{14} - n'_{15} = (w_{1c} + w_{1d}) \\
w'_{2b} &= + n'_{22} - n'_{23} = (w_{2b} + w_{2e}) & w'_{2c} &= + n'_{24} - n'_{25} = (w_{2c} + w_{2d}) \\
w'_{3b} &= - n'_{32} + n'_{33} = (w_{3b} - w_{3e}) & w'_{3c} &= + n'_{34} - n'_{35} = (w_{3c} + w_{3d}) \\
w'_{4b} &= - n'_{42} + n'_{43} = (w_{4b} - w_{4e}) & w'_{4c} &= + n'_{44} - n'_{45} = (w_{4c} + w_{4d}) \\
w'_{5b} &= - n'_{52} + n'_{53} = (w_{5b} - w_{5e}) & w'_{5c} &= - n'_{54} + n'_{55} = (w_{5c} - w_{5d}).
\end{aligned} \qquad (127)$$

---

* Between the elements of this 5×5 matrix for the left-hand half of the frame and the elements of the 10×10 matrix for the whole frame, as mentioned on page 27, there exist, according to the "load transposition method" the following relationships:

$$n'_{11} = n_{11} - n_{1,10}, \quad n'_{12} = n_{12} - n_{19} \text{ etc., up to} \quad n'_{55} = n_{55} - n_{56}.$$

** The members enclosed in parentheses have been formed with the values s and w of the 10×10 matrix.

Loading Condition 1a

Horizontal member carrying vertical loading of arbitrary magnitude, but symmetrical about center of frame

$$(\Re_5 = \mathfrak{L}_1, \quad \mathfrak{L}_5 = \Re_1; \quad \Re_4 = \mathfrak{L}_2, \quad \mathfrak{L}_4 = \Re_2; \quad \Re_3 = \mathfrak{L}_3).$$

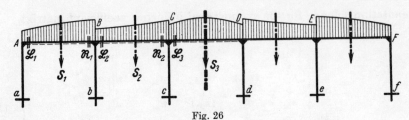

Fig. 26

Beam loading diagram with load terms $\mathfrak{L}$ and $\Re$ for left-hand half of frame

Beam moments: $(L_\nu = \mathfrak{L}_\nu \, k_\nu \quad \text{and} \quad R_\nu = \Re_\nu \, k_\nu):$

	L_1	R_1	L_2	R_2	L_3
$X'_1 = M_{A1} = M_{F5} =$	$-n'_{11}$	$+n'_{12}$	$+n'_{13}$	$-n'_{14}$	$-n'_{15}$
$X'_2 = M_{B1} = M_{E5} =$	$+n'_{21}$	$-n'_{22}$	$-n'_{23}$	$+n'_{24}$	$+n'_{25}$
$X'_3 = M_{B2} = M_{E4} =$	$+n'_{31}$	$-n'_{32}$	$-n'_{33}$	$+n'_{34}$	$+n'_{35}$
$X'_4 = M_{C2} = M_{D4} =$	$-n'_{41}$	$+n'_{42}$	$+n'_{43}$	$-n'_{44}$	$-n'_{45}$
$X'_5 = M_{C3} = M_{D3} =$	$-n'_{51}$	$+n'_{52}$	$+n'_{53}$	$-n'_{54}$	$-n'_{55}.$

(128)

Column moments:

at top:

$$M_{Aa} = -M_{Ff} = + X'_1$$
$$M_{Bb} = -M_{Ee} = -X'_2 + X'_3$$
$$M_{Cc} = -M_{Dd} = -X'_4 + X'_5;$$

at base:

$$M_a = -M_f = -M_{Aa} \cdot \varepsilon_a$$
$$M_b = -M_e = -M_{Bb} \cdot \varepsilon_b$$
$$M_c = -M_d = -M_{Cc} \cdot \varepsilon_c.$$

(129) and (130)

Special case: symmetrical bay loading ($R_1 = L_1$ and $R_2 = L_2$):

$$\begin{aligned}
X'_1 = M_{A1} = M_{F5} &= -L_1 \cdot s'_{11} + L_2 \cdot s'_{12} - L_3 \cdot n'_{15} \\
X'_2 = M_{B1} = M_{E5} &= -L_1 \cdot s'_{21} - L_2 \cdot s'_{22} + L_3 \cdot n'_{25} \\
X'_3 = M_{B2} = M_{E4} &= -L_1 \cdot s'_{31} - L_2 \cdot s'_{32} + L_3 \cdot n'_{35} \\
X'_4 = M_{C2} = M_{D4} &= +L_1 \cdot s'_{41} - L_2 \cdot s'_{42} - L_3 \cdot n'_{45} \\
X'_5 = M_{C3} = M_{D3} &= +L_1 \cdot s'_{51} - L_2 \cdot s'_{52} - L_3 \cdot n'_{55}.
\end{aligned}$$

(131)

Loading Conditions 2a and 3a

All the columns carrying arbitrary horizontal loading (2a) and external rotational moments applied at all the beam joints (3a), but load arrangement as a whole is symmetrical about center of frame

$(\mathfrak{L}_f = -\mathfrak{L}_a, \quad \mathfrak{R}_f = -\mathfrak{R}_a; \quad \mathfrak{L}_e = -\mathfrak{L}_b, \quad \mathfrak{R}_e = -\mathfrak{R}_b; \quad \mathfrak{L}_d = -\mathfrak{L}_c, \quad \mathfrak{R}_d = -\mathfrak{R}_c;$
$D_F = -D_A \quad D_E = -D_B \quad D_D = -D_C)$

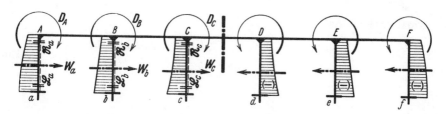

Fig. 27
Column loading diagram with load terms $\mathfrak{L}$ and $\mathfrak{R}$ for the columns, and external rotational moments D, for left-hand half of frame

Joint load terms $(L_n = \mathfrak{L}_n k_n \quad \text{and} \quad R_n = \mathfrak{R}_n k_n)$:

$$\mathfrak{B}_N = D_N \cdot b_n + L_n \cdot \varepsilon_n - R_n \qquad (132)$$
$$\text{for} \quad N = A, B, C \quad \text{and} \quad n = a, b, c.$$

Beam moments:

	$\mathfrak{B}_A$	$\mathfrak{B}_B$	$\mathfrak{B}_C$
$X_1' = M_{A1} = M_{F5} =$	$+n_{11}'$	$+w_{1b}'$	$-w_{1c}'$
$X_2' = M_{B1} = M_{E5} =$	$-n_{21}'$	$-w_{2b}'$	$+w_{2c}'$
$X_3' = M_{B2} = M_{E4} =$	$-n_{31}'$	$+w_{3b}'$	$+w_{3c}'$
$X_4' = M_{C2} = M_{D4} =$	$+n_{41}'$	$-w_{4b}'$	$-w_{4c}'$
$X_5' = M_{C3} = M_{D3} =$	$+n_{51}'$	$-w_{5b}'$	$+w_{5c}'.$

(133)

Column moments:

at top: at base:

$M_{Aa} = -M_{Ff} = -D_A \qquad\quad + X_1' \qquad M_a = -M_f = -(\mathfrak{L}_a + M_{Aa})\,\varepsilon_a$
$M_{Bb} = -M_{Ee} = -D_B - X_2' + X_3' \qquad M_b = -M_e = -(\mathfrak{L}_b + M_{Bb})\,\varepsilon_b$
$M_{Cc} = -M_{Dd} = -D_C - X_4' + X_5'; \qquad M_c = -M_d = -(\mathfrak{L}_c + M_{Cc})\,\varepsilon_c.$

(134) and (135)

b) Anti-symmetrical load arrangements

Coefficients

Flexibility coefficients of all the members, according to formulas (1), for $\nu = 1, 2, 3$ and $n = a, b, c$

Flexibilities of the columns, according to formula (2), for $n = a, b, c$

Diagonal coefficients:

$$\left.\begin{array}{lll} O_1 = b_a + 2k_1 & O_3 = b_b + 2k_2 & O_5'' = b_c + k_3 \\ O_2 = 2k_1 + b_b & O_4 = 2k_2 + b_c & \end{array}\right\} \quad (136)$$

Moment carry-over factors and auxiliary coefficients:

$$\left.\begin{array}{ll} \begin{array}{l}\text{backward reduction:} \\[4pt] \sigma_{ct} = \dfrac{b_c}{r_5''} \qquad r_5'' = O_5'' \\[6pt] \iota_{2t} = \dfrac{k_2}{r_4''} \qquad r_4'' = O_4 - b_c \cdot \sigma_{ct} \\[6pt] \sigma_{bt} = \dfrac{b_b}{r_3''} \qquad r_3'' = O_3 - k_2 \cdot \iota_{2t} \\[6pt] \iota_{1t} = \dfrac{k_1}{r_2''} \qquad r_2'' = O_2 - b_b \cdot \sigma_{bt} \\[6pt] \qquad\qquad\;\; r_1'' = O_1 - k_1 \cdot \iota_{1t} . \end{array} & \begin{array}{l}\text{forward reduction:} \\[4pt] \qquad\qquad\; v_1 = O_1 \\[6pt] \iota_1' = \dfrac{k_1}{v_1} \\[6pt] \qquad\qquad\; v_2 = O_2 - k_1 \cdot \iota_1' \\[6pt] \sigma_b' = \dfrac{b_b}{v_2} \\[6pt] \qquad\qquad\; v_3 = O_3 - b_b \cdot \sigma_b' \\[6pt] \iota_2' = \dfrac{k_2}{v_3} \\[6pt] \qquad\qquad\; v_4 = O_4 - k_2 \cdot \iota_2' \\[6pt] \sigma_c' = \dfrac{b_c}{v_4} \\[6pt] \qquad\qquad\; v_5'' = O_5'' - b_c \cdot \sigma_c' . \end{array} \end{array}\right\} \quad \begin{array}{c}(137) \\ \text{and} \\ (138)\end{array}$$

Principal influence coefficients by recursion:

$$\left.\begin{array}{ll} \begin{array}{l}\text{forward:} \\[4pt] n_{11}'' = \dfrac{1}{r_1''} \\[8pt] n_{22}'' = \dfrac{1}{r_2''} + n_{11}'' \cdot \iota_{1t}^2 \\[8pt] n_{33}'' = \dfrac{1}{r_3''} + n_{22}'' \cdot \sigma_{bt}^2 \\[8pt] n_{44}'' = \dfrac{1}{r_4''} + n_{33}'' \cdot \iota_{2t}^2 \\[8pt] n_{55}'' = \dfrac{1}{r_5''} + n_{44}'' \cdot \sigma_{ct}^2 . \end{array} & \begin{array}{l}\text{backward:} \\[4pt] n_{55}'' = \dfrac{1}{v_5''} \\[8pt] n_{44}'' = \dfrac{1}{v_4} + n_{55}'' \cdot \sigma_c'^2 \\[8pt] n_{33}'' = \dfrac{1}{v_3} + n_{44}'' \cdot \iota_2'^2 \\[8pt] n_{22}'' = \dfrac{1}{v_2} + n_{33}'' \cdot \sigma_b'^2 \\[8pt] n_{11}'' = \dfrac{1}{v_1} + n_{22}'' \cdot \iota_1'^2 . \end{array} \end{array}\right\} \quad \begin{array}{c}(139) \\ \text{and} \\ (140)\end{array}$$

Principal influence coefficients direct

$$n''_{11} = \frac{1}{r''_1}; \qquad n''_{ii} = \frac{1}{r''_i + v_i - O_i}, \qquad (i = 2, 3, 4); \qquad n''_{55} = \frac{1}{v''_5}. \qquad (141)$$

Computation scheme for the symmetrical influence coefficient matrix*

$$\left. \begin{array}{l} \cdot\, \iota_{1t} \\ \cdot\, \sigma_{bt} \\ \cdot\, \iota_{2t} \\ \cdot\, \sigma_{ct} \end{array} \right| \begin{array}{ccccc} (n''_{11}) & n''_{12} & n''_{13} & n''_{14} & n''_{15} \\ \downarrow & \uparrow & \uparrow & \uparrow & \uparrow \\ n''_{21} & (n''_{22}) & n''_{23} & n''_{24} & n''_{25} \\ \downarrow & \downarrow & \uparrow & \uparrow & \uparrow \\ n''_{31} & n''_{32} & (n''_{33}) & n''_{34} & n''_{35} \\ \downarrow & \downarrow & \downarrow & \uparrow & \uparrow \\ n''_{41} & n''_{42} & n''_{43} & (n''_{44}) & n''_{45} \\ \downarrow & \downarrow & \downarrow & \downarrow & \uparrow \\ n''_{51} & n''_{52} & n''_{53} & n''_{54} & (n''_{55}) \end{array} \right| \begin{array}{l} \cdot\, \iota'_1 \\ \cdot\, \sigma'_b \\ \cdot\, \iota'_2 \\ \cdot\, \sigma'_c \end{array} \qquad (142)$$

Composite influence coefficients** :

$$\begin{aligned}
s''_{11} &= +n''_{11} - n''_{12} = (s_{11} - s_{15}) & s''_{12} &= +n''_{13} - n''_{14} = (s_{12} - s_{14}) \\
s''_{21} &= -n''_{21} + n''_{22} = (s_{21} + s_{25}) & s''_{22} &= +n''_{23} - n''_{24} = (s_{22} - s_{24}) \\
s''_{31} &= -n''_{31} + n''_{32} = (s_{31} + s_{35}) & s''_{32} &= +n''_{33} - n''_{34} = (s_{32} - s_{34}) \\
s''_{41} &= -n''_{41} + n''_{42} = (s_{41} + s_{45}) & s''_{42} &= -n''_{43} + n''_{44} = (s_{42} + s_{44}) \\
s''_{51} &= -n''_{51} + n''_{52} = (s_{51} + s_{55}) & s''_{52} &= -n''_{53} + n''_{54} = (s_{52} + s_{54});
\end{aligned} \qquad (143)$$

$$\begin{aligned}
w''_{1b} &= +n''_{12} - n''_{13} = (w_{1b} - w_{1e}) & w''_{1c} &= +n''_{14} - n''_{15} = (w_{1c} - w_{1d}) \\
w''_{2b} &= +n''_{22} - n''_{23} = (w_{2b} - w_{2e}) & w''_{2c} &= +n''_{24} - n''_{25} = (w_{2c} - w_{2d}) \\
w''_{3b} &= -n''_{32} + n''_{33} = (w_{3b} + w_{3e}) & w''_{3c} &= +n''_{34} - n''_{35} = (w_{3c} - w_{3d}) \\
w''_{4b} &= -n''_{42} + n''_{43} = (w_{4b} + w_{4e}) & w''_{4c} &= +n''_{44} - n''_{45} = (w_{4c} - w_{4d}) \\
w''_{5b} &= -n''_{52} + n''_{53} = (w_{5b} + w_{5e}) & w''_{5c} &= -n''_{54} + n''_{55} = (w_{5c} + w_{5d}).
\end{aligned} \qquad (144)$$

---

* Between the elements of this 5 × 5 matrix for the left-hand half of the frame and the elements of the 10 × 10 matrix for the whole frame, as mentioned on page 11, there exist, according to the "load transposition method"; the following relationships:

$$n''_{11} = n_{11} + n_{1,10}, \qquad n''_{12} = n_{12} + n_{19} \quad \text{etc., up to} \quad n''_{55} = n_{55} + n_{56}.$$

** The members enclosed in parentheses have been formed with the values s and w of the 10 × 10 matrix.

Loading Condition 1b

Horizontal member carrying vertical loading of arbitrary magnitude, but anti-symmetrical about center of frame

$(\mathfrak{R}_5 = -\mathfrak{L}_1, \quad \mathfrak{L}_5 = -\mathfrak{R}_1; \quad \mathfrak{R}_4 = -\mathfrak{L}_2, \quad \mathfrak{L}_4 = -\mathfrak{R}_2; \quad \mathfrak{R}_3 = -\mathfrak{L}_3)$

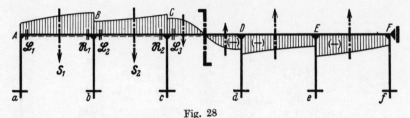

Fig. 28

Beam loading diagram with load terms $\mathfrak{L}$ and $\mathfrak{R}$ for left-hand half of frame

Beam moments: $L_\nu = \mathfrak{L}_\nu k_\nu$ and $R_\nu = \mathfrak{R}_\nu k_\nu$:

	L_1	R_1	L_2	R_2	L_3
$X_1'' = M_{A1} = -M_{F5} =$	$-n_{11}''$	$+n_{12}''$	$+n_{13}''$	$-n_{14}''$	$-n_{15}''$
$X_2'' = M_{B1} = -M_{E5} =$	$+n_{21}''$	$-n_{22}''$	$-n_{23}''$	$+n_{24}''$	$+n_{25}''$
$X_3'' = M_{B2} = -M_{E4} =$	$+n_{31}''$	$-n_{32}''$	$-n_{33}''$	$+n_{34}''$	$+n_{35}''$
$X_4'' = M_{C2} = -M_{D4} =$	$-n_{41}''$	$+n_{42}''$	$+n_{43}''$	$-n_{44}''$	$-n_{45}''$
$X_5'' = M_{C3} = -M_{D3} =$	$-n_{51}''$	$+n_{52}''$	$+n_{53}''$	$-n_{54}''$	$-n_{55}''.$

(145)

Column moments:

at top:

$M_{Aa} = M_{Ff} = \quad\quad + X_1''$
$M_{Bb} = M_{Ee} = -X_2'' + X_3''$
$M_{Cc} = M_{Dd} = -X_4'' + X_5'';$

at base:

$M_a = M_f = -M_{Aa} \cdot \varepsilon_a$
$M_b = M_e = -M_{Bb} \cdot \varepsilon_b$
$M_c = M_d = -M_{Cc} \cdot \varepsilon_c.$

(146) and (147)

Special case: symmetrical bay loading ($R_1 = L_1$ and $R_2 = L_2$):

$$X_1'' = M_{A1} = -M_{F5} = -L_1 \cdot s_{11}'' + L_2 \cdot s_{12}''$$
$$X_2'' = M_{B1} = -M_{E5} = -L_1 \cdot s_{21}'' - L_2 \cdot s_{22}''$$
$$X_3'' = M_{B2} = -M_{E4} = -L_1 \cdot s_{31}'' - L_2 \cdot s_{32}''$$
$$X_4'' = M_{C2} = -M_{D4} = +L_1 \cdot s_{41}'' - L_2 \cdot s_{42}''$$
$$X_5'' = M_{C3} = -M_{D3} = +L_1 \cdot s_{51}'' - L_2 \cdot s_{52}''.$$

(148)

Loading Conditions 2b and 3b

All the columns carrying arbitrary horizontal loading (2b) and external rotational moments applied to all the beam joints (3b), but load arrangement as a whole is anti-symmetrical about center of frame

$$(\mathfrak{L}_f = \mathfrak{L}_a, \quad \mathfrak{R}_f = \mathfrak{R}_a; \quad \mathfrak{L}_e = \mathfrak{L}_b, \quad \mathfrak{R}_e = \mathfrak{R}_b; \quad \mathfrak{L}_d = \mathfrak{L}_c, \quad \mathfrak{R}_d = \mathfrak{R}_c;$$
$$D_F = D_A \quad D_E = D_B \quad D_D = D_C)$$

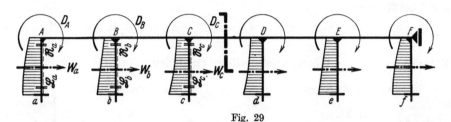

Fig. 29

Column loading diagram with load terms $\mathfrak{L}$ and $\mathfrak{R}$ for the columns, and external rotational moments D, for left-hand half of frame

Joint load terms: $(L_n = \mathfrak{L}_n k_n$ and $R_n = \mathfrak{R}_n k_n)$:

$$\left. \begin{array}{l} \mathfrak{B}_N = D_N \cdot b_n + L_n \cdot \varepsilon_n - R_n \\ \text{for} \quad N = A, B, C, \quad \text{and} \quad n = a, b, c. \end{array} \right\} \quad (149)$$

Beam moments:

	$\mathfrak{B}_A$	$\mathfrak{B}_B$	$\mathfrak{B}_C$
$X_1'' = M_{A1} = -M_{F5} =$	$+n_{11}''$	$+w_{1b}''$	$-w_{1c}''$
$X_2'' = M_{B1} = -M_{E5} =$	$-n_{21}''$	$-w_{2b}''$	$+w_{2c}''$
$X_3'' = M_{B2} = -M_{E4} =$	$-n_{31}''$	$+w_{3b}''$	$+w_{3c}''$
$X_4'' = M_{C2} = -M_{D4} =$	$+n_{41}''$	$-w_{4b}''$	$-w_{4c}''$
$X_5'' = M_{C3} = -M_{D3} =$	$+n_{51}''$	$-w_{5b}''$	$+w_{5c}''.$

(150)

Column moments:

at top: \hspace{2cm} at base:

$$\left. \begin{array}{ll} M_{Aa} = M_{Ff} = -D_A \qquad + X_1'' & M_a = M_f = -(\mathfrak{L}_a + M_{Aa})\varepsilon_a \\ M_{Bb} = M_{Ee} = -D_B - X_2'' + X_3'' & M_b = M_e = -(\mathfrak{L}_b + M_{Bb})\varepsilon_b \\ M_{Cc} = M_{Dd} = -D_C - X_4'' + X_5''; & M_c = M_d = -(\mathfrak{L}_c + M_{Cc})\varepsilon_c. \end{array} \right\} \begin{array}{l} (151) \\ \text{and} \\ (152) \end{array}$$

II. Treatment by the Deformation Method

a) Coefficients for symmetrical load arrangements

Stiffness coefficients of all the members, according to formulas (20), for $\nu = 1, 2, 3$ and $n = a, b, c$

Stiffness of the columns, according to formulas (21), for $n = a, b, c$

Joint coefficients:

$$K_A = 3S_a + 2K_1 \quad K_B = 2K_1 + 3S_b + 2K_2 \quad K'_C = 2K_2 + 3S_c + K_3. \quad (153)$$

Rotation carry-over factors and auxiliary coefficients:

backward reduction: | forward reduction:

$$j_{2s} = \frac{K_2}{r'_c} \quad \begin{aligned} r'_c &= K'_C \\ r'_b &= K_B - K_2 \cdot j_{2s} \end{aligned} \qquad j'_1 = \frac{K_1}{v_a} \quad \begin{aligned} v_a &= K_A \\ v_b &= K_B - K_1 \cdot j'_1 \end{aligned} \quad \begin{matrix}(154)\\ \text{and} \\ (155)\end{matrix}$$

$$j_{1s} = \frac{K_1}{r'_b} \quad r'_a = K_A - K_1 \cdot j_{1s} \qquad j'_2 = \frac{K_2}{v_b} \quad v'_c = K'_C - K_2 \cdot j'_2.$$

Principal influence coefficients by recursion:

forward: | backward:

$$\begin{aligned} u'_{aa} &= 1/r'_a \\ u'_{bb} &= 1/r'_b + u'_{aa} \cdot j^2_{1s} \\ u'_{cc} &= 1/r'_c + u'_{bb} \cdot j^2_{2s}. \end{aligned} \qquad \begin{aligned} u'_{cc} &= 1/v'_c \\ u'_{bb} &= 1/v_b + u'_{cc} \cdot j'^2_2 \\ u'_{aa} &= 1/v_a + u'_{bb} \cdot j'^2_1. \end{aligned} \quad \begin{matrix}(156)\\ \text{and} \\ (157)\end{matrix}$$

Principal influence coefficients direct

$$u'_{aa} = \frac{1}{r'_a} \qquad u'_{bb} = \frac{1}{r'_b + v_b - K_B} \qquad u'_{cc} = \frac{1}{v'_c}. \quad (158)$$

Computation scheme for the symmetrical influence coefficient matrix*

$$\begin{array}{c|ccc|c} & (u'_{aa}) & u'_{ab} & u'_{ac} & \\ \cdot j_{1s} & \downarrow & \uparrow & \uparrow & \cdot j'_1 \\ & u'_{ba} & (u'_{bb}) & u'_{bc} & \\ \cdot j_{2s} & \downarrow & \downarrow & \uparrow & \cdot j'_2 \\ & u'_{ca} & u'_{cb} & (u'_{cc}) & \end{array} \quad (159)$$

---

* Between the elements of this 3 × 3 matrix for the left-hand half of the frame and the elements of the 6 × 6 matrix for the whole frame, as indicated on page 27, there exist, according to the "load transposition method", the following relationships:

$$u'_{aa} = u_{aa} + u_{af} \quad \text{etc., up to} \quad u'_{cc} = u_{cc} + u_{cd}.$$

b) Coefficients for anti-symmetrical load arrangements

Stiffness coefficients of all the members, acccording to formulas (20), for $\nu = 1, 2, 3$ and $n = a, b, c$

Stiffness of the columns, according to formulas (21) for $n = a, b, c$

Joint coefficients:

$$K_A = 3S_a + 2K_1 \quad K_B = 2K_1 + 3S_b + 2K_2 \quad K_C'' = 2K_2 + 3S_c + 3\mathbf{K_3}. \quad (160)$$

Rotation carry-over factors and auxiliary coefficients:

$$\left.\begin{array}{l}
\text{backward reduction:} \\
\quad r_c'' = K_C'' \\
j_{2t} = \dfrac{K_2}{r_c''} \\
\quad r_b'' = K_B - K_2 \cdot j_{2t} \\
j_{1t} = \dfrac{K_1}{r_b''} \\
\quad r_a'' = K_A - K_1 \cdot j_{1t}.
\end{array} \;\middle|\; \begin{array}{l}
\text{forward reduction:} \\
\quad v_a = K_A \\
j_1' = \dfrac{K_1}{v_a} \\
\quad v_b = K_B - K_1 \cdot j_1' \\
j_2' = \dfrac{K_2}{v_b} \\
\quad v_c'' = K_C'' - K_2 \cdot j_2'.
\end{array}\right\} \begin{array}{c} (161) \\ \text{and} \\ (162) \end{array}$$

Principal influence coefficients by recursion:

$$\left.\begin{array}{l}
\text{forward:} \\
u_{aa}'' = 1/r_a'' \\
u_{bb}'' = 1/r_b'' + u_{aa}'' \cdot j_{1t}^2 \\
u_{cc}'' = 1/r_c'' + u_{bb}'' \cdot j_{2t}^2.
\end{array} \;\middle|\; \begin{array}{l}
\text{backward:} \\
u_{cc}'' = 1/v_c'' \\
u_{bb}'' = 1/v_b + u_{cc}'' \cdot j_2'^{2} \\
u_{aa}'' = 1/v_a + u_{bb}'' \cdot j_1'^{2}.
\end{array}\right\} \begin{array}{c} (163) \\ \text{and} \\ (164) \end{array}$$

Principal influence coefficients direct

$$u_{aa}'' = \frac{1}{r_a''} \qquad u_{bb}'' = \frac{1}{r_b'' + v_b - K_B} \qquad u_{cc}'' = \frac{1}{v_c''}. \quad (165)$$

Computation scheme for the symmetrical influence coefficient matrix*

$$\left.\begin{array}{c|ccc|c}
 & (u_{aa}'') & u_{ab}'' & u_{ac}'' & \\
\cdot j_{1t} & \downarrow & \uparrow & \uparrow & \cdot j_1' \\
 & u_{ba}'' & (u_{bb}'') & u_{bc}'' & \\
\cdot j_{2t} & \downarrow & \downarrow & \uparrow & \cdot j_2' \\
 & u_{ca}'' & u_{cb}'' & (u_{cc}'') & \\
\end{array}\right\} \quad (166)$$

---

* Between the elements of this 3×3 matrix for the left-hand half of the frame and the elements of the 6×6 matrix for the whole frame, as indicated on page 27, there exist, according to the "load transposition method", the following relationships:
$$u_{aa}'' = u_{aa} - u_{af} \text{ etc., up to } u_{cc}'' = u_{cc} - u_{cd}.$$

Loading Condition 4a

Overall loading condition (superposition of Loading Conditions 1a, 2a and 3a): load arrangement as a whole symmetrical about center of frame

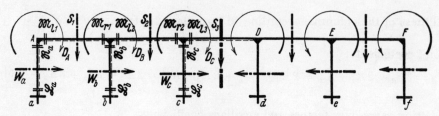

Fig. 30

Loading diagram with fixed-end moments $\mathfrak{M}_l$ and $\mathfrak{M}_r$ for the beam spans, load terms $\mathfrak{L}$ and $\mathfrak{R}$ for the columns, and external rotational moments D, for left-hand half of frame.

Column restraining moments at the beam joints:

$$\mathfrak{C}_{rn} = -\frac{\mathfrak{R}_n - \mathfrak{L}_n \cdot \varepsilon_n}{2 - \varepsilon_n} \quad \text{for} \quad n = a, b, c. \tag{167}$$

Joint load terms

$$\left. \begin{aligned} \mathfrak{K}_A &= \phantom{\mathfrak{M}_{r1}} - \mathfrak{M}_{l1} + \mathfrak{C}_{ra} + D_A \\ \mathfrak{K}_B &= \mathfrak{M}_{r1} - \mathfrak{M}_{l2} + \mathfrak{C}_{rb} + D_B \\ \mathfrak{K}_C &= \mathfrak{M}_{r2} - \mathfrak{M}_{l3} + \mathfrak{C}_{rc} + D_C. \end{aligned} \right\} \tag{168}$$

auxiliary moments:

$$\left. \begin{aligned} Y'_A &= + \mathfrak{K}_A \cdot u'_{aa} - \mathfrak{K}_B \cdot u'_{ab} + \mathfrak{K}_C \cdot u'_{ac} \\ Y'_B &= - \mathfrak{K}_A \cdot u'_{ba} + \mathfrak{K}_B \cdot u'_{bb} - \mathfrak{K}_C \cdot u'_{bc} \\ Y'_C &= + \mathfrak{K}_A \cdot u'_{ca} - \mathfrak{K}_B \cdot u'_{cb} + \mathfrak{K}_C \cdot u'_{cc}. \end{aligned} \right\} \tag{169}$$

Beam moments:

$$\left. \begin{aligned} M_{A1} &= M_{F5} = \mathfrak{M}_{l1} + (2\,Y'_A + Y'_B)\,K_1 \\ M_{B2} &= M_{E4} = \mathfrak{M}_{l2} + (2\,Y'_B + Y'_C)\,K_2 \\ M_{C3} &= M_{D3} = \mathfrak{M}_{l3} + Y'_C \cdot K_3; \\ & \qquad M_{B1} = M_{E5} = \mathfrak{M}_{r1} - (Y'_A + 2\,Y'_B)\,K_1 \\ & \qquad M_{C2} = M_{D4} = \mathfrak{M}_{r2} - (Y'_B + 2\,Y'_C)\,K_2. \end{aligned} \right\} \tag{170}$$

Column moments:

at top:
$$M_{Aa} = -M_{Ff} = \mathfrak{C}_{ra} - Y'_A \cdot 3S_a$$
$$M_{Bb} = -M_{Ee} = \mathfrak{C}_{rb} - Y'_B \cdot 3S_b \quad (171)$$
$$M_{Cc} = -M_{Dd} = \mathfrak{C}_{rc} - Y'_C \cdot 3S_c;$$

at base:
$$M_a = -M_f = -(\mathfrak{L}_a + M_{Aa})\,\varepsilon_a$$
$$M_b = -M_e = -(\mathfrak{L}_b + M_{Bb})\,\varepsilon_b \quad (171\text{a})$$
$$M_c = -M_d = -(\mathfrak{L}_c + M_{Cc})\,\varepsilon_c.$$

Check relationships:

$$D_B + M_{Bb} + M_{B1} - M_{B2} = 0$$
$$D_A + M_{Aa} - M_{A1} = 0 \qquad\qquad\qquad\qquad (172)$$
$$D_C + M_{Cc} + M_{C2} - M_{C3} = 0.$$

Loading Condition 4b

Overall loading condition (superposition of Loading Conditions 1b, 2b and 3b): load arrangement as a whole anti-symmetrical about center of frame

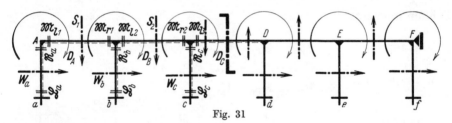

Fig. 31

Loading diagram with fixed-end moments $\mathfrak{M}_l$ and $\mathfrak{M}_r$ for the beam spans, load terms $\mathfrak{L}$ and $\mathfrak{R}$ for the columns, and external rotational moments D, for left-hand half of frame

Column restraining moments at the beam joints:

$$\mathfrak{C}_{rn} = -\frac{\mathfrak{R}_n - \mathfrak{L}_n \cdot \varepsilon_n}{2 - \varepsilon_n} \quad \text{for} \quad n = a, b, c. \quad (173)$$

Joint load terms:

$$\mathfrak{K}_A = \qquad -\mathfrak{M}_{l1} + \mathfrak{C}_{ra} + D_A$$
$$\mathfrak{K}_B = \mathfrak{M}_{r1} - \mathfrak{M}_{l2} + \mathfrak{C}_{rb} + D_B \quad (174)$$
$$\mathfrak{K}_C = \mathfrak{M}_{r2} - \mathfrak{M}_{l3} + \mathfrak{C}_{rc} + D_C.$$

auxiliary moments:

$$\left.\begin{aligned}Y_A'' &= + \mathfrak{K}_A \cdot u_{aa}'' - \mathfrak{K}_B \cdot u_{ab}'' + \mathfrak{K}_C \cdot u_{ac}'' \\ Y_B'' &= - \mathfrak{K}_A \cdot u_{ba}'' + \mathfrak{K}_B \cdot u_{bb}'' - \mathfrak{K}_C \cdot u_{bc}'' \\ Y_C'' &= + \mathfrak{K}_A \cdot u_{ca}'' - \mathfrak{K}_B \cdot u_{cb}'' + \mathfrak{K}_C \cdot u_{cc}''. \end{aligned}\right\} \quad (175)$$

Beam moments:

$$\left.\begin{aligned}M_{A1} &= - M_{F5} = \mathfrak{M}_{l1} + (2\,Y_A'' + Y_B'')\,K_1 \\ M_{B2} &= - M_{E4} = \mathfrak{M}_{l2} + (2\,Y_B'' + Y_C'')\,K_2 \\ M_{C3} &= - M_{D3} = \mathfrak{M}_{l3} + Y_C'' \cdot 3\,K_3; \\ M_{B1} &= - M_{E5} = \mathfrak{M}_{r1} - (Y_A'' + 2\,Y_B'')\,K_1 \\ M_{C2} &= - M_{D4} = \mathfrak{M}_{r2} - (Y_B'' + 2\,Y_C'')\,K_2. \end{aligned}\right\} \quad (176)$$

Column moments:

at top:
$$\left.\begin{aligned}M_{Aa} &= M_{Ff} = \mathfrak{C}_{ra} - Y_A'' \cdot 3\,S_a \\ M_{Bb} &= M_{Ee} = \mathfrak{C}_{rb} - Y_B'' \cdot 3\,S_b \\ M_{Cc} &= M_{Dd} = \mathfrak{C}_{rc} - Y_C'' \cdot 3\,S_c; \end{aligned}\right\} \quad (177)$$

at base:
$$\left.\begin{aligned}M_a &= M_f = -(\mathfrak{L}_a + M_{Aa})\,\varepsilon_a \\ M_b &= M_e = -(\mathfrak{L}_b + M_{Bb})\,\varepsilon_b \\ M_c &= M_d = -(\mathfrak{L}_c + M_{Cc})\,\varepsilon_c. \end{aligned}\right\} \quad (177\mathrm{a})$$

Check relationships:

$$\left.\begin{aligned}D_A + M_{Aa} \qquad\qquad - M_{A1} &= 0 \\ D_B + M_{Bb} + M_{B1} - M_{B2} &= 0 \\ D_C + M_{Cc} + M_{C2} - M_{C3} &= 0. \end{aligned}\right\} \quad (178)$$

Horizontal thrusts, column shears and lateral restraining force for all loading conditions

For the asymmetrical Loading Conditions 1 and 3*:

Column thrusts and shears:

$$H_n = +Q_n = \frac{M_{Nn}(1+\varepsilon_n)}{h_n}; \quad \text{for } n = a, b, c, d, e, f \text{ and } N = A, B, C, D, E, F. \tag{179}$$

Lateral restraining force:

$$H = -(H_a + H_b + H_c + H_d + H_e + H_f). \tag{180}$$

For the asymmetrical Loading Conditions 2 and 4*:

Column thrusts and shears:

$$T_n = \frac{M_{Nn} - M_n}{h_n}; \quad H_n = Q_{ln} = \frac{\mathfrak{S}_{rn}}{h_n} + T_n, \quad Q_{rn} = -\frac{\mathfrak{S}_{ln}}{h_n} + T_n; \tag{181}$$

$$\text{for } n = a, b, c, d, e, f \text{ and } N = A, B, C, D, E, F.$$

Lateral restraining force:

$$H = -(Q_{ra} + Q_{rb} + Q_{rc} + Q_{rd} + Q_{re} + Q_{rf}). \tag{182}$$

For the symmetrical Loading Conditions 1a and 3a:

H_n as equation (179); for $n = a, b, c,$ and $N = A, B, C,$ (183)

Lateral restraining force $H = 0$ (184)

For the symmetrical Loading Conditions 2a and 4a:

$T_n; H_n = Q_{ln}, Q_{rn}$ as equation (181); for $n = a, b, c$ and $N = A, B, C.$ (185)

Lateral restraining force $H = 0$ (186)

For the anti-symmetrical Loading Conditions 1b and 3b:

H_n as equation (179); for $n = a, b, c$ and $N = A, B, C.$ (187)

Lateral restraining force $H = -2(H_a + H_b + H_c).$ (188)

For the anti-symmetrical Loading Conditions 2b and 4b:

$T_n; H_n = Q_{ln}, Q_{rn}$ as equation (181); for $n = a, b, c$ and $N = A, B, C.$ (189)

Lateral restraining force $H = -2(Q_{ra} + Q_{rb} + Q_{rc}).$ (190)

---

* These loading conditions, which have not been given for Frame Shape 5, are merely indicated on page 27.

Frame Shape 5v

(Frame Shape 5 with laterally unrestrained horizontal member)

Note: Irrespective of symmetry, Frame Shape 5v can be analysed with the aid of the formulas relating to Frame Shape 1v (see page 9/10). The computation procedure is nevertheless again indicated below, as a self-contained whole, having due regard to the anti-symmetry properties.

First step: Calculation of all the requisite asymmetrical or anti-symmetrical loading conditions as well as the appropriate lateral restraining force H according to Frame Shape 5. (The symmetrical Loading Conditions need not be considered, because for these $H = 0$).

Second step: Determination of the effect of a horizontal point load W, acting at the level of the horizontal member, with the aid of Loading Condition 5 for Frame Shape 5v, as indicated below.

Third step: Determination of the final moments and forces for Frame Shape 5v by superposition of the results of the first step and those of the second step for $W = H$.

Loading Condition 5

Horizontal point load W at level of horizontal member

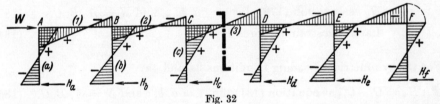

Fig. 32

Loading diagram and anti-symmetrical diagram of bending moments. (For clarity, only the reactions H_n at the column feet are shown).

Note: The *collective subscripts* have the following meaning:

$$\nu = 1, 2, 3; \qquad N = A, B, C; \qquad n = a, b, c.$$

Column ratios
with arbitrary comparison height h_k $\left.\right\}$ $\gamma_n = \dfrac{h_k}{h_n}.$ (191)

Specific bending moments m (instead of M):
either according to *Method I/b* from the formulas (150), (151) and (152)—i.e., the moments $X_i'' = m_{N\nu}$, m_{Nn} and m_n—for:

$$\mathfrak{B}_N = \gamma_n(1 + \varepsilon_n) \quad \text{and} \quad \mathfrak{L}_n = \gamma_n/k_n; \qquad (192)$$

or according to *Method II/b* from the formulas (175), (176), (177) and (178)— i.e., the moments y_N'' (instead of Y_N''), m_{Nv}, m_{Nn} and m_n—for:

$$\mathfrak{K}_N = \mathfrak{C}_{rn} = S_n \gamma_n (1 + \varepsilon_n) \quad \text{and} \quad \mathfrak{L}_n = \gamma_n K_n. \tag{193}$$

Note: All the other load terms in bold type in the formulas indicated should be omitted.

Specific h_k-fold column shears:

$$p_n = (m_{Nn} - m_n) \gamma_n; \tag{194}$$

Corresponding specific h_k-fold applied load

$$p = 2(p_a + p_b + p_c). \tag{195}$$

Conversion moment for a given load W:

$$U = \frac{W h_k}{p}. \tag{196}$$

Final moments and forces for the given concentrated load W:

Beam moments:

$$\begin{aligned} M_{A1} &= -M_{F5} = U \cdot m_{A1} & M_{B2} &= -M_{E4} = U \cdot m_{B2} & M_{C3} &= -M_{D3} = U \cdot m_{C3} \\ M_{B1} &= -M_{E5} = U \cdot m_{B1}; & M_{C2} &= -M_{D4} = U \cdot m_{C2}; & & \end{aligned} \tag{197}$$

Column moments:

$$\begin{aligned} M_{Aa} &= M_{Ff} = U \cdot m_{Aa} & M_{Bb} &= M_{Ee} = U \cdot m_{Bb} & M_{Cc} &= M_{Dd} = U \cdot m_{Cc}; \\ M_a &= M_f = U \cdot m_a & M_b &= M_e = U \cdot m_b & M_c &= M_d = U \cdot m_c. \end{aligned} \tag{198}$$

Column shears and horizontal thrusts for left-hand half of frame:

$$Q_n = H_n = \frac{U \cdot p_n}{h_k} = W \cdot \frac{p_n}{p}. \tag{199}$$

Supplementary note: The computation procedure given here for the "influence of a horizontal concentrated load W acting at the level of the laterally unrestrained horizontal member", for the symmetrical Frame Shape 5v, is fundamentally applicable also to all other single-story symmetrical frames having an even number of columns. It should only be noted that the subscript n in each case relates to all the rigidly connected columns in the left-hand half of the frame and that the subscript N relates to all the corresponding beam joints.

— 44 —

Frame Shape 6
Symmetrical single-story three-bay frame with laterally restrained horizontal member and four elastically restrained columns
(Symmetrical form of Frame Shape 1)

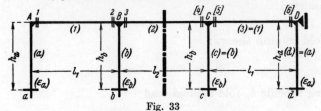

Fig. 33
Frame shape, dimensions and symbols

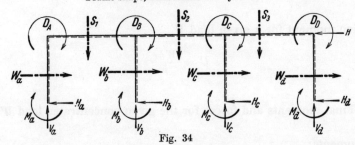

Fig. 34
Definition of positive direction for all external loads on joints and members, for all column reactions, and for the lateral restraining force

All the dimensions and coefficients for the right-hand half of the frame are the same as those for the left-hand half. In the case of the columns the broken line is always placed on the right-hand side of each column, despite the symmetry of the frame.

Note: For Frame Shape 6 *the coefficients and the Loading Conditions 1 – 4 for Frame Shape 1* may be employed. On account of symmetry, the forward reduction (5) becomes the same as the backward reduction (4) [and similarly (24) becomes the same as (23)], and the backward recursion (7) becomes the same as the forward recursion (6) [and similarly (26) becomes the same as (25)]. Furthermore, the square matrix of influence coefficients (9), or (28), becomes bisymmetrical, i.e., symmetrical about the principal diagonal and the secondary diagonal. (If the *"load transposition method"* is employed, however, symmetrical and anti-symmetrical loading conditions together with the appropriate coefficients may also be calculated directly, i.e., with reference to Frame Shape 1, as indicated below.*)

---

* It would, alternatively, be possible simply to refer the reader to Frame Shape 5, inasmuch as the formulas for that frame are generally valid for symmetrical continuous frames with any odd number of bays (see the note "Important" on page 27). However, in view of the practical importance of the symmetrical three-bay frame, the formulas for it are nevertheless given separately.

I. Treatment by the Method of Forces

a) Symmetrical load arrangements

Coefficients

Flexibility coefficients of all the members and flexibilities of the columns, according to formulas (1) and (2), for $\nu = 1, 2$ and $n = a, b$.

Diagonal coefficients:

$$O_1 = b_a + 2k_1 \qquad O_2 = 2k_1 + b_b \qquad O_3' = b_b + 3k_2. \tag{200}$$

Moment carry-over factors and auxiliary coefficients:

$$
\left.\begin{array}{l|l}
\text{backward reduction:} & \text{forward reduction:} \\[4pt]
\begin{aligned}
\sigma_{bs} &= \frac{b_b}{r_3'} \\
\iota_{1s} &= \frac{k_1}{r_2'}
\end{aligned}
\quad
\begin{aligned}
r_3' &= O_3' \\
r_2' &= O_2 - b_b \cdot \sigma_{bs} \\
r_1' &= O_1 - k_1 \cdot \iota_{1s}.
\end{aligned}
&
\begin{aligned}
\iota_1' &= \frac{k_1}{v_1} \\
\sigma_b' &= \frac{b_b}{v_2}
\end{aligned}
\quad
\begin{aligned}
v_1 &= O_1 \\
v_2 &= O_2 - k_1 \cdot \iota_1' \\
v_3' &= O_3' - b_b \cdot \sigma_b'.
\end{aligned}
\end{array}\right\} \begin{array}{c} (201) \\ \text{and} \\ (202) \end{array}
$$

Principal influence coefficients by recursion:

$$
\left.\begin{array}{l|l}
\text{forward:} & \text{backward:} \\[4pt]
\begin{aligned}
n_{11}' &= 1/r_1' \\
n_{22}' &= 1/r_2' + n_{11}' \cdot \iota_{1s}^2 \\
n_{33}' &= 1/r_3' + n_{22}' \cdot \sigma_{bs}^2.
\end{aligned}
&
\begin{aligned}
n_{33}' &= 1/v_3' \\
n_{22}' &= 1/v_2 + n_{33}' \cdot \sigma_b'^2 \\
n_{11}' &= 1/v_1 + n_{22}' \cdot \iota_1'^2.
\end{aligned}
\end{array}\right\} \begin{array}{c} (203) \\ \text{and} \\ (204) \end{array}
$$

Principal influence coefficients direct

$$n_{11}' = \frac{1}{r_1'} \qquad n_{22}' = \frac{1}{r_2' + v_2 - O_2} \qquad n_{33}' = \frac{1}{v_3'}. \tag{205}$$

Computation scheme for the symmetrical influence coefficient matrix*

$$
\left.\begin{array}{c|ccc|c}
 & (n_{11}') & n_{12}' & n_{13}' & \\
\cdot \iota_{1s} & \downarrow & \uparrow & \uparrow & \cdot \iota_1' \\
 & n_{21}' & (n_{22}') & n_{23}' & \\
\cdot \sigma_{bs} & \downarrow & \downarrow & \uparrow & \cdot \sigma_b' \\
 & n_{31}' & n_{32}' & (n_{33}') & \\
\end{array}\right\} (206)
$$

---

* With the aid of the "load transposition method" the elements of matrix (206) may also be obtained from matrix (9), as follows:

$$n_{11}' = n_{11} - n_{16}, \qquad n_{12}' = n_{12} - n_{15} \text{ etc., up to } n_{33}' = n_{33} - n_{34}.$$

Loading Condition 1a

Horizontal member carrying vertical loading of arbitrary magnitude, but symmetrical about center of frame

$$(\mathfrak{R}_3 = \mathfrak{L}_1, \quad \mathfrak{L}_3 = \mathfrak{R}_1; \quad \mathfrak{R}_2 = \mathfrak{L}_2)$$

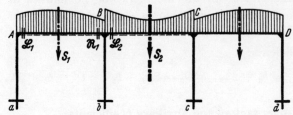

Fig. 35

Beam loading diagram with load terms $\mathfrak{L}$ and $\mathfrak{R}$ for left-hand half of frame

Beam moments ($L_1 = \mathfrak{L}_1 k_1$, $R_1 = \mathfrak{R}_1 k_1$; $L_2 = \mathfrak{L}_2 k_2$):

	L_1	R_1	L_2
$X_1' = M_{A1} = M_{D3} =$	$-n_{11}'$	$+n_{12}'$	$+n_{13}'$
$X_2' = M_{B1} = M_{C3} =$	$+n_{21}'$	$-n_{22}'$	$-n_{23}'$
$X_3' = M_{B2} = M_{C2} =$	$+n_{31}'$	$-n_{32}'$	$-n_{33}'$.

(207)

Column moments:

at top: at base:

$$M_{Aa} = -M_{Dd} = \quad\quad +X_1' \quad\quad M_a = -M_d = -M_{Aa} \cdot \varepsilon_a$$

$$M_{Bb} = -M_{Cc} = -X_2' + X_3'; \quad\quad M_b = -M_c = -M_{Bb} \cdot \varepsilon_b.$$

(208) and (209)

Special case: symmetrical bay loading ($R_1 = L_1$):

$$\begin{aligned}
X_1' = M_{A1} = M_{D3} &= -L_1 \cdot s_{11}' + L_2 \cdot n_{13}' \\
X_2' = M_{B1} = M_{C3} &= -L_1 \cdot s_{21}' - L_2 \cdot n_{23}' \\
X_3' = M_{B2} = M_{C2} &= -L_1 \cdot s_{31}' - L_2 \cdot n_{33}';
\end{aligned}$$

(210)

where*:

$$\begin{aligned}
s_{11}' &= +n_{11}' - n_{12}' = (s_{11} + s_{13}) \\
s_{21}' &= -n_{21}' + n_{22}' = (s_{21} - s_{23}) \\
s_{31}' &= -n_{31}' + n_{32}' = (s_{31} - s_{33}).
\end{aligned}$$

(211)

---

* The members enclosed in parentheses have been formed with the values s (10) of page 3.

Loading Conditions 2a and 3a

All the columns carrying arbitrary horizontal loading (2a), and external rotational moments applied at all the beam joints (3a), but load arrangement as a whole is symmetrical about center of frame

$(\mathfrak{L}_d = -\mathfrak{L}_a, \mathfrak{R}_d = -\mathfrak{R}_a; \mathfrak{L}_c = -\mathfrak{L}_b, \mathfrak{R}_c = -\mathfrak{R}_b; D_D = -D_A, D_C = -D_B)$

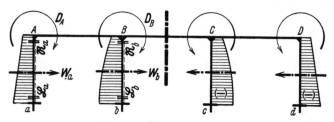

Fig. 36

Column loading diagram with load terms $\mathfrak{L}$ and $\mathfrak{R}$ for the columns, and external rotational moments D, for left-hand half of frame

Joint load terms: $(L_n = \mathfrak{L}_n k_n$ and $R_n = \mathfrak{R}_n k_n)$:

$$\mathfrak{B}_A = D_A \cdot b_a + L_a \cdot \varepsilon_a - R_a \qquad \mathfrak{B}_B = D_B \cdot b_b + L_b \cdot \varepsilon_b - R_b. \qquad (212)$$

Beam moments:

$$\begin{aligned} X'_1 = M_{A1} = M_{D3} = &+\mathfrak{B}_A \cdot n'_{11} + \mathfrak{B}_B \cdot w'_{1b} \\ X'_2 = M_{B1} = M_{C3} = &-\mathfrak{B}_A \cdot n'_{21} - \mathfrak{B}_B \cdot w'_{2b} \\ X'_3 = M_{B2} = M_{C2} = &-\mathfrak{B}_A \cdot n'_{31} + \mathfrak{B}_B \cdot w'_{3b}; \end{aligned} \qquad (213)$$

where*

$$\begin{aligned} w'_{1b} &= +n'_{12} - n'_{13} = (w_{1b} + w_{1c}) \\ w'_{2b} &= +n'_{22} - n'_{23} = (w_{2b} + w_{2c}) \\ w'_{3b} &= -n'_{32} + n'_{33} = (w_{3b} - w_{3c}). \end{aligned} \qquad (214)$$

Column moments:

at top:
$$\begin{aligned} M_{Aa} &= -M_{Dd} = -D_A + X'_1 \\ M_{Bb} &= -M_{Cc} = -D_B - X'_2 + X'_3; \end{aligned} \qquad (215)$$

at base:
$$\begin{aligned} M_a &= -M_d = -(\mathfrak{L}_a + M_{Aa})\varepsilon_a \\ M_b &= -M_c = -(\mathfrak{L}_b + M_{Bb})\varepsilon_b. \end{aligned} \qquad (216)$$

---

* The members enclosed in parentheses have been formed with the values w (11) of page 3.

b) Anti-symmetrical load arrangements

Coefficients

Flexibility coefficients of all the members and flexibilities of the columns, according to formulas (1) and (2), for $\nu = 1, 2$ and $n = a, b$

Diagonal coefficients:

$$O_1 = b_a + 2k_1 \qquad O_2 = 2k_1 + b_b \qquad O_3'' = b_b + k_2. \tag{217}$$

Moment carry-over factors and auxiliary coefficients:

$$\begin{array}{l|l}
\text{backward reduction:} & \text{forward reduction:} \\
\sigma_{bt} = \dfrac{b_b}{r_3''} \quad \begin{array}{l} r_3'' = O_3'' \\ r_2'' = O_2 - b_b \cdot \sigma_{bt} \end{array} & \iota_1' = \dfrac{k_1}{v_1} \quad \begin{array}{l} v_1 = O_1 \\ v_2 = O_2 - k_1 \cdot \iota_1' \end{array} \\
\iota_{1t} = \dfrac{k_1}{r_2''} \quad r_1'' = O_1 - k_1 \cdot \iota_{1t}. & \sigma_b' = \dfrac{b_b}{v_2} \quad v_3'' = O_3'' - b_b \cdot \sigma_b'.
\end{array} \quad \begin{matrix}(218)\\ \text{and}\\ (219)\end{matrix}$$

Principal influence coefficients by recursion:

$$\begin{array}{l|l}
\text{forward:} & \text{backward:} \\
n_{11}'' = 1/r_1'' & n_{33}'' = 1/v_3'' \\
n_{22}'' = 1/r_2'' + n_{11}'' \cdot \iota_{1t}^2 & n_{22}'' = 1/v_2 + n_{33}'' \cdot \sigma_b'^2 \\
n_{33}'' = 1/r_3'' + n_{22}'' \cdot \sigma_{bt}^2. & n_{11}'' = 1/v_1 + n_{22}'' \cdot \iota_1'^2.
\end{array} \quad \begin{matrix}(220)\\ \text{and}\\ (221)\end{matrix}$$

Principal influence coefficients direct

$$n_{11}'' = \frac{1}{r_1''} \qquad n_{22}'' = \frac{1}{r_2'' + v_2 - O_2} \qquad n_{33}'' = \frac{1}{v_3''}. \tag{222}$$

Computation scheme for the symmetrical influence coefficient matrix*

$$\left. \begin{array}{c|ccc|c}
 & (n_{11}'') & n_{12}'' & n_{13}'' & \\
\cdot \iota_{1t} & \downarrow \uparrow & \uparrow & \uparrow & \cdot \iota_1' \\
 & n_{21}'' & (n_{22}'') & n_{23}'' & \\
\cdot \sigma_{bt} & \downarrow & \downarrow & \uparrow & \cdot \sigma_b' \\
 & n_{31}'' & n_{32}'' & (n_{33}'') &
\end{array} \right\} \tag{223}$$

---

* With the aid of the "load transposition method" the elements of the matrix (223) may also be obtained from matrix (9), as follows:

$$n_{11}'' = n_{11} + n_{16}, \quad n_{12}'' = n_{12} + n_{15} \quad \text{etc., up to} \quad n_{33}'' = n_{33} + n_{34}.$$

Loading Condition 1b

Horizontal member carrying vertical loading of arbitrary magnitude, but anti-symmetrical about center of frame

$$(\mathfrak{R}_3 = -\mathfrak{L}_1, \quad \mathfrak{L}_3 = -\mathfrak{R}_1; \quad \mathfrak{R}_2 = -\mathfrak{L}_2)$$

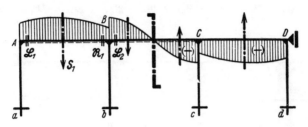

Fig. 37
Beam loading diagram with load terms $\mathfrak{L}$ and $\mathfrak{R}$ for left-hand half of frame

Beam moments $(L_1 = \mathfrak{L}_1 k_1, \quad R_1 = \mathfrak{R}_1 k_1; \quad L_2 = \mathfrak{L}_2 k_2)$:

$$
\begin{array}{l|ccc}
 & L_1 & R_1 & L_2 \\
\hline
X_1'' = M_{A1} = -M_{D3} = & -n_{11}'' & +n_{12}'' & +n_{13}'' \\
X_2'' = M_{B1} = -M_{C3} = & +n_{21}'' & -n_{22}'' & -n_{23}'' \\
X_3'' = M_{B2} = -M_{C2} = & +n_{31}'' & -n_{32}'' & -n_{33}''.
\end{array}
\quad (224)
$$

Column moments:

at top: $\qquad\qquad$ at base: $\qquad$ (225)

$$M_{Aa} = M_{Dd} = \quad +X_1'' \qquad M_a = M_d = -M_{Aa} \cdot \varepsilon_a$$
$$M_{Bb} = M_{Cc} = -X_2'' + X_3''; \qquad M_b = M_c = -M_{Bb} \cdot \varepsilon_b.$$

and (226)

Special case: symmetrical bay loading $\qquad R_1 = L_1$):

$$
\begin{aligned}
X_1'' &= M_{A1} = -M_{D3} = -L_1 \cdot s_{11}'' \\
X_2'' &= M_{B1} = -M_{C3} = -L_1 \cdot s_{21}'' \\
X_3'' &= M_{B2} = -M_{C2} = -L_1 \cdot s_{31}'';
\end{aligned}
\quad (227)
$$

where*

$$
\begin{aligned}
s_{11}'' &= +n_{11}'' - n_{12}'' = (s_{11} - s_{13}) \\
s_{21}'' &= -n_{21}'' + n_{22}'' = (s_{21} + s_{23}) \\
s_{31}'' &= -n_{31}'' + n_{32}'' = (s_{31} + s_{33}).
\end{aligned}
\quad (228)
$$

---

* The members enclosed in parentheses have been formed with the values s (10) of page 3.

Loading Conditions 2b and 3b

All the columns carrying arbitrary horizontal loading (2b), and external rotational moments applied to all the beam joints (3b), but load arrangement as a whole is anti-symmetrical about center of frame

$$(\mathfrak{L}_d = \mathfrak{L}_a, \quad \mathfrak{R}_d = \mathfrak{R}_a; \quad \mathfrak{L}_c = \mathfrak{L}_b, \quad \mathfrak{R}_c = \mathfrak{R}_b; \quad D_D = D_A, \quad D_C = D_B)$$

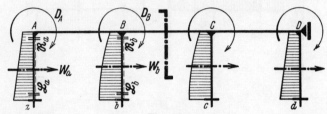

Fig. 38

Column loading diagram with load terms $\mathfrak{L}$ and $\mathfrak{R}$ for the columns, and external rotational moments D, for left-hand half of frame

Joint load terms: $(L_n = \mathfrak{L}_n k_n \quad \text{and} \quad R_n = \mathfrak{R}_n k_n)$:

$$\mathfrak{B}_A = D_A \cdot b_a + L_a \cdot \varepsilon_a - R_a \qquad \mathfrak{B}_B = D_B \cdot b_b + L_b \cdot \varepsilon_b - R_b. \qquad (229)$$

Beam moments:

$$\left. \begin{aligned} X_1'' &= M_{A1} = -M_{D3} = +\mathfrak{B}_A \cdot n_{11}'' + \mathfrak{B}_B \cdot w_{1b}'' \\ X_2'' &= M_{B1} = -M_{C3} = -\mathfrak{B}_A \cdot n_{21}'' - \mathfrak{B}_B \cdot w_{2b}'' \\ X_3'' &= M_{B2} = -M_{C2} = -\mathfrak{B}_A \cdot n_{31}'' + \mathfrak{B}_B \cdot w_{3b}'' \end{aligned} \right\} \qquad (230)$$

where*:

$$\left. \begin{aligned} w_{1b}'' &= +n_{12}'' - n_{13}'' = (w_{1b} - w_{1c}) \\ w_{2b}'' &= +n_{22}'' - n_{23}'' = (w_{2b} - w_{2c}) \\ w_{3b}'' &= -n_{32}'' + n_{33}'' = (w_{3b} + w_{3c}) \end{aligned} \right\} \qquad (231)$$

Column moments:

at top:
$$\left. \begin{aligned} M_{Aa} &= M_{Dd} = -D_A \quad\quad\quad + X_1'' \\ M_{Bb} &= M_{Cc} = -D_B - X_2'' + X_3''; \end{aligned} \right\} \qquad (232)$$

at base:
$$\left. \begin{aligned} M_a &= M_d = -(\mathfrak{L}_a + M_{Aa})\, \varepsilon_a \\ M_b &= M_c = -(\mathfrak{L}_b + M_{Bb})\, \varepsilon_b. \end{aligned} \right\} \qquad (233)$$

---

* The members enclosed in parentheses have been formed with the values w (11) of page 3 .

II. Treatment by the Deformation Method

Coefficients

Stiffness coefficients of all the members and stiffnesses of the columns, according to formulas (20) and (21), for $\nu = 1, 2$ and $n = a, b$

a) For symmetrical load arrangements

Joint coefficients:
$$K_A = 3S_a + 2K_1 \qquad K'_B = 2K_1 + 3S_b + \mathbf{K_2}. \tag{234}$$

Rotation carry-over factors
$$j_{1s} = K_1/K'_B \qquad j'_1 = K_1/K_A. \tag{235}$$

Influence coefficients*:
$$\left.\begin{array}{c} u'_{aa} = \dfrac{1}{K_A - K_1 \cdot j_{1s}} \qquad u'_{bb} = \dfrac{1}{K'_B - K_1 \cdot j'_1} \\ u'_{ab} = u'_{ba} = u'_{aa} \cdot j_{1s} = u'_{bb} \cdot j'_1. \end{array}\right\} \tag{236}$$

b) For anti-symmetrical load arrangements

Joint coefficients:
$$K_A = 3S_a + 2K_1 \qquad K''_B = 2K_1 + 3S_b + \mathbf{3K_2}. \tag{237}$$

Rotation carry-over factors
$$j_{1t} = K_1/K''_B \qquad j'_1 = K_1/K_A. \tag{238}$$

Influence coefficients*:
$$\left.\begin{array}{c} u''_{aa} = \dfrac{1}{K_A - K_1 \cdot j_{1t}} \qquad u''_{bb} = \dfrac{1}{K''_B - K_1 \cdot j'_1} \\ u''_{ab} = u''_{ba} = u''_{aa} \cdot j_{1t} = u''_{bb} \cdot j'_1. \end{array}\right\} \tag{239}$$

---

* With the aid of the "load transposition method" the influence coefficients (236) and (239) may also be obtained from matrix (28), as follows:

$$\begin{aligned} u'_{aa} &= u_{aa} + u_{ad} & u'_{bb} &= u_{bb} + u_{bc} \\ u'_{ab} &= u'_{ba} = u_{ab} + u_{ac} &= u_{ba} + u_{bd}; \end{aligned}$$

and

$$\begin{aligned} u''_{aa} &= u_{aa} - u_{ad} & u''_{bb} &= u_{bb} - u_{bc} \\ u''_{ab} &= u''_{ba} = u_{ab} - u_{ac} &= u_{ba} - u_{bd}. \end{aligned}$$

Loading Condition 4a

Overall loading condition (superposition of Loading Conditions 1a, 2a and 3a): load arrangement as a whole symmetrical about center of frame

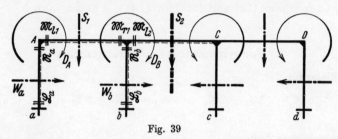

Fig. 39

Loading diagram with fixed-end moments $\mathfrak{M}_l$ and $\mathfrak{M}_r$ for the beam spans, load terms $\mathfrak{L}$ and $\mathfrak{R}$ for the columns, and external rotational moments D, for left-hand half of frame

Column restraining moments at the beam joints:

$$\mathfrak{C}_{ra} = -\frac{\mathfrak{R}_a - \mathfrak{L}_a \cdot \varepsilon_a}{2 - \varepsilon_a} \qquad \mathfrak{C}_{rb} = -\frac{\mathfrak{R}_b - \mathfrak{L}_b \cdot \varepsilon_b}{2 - \varepsilon_b}. \qquad (240)$$

Joint load terms

$$\mathfrak{K}_A = -\mathfrak{M}_{l1} + \mathfrak{C}_{ra} + D_A \qquad \mathfrak{K}_B = \mathfrak{M}_{r1} - \mathfrak{M}_{l2} + \mathfrak{C}_{rb} + D_B. \qquad (241)$$

auxiliary moments:

$$Y'_A = +\mathfrak{K}_A \cdot u'_{aa} - \mathfrak{K}_B \cdot u'_{ab} \qquad Y'_B = -\mathfrak{K}_A \cdot u'_{ba} + \mathfrak{K}_B \cdot u'_{bb}. \qquad (242)$$

Beam moments:

$$\left.\begin{array}{l} M_{A1} = M_{D3} = \mathfrak{M}_{l1} + (2Y'_A + Y'_B)K_1 \\ M_{B1} = M_{C3} = \mathfrak{M}_{r1} - (Y'_A + 2Y'_B)K_1 \qquad M_{B2} = M_{C2} = \mathfrak{M}_{l2} + Y'_B \cdot K_2 \end{array}\right\} \quad (243)$$

Column moments:

at top: \qquad\qquad\qquad\qquad at base: \qquad\qquad (244)

$$M_{Aa} = -M_{Dd} = \mathfrak{C}_{ra} - Y'_A \cdot 3S_a \qquad M_a = -M_d = -(\mathfrak{L}_a + M_{Aa})\varepsilon_a \quad \text{and}$$
$$M_{Bb} = -M_{Cc} = \mathfrak{C}_{rb} - Y'_B \cdot 3S_b; \qquad M_b = -M_c = -(\mathfrak{L}_b + M_{Bb})\varepsilon_b. \quad (245)$$

Check relationships:

$$D_A + M_{Aa} - M_{A1} = 0 \qquad D_B + M_{Bb} + M_{B1} - M_{B2} = 0. \qquad (246)$$

Loading Condition 4b

Overall loading condition (superposition of Loading Conditions 1b, 2b and 3b): load arrangement as a whole anti-symmetrical about center of frame

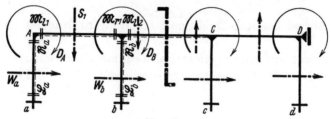

Fig. 40

Loading diagram with fixed-end moments $\mathfrak{M}_l$ and $\mathfrak{M}_r$ for the beam spans, load terms $\mathfrak{L}$ and $\mathfrak{R}$ for the columns, and external rotational moments D, for left-hand half of frame

Column restraining moments at the beam joints:

$$\mathfrak{C}_{ra} = -\frac{\mathfrak{R}_a - \mathfrak{L}_a \cdot \varepsilon_a}{2 - \varepsilon_a} \qquad \mathfrak{C}_{rb} = -\frac{\mathfrak{R}_b - \mathfrak{L}_b \cdot \varepsilon_b}{2 - \varepsilon_b}. \qquad (247)$$

Joint load terms

$$\mathfrak{R}_A = -\mathfrak{M}_{l1} + \mathfrak{C}_{ra} + D_A \qquad \mathfrak{R}_B = \mathfrak{M}_{r1} - \mathfrak{M}_{l2} + \mathfrak{C}_{rb} + D_B. \qquad (248)$$

auxiliary moments:

$$Y''_A = +\mathfrak{R}_A \cdot u''_{aa} - \mathfrak{R}_B \cdot u''_{ab} \qquad Y''_B = -\mathfrak{R}_A \cdot u''_{ba} + \mathfrak{R}_B \cdot u''_{bb}. \qquad (249)$$

Beam moments:

$$\left. \begin{aligned} M_{A1} &= -M_{D3} = \mathfrak{M}_{l1} + (2Y''_A + Y''_B)K_1 \\ M_{B1} &= -M_{C3} = \mathfrak{M}_{r1} - (Y''_A + 2Y''_B)K_1 \qquad M_{B2} = -M_{C2} = \mathfrak{M}_{l2} + Y''_B \cdot 3K_2 \end{aligned} \right\} \quad (250)$$

Column moments:

at top: at base:

$$\left. \begin{aligned} M_{Aa} &= M_{Dd} = \mathfrak{C}_{ra} - Y''_A \cdot 3S_a & M_a &= M_d = -(\mathfrak{L}_a + M_{Aa})\varepsilon_a \\ M_{Bb} &= M_{Cc} = \mathfrak{C}_{rb} - Y''_B \cdot 3S_b; & M_b &= M_c = -(\mathfrak{L}_b + M_{Bb})\varepsilon_b. \end{aligned} \right\} \quad \begin{matrix}(251)\\ \text{and}\\ (252)\end{matrix}$$

Check relationships:

$$D_A + M_{Aa} - M_{A1} = 0 \qquad D_B + M_{Bb} + M_{B1} - M_{B2} = 0. \qquad (253)$$

— 54 —

(Frame Shape 6—cont.)

Horizontal thrusts, column shears and lateral restraining force for all Loading Conditions

For asymmetrical Loading Conditions: according to formulas (36)—(39).

For symmetrical and anti-symmetrical Loading Conditions: according to formulas (183)—(190), for the collective subscripts $N = A, B$ and $n = a, b$.

Frame Shape 6v

(Frame Shape 6 with laterally unrestrained horizontal member)

Calculation procedure:

either exactly as for Frame Shape 1v (see page 9/10), having due regard to the properties of anti-symmetry—or as for Frame Shape 5v (see page 42/43) for the collective subscripts $\nu = 1, 2; N = A, B; n = a, b$.

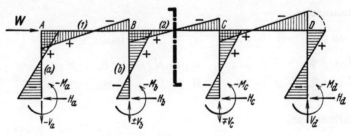

Fig. 41
Loading Condition 5 with anti-symmetrical diagram of bending moments and support reactions

Frame Shape 7

Symmetrical single-story single-bay frame with laterally restrained horizontal member and two elastically restrained vertical columns

(Symmetrical form of Frame Shape 2)

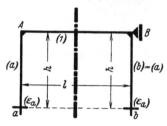
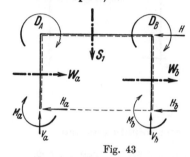

Fig. 42
Frame shape, dimensions and symbols

Fig. 43
Definition of positive direction for all external loads on joints and members, for all column reactions, and for the lateral restraining force

---

All the dimensions and coefficients for the right-hand half of the frame are the same as those for the left-hand half. In the case of the columns the broken line is always placed on the right-hand side of each column, despite the symmetry of the frame.

Note: For Frame Shape 7 the formulas that have been given for *Frame Shape 2* may be employed. However, various simplifications can be introduced thanks to the symmetry of the frame. Having regard to the practical importance of Frame Shape 7, all the relevant formulas have therefore been reproduced as a self-contained whole under the present heading.

Important: In the loading conditions considered below, the left-hand diagram in each case represents the loading, while the right-hand diagram represents the bending moment diagram and all the support reactions.

I. Treatment by the Method of Forces

Coefficients, where $J_k = J_1$ and $l_k = l$, hence $k_1 = 1$:

$$k_a = \frac{J_1}{J_a} \cdot \frac{h}{l}; \qquad b_a = k_a(2 - \varepsilon_a), \quad (\varepsilon_a \text{ is assumed to be known}).$$

$$N_1 = b_a + 1 \qquad N_2 = b_a + 2 \qquad N_3 = b_a + 3. \tag{254}$$

Loading Condition 1

Beam carrying arbitrary vertical loading

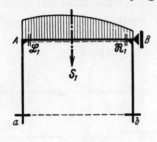
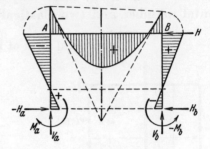

Fig. 44 Fig. 45

Beam and column moments

$$M_{A1} = +M_{Aa} = \frac{-\mathfrak{L}_1 N_2 + \mathfrak{R}_1}{N_1 N_3} \qquad M_{B1} = -M_{Bb} = \frac{+\mathfrak{L}_1 - \mathfrak{R}_1 N_2}{N_1 N_3};$$
$$M_a = -M_{Aa} \cdot \varepsilon_a \qquad\qquad M_b = -M_{Bb} \cdot \varepsilon_a. \quad\Bigg\} \quad (255)$$

Reactions at supports

$$H_a = \frac{M_{Aa}(1+\varepsilon_a)}{h} \quad H_b = \frac{M_{Bb}(1+\varepsilon_a)}{h}; \quad H = \frac{(\mathfrak{L}_1 - \mathfrak{R}_1)(1+\varepsilon_a)}{hN_1};$$
$$V_a = \frac{\mathfrak{S}_{r1}}{l} + \frac{(\mathfrak{L}_1 - \mathfrak{R}_1)}{lN_1} \quad V_b = \frac{\mathfrak{S}_{l1}}{l} - \frac{(\mathfrak{L}_1 - \mathfrak{R}_1)}{lN_1}. \quad\Bigg\} \quad (256)$$

Loading Condition 1a

Beam carrying arbitrary symmetrical vertical loading
(Special case of Loading Condition 1, with $\mathfrak{R}_1 = \mathfrak{L}_1$)

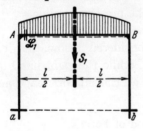
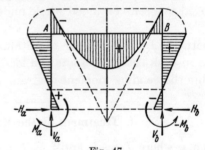

Fig. 46 Fig. 47

$$M_{A1} = +M_{Aa} = M_{B1} = -M_{Bb} = -\frac{\mathfrak{L}_1}{N_3} \qquad M_a = -M_b = -M_{Aa} \cdot \varepsilon_a. \quad (257)$$

$$H_b = -H_a = \frac{\mathfrak{L}_1(1+\varepsilon_a)}{hN_3}; \quad H = 0; \quad V_a = V_b = \frac{S_1}{2}. \quad (258)$$

Loading Condition 1b

Beam carrying arbitrary anti-symmetrical vertical loading
(Special case of Loading Condition 1, with $\Re_1 = -\mathfrak{L}_1$)

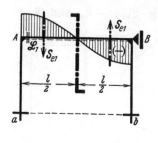
Fig. 48

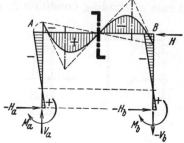
Fig. 49

$$M_{A1} = +M_{Aa} = -M_{B1} = +M_{Bb} = -\frac{\mathfrak{L}_1}{N_1} \qquad M_a = M_b = -M_{Aa} \cdot \varepsilon_a. \quad (259)$$

$$H_a = H_b = -\frac{\mathfrak{L}_1}{hN_1}; \qquad H = \frac{2\mathfrak{L}_1}{hN_1}; \qquad V_a = -V_b = \frac{\mathfrak{S}_{r1}}{l} + \frac{2\mathfrak{L}_1}{lN_1}. \quad (260)$$

Loading Condition 2

Both columns carrying arbitrary horizontal loading

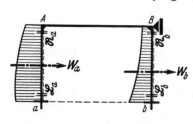
Fig. 50

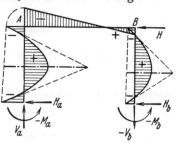
Fig. 51

Joint load terms:

$$\mathfrak{B}_A = (\Re_a - \mathfrak{L}_a \cdot \varepsilon_a) k_a \qquad \mathfrak{B}_B = (\Re_b - \mathfrak{L}_b \cdot \varepsilon_a) k_a. \quad (261)$$

Beam and column moments

$$\left. \begin{array}{ll} M_{A1} = +M_{Aa} = -\dfrac{\mathfrak{B}_A N_2 + \mathfrak{B}_B}{N_1 N_3} & M_{B1} = -M_{lb} = +\dfrac{\mathfrak{B}_A + \mathfrak{B}_B N_2}{N_1 N_3}; \\ M_a = -(\mathfrak{L}_a + M_{Aa}) \varepsilon_a & M_b = -(\mathfrak{L}_b + M_{Bb}) \varepsilon_a. \end{array} \right\} \quad (262)$$

Reactions at supports

All forces H as given in equations (73) and (74), where $(h_a = h_b) = h$.

$$V_a = -V_b = \frac{M_{B1} - M_{A1}}{l} = \frac{\mathfrak{B}_A + \mathfrak{B}_B}{lN_1}. \quad (263)$$

Loading Condition 2a

Both columns carrying arbitrary equal horizontal loading, symmetrical about center of frame

(Special case of Loading Condition 2, with $\mathfrak{L}_b = -\mathfrak{L}_a$ and $\mathfrak{R}_b = -\mathfrak{R}_a$)

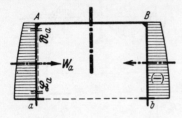

Fig. 52   Fig. 53

$$\left. \begin{array}{l} M_{A1} = +M_{Aa} = M_{B1} = -M_{Bb} = -\dfrac{\mathfrak{B}_A}{N_3}; \\[6pt] M_b = -M_a = (\mathfrak{L}_a + M_{Aa})\,\varepsilon_a. \end{array} \right\} \quad (264)$$

$$H_a = -H_b = \frac{\mathfrak{S}_{ra}}{h} + \frac{M_{Aa} - M_a}{h}; \quad H = 0; \quad V_a = V_b = 0. \quad (265)$$

Loading Condition 2b

Both columns carrying arbitrary equal horizontal loading, anti-symmetrical about center of frame

(Special case of Loading Condition 2, with $\mathfrak{L}_b = \mathfrak{L}_a$ and $\mathfrak{R}_b = \mathfrak{R}_a$)

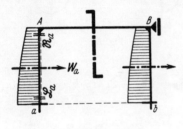 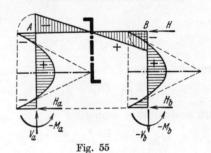

Fig. 54   Fig. 55

$$\left. \begin{array}{l} M_{A1} = +M_{Aa} = -M_{B1} = +M_{Bb} = -\dfrac{\mathfrak{B}_A}{N_1}, \quad V_a = -V_b = \dfrac{2\mathfrak{B}_A}{lN_1}. \\[6pt] M_a = M_b = -(\mathfrak{L}_a + M_{Aa})\,\varepsilon_a. \end{array} \right\} \quad (266)$$

$$H_a = H_b = \frac{\mathfrak{S}_{ra}}{h} + \frac{M_{Aa} - M_a}{h} \qquad H = 2\left(\frac{\mathfrak{S}_{la}}{h} - \frac{M_{Aa} - M_a}{h}\right). \quad (267)$$

Loading Condition 3

Two rotational moments of different magnitude applied at the corners

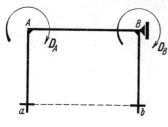
Fig. 56

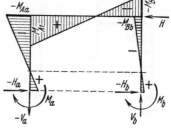
Fig. 57

Beam and column moments

$$M_{A1} = + \frac{(D_A N_2 + D_B) b_a}{N_1 N_3} \qquad M_{B1} = - \frac{(D_A + D_B N_2) b_a}{N_1 N_3};$$
$$M_{Aa} = + M_{A1} - D_A \qquad M_{Bb} = - M_{B1} - D_B \qquad (268)$$
$$M_a = - M_{Aa} \cdot \varepsilon_a; \qquad M_b = - M_{Bb} \cdot \varepsilon_a.$$

Reactions at supports

$$H_a = \frac{M_{Aa}(1 + \varepsilon_a)}{h} \qquad H_b = \frac{M_{Bb}(1 + \varepsilon_a)}{h}; \qquad H = -H_a - H_b;$$
$$V_b = -V_a = \frac{M_{A1} - M_{B1}}{l} = \frac{(D_A + D_B) b_a}{l N_1}. \qquad (269)$$

Loading Condition 3a

Symmetrical pair of rotational moments applied at the corners

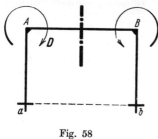
Fig. 58

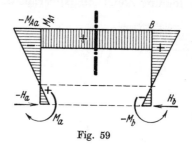
Fig. 59

$$M_{A1} = M_{B1} = + \frac{D b_a}{N_3} \qquad M_{Aa} = - M_{Bb} = - \frac{3 D}{N_3}; \qquad M_a = - M_b = - M_{Aa} \cdot \varepsilon_a;$$
(270)
$$(M_{A1} - M_{Aa} = D). \qquad H_b = -H_a = \frac{M_{Bb}(1 + \varepsilon_a)}{h}; \qquad H = 0; \qquad V_a = V_b = 0.$$
(271)

Loading Condition 3b

Anti-symmetrical pair of rotational moments applied at the corners

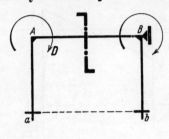

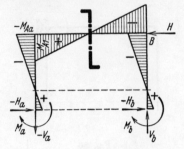

Fig. 60   Fig. 61

$$M_{A1} = -M_{B1} = +\frac{D b_a}{N_1} \qquad M_{Aa} = M_{Eb} = -\frac{D}{N_1}; \qquad \begin{array}{l} M_a = M_b = -M_{Aa} \cdot \varepsilon_a \\ (M_{A1} - M_{Aa} = D). \end{array} \quad (272)$$

$$H_a = H_b = \frac{M_{Aa}(1+\varepsilon_a)}{h} \qquad H = -2H_a; \qquad V_b = -V_a = \frac{2M_{A1}}{l}. \qquad (273)$$

II. Treatment by the Deformation Method

Coefficients, where $J_k = J_1$ and $l_k = l$, and hence $K_1 = 1$:

$$K_a = \frac{J_a}{J_1} \cdot \frac{l}{h}; \qquad S_a = \frac{K_a}{2-\varepsilon_a}, \; (\varepsilon_a \text{ is assumed to be known}).$$
$$U_1 = 3S_a + 1 \qquad U_2 = 3S_a + 2 \qquad U_3 = 3S_a + 3. \qquad (274)$$

Loading Condition 4

Overall loading condition (superposition of Loading Conditions 1, 2 and 3)

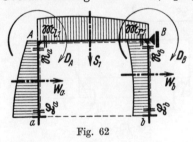

Fig. 62

All the formulas (except those for the values Y) are the same as for Frame Shape 2, page 14, writing $\varepsilon_b = \varepsilon_a$ and $S_b = S_a$

Auxiliary moments:

$$Y_A = \frac{+\mathfrak{K}_A U_2 - \mathfrak{K}_B}{U_1 U_3} \qquad Y_B = \frac{-\mathfrak{K}_A + \mathfrak{K}_B U_2}{U_1 U_3}. \qquad (275)$$

All forces H as given in equations (73) and (74), where $(h_a = h_b) = h$.

$$V_a = \frac{\mathfrak{S}_{r1}}{l} + \frac{M_{B1} - M_{A1}}{l} \qquad V_b = \frac{\mathfrak{S}_{l1}}{l} - \frac{M_{B1} - M_{A1}}{l}. \qquad (276)$$

Loading Condition 4a

Symmetrical overall loading condition—(1a, 2a and 3a)

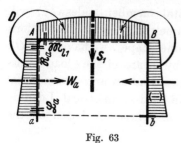

Fig. 63

Column restraining moment

$$\mathfrak{C}_{ra} = -\frac{\mathfrak{R}_a - \mathfrak{L}_a \cdot \varepsilon_a}{2 - \varepsilon_a}. \qquad (277)$$

Auxiliary moment

$$Y'_A = \frac{-\mathfrak{M}_{l1} + \mathfrak{C}_{ra} + D}{U_1}. \qquad (278)$$

Beam and column moments

$$\left.\begin{array}{l} M_{A1} = M_{B1} = \mathfrak{M}_{l1} + Y'_A; \\ M_{Aa} = -M_{Bb} = \mathfrak{C}_{ra} - Y'_A \cdot 3 S_a \qquad M_a = -M_b = -(\mathfrak{L}_a + M_{Aa})\,\varepsilon_a. \end{array}\right\} \quad (279)$$

Reactions at supports

$$H_a = -H_b = \frac{\mathfrak{C}_{ra}}{h} + \frac{M_{Aa} - M_a}{h}; \qquad H = 0; \qquad V_a = V_b = \frac{S_1}{2}. \qquad (280)$$

Loading Condition 4b

Anti-symmetrical overall loading condition—(1b, 2b and 3b)

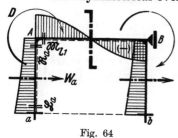

Fig. 64

Column restraining moment

exactly as formula (277).

Auxiliary moment

$$Y''_A = \frac{-\mathfrak{M}_{l1} + \mathfrak{C}_{ra} + D}{U_3}. \qquad (281)$$

Beam and column moments

$$\left.\begin{array}{l} M_{A1} = -M_{B1} = \mathfrak{M}_{l1} + 3\,Y''_A; \\ M_{Aa} = M_{Bb} = \mathfrak{C}_{ra} - Y''_A \cdot 3 S_a \qquad M_a = M_b = -(\mathfrak{L}_a + M_{Aa})\,\varepsilon_a. \end{array}\right\} \quad (282)$$

Reactions at supports

$$\left.\begin{array}{l} H_a = H_b = \dfrac{\mathfrak{C}_{ra}}{h} + \dfrac{M_{Aa} - M_a}{h}; \qquad H = 2(\boldsymbol{W}_a - H_a) = 2\left(\dfrac{\mathfrak{S}_{la}}{h} - \dfrac{M_{Aa} - M_a}{h}\right); \\ V_a = -V_b = \dfrac{\mathfrak{S}_{r1}}{l} - \dfrac{2 M_{A1}}{l}. \end{array}\right\} \quad (283)$$

— 62 —

Frame Shape 7v
(Frame Shape 7 with laterally unrestrained horizontal member)

Note: Lateral displacement of the horizontal member (sidesway) will occur only with asymmetrical and anti-symmetrical loading conditions (with all symmetrical loading conditions for Frame Shape 7 we have $H = O$). As every asymmetrical loading condition can, by means of the "load transposition method," be split up into a symmetrical and an anti-symmetrical loading condition, only the formulas for the anti-symmetrical Loading Conditions 5, 1b, 2b and 3b are directly given. These are based on the Method of Forces (Method I).

As in the previous case (Frame Shape 7), the left-hand diagram again in each case represents the loading, while the right-hand diagram represents the bending moment and all the support reactions.

Coefficients, where $J_k = J_1$ and $l_k = l$, and hence $k_1 = 1$: (284)

$$k_a = \frac{J_1}{J_a} \cdot \frac{h}{l}; \qquad N^v = \left(\frac{1}{\varepsilon_a} + 4\right) k_a + 1, \quad (\varepsilon_a \text{ is assumed to be known})$$

Loading Condition 5
Horizontal concentrated load W at level of horizontal member

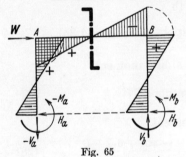

Fig. 65

Beam and column moments

$$\left.\begin{array}{l} M_{A1} = +\dfrac{Wh}{2} \cdot \dfrac{(1/\varepsilon_a + 1)k_a}{N^v}; \\ (M_{Aa} = M_{Bb} = -M_{B1}) = M_{A1}; \\ M_a = M_b = -Wh/2 + M_{Aa}. \end{array}\right\} \quad (285)$$

Beam and column moments

$$H_a = H_b = W/2 \qquad V_b = -V_a = 2M_{A1}/l. \quad (286)$$

Loading Condition 3b
Anti-symmetrical pair of rotational moments applied at the corners

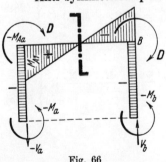

Fig. 66

Beam and column moments

$$\left.\begin{array}{l} M_{Aa} = M_a = M_{Bb} = M_b = -\dfrac{D}{N^v}; \\ M_{A1} = -M_{B1} = D + M_{Aa}. \end{array}\right\} \quad (287)$$

Reactions at supports $H_a = H_b = 0; \qquad V_b = -V_a = 2M_{A1}/l.$ (288)

Loading Condition 1b

Beam carrying arbitrary anti-symmetrical vertical loading

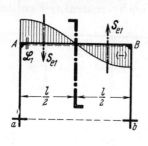

Fig. 67   Fig. 68

Beam and column moments

$$M_{A1} = M_{Aa} = M_a = -M_{B1} = M_{Bb} = M_b = -\frac{\mathfrak{L}_1}{N^v}. \qquad (289)$$

Reactions at supports

$$H_a = H_b = 0; \qquad V_a = -V_b = \frac{\mathfrak{S}_{r1}}{l} - \frac{2 M_{A1}}{l}. \qquad (290)$$

Loading Condition 2b

Both columns carrying arbitrary horizontal loading, anti-symmetrical about center of frame

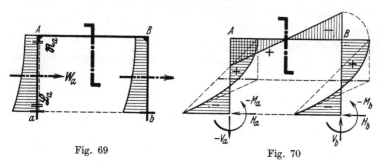

Fig. 69   Fig. 70

Beam and column moments

$$\left. \begin{array}{c} M_{A1} = M_{Aa} = -M_{B1} = M_{Bb} = +\left[\mathfrak{S}_{la}\left(\dfrac{1}{\varepsilon_a}+1\right) - (\mathfrak{L}_a + \mathfrak{R}_a)\right] \cdot \dfrac{k_a}{N^v}; \\ M_a = M_b = -\mathfrak{S}_{la} + M_{Aa}. \end{array} \right\} \quad (291)$$

Reactions at supports

$$H_a = H_b = \boldsymbol{W}_a; \qquad V_b = -V_a = \frac{2 M_{A1}}{l}. \qquad (292)$$

Frame Shape 8

Symmetrical single-story six-bay frame with laterally restrained horizontal member and seven elastically restrained columns

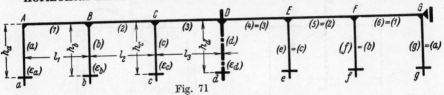

Fig. 71
Frame shape, dimensions and symbols

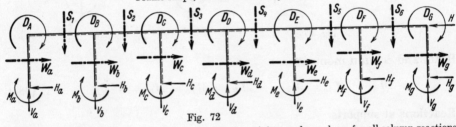

Fig. 72
Definition of positive direction for all external loads on joints and members, for all column reactions, and for the lateral restraining force

> All the dimensions and coefficients for the right-hand half of the frame are the same as those for the left-hand half. In the case of the columns the broken line is always placed on the right-hand side of each column, despite the symmetry of the frame.

Note: For Frame Shape 8 the *coefficients* and the *Loading Conditions* 1 – 4 for *Frame Shape* 3, enlarged by the addition of two bays, may be employed. On account of symmetry, the forward reduction becomes the same as the backward reduction, and the backward recursion becomes the same as the forward recursion, in both Method I and Method II. Furthermore, the *square matrices of influence coefficients* (with 12×12 elements in Method I and 7×7 elements in Method II) are bisymmetrical, i.e., symmetrical about the principal diagonal and the secondary diagonal. If the *"load transposition method"* is employed, however, it is also possible to start from the following symmetrical and anti-symmetrical loading conditions, which in the present case are given only for the Deformation Method (Method II).*

> **Important**
>
> All the instructions, formula sequences and formula matrices given for Frame Shape 8 can readily be extended to suit symmetrical frames of similar type having an even number of bays.

---

* The treatment by the Method of Forces (Method I) has been omitted in order to save space. The formulas for a similar type of frame according to that method are given under Frame Shape 9 (page 72).

a) Coefficients for symmetrical load arrangements

Stiffness coefficients of all the members, according to formulas (94), for $\nu = 1, 2, 3$ and $n = a, b, c$

Stiffnesses of the columns, according to formulas (95), for $n = a, b, c$

Joint coefficients:

$$K_A = 3S_a + 2K_1 \quad K_B = 2K_1 + 3S_b + 2K_2 \quad K_C = 2K_2 + 3S_c + 2K_3. \quad (293)$$

Rotation carry-over factors and auxiliary coefficients:

$$\begin{array}{l|l}
\text{backward reduction:} & \text{forward reduction:} \\
\begin{aligned} j_{2s} &= \frac{K_2}{r_c'} & r_c' &= K_C \\ & & r_b' &= K_B - K_2 \cdot j_{2s} \\ j_{1s} &= \frac{K_1}{r_b'} \\ & & r_a' &= K_A - K_1 \cdot j_{1s}. \end{aligned} &
\begin{aligned} & & v_a &= K_A \\ j_1' &= \frac{K_1}{v_a} \\ & & v_b &= K_B - K_1 \cdot j_1' \\ j_2' &= \frac{K_2}{v_b} \\ & & v_c &= K_C - K_2 \cdot j_2'. \end{aligned}
\end{array} \quad \begin{matrix}(294)\\ \text{and}\\ (295)\end{matrix}$$

Principal influence coefficients by recursion:

$$\begin{array}{l|l}
\text{forward:} & \text{backward:} \\
\begin{aligned} u_{aa}' &= 1/r_a' \\ u_{bb}' &= 1/r_b' + u_{aa}' \cdot j_{1s}^2 \\ u_{cc}' &= 1/r_c' + u_{bb}' \cdot j_{2s}^2. \end{aligned} &
\begin{aligned} u_{cc}' &= 1/v_c \\ u_{bb}' &= 1/v_b + u_{cc}' \cdot j_2'^2 \\ u_{aa}' &= 1/v_a + u_{bb}' \cdot j_1'^2. \end{aligned}
\end{array} \quad \begin{matrix}(296)\\ \text{and}\\ (297)\end{matrix}$$

Principal influence coefficients direct

$$u_{aa}' = \frac{1}{r_a'} \qquad u_{bb}' = \frac{1}{r_b' + v_b - K_B} \qquad u_{cc}' = \frac{1}{v_c}. \quad (298)$$

Computation scheme for the symmetrical influence coefficient matrix*

$$\left.\begin{array}{c|ccc|c}
& (u_{aa}') & u_{ab}' & u_{ac}' & \\
\cdot j_{1s} & \downarrow & \uparrow & \uparrow & \cdot j_1' \\
& u_{ba}' & (u_{bb}') & u_{bc}' & \\
\cdot j_{2s} & \downarrow & \downarrow & \uparrow & \cdot j_2' \\
& u_{ca}' & u_{cb}' & (u_{cc}') &
\end{array}\right\} \quad (299)$$

---

*Between the elements of this 3 × 3 matrix for the left-hand half of the frame and the elements of the 7 × 7 matrix for the whole frame, as indicated on page 64, there exist, according to the "load transposition method", the following relationships:

$$u_{aa}' = u_{aa} - u_{ay} \text{ etc., up to } u_{cc}' = u_{cc} - u_{ce}.$$

(The values u' with the subscripts d for the center column vanish)

b) Coefficients for anti-symmetrical load arrangements

Stiffness coefficients of all the members, according to formulas (94), for $\nu = 1, 2, 3$ and $n = a, b, c, d$

Stiffnesses of the columns, according to formulas (95) for $n = a, b, c, d$

Joint coefficients:

$$\left.\begin{aligned}K_A &= 3S_a + 2K_1 & K_B &= 2K_1 + 3S_b + 2K_2 \\ K_C &= 2K_2 + 3S_c + 2K_3 & K_D'' &= 2K_3 + 3S_d/2.\end{aligned}\right\} \quad (300)$$

Rotation carry-over factors and auxiliary coefficients:

backward reduction: | forward reduction:

$$\left.\begin{array}{ll|ll} j_{3t} = \dfrac{K_3}{r_d''} & r_d'' = K_D'' & j_1' = \dfrac{K_1}{v_a} & v_a = K_A \\ & r_c'' = K_C - K_3 \cdot j_{3t} & & v_b = K_B - K_1 \cdot j_1' \\ j_{2t} = \dfrac{K_2}{r_c''} & r_b'' = K_B - K_2 \cdot j_{2t} & j_2' = \dfrac{K_2}{v_b} & v_c = K_C - K_2 \cdot j_2' \\ j_{1t} = \dfrac{K_1}{r_b''} & r_a'' = K_A - K_1 \cdot j_{1t}. & j_3' = \dfrac{K_3}{v_c} & v_d'' = K_D'' - K_3 \cdot j_3'.\end{array}\right\} \quad \begin{array}{c}(301)\\ \text{and}\\ (302)\end{array}$$

Principal influence coefficients by recursion:

forward: | backward:

$$\left.\begin{array}{ll|ll} u_{aa}'' &= 1/r_a'' & u_{dd}'' &= 1/v_d'' \\ u_{bb}'' &= 1/r_b'' + u_{aa}'' \cdot j_{1t}^2 & u_{cc}'' &= 1/v_c + u_{dd}'' \cdot j_3'^2 \\ u_{cc}'' &= 1/r_c'' + u_{bb}'' \cdot j_{2t}^2 & u_{bb}'' &= 1/v_b + u_{cc}'' \cdot j_2'^2 \\ u_{dd}'' &= 1/r_d'' + u_{cc}'' \cdot j_{3t}^2. & u_{aa}'' &= 1/v_a + u_{bb}'' \cdot j_1'^2.\end{array}\right\} \quad \begin{array}{c}(303)\\ \text{and}\\ (304)\end{array}$$

Principal influence coefficients direct

$$u_{aa}'' = \dfrac{1}{r_a''} \quad u_{bb}'' = \dfrac{1}{r_b'' + v_b - K_B} \quad u_{cc}'' = \dfrac{1}{r_c'' + v_c - K_C} \quad u_{dd}'' = \dfrac{1}{v_{dd}''}. \quad (305)$$

Computation scheme for the symmetrical influence coefficient matrix*

$$\left.\begin{array}{c|cccc|c} \cdot j_{1t} & (u_{aa}'') & u_{ab}'' & u_{ac}'' & u_{ad}'' & \cdot j_1' \\ & \downarrow & \uparrow & \uparrow & \uparrow & \\ & u_{ba}'' & (u_{bb}'') & u_{bc}'' & u_{bd}'' & \\ \cdot j_{2t} & \downarrow & \downarrow & \uparrow & \uparrow & \cdot j_2' \\ & u_{ca}'' & u_{cb}'' & (u_{cc}'') & u_{cd}'' & \\ \cdot j_{3t} & \downarrow & \downarrow & \downarrow & \uparrow & \cdot j_3' \\ & u_{da}'' & u_{db}'' & u_{dc}'' & (u_{dd}'') & \end{array}\right\} \quad (306)$$

---

* See footnote on next page.

Loading Condition 4a

Symmetrical loading condition

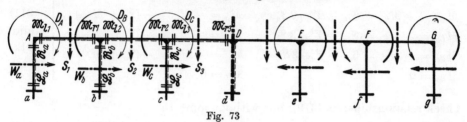

Fig. 73

Loading diagram with fixed-end moments $\mathfrak{M}_l$ and $\mathfrak{M}_r$ for the beam spans, load terms $\mathfrak{L}$ and $\mathfrak{R}$ for the columns, and external rotational moments D, for left-hand half of frame
(Symmetrical loading of the central column d and central joint D not shown)

Column restraining moments at the beam joints:

$$\mathfrak{C}_{rn} = -\frac{\mathfrak{R}_n - \mathfrak{L}_n \cdot \varepsilon_n}{2 - \varepsilon_n}, \quad \text{for} \quad n = a, b, c. \tag{307}$$

Joint load terms:

$$\left. \begin{array}{l} \mathfrak{K}_A = -\mathfrak{M}_{l1} + \mathfrak{C}_{ra} + D_A \qquad \mathfrak{K}_B = \mathfrak{M}_{r1} - \mathfrak{M}_{l2} + \mathfrak{C}_{rb} + D_B \\ \qquad\qquad\qquad\qquad \mathfrak{K}_C = \mathfrak{M}_{r2} - \mathfrak{M}_{l3} + \mathfrak{C}_{rc} + D_C . \end{array} \right\} \tag{308}$$

Auxiliary moments:

$$\left. \begin{array}{l} Y'_A = +\mathfrak{K}_A \cdot u'_{aa} - \mathfrak{K}_B \cdot u'_{ab} + \mathfrak{K}_C \cdot u'_{ac} \\ Y'_B = -\mathfrak{K}_A \cdot u'_{ba} + \mathfrak{K}_B \cdot u'_{bb} - \mathfrak{K}_C \cdot u'_{bc} \\ Y'_C = +\mathfrak{K}_A \cdot u'_{ca} - \mathfrak{K}_B \cdot u'_{cb} + \mathfrak{K}_C \cdot u'_{cc} . \end{array} \right\} \tag{309}$$

Beam moments:

$$\left. \begin{array}{l} M_{A1} = M_{G6} = \mathfrak{M}_{l1} + (2Y'_A + Y'_B) K_1 \\ M_{B2} = M_{F5} = \mathfrak{M}_{l2} + (2Y'_B + Y'_C) K_2 \\ M_{C3} = M_{E4} = \mathfrak{M}_{l3} + 2Y'_C \cdot K_3; \\ \qquad M_{B1} = M_{F6} = \mathfrak{M}_{r1} - (Y'_A + 2Y'_B) K_1 \\ \qquad M_{C2} = M_{E5} = \mathfrak{M}_{r2} - (Y'_B + 2Y'_C) K_2 \\ \qquad M_{D3} = M_{D4} = \mathfrak{M}_{r3} - Y'_C \cdot K_3 . \end{array} \right\} \tag{310}$$

---

* Between the elements of the 4 × 4 matrix (306) for the left-hand half of the frame and the elements of the 7 × 7 matrix for the whole frame, mentioned on page 64, there exist, according to the "load transposition method", the following relationships:

$$u''_{aa} = u_{aa} + u_{ag} \quad \text{etc., up to} \quad u''_{cc} = u_{cc} + u_{ce}, \quad u''_{dd} = 2 u_{dd}.$$

(The values u'' of the d-column of (306) become twice as large as the values u of the d-column of the 7 × 7 matrix).

Column moments:

at top:

$$M_{Aa} = -M_{Gg} = \mathfrak{C}_{ra} - Y'_A \cdot 3S_a$$
$$M_{Bb} = -M_{Ff} = \mathfrak{C}_{rb} - Y'_B \cdot 3S_b$$
$$M_{Cc} = -M_{Ee} = \mathfrak{C}_{rc} - Y'_C \cdot 3S_c$$
$$M_{Dd} = 0;$$

at base:

$$M_a = -M_g = -(\mathfrak{L}_a + M_{Aa})\varepsilon_a$$
$$M_b = -M_f = -(\mathfrak{L}_b + M_{Bb})\varepsilon_b$$
$$M_c = -M_e = -(\mathfrak{L}_c + M_{Cc})\varepsilon_c$$
$$M_d = 0.$$

(311) and (312)

Check relationships: as (319), but without joint D.

Loading Condition 4b

Anti-symmetrical overall loading condition

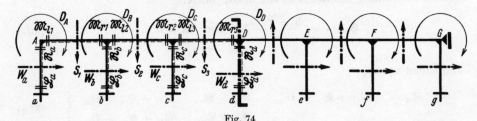

Fig. 74

Loading diagram with fixed-end moments $\mathfrak{M}_l$ and $\mathfrak{M}_r$ for the beam spans, load terms $\mathfrak{L}$ and $\mathfrak{R}$ for the columns, and external rotational moments $\boldsymbol{D}$, for left-hand half of frame, including central column and central joint

Column restraining moments at the beam joints:

$$\mathfrak{C}_{rn} = -\frac{\mathfrak{R}_n - \mathfrak{L}_n \cdot \varepsilon_n}{2 - \varepsilon_n}, \quad \text{for} \quad n = a, b, c, d. \tag{313}$$

Joint load terms:

$$\begin{aligned}
\mathfrak{R}_A &= \quad\quad -\mathfrak{M}_{l1} + \mathfrak{C}_{ra} + D_A & \mathfrak{R}_C &= \mathfrak{M}_{r2} - \mathfrak{M}_{l3} + \mathfrak{C}_{rc} + D_C \\
\mathfrak{R}_B &= \mathfrak{M}_{r1} - \mathfrak{M}_{l2} + \mathfrak{C}_{rb} + D_B & \mathfrak{R}''_D &= \mathfrak{M}_{r3} + \mathfrak{C}_{rd}/2 + D_D/2.
\end{aligned} \tag{314}$$

Auxiliary moments:

	$\mathfrak{R}_A$	$\mathfrak{R}_B$	$\mathfrak{R}_C$	$\mathfrak{R}''_D$
$Y''_A =$	$+u''_{aa}$	$-u''_{ab}$	$+u''_{ac}$	$-u''_{ad}$
$Y''_B =$	$-u''_{ba}$	$+u''_{bb}$	$-u''_{bc}$	$+u''_{bd}$
$Y''_C =$	$+u''_{ca}$	$-u''_{cb}$	$+u''_{cc}$	$-u''_{cd}$
$Y''_D =$	$-u''_{da}$	$+u''_{db}$	$-u''_{dc}$	$+u''_{dd}.$

(315)

Beam moments:

$$\left.\begin{aligned}
M_{A1} &= -M_{G6} = \mathfrak{M}_{l1} + (2Y_A'' + Y_B'') K_1 \\
M_{B2} &= -M_{F5} = \mathfrak{M}_{l2} + (2Y_B'' + Y_C'') K_2 \\
M_{C3} &= -M_{E4} = \mathfrak{M}_{l3} + (2Y_C'' + Y_D'') K_3; \\
M_{B1} &= -M_{F6} = \mathfrak{M}_{r1} - (Y_A'' + 2Y_B'') K_1 \\
M_{C2} &= -M_{E5} = \mathfrak{M}_{r2} - (Y_B'' + 2Y_C'') K_2 \\
M_{D3} &= -M_{D4} = \mathfrak{M}_{r3} - (Y_C'' + 2Y_D'') K_3.
\end{aligned}\right\} \quad (316)$$

Column moments:

at top: at base:

$$\left.\begin{array}{ll}
M_{Aa} = M_{Gg} = \mathfrak{C}_{ra} - Y_A'' \cdot 3S_a & M_a = M_g = -(\mathfrak{L}_a + M_{Aa})\,\varepsilon_a \\
M_{Bb} = M_{Ff} = \mathfrak{C}_{rb} - Y_B'' \cdot 3S_b & M_b = M_f = -(\mathfrak{L}_b + M_{Bb})\,\varepsilon_b \\
M_{Cc} = M_{Ee} = \mathfrak{C}_{rc} - Y_C'' \cdot 3S_c & M_c = M_e = -(\mathfrak{L}_c + M_{Cc})\,\varepsilon_c \\
M_{Dd} = \mathfrak{C}_{rd} - Y_D'' \cdot 3S_d; & M_d = -(\mathfrak{L}_d + M_{Dd})\,\varepsilon_d.
\end{array}\right\} \begin{array}{c}(317)\\ \text{and} \\ (318)\end{array}$$

Check relationships:

$$\left.\begin{aligned}
D_A + M_{Aa} \phantom{+ M_{B1}} - M_{A1} &= 0 \\
D_B + M_{Bb} + M_{B1} - M_{B2} &= 0 \\
D_C + M_{Cc} + M_{C2} - M_{C3} &= 0 \\
D_D + M_{Dd} + 2M_{D3} \phantom{- M_{A1}} &= 0.
\end{aligned}\right\} \quad (319)$$

Horizontal thrusts, column shears and lateral restraining force for all loading conditions

For asymmetrical loading conditions, according to formulas (179)/(180) or (181)/(182), for $n = a, b, c, d, e, f, g$ and $N = A, B, C, D, E, F, G$.

For symmetrical loading conditions, according to formulas (183)/(184) or (185)/(186), for $n = a, b, c$ and $N = A, B, C$.

For anti-symmetrical loading conditions, according to formula (187) or (189), for $n = a, b, c, d$ and $N = A, B, C, D$. Furthermore:

$$H = -2(H_a + H_b + H_c) - H_d \quad \text{or} \quad H = -2(Q_{ra} + Q_{rb} + Q_{rc}) - Q_{rd}. \quad (320)$$

Frame Shape 8v
(Frame Shape 8 with laterally unrestrained horizontal member)

Note: Irrespective of symmetry, Frame Shape 8v can be analysed with the aid of the formulas relating to Frame Shape 1v (see page 9/10). The computation procedure is nevertheless again indicated below, as a self-contained whole, having due regard to the anti-symmetry properties.

First step: Calculation of all the requisite asymmetrical or anti-symmetrical loading conditions as well as the appropriate lateral restraining force H according to Frame Shape 8. (The symmetrical Loading Conditions need not be considered, because for these $H = 0$).

Second step: Determination of the effect of a horizontal concentrated load W, acting at the level of the horizontal member, with the aid of Loading Condition 5 for Frame Shape 8v, as indicated below.

Third step: Determination of the final moments and forces for Frame Shape 8v by superposition of the results of the first step and those of the second step for $W = H$.

Loading Condition 5

Horizontal concentrated load W at level of horizontal member

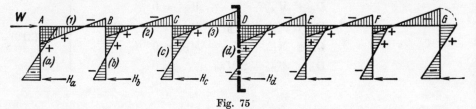

Fig. 75

Loading diagram and anti-symmetrical diagram of bending moments. (For clarity, only the reactions H_n at the column bases are shown)

Note: The *collective subscripts* have the following meanings:
$$\nu = 1, 2, 3; \qquad N = A, B, C, [D]; \qquad n = a, b, c, [d].$$

Column ratios
with arbitrary comparison height h_k $\qquad \gamma_n = \dfrac{h_k}{h_n}.$ (321)

Specific bending moments m (instead of M):
either according to Method I, Loading Condition 2b*), the moments m_{Nn} and m_n — for:

$$\mathfrak{B}_N = \gamma_n(1 + \varepsilon_n), \quad [\mathfrak{B}''_D = \gamma_d(1 + \varepsilon_d)/2] \quad \text{and} \quad \mathfrak{L}_n = \gamma_n/k_n; \qquad (322)$$

---
* Not given for Frame Shape 8, but see Loading Condition 2b for Frame Shape 9, which represents a similar case.

or according to *Method II/b*, from the formulas (316) – (319)—i.e., the moments y''_N (instead of Y''_N), m_{Nv}, m_{Nn} and m_n—for:

$$\mathfrak{K}_N = \mathfrak{C}_{rn} = S_n \gamma_n (1 + \varepsilon_n), \quad [\mathfrak{K}''_D = \mathfrak{C}_{rd}/2] \quad \text{and} \quad \mathfrak{L}_n = \gamma_n K_n. \tag{323}$$

Note: All the other load terms in bold type in the formulas indicated should be omitted.

Specific h_k-fold column shears:

$$p_n = (m_{Nn} - m_n)\gamma_n; \tag{324}$$

Corresponding specific h_k-fold applied load

$$p = 2(p_a + p_b + p_c) + p_d. \tag{325}$$

Conversion moment for a given load W:

$$U = \frac{W h_k}{p}. \tag{326}$$

Final moments and forces for the given concentrated load W:

Beam moments:

$$\left. \begin{array}{ll} M_{A1} = -M_{G6} = U \cdot m_{A1} \\ M_{B1} = -M_{F6} = U \cdot m_{B1} \end{array} \text{etc., up to} \begin{array}{l} M_{C3} = -M_{E4} = U \cdot m_{C3} \\ M_{D3} = -M_{D4} = U \cdot m_{D3}; \end{array} \right\} \tag{327}$$

Column moments:

$$\left. \begin{array}{ll} M_{Aa} = M_{Gg} = U \cdot m_{Aa} \\ M_a = M_g = U \cdot m_a \end{array} \text{etc., up to} \begin{array}{l} M_{Dd} = U \cdot m_{Dd} \\ M_d = U \cdot m_d. \end{array} \right\} \tag{328}$$

Column shears and horizontal thrusts for left-hand half of frame:

$$Q_n = H_n = \frac{U \cdot p_n}{h_k} = W \cdot \frac{p_n}{p}. \tag{329}$$

Supplementary note: The computation procedure given here for the "influence of horizontal concentrated load W acting at the level of the laterally unrestrained horizontal member", for the symmetrical Frame Shape 8v, is fundamentally applicable also to all other single-story symmetrical frames having an odd number of columns. It should only be noted that the subscript n in each case relates to all the rigidly connected columns and that the subscript N relates to all the corresponding beam joints.

Frame Shape 9

Symmetrical single-story four-bay frame with laterally restrained horizontal member and five elastically restrained columns

(Symmetrical form of Frame Shape 3)

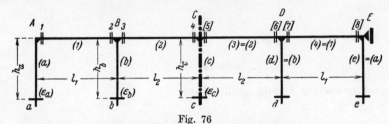

Fig. 76
Frame shape, dimensions and symbols

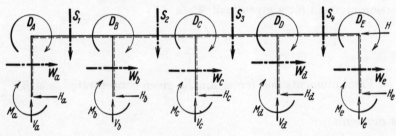

Fig. 77
Definition of positive direction for all external loads on joints and members, for all column reactions, and for the lateral restraining force

---

All the dimensions and coefficients for the right-hand half of the frame are the same as those for the left-hand half. In the case of the columns the broken line is always placed on the right-hand side of each column, despite the symmetry of the frame.

Note: For Frame Shape 9 the *coefficients* and the *Loading Conditions 1—4 for Frame Shape 3* may be employed. On account of symmetry, the reduction sequences (78)/(79) and (97)/(98) and the recursion sequences (80)/(81) and (99)/(100), respectively, become equal. Furthermore, the square matrices of influence coefficients (83) and (102) are bisymmetrical, i.e., symmetrical about the principal diagonal and the secondary diagonal. If the *"load transposition method"* is employed, however, it is also possible to start from the following symmetrical and anti-symmetrical loading conditions, which in the present case are given only for the Method of Forces (Method I).*

---

* The treatment by the Deformation Method (Method II) has been omitted in order to save space. The formulas for a similar type of frame according to that method are given under Frame Shape 8 (pages 64 to 69).

> **Important**
>
> All the instructions, formula sequences and formula matrices given for Frame Shape 9 can readily be extended to suit symmetrical frames of similar type having an even number of bays.

a) Coefficients for symmetrical load arrangements

Flexibility coefficients of all the members, according to formulas (75), for $\nu = 1, 2$ and $n = a, b$.

Flexibilities of the columns, according to formulas (76), for $n = a, b$.

Diagonal coefficients:

$$O_1 = b_a + 2k_1 \qquad O_2 = 2k_1 + b_b \qquad O_3 = b_b + 2k_2; \qquad (O_4' = 2k_2). \tag{330}$$

Moment carry-over factors and auxiliary coefficients:

$$
\left.
\begin{array}{l|l}
\text{backward reduction:} & \text{forward reduction:} \\[4pt]
\iota_{2s} = \dfrac{1}{2} & v_1 = O_1 \\
\qquad r_3' = O_3 - k_2 \cdot \dfrac{1}{2} & \iota_1' = \dfrac{k_1}{v_1} \\
\sigma_{bs} = \dfrac{b_b}{r_3'} & \qquad v_2 = O_2 - k_1 \cdot \iota_1' \\
\qquad r_2' = O_2 - b_b \cdot \sigma_{bs} & \sigma_b' = \dfrac{b_b}{v_2} \\
\iota_{1s} = \dfrac{k_1}{r_2'} & \qquad v_3 = O_3 - b_b \cdot \sigma_b' \\
\qquad r_1' = O_1 - k_1 \cdot \iota_{1s}. & \iota_2' = \dfrac{k_2}{v_3}.
\end{array}
\right\}
\begin{array}{l}(331)\\ \text{and}\\ (332)\end{array}
$$

Principal influence coefficients by recursion:

$$
\left.
\begin{array}{l|l}
\text{forward:} & \text{backward:} \\[4pt]
n_{11}' = \dfrac{1}{r_1'} & n_{44}' = \dfrac{1}{k_2(2 - \iota_2')} \\
n_{22}' = \dfrac{1}{r_2'} + n_{11}' \cdot \iota_{1s}^2 & n_{33}' = \dfrac{1}{v_3} + n_{44}' \cdot \iota_2'^2 \\
n_{33}' = \dfrac{1}{r_3'} + n_{22}' \cdot \sigma_{bs}^2 & n_{22}' = \dfrac{1}{v_2} + n_{33}' \cdot \sigma_b'^2 \\
n_{44}' = \dfrac{1}{2k_2} + n_{33}' \cdot \dfrac{1}{4}. & n_{11}' = \dfrac{1}{v_1} + n_{22}' \cdot \iota_1'^2.
\end{array}
\right\}
\begin{array}{l}(333)\\ \text{and}\\ (334)\end{array}
$$

Computation scheme for the symmetrical influence coefficient matrix*

$$
\left.
\begin{array}{r|cccc|l}
 & (n_{11}') & n_{12}' & n_{13}' & n_{14}' & \\
\cdot\, \iota_{1s} & \downarrow & \uparrow & \uparrow & \uparrow & \cdot\, \iota_1' \\
 & n_{21}' & (n_{22}') & n_{23}' & n_{24}' & \\
\cdot\, \sigma_{bs} & \downarrow & \downarrow & \uparrow & \uparrow & \cdot\, \sigma_b' \\
 & n_{31}' & n_{32}' & (n_{33}') & n_{34}' & \\
\cdot\, \dfrac{1}{2} & \downarrow & \downarrow & \downarrow & \uparrow & \cdot\, \iota_2' \\
 & n_{41}' & n_{42}' & n_{43}' & (n_{44}') & \\
\end{array}
\right\}
(335)
$$

---

* Between the elements of the matrix (335) and the elements of the matrix (83) there exist, according to the "load transposition method", the following relationships:

$$n_{11}' = n_{11} + n_{18}, \qquad n_{12}' = n_{12} + n_{17} \quad \text{etc., up to} \quad n_{44}' = n_{44} + n_{45}.$$

Principal influence coefficients direct

$$n'_{11} = \frac{1}{r'_1}; \quad n'_{ii} = \frac{1}{r'_i + v_i - O_i}, \quad (i = 2, 3); \quad n'_{44} = \frac{1}{k_2(2 - \iota'_2)}. \quad (336)$$

Composite influence coefficients

$$\begin{aligned}
s'_{11} &= + n'_{11} - n'_{12} = (s_{11} - s_{14}) & s'_{12} &= & n'_{14} &= (s_{12} - s_{13}) \\
& & s'_{22} &= & n'_{24} &= (s_{22} - s_{23}) \\
s'_{21} &= - n'_{21} + n'_{22} = (s_{21} + s_{24}) & s'_{32} &= & n'_{34} &= (s_{32} - s_{33}) \\
s'_{31} &= - n'_{31} + n'_{32} = (s_{31} + s_{34}) \\
s'_{41} &= - n'_{41} + n'_{42} = (s_{41} + s_{44}) & s'_{42} &= - n'_{43} + n'_{44} = (s_{42} + s_{43});
\end{aligned} \quad (337)$$

$$\begin{aligned}
w'_{1b} &= + n'_{12} - n'_{13} = (w_{1b} - w_{1d}) \\
w'_{2b} &= + n'_{22} - n'_{23} = (w_{2b} - w_{2d}) \\
w'_{3b} &= - n'_{32} + n'_{33} = (w_{3b} + w_{3d}) \\
w'_{4b} &= - n'_{42} + n'_{43} = (w_{4b} + w_{4d}).
\end{aligned} \quad (338)$$

b) Coefficients for anti-symmetrical load arrangements

Flexibility coefficients of all the members, according to formulas (75), for $\nu = 1, 2$ and $n = a, b, c$.

Flexibilities of the columns, according to formulas (76), for $n = a, b, c$

Diagonal coefficients:

$$O_1 = b_a + 2k_1 \quad O_2 = 2k_1 + b_b \quad O_3 = b_b + 2k_2 \quad O''_4 = 2k_2 + 2b_c. \quad (339)$$

Moment carry-over factors and auxiliary coefficients:

$$\begin{array}{c|c}
\text{backward reduction:} & \text{forward reduction:} \\
\begin{aligned}
& & r''_4 &= O''_4 \\
\iota_{2t} &= \frac{k_2}{r''_4} & & \\
& & r''_3 &= O_3 - k_2 \cdot \iota_{2t} \\
\sigma_{bt} &= \frac{b_b}{r''_3} & & \\
& & r''_2 &= O_2 - b_b \cdot \sigma_{bt} \\
\iota_{1t} &= \frac{k_1}{r''_2} & & \\
& & r''_1 &= O_1 - k_1 \cdot \iota_{1t}.
\end{aligned}
&
\begin{aligned}
v_1 &= O_1 \\
\iota'_1 &= \frac{k_1}{v_1} & & \\
& & v_2 &= O_2 - k_1 \cdot \iota'_1 \\
\sigma'_b &= \frac{b_b}{v_2} & & \\
& & v_3 &= O_3 - b_b \cdot \sigma'_b \\
\iota'_2 &= \frac{k_2}{v_3} & & \\
& & v''_4 &= O''_4 - k_2 \cdot \iota'_2.
\end{aligned}
\end{array} \quad \begin{matrix}(340) \\ \text{and} \\ (341)\end{matrix}$$

Principal influence coefficients direct

$$n''_{11} = \frac{1}{r''_1} \quad n''_{22} = \frac{1}{r'_2 + v_2 - O_2} \quad n''_{33} = \frac{1}{r'_3 + v_3 - O_3} \quad n''_{44} = \frac{1}{v''_4}. \quad (342)$$

---

* The members enclosed in parentheses have been formed with the values s and w (84)/(85).

Principal influence coefficients by recursion:

$$\left. \begin{array}{ll} \text{forward:} & \text{backward:} \\[4pt] n''_{11} = \dfrac{1}{r''_1} & n''_{44} = \dfrac{1}{v''_4} \\[6pt] n''_{22} = \dfrac{1}{r''_2} + n''_{11} \cdot \iota'^2_{1t} & n''_{33} = \dfrac{1}{v_3} + n''_{44} \cdot \iota'^2_2 \\[6pt] n''_{33} = \dfrac{1}{r''_3} + n''_{22} \cdot \sigma^2_{bt} & n''_{22} = \dfrac{1}{v_2} + n''_{33} \cdot \sigma'^2_b \\[6pt] n''_{44} = \dfrac{1}{r''_4} + n''_{33} \cdot \iota'^2_{2t}. & n''_{11} = \dfrac{1}{v_1} + n''_{22} \cdot \iota'^2_1. \end{array} \right\} \begin{array}{c} (343) \\ \text{and} \\ (344) \end{array}$$

Computation scheme for the symmetrical influence coefficient matrix*

$$\left. \begin{array}{c} \\ \cdot\, \iota_{1t} \\ \\ \cdot\, \sigma_{bt} \\ \\ \cdot\, \iota_{2t} \end{array} \begin{vmatrix} (n''_{11}) & n''_{12} & n''_{13} & n''_{14} \\ \downarrow & \uparrow & \uparrow & \uparrow \\ n''_{21} & (n''_{22}) & n''_{23} & n''_{24} \\ \downarrow & \downarrow & \uparrow & \uparrow \\ n''_{31} & n''_{32} & (n''_{33}) & n''_{34} \\ \downarrow & \downarrow & \downarrow & \uparrow \\ n''_{41} & n''_{42} & n''_{43} & (n''_{44}) \end{vmatrix} \begin{array}{c} \cdot\, \iota'_1 \\ \\ \cdot\, \sigma'_b \\ \\ \cdot\, \iota'_2 \end{array} \right\} (345)$$

Composite influence coefficients**

$$\left. \begin{array}{ll} s''_{11} = +n''_{11} - n''_{12} = (s_{11} + s_{14}) & s''_{12} = +n''_{13} - n''_{14} = (s_{12} + s_{13}) \\[4pt] s''_{21} = -n''_{21} + n''_{22} = (s_{21} - s_{24}) & s''_{22} = +n''_{23} - n''_{24} = (s_{22} + s_{23}) \\[4pt] s''_{31} = -n''_{31} + n''_{32} = (s_{31} - s_{34}) & s''_{32} = +n''_{33} - n''_{34} = (s_{32} + s_{33}) \\[4pt] s''_{41} = -n''_{41} + n''_{42} = (s_{41} - s_{44}) & s''_{42} = -n''_{43} + n''_{44} = (s_{42} - s_{43}); \end{array} \right\} (346)$$

$$\left. \begin{array}{l} w''_{1b} = +n''_{12} - n''_{13} = (w_{1b} + w_{1d}) \\[4pt] w''_{2b} = +n''_{22} - n''_{23} = (w_{2b} + w_{2d}) \\[4pt] w''_{3b} = -n''_{32} + n''_{33} = (w_{3b} - w_{3d}) \\[4pt] w''_{4b} = -n''_{42} + n''_{43} = (w_{4b} - w_{4d}). \end{array} \right\} (347)$$

---

* Between the elements of the matrix (345) and the elements of the matrix (83) there exist, according to the "load transposition method", the following relationships:

$$n''_{11} = n_{11} - n_{18}, \quad n''_{12} = n_{12} - n_{17} \text{ etc., up to } n''_{44} = n_{44} - n_{45}.$$

** The members enclosed in parentheses have been formed with the values s and w (84)/(85).

— 76 —

Loading Condition 1a

Horizontal member carrying vertical loading of arbitrary magnitude, symmetrical about center of frame

$$(\Re_4 = \mathfrak{L}_1, \quad \mathfrak{L}_4 = \Re_1; \quad \Re_3 = \mathfrak{L}_2, \quad \mathfrak{L}_3 = \Re_2)$$

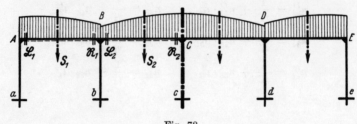

Fig. 78

Beam loading diagram with load terms $\mathfrak{L}$ and $\Re$ for left-hand half of frame

Beam moments ($L_\nu = \mathfrak{L}_\nu k_\nu$ and $R_\nu = \Re_\nu k_\nu$):

$$\left. \begin{array}{l|cccc}
 & L_1 & R_1 & L_2 & R_2 \\
\hline
X'_1 = M_{A1} = M_{E4} = & -n'_{11} & +n'_{12} & +n'_{13} & -n'_{14} \\
X'_2 = M_{B1} = M_{D4} = & +n'_{21} & -n'_{22} & -n'_{23} & +n'_{24} \\
X'_3 = M_{B2} = M_{D3} = & +n'_{31} & -n'_{32} & -n'_{33} & +n'_{34} \\
X'_4 = M_{C2} = M_{C3} = & -n'_{41} & +n'_{42} & +n'_{43} & -n'_{44}
\end{array} \right\} \quad (348)$$

Column moments:

at top:

$$\left. \begin{array}{ll}
M_{Aa} = -M_{Ee} = \quad +X'_1 & \quad M_a = -M_e = -M_{Aa} \cdot \varepsilon_a \\
M_{Bb} = -M_{Dd} = -X'_2 + X'_3 & \quad M_b = -M_d = -M_{Bb} \cdot \varepsilon_b \\
M_{Cc} = 0; & \quad M_c = 0.
\end{array} \right\} \quad \begin{array}{l}(349) \\ \text{and} \\ (350)\end{array}$$

at base:

Special case: symmetrical bay loading ($R_1 = L_1$ and $R_2 = L_2$):

$$\left. \begin{array}{l}
X'_1 = M_{A1} = M_{E4} = -L_1 \cdot s'_{11} + L_2 \cdot s'_{12} \\
X'_2 = M_{B1} = M_{D4} = -L_1 \cdot s'_{21} - L_2 \cdot s'_{22} \\
X'_3 = M_{B2} = M_{D3} = -L_1 \cdot s'_{31} - L_2 \cdot s'_{32} \\
X'_4 = M_{C2} = M_{C3} = +L_1 \cdot s'_{41} - L_2 \cdot s'_{42}.
\end{array} \right\} \quad (351)$$

Loading Conditions 2a and 3a

All the columns carrying arbitrary horizontal loading (2a), and external rotational moments applied at all the beam joints (3a), but load arrangement as a whole is symmetrical about center of frame

$(\mathfrak{L}_e = -\mathfrak{L}_a, \ \mathfrak{R}_e = -\mathfrak{R}_a; \ \mathfrak{L}_d = -\mathfrak{L}_b, \ \mathfrak{R}_d = -\mathfrak{R}_b; \ D_E = -D_A, \ D_D = -D_B)$

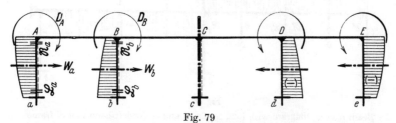

Fig. 79
Column loading diagram with load terms $\mathfrak{L}$ and $\mathfrak{R}$ for the columns, and external rotational moments D, for left-hand half of frame
(Symmetrical loading of the central column d and the central joint D not shown)

Joint load terms: $(L_n = \mathfrak{L}_n k_n \ \text{and} \ R_n = \mathfrak{R}_n k_n)$:

$$\mathfrak{B}_A = D_A \cdot b_a + L_a \cdot \varepsilon_a - R_a \qquad \mathfrak{B}_B = D_B \cdot b_b + L_b \cdot \varepsilon_b - R_b. \qquad (352)$$

Beam moments:

$$\left. \begin{aligned} X_1' &= M_{A1} = M_{E4} = +\mathfrak{B}_A \cdot n_{11}' + \mathfrak{B}_B \cdot w_{1b}' \\ X_2' &= M_{B1} = M_{D4} = -\mathfrak{B}_A \cdot n_{21}' - \mathfrak{B}_B \cdot w_{2b}' \\ X_3' &= M_{B2} = M_{D3} = -\mathfrak{B}_A \cdot n_{31}' + \mathfrak{B}_B \cdot w_{3b}' \\ X_4' &= M_{C2} = M_{C3} = +\mathfrak{B}_A \cdot n_{41}' - \mathfrak{B}_B \cdot w_{4b}' \end{aligned} \right\} \qquad (353)$$

Column moments:

at top:
$$\left. \begin{aligned} M_{Aa} &= -M_{Ee} = -D_A && + X_1' \\ M_{Bb} &= -M_{Dd} = -D_B &&- X_2' + X_3' \\ M_{Cc} &= 0; \end{aligned} \right\} \qquad (354)$$

at base:
$$\left. \begin{aligned} M_a &= -M_e = -(\mathfrak{L}_a + M_{Aa})\varepsilon_a \\ M_b &= -M_d = -(\mathfrak{L}_b + M_{Bb})\varepsilon_b \\ M_c &= 0. \end{aligned} \right\} \qquad (355)$$

Loading Condition 1b

Horizontal member carrying vertical loading of arbitrary magnitude, antisymmetrical about center of frame

$$(\mathfrak{R}_4 = -\mathfrak{L}_1, \quad \mathfrak{L}_4 = -\mathfrak{R}_1; \quad \mathfrak{R}_3 = -\mathfrak{L}_2, \quad \mathfrak{L}_3 = -\mathfrak{R}_2)$$

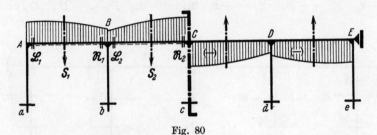

Fig. 80
Beam loading diagram with load terms $\mathfrak{L}$ and $\mathfrak{R}$ for left-hand half of frame

Beam moments ($L_\nu = \mathfrak{L}_\nu k_\nu$ and $R_\nu = \mathfrak{R}_\nu k_\nu$):

	L_1	R_1	L_2	R_2
$X_1'' = M_{A1} = -M_{E4} =$	$-n_{11}''$	$+n_{12}''$	$+n_{13}''$	$-n_{14}''$
$X_2'' = M_{B1} = -M_{D4} =$	$+n_{21}''$	$-n_{22}''$	$-n_{23}''$	$+n_{24}''$
$X_3'' = M_{B2} = -M_{D3} =$	$+n_{31}''$	$-n_{32}''$	$-n_{33}''$	$+n_{34}''$
$X_4'' = M_{C2} = -M_{C3} =$	$-n_{41}''$	$+n_{42}''$	$+n_{43}''$	$-n_{44}''$

(356)

Column moments:

at top: at base:

$$\begin{aligned}
M_{Aa} &= M_{Ee} = +X_1'' & M_a &= M_e = -M_{Aa} \cdot \varepsilon_a \\
M_{Bb} &= M_{Dd} = -X_2'' + X_3'' & M_b &= M_d = -M_{Bb} \cdot \varepsilon_b \\
M_{Cc} &= -2X_4''; & M_c &= -M_{Cc} \cdot \varepsilon_c.
\end{aligned}$$

(357) and (358)

Special case: symmetrical bay loading ($R_1 = L_1$ and $R_2 = L_2$)

$$\begin{aligned}
X_1'' &= M_{A1} = -M_{E4} = -L_1 \cdot s_{11}'' + L_2 \cdot s_{12}'' \\
X_2'' &= M_{B1} = -M_{D4} = -L_1 \cdot s_{21}'' - L_2 \cdot s_{22}'' \\
X_3'' &= M_{B2} = -M_{D3} = -L_1 \cdot s_{31}'' - L_2 \cdot s_{32}'' \\
X_4'' &= M_{C2} = -M_{C3} = +L_1 \cdot s_{41}'' - L_2 \cdot s_{42}''.
\end{aligned}$$

(359)

Loading Conditions 2b and 3b

All the columns carrying arbitrary horizontal loading (2b) and external rotational moments applied to all the beam joints (3b), but load arrangement as a whole is anti-symmetrical about center of frame

$$(\mathfrak{L}_e = \mathfrak{L}_a, \quad \mathfrak{R}_e = \mathfrak{R}_a; \quad \mathfrak{L}_d = \mathfrak{L}_b, \quad \mathfrak{R}_d = \mathfrak{R}_b; \quad D_E = D_A, \quad D_D = D_B)$$

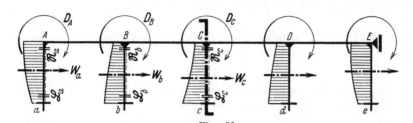

Fig. 81

Column loading diagram with load terms $\mathfrak{L}$ and $\mathfrak{R}$ for the columns, and external rotational moments D, for left-hand half of frame, including central column and central joint

Joint load terms: $(L_n = \mathfrak{L}_n k_n \text{ and } R_n = \mathfrak{R}_n k_n)$:

$$\left. \begin{array}{c} \mathfrak{B}_N = D_N \cdot b_n + L_n \cdot \varepsilon_n - R_n; \qquad \mathfrak{B}''_C = \mathfrak{B}_C/2; \\ \text{for} \quad N = A, B, C \quad \text{and} \quad n = a, b, c. \end{array} \right\} \quad (360)$$

Beam moments:

	$\mathfrak{B}_A$	$\mathfrak{B}_B$	$\mathfrak{B}''_C$
$X''_1 = M_{A1} = -M_{E4} =$	$+n''_{11}$	$+w''_{1b}$	$-n''_{14}$
$X''_2 = M_{B1} = -M_{D4} =$	$-n''_{21}$	$-w''_{2b}$	$+n''_{24}$
$X''_3 = M_{B2} = -M_{D3} =$	$-n''_{31}$	$+w''_{3b}$	$+n''_{34}$
$X''_4 = M_{C2} = -M_{C3} =$	$+n''_{41}$	$-w''_{4b}$	$-n''_{44}.$

(361)

Column moments:

at top:
$$\left. \begin{array}{l} M_{Aa} = M_{Ee} = -D_A \qquad\quad + X''_1 \\ M_{Bb} = M_{Dd} = -D_B - X''_2 + X''_3 \\ M_{Cc} = -D_C - 2X''_4; \end{array} \right\} \quad (362)$$

at base:
$$\left. \begin{array}{l} M_a = M_e = -(\mathfrak{L}_a + M_{Aa})\,\varepsilon_a \\ M_b = M_d = -(\mathfrak{L}_b + M_{Bb})\,\varepsilon_b \\ M_c \quad\;\, = -(\mathfrak{L}_c + M_{Cc})\,\varepsilon_c. \end{array} \right\} \quad (363)$$

(Frame Shape 9—cont.)

Horizontal thrusts, column shears and lateral restraining force for all loading conditions

For asymmetrical loading conditions, according to formulas (179)/(180) or (181)/(182), for $n = a, b, c, d, e$ and $N = A, B, C, D, E$.

For symmetrical loading conditions, according to formulas (183)/(184) or (185)/(186), for $n = a, b$ and $N = A, B$.

For anti-symmetrical loading conditions, according to formulas (187) or (189), for $n = a, b, c$ and $N = A, B, C$. Furthermore:

$$H = -2(H_a + H_b) - H_c \quad \text{or} \quad H = -2(Q_{ra} + Q_{rb}) - Q_{rc}. \qquad (364)$$

Frame Shape 9v

(Frame Shape 9 with laterally unrestrained horizontal member)

Procedure

either exactly as for Frame Shape 3v, see page 24/25, having due regard to anti-symmetry properties,

or, alternatively, similar to procedure for Frame Shape 8v, see pages 70-71, for the collective subscripts $v = 1, 2$; and $N = A, B, C$.

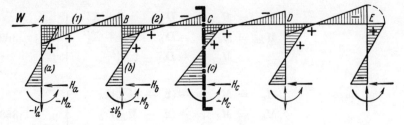

Fig. 82
Loading Condition 5 with anti-symmetrical diagram of bending moments and column reactions

Frame Shape 10

Symmetrical single-story two-bay frame with laterally restrained horizontal member and three elastically restrained columns

(Symmetrical form of Frame Shape 4)

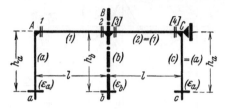

Fig. 83
Frame shape, dimensions and symbols

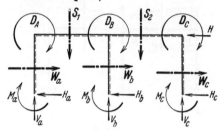

Fig. 84
Definition of positive direction for all external loads on joints and members, for all column reactions, and for the lateral restraining force

> All the dimensions and coefficients for the right-hand half of the frame are the same as those for the left-hand half. In the case of the columns the broken line is always placed on the right-hand side of each column, despite the symmetry of the frame.

Note: Frame Shape 10 could, in principle, be analyzed with the aid of the formulas for *Frame Shape* 4. On account of symmetry, the backward reduction becomes the same as the forward reduction, and the backward recursion becomes the same as the forward recursion (for both Method I and Method II). Furthermore, the square matrices of influence coefficients (with 4×4 elements in Method I, and with 3×3 elements in Method II) become bisymmetrical, i.e., symmetrical about the principal diagonal and the secondary diagonal. If the *"load transposition method"* is employed, however, it is also possible to start from the following symmetrical and antisymmetrical loading conditions*.

I. Treatment by the Method of Forces

(For $J_k = J_1$ and $l_k = l$, hence $k_1 = 1$)

a) Coefficients for symmetrical load arrangements

$$\left.\begin{array}{l} k_a = \dfrac{J_1}{J_a} \cdot \dfrac{h_a}{l}, \quad b_a = k_a(2-\varepsilon_a); \quad \iota_1' = \dfrac{1}{b_a+2}, \quad b_1' = 2 - \iota_1'; \\[2mm] n_{11}' = \dfrac{1}{b_a+3/2} \quad n_{22}' = \dfrac{1}{b_1'} \quad (n_{12}' = n_{21}') = \left(\dfrac{n_{11}'}{2} = n_{22}' \cdot \iota_1'\right). \end{array}\right\} \quad (365)$$

---

* It would, alternatively, be possible simply to refer to Frame Shape 9 (for Method I) or to Frame Shape 8 (for Method II), inasmuch as the formulas for those cases are generally valid for symmetrical continuous frames with any even number of bays. However, in view of the practical importance of the symmetrical two-bay frame, the formulas for it are nevertheless given separately.

Loading Condition 1a
Both spans of horizontal member carrying vertical loading of arbitrary magnitude symmetrical about center of frame
$(\mathfrak{R}_2 = \mathfrak{L}_1, \quad \mathfrak{L}_2 = \mathfrak{R}_1)$

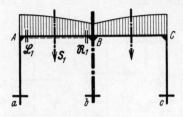

Fig. 85
Beam loading diagram with load terms $\mathfrak{L}$ and $\mathfrak{R}$ for left-hand half of frame

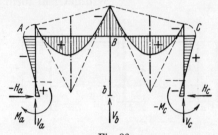

Fig. 86
Symmetrical diagram of bending moments and reactions at supports

Beam moments ($L_1 = \mathfrak{L}_1$ and $R_1 = \mathfrak{R}_1$ since $k_1 = 1$)*:

$$\left. \begin{aligned} X_1' = M_{A1} = M_{C2} = -L_1 \cdot n_{11}' + R_1 \cdot n_{12}' = -\frac{L_1 - R_1/2}{b_a + 3/2} \\ X_2' = M_{B1} = M_{B2} = +L_1 \cdot n_{21}' - R_1 \cdot n_{22}' = -\frac{R_1 - L_1 \cdot \iota_1'}{b_1'} \end{aligned} \right\} \quad (366)$$

Column moments:

$$\left. \begin{aligned} \text{at top:} \quad & M_{Aa} = -M_{Cc} = +X_1' & M_{Bb} = 0; \\ \text{at base:} \quad & M_a = -M_c = -M_{Aa} \cdot \varepsilon_a & M_b = 0. \end{aligned} \right\} \quad (367)$$

Special case: symmetrical bay loading ($R_1 = L_1$):

$$\left. \begin{aligned} X_1' = M_{A1} = M_{C2} = -L_1 \cdot s_{11}' = -\frac{L_1/2}{b_a + 3/2} \\ X_2' = M_{B1} = M_{B2} = -L_1 \cdot s_{21}' = -\frac{L_1(1 - \iota_1')}{b_1'}; \end{aligned} \right\} \quad (368)$$

where**:

$$\left. \begin{aligned} s_{11}' = +n_{11}' - n_{12}' = (s_{11} - s_{12}) \\ s_{21}' = -n_{21}' + n_{22}' = (s_{21} + s_{22}). \end{aligned} \right\} \quad (369)$$

---

* For completeness it should be pointed out that between the elements of the n-matrix in (366) and the elements of the 4×4 matrix relating to Frame Shape 4 there exist the following relationships:
$n_{11}' = n_{11} + n_{14}, \quad (n_{12}' = n_{21}') = (n_{12} + n_{13} = n_{21} + n_{24}), \quad n_{22}' = n_{22} + n_{23}.$

** The members enclosed in parentheses have been formed with the values s of Frame Shape 4.

Loading Conditions 2a and 3a

Both external columns carrying arbitrary horizontal loading, symmetrical about the center of the frame (2a), and symmetrical pair of rotational moments applied to the two corners of the frame (3a)

$$(\mathfrak{L}_c = -\mathfrak{L}_a, \quad \mathfrak{R}_c = -\mathfrak{R}_a; \quad D_C = -D_A)$$

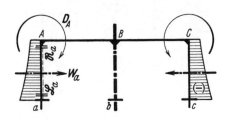
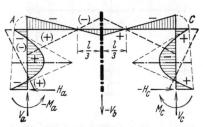

Fig. 87
Column loading diagram with load terms and for the left-hand column, and external rotational moments

Fig. 88
Symmetrical diagram of bending moments for Loading Condition 2a (full lines) and Loading Condition 3a (dotted); also reactions at supports

Joint load term:

$$\mathfrak{B}_A = D_A \cdot b_a + (\mathfrak{L}_a \cdot \varepsilon_a - \mathfrak{R}_a) \, k_a. \tag{370}$$

Beam moments:

$$\left. \begin{aligned} X_1' &= M_{A1} = M_{C2} = +\mathfrak{B}_A \cdot n_{11}' = +\frac{\mathfrak{B}_A}{b_a + 3/2} \\ X_2' &= M_{B1} = M_{B2} = -\mathfrak{B}_A \cdot n_{21}' = -X_1'/2. \end{aligned} \right\} \tag{371}$$

Column moments:

at top: $M_{Aa} = -M_{Cc} = -D_A + X_1' \qquad M_{Bb} = 0$

at base: $M_a = -M_c = -(\mathfrak{L}_a + M_{Aa})\varepsilon_a \qquad M_b = 0.$ $\qquad \Big\} \tag{372}$

---

b) Coefficients for anti-symmetrical load arrangements:

$$\left. \begin{aligned} k_a &= \frac{J_1}{J_a} \cdot \frac{h_a}{l} \qquad k_b = \frac{J_1}{J_b} \cdot \frac{h_b}{l}; \qquad \begin{array}{l} b_a = k_a(2 - \varepsilon_a) \\ b_b = k_b(2 - \varepsilon_b); \end{array} \\ \iota_{1t} &= \frac{1}{2 + 2b_b} \qquad b_{1t} = (2 - \iota_{1t}) \qquad \iota_1' = \frac{1}{b_a + 2} \qquad b_1' = (2 - \iota_1'); \\ n_{11}' &= \frac{1}{b_a + b_{1t}} \qquad n_{22}'' = \frac{1}{b_1' + 2b_b} \qquad (n_{12}'' = n_{21}'') = n_{11}'' \cdot \iota_{1t} = n_{22}'' \cdot \iota_1'. \end{aligned} \right\} \tag{373}$$

Loading Condition 1b

Both spans of horizontal member carrying vertical loading of arbitrary magnitude, anti-symmetrical about center of frame

$(\Re_2 = -\mathfrak{L}_1, \quad \mathfrak{L}_2 = -\Re_1)$

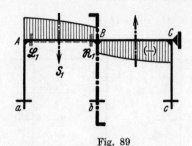

Fig. 89
Beam loading diagram with load terms $\mathfrak{L}$ and $\Re$ for left-hand half of frame

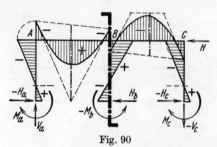

Fig. 90
Anti-symmetrical diagram of bending moments and reactions at supports

Beam moments ($L_1 = \mathfrak{L}_1$ and $R_1 = \Re_1$ since $k_1 = 1$)*

$$\left.\begin{aligned}X_1'' &= M_{A1} = -M_{C2} = -L_1 \cdot n_{11}'' + R_1 \cdot n_{12}'' = -\frac{L_1 - R_1 \cdot \iota_{1t}}{b_a + b_{1t}} \\ X_2'' &= M_{B1} = -M_{B2} = +L_1 \cdot n_{21}'' - R_1 \cdot n_{22}'' = -\frac{R_1 - L_1 \cdot \iota_1'}{b_1' + 2b_b}\end{aligned}\right\} \quad (374)$$

Column moments:

at top: $\quad M_{Aa} = M_{Cc} = +X_1'' \qquad M_{Bb} = -2 X_2'';$

at base: $\quad M_a = M_c = -M_{Aa} \cdot \varepsilon_a \qquad M_b = -M_{Bb} \cdot \varepsilon_b.$ $\qquad (375)$

Special case: symmetrical bay loading ($R_1 = L_1$):

$$\left.\begin{aligned}X_1'' &= M_{A1} = -M_{C2} = -L_1 \cdot s_{11}'' = -\frac{L_1(1 - \iota_{1t})}{b_a + b_{1t}} \\ X_2'' &= M_{B1} = -M_{B2} = -L_1 \cdot s_{21}'' = -\frac{L_1(1 - \iota_1')}{b_1' + 2b_b};\end{aligned}\right\} \quad (376)$$

where**:
$$\left.\begin{aligned}s_{11}'' &= +n_{11}'' - n_{12}'' = (s_{11} + s_{12}) \\ s_{21}'' &= -n_{21}'' + n_{22}'' = (s_{21} - s_{22}).\end{aligned}\right\} \quad (377)$$

---

* For completeness it should be pointed out that between the elements of the n-matrix in (374) and the elements of the 4×4 matrix relating to Frame Shape 4 there exist the following relationships:

$\quad n_{11}'' = n_{11} - n_{14}, \quad (n_{12}'' = n_{21}'') = (n_{12} - n_{13} = n_{21} - n_{24}), \quad n_{22}'' = n_{22} - n_{23}.$

** The members enclosed in parentheses have been formed with the values s of Frame Shape 4.

Loading Conditions 2b and 3b

All the columns carrying arbitrary horizontal loading (2b) and external rotational moments applied to all the beam joints (3b), but load arrangement as a whole is anti-symmetrical about center of frame

$$(\mathfrak{L}_c = \mathfrak{L}_a, \quad \mathfrak{R}_c = \mathfrak{R}_a; \quad D_C = D_A)$$

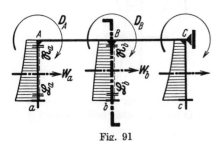

Fig. 91
Column loading diagram with load terms $\mathfrak{L}$ and $\mathfrak{R}$ for the columns, and external rotational moments D, for left-hand half of frame

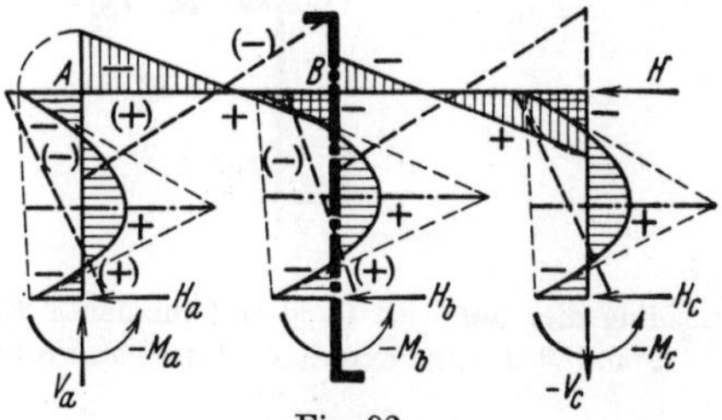

Fig. 92
Anti-symmetrical diagram of bending moments for Loading Condition 2b (full lines) and Loading Condition 3b (dotted); also reactions at supports

Joint load terms: $(L_n = \mathfrak{L}_n k_n \text{ and } R_n = \mathfrak{R}_n k_n)$:

$$\mathfrak{B}_A = D_A \cdot b_a + L_a \cdot \varepsilon_a - R_a \qquad \mathfrak{B}_B'' = (D_B \cdot b_b + L_b \cdot \varepsilon_b - R_b)/2 \quad (378)$$

Beam moments:

$$\left. \begin{aligned} X_1'' &= M_{A1} = -M_{C2} = +\mathfrak{B}_A \cdot n_{11}'' + \mathfrak{B}_B'' \cdot n_{12}'' = +\frac{\mathfrak{B}_A + \mathfrak{B}_B'' \cdot \iota_{1t}}{b_a + b_{1t}} \\ X_2'' &= M_{B1} = -M_{B2} = -\mathfrak{B}_A \cdot n_{21}'' - \mathfrak{B}_B'' \cdot n_{22}'' = -\frac{\mathfrak{B}_A \cdot \iota_1' + \mathfrak{B}_B''}{b_1' + 2b_b} \cdot \end{aligned} \right\} \quad (379)$$

Column moments:

$$\left. \begin{aligned} &\text{at top:} && \text{at base:} \\ M_{Aa} &= M_{Cc} = -D_A + X_1'' & M_a &= M_c = -(\mathfrak{L}_a + M_{Aa})\varepsilon_a \\ M_{Bb} &= -D_B - 2X_2''; & M_b &= -(\mathfrak{L}_b + M_{Bb})\varepsilon_b. \end{aligned} \right\} \quad (380)$$

II. Treatment by the Deformation Method

(For $J_k = J_1$ and $l_k = l$, hence $K_1 = 1$)

a) Coefficients for symmetrical load arrangements

$$K_a = \frac{J_a}{J_1} \cdot \frac{l}{h_a} = \frac{1}{k_a} \qquad S_a = \frac{K_a}{2 - \varepsilon_a} = \frac{1}{b_a}; \qquad K_A = 3S_a + 2. \quad (381)$$

Loading Condition 4a

Symmetrical overall loading condition
(Superposition of Loading Conditions 1a, 2a and 3a)

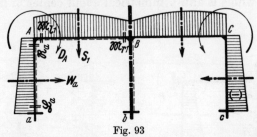

Fig. 93

Loading diagram with fixed-end moments $\mathfrak{M}_l$ and $\mathfrak{M}_r$ for the horizontal member, load terms $\mathfrak{L}$ and $\mathfrak{R}$ for the external column, and rotational moment D, for left-hand half of frame

Column restraining moment:

$$\mathfrak{E}_{ra} = -\frac{\mathfrak{R}_a - \mathfrak{L}_a \cdot \varepsilon_a}{2 - \varepsilon_a}. \tag{382}$$

Auxiliary moment:

$$Y'_A = \frac{-\mathfrak{M}_{l1} + \mathfrak{E}_{ra} + D_A}{K_A}. \tag{383}$$

Beam moments:

$$M_{A1} = M_{C2} = \mathfrak{M}_{l1} + 2 Y'_A \qquad M_{B1} = M_{B2} = \mathfrak{M}_{r1} - Y'_A. \tag{384}$$

Column moments:

$$\left.\begin{array}{ll}
\text{at top:} & M_{Aa} = -M_{Cc} = \mathfrak{E}_{ra} - Y'_A \cdot 3 S_a \qquad M_{Bb} = 0; \\
\text{at base:} & M_a = -M_c = -(\mathfrak{L}_a + M_{Aa})\varepsilon_a \qquad M_b = 0.
\end{array}\right\} \tag{385}$$

Check relationship:

$$D_A + M_{Aa} - M_{A1} = 0. \tag{386}$$

b) Coefficients for anti-symmetrical load arrangements:

$$\left.\begin{array}{l}
K_a = \dfrac{J_a}{J_1} \cdot \dfrac{l}{h_a} \qquad K_b = \dfrac{J_b}{J_1} \cdot \dfrac{l}{h_b}; \qquad S_a = \dfrac{K_a}{2 - \varepsilon_a} \qquad S_b = \dfrac{K_b}{2 - \varepsilon_b}; \\
K_A = 3 S_a + 2 \qquad K''_B = 2 + 3 S_b/2; \qquad j_{1t} = 1/K''_B \qquad j'_1 = 1/K_A; \\
\qquad S_{1t} = (2 - j_{1t}) \qquad S'_1 = (2 - j'_1). \\
u''_{aa} = \dfrac{1}{K_A - j_{1t}} = \dfrac{1}{3 S_a + S_{1t}} \qquad u''_{bb} = \dfrac{1}{K''_B - j'_1} = \dfrac{1}{S' + 3 S_b/2}; \\
\qquad (u''_{ab} = u''_{ba}) = (u''_{aa} \cdot j_{1t} = u''_{bb} \cdot j'_1).
\end{array}\right\} \tag{387}$$

Loading Condition 4b

Anti-symmetrical overall loading condition
(Superposition of Loading Conditions 1b, 2b and 3b)

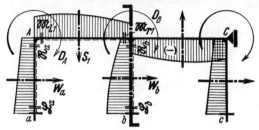

Fig. 94

Loading diagram with fixed-end moments $\mathfrak{M}_l$ and $\mathfrak{M}_r$ for the horizontal member, load terms $\mathfrak{L}$ and $\mathfrak{R}$ for the columns, and external rotational moments $\boldsymbol{D}$, for left-hand half of frame, including central column and central joint

Column restraining moments at the beam joints:

$$\mathfrak{C}_{ra} = -\frac{\mathfrak{R}_a - \mathfrak{L}_a \cdot \varepsilon_a}{2 - \varepsilon_a} \qquad \mathfrak{C}_{rb} = -\frac{\mathfrak{R}_b - \mathfrak{L}_b \cdot \varepsilon_b}{2 - \varepsilon_a}. \tag{388}$$

Joint load terms:

$$\mathfrak{K}_A = -\mathfrak{M}_{l1} + \mathfrak{C}_{ra} + \boldsymbol{D}_A \qquad \mathfrak{K}_B'' = +\mathfrak{M}_{r1} + \mathfrak{C}_{rb}/2 + \boldsymbol{D}_B/2. \tag{389}$$

auxiliary moments:

$$\left.\begin{aligned}
Y_A'' &= +\mathfrak{K}_A \cdot u_{aa}'' - \mathfrak{K}_B'' \cdot u_{ab}'' = +\frac{\mathfrak{K}_A - \mathfrak{K}_B'' \cdot j_{1t}}{3S_a + S_{1t}} \\
Y_B'' &= -\mathfrak{K}_A \cdot u_{ba}'' + \mathfrak{K}_B'' \cdot u_{bb}'' = +\frac{\mathfrak{K}_B'' - \mathfrak{K}_A \cdot j_1}{S_1' + 3S_b/2}.
\end{aligned}\right\} \tag{390}$$

Beam moments:

$$\left.\begin{aligned}
M_{A1} &= -M_{C2} = \mathfrak{M}_{l1} + (2Y_A'' + Y_B'') \\
M_{B1} &= -M_{B2} = \mathfrak{M}_{r1} - (Y_A'' + 2Y_B'').
\end{aligned}\right\} \tag{391}$$

Column moments:

at top: $\quad M_{Aa} = M_{Cc} = \mathfrak{C}_{ra} - Y_A'' \cdot 3S_a \qquad M_{Bb} = \mathfrak{C}_{rb} - Y_B'' \cdot 3S_b;$

at base: $\quad M_a = M_c = -(\mathfrak{L}_a + M_{Aa})\varepsilon_a \qquad M_b = -(\mathfrak{L}_b + M_{Bb})\varepsilon_b.$ (392)

Check relationships:

$$\boldsymbol{D}_A + M_{Aa} - M_{A1} = 0 \qquad \boldsymbol{D}_B + M_{Bb} + 2M_{B1} = 0. \tag{393}$$

(Frame Shape 10—cont.)

Horizontal thrusts, column shears and lateral restraining force for all loading conditions

For asymmetrical loading conditions, according to formulas (179)/(180) or (181)/(182), for $n = a, b, c$ and $N = A, B, C$.

For symmetrical loading conditions, according to formulas (183)/(184) or (185)/(186), for $n = a$ and $N = A$.

For anti-symmetrical loading conditions, according to formulas (187) or (189), for $n = a, b$ and $N = A, B$. Furthermore:

$$H = -2H_a - H_b \quad \text{or} \quad H = -2Q_{ra} - Q_{rb}. \tag{394}$$

Frame Shape 10v

(Frame Shape 10 with laterally unrestrained horizontal member)

Procedure

either similar to procedure for Frame Shape 3v, see pages 24-25, for the collective subscripts $\nu = 1, 2, n = a, b, c$ and $N = A, B, C$, having due regard to anti-symmetry properties,

or, alternatively, similar to procedure for Frame Shape 8v, see pages 70-71, for the collective subscripts $\nu = 1, n = a, b,$ and $N = A, B$.

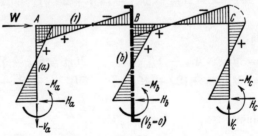

Fig. 95

Loading Condition 5 with anti-symmetrical diagram of bending moments and column reactions

Frame Shape 11

Single-story five-bay frame with laterally restrained horizontal member, elastically restrained at each end, and four elastically restrained columns

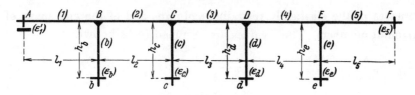

Fig. 96
Frame shape, dimensions and symbols

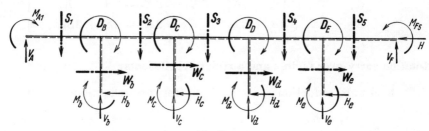

Fig. 97
Definition of positive direction for all external loads on joints and members, for all reactions at supports, and for the lateral restraining force

---

Important

All the formula sequences and formula matrices given for Frame Shape 11 can readily be extended to suit frames of similar type having any number of bays.

Note: The *collective subscripts* have the following meaning:

$\nu = 1, 2, 3, 4, 5$ denoting the bays (or beam spans)
$n = b, c, d, e$ denoting the columns
$N = B, C, D, E$ denoting the joints

The problem will be treated by the Deformation Method (Method II) only. For a treatment of similar continuous frames with any number of bays by the Method of Forces (Method I) see Frame Shape 12, pages 94-98, or Frame Shape 13, pages 99-103.

Coefficients:

Stiffness coefficients for all individual members:

$$K_\nu = \frac{J_\nu}{J_k} \cdot \frac{l_k}{l_\nu} = \frac{1}{k_\nu}\,; \qquad K_n = \frac{J_n}{J_k} \cdot \frac{l_k}{h_n} = \frac{1}{k_n}\,. \qquad (395)$$

Stiffnesses of the elastically restrained columns and of the end spans of the horizontal member: The ϵ values are assumed to be known:

$$S_1 = \frac{K_1}{2 - \varepsilon_1'} = \frac{1}{b_1'}\,; \qquad S_n = \frac{K_n}{2 - \varepsilon_n} = \frac{1}{b_n}\,; \qquad S_5 = \frac{K_5}{2 - \varepsilon_5} = \frac{1}{b_5}\,. \qquad (396)$$

Joint coefficients:

$$\begin{aligned} K_B &= 3S_1 + 3S_b + 2K_2 & K_D &= 2K_3 + 3S_d + 2K_4 \\ K_C &= 2K_2 + 3S_c + 2K_3 & K_E &= 2K_4 + 3S_e + 3S_5 \end{aligned} \qquad (397)$$

Rotation carry-over factors and auxiliary coefficients:

backward reduction: | forward reduction:

$$\begin{aligned} & & r_e = K_E & & & & v_b = K_B & \\ j_4 &= \frac{K_4}{r_e} & & & j_2' &= \frac{K_2}{v_b} & & \\ & & r_d = K_D - K_4 \cdot j_4 & & & & v_c = K_C - K_2 \cdot j_2' & \\ j_3 &= \frac{K_3}{r_d} & & & j_3' &= \frac{K_3}{v_c} & & \\ & & r_c = K_C - K_3 \cdot j_3 & & & & v_d = K_D - K_3 \cdot j_3' & \\ j_2 &= \frac{K_2}{r_c} & & & j_4' &= \frac{K_4}{v_d} & & \\ & & r_b = K_B - K_2 \cdot j_2\,. & & & & v_e = K_E - K_4 \cdot j_4'\,. & \end{aligned} \qquad \begin{matrix}(398)\\ \text{and}\\ (399)\end{matrix}$$

Principal influence coefficients by recursion:

forward: | backward:

$$\begin{aligned} u_{bb} &= 1/r_b & u_{ee} &= 1/v_e \\ u_{cc} &= 1/r_c + u_{bb} \cdot j_2^2 & u_{dd} &= 1/v_d + u_{ee} \cdot j_4'^2 \\ u_{dd} &= 1/r_d + u_{cc} \cdot j_3^2 & u_{cc} &= 1/v_c + u_{dd} \cdot j_3'^2 \\ u_{ee} &= 1/r_e + u_{dd} \cdot j_4^2\,. & u_{bb} &= 1/v_b + u_{cc} \cdot j_2'^2\,. \end{aligned} \qquad \begin{matrix}(400)\\ \text{and}\\ (401)\end{matrix}$$

Principal influence coefficients direct

$$u_{bb} = \frac{1}{r_b} \quad u_{cc} = \frac{1}{r_c + v_c - K_C} \quad u_{dd} = \frac{1}{r_d + v_d - K_D} \quad u_{ee} = \frac{1}{v_e}\,. \qquad (402)$$

Computation scheme for the symmetrical influence coefficient matrix

$$
\begin{array}{c|cccc|c}
\cdot j_2 & (u_{bb}) & u_{bc} & u_{bd} & u_{be} & \cdot j_2' \\
& \downarrow & \uparrow & \uparrow & \uparrow & \\
& u_{cb} & (u_{cc}) & u_{cd} & u_{ce} & \\
\cdot j_3 & \downarrow & \downarrow & \uparrow & \uparrow & \cdot j_3' \\
& u_{db} & u_{dc} & (u_{dd}) & u_{de} & \\
\cdot j_4 & \downarrow & \downarrow & \downarrow & \uparrow & \cdot j_4' \\
& u_{eb} & u_{ec} & u_{ed} & (u_{ee}) & \\
\end{array}
\quad (403)
$$

Loading Condition 4
Overall loading condition (superposition of Loading Conditions 1, 2 and 3)

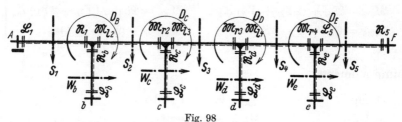

Fig. 98

Loading diagram with fixed-end moments $\mathfrak{M}_l$ and $\mathfrak{M}_r$ for the internal spans, load terms $\mathfrak{L}$ and $\mathfrak{R}$ for the semi-rigidly fixed end spans and columns, and external rotational moments D

Restraining moments of the end spans of the horizontal member at the joints of the frame:

$$\mathfrak{E}_{r1} = -\frac{\mathfrak{R}_1 - \mathfrak{L}_1 \cdot \varepsilon_1'}{2 - \varepsilon_1'} ; \qquad \mathfrak{E}_{l5} = -\frac{\mathfrak{L}_5 - \mathfrak{R}_5 \cdot \varepsilon_5}{2 - \varepsilon_5} . \qquad (404)$$

Restraining moments of the columns at the joints of the frame:

$$\mathfrak{E}_{rn} = -\frac{\mathfrak{R}_n - \mathfrak{L}_n \cdot \varepsilon_n}{2 - \varepsilon_n} . \qquad (405)$$

Joint load terms:

$$\begin{aligned}
\mathfrak{K}_B &= \mathfrak{E}_{r1} - \mathfrak{M}_{l2} + \mathfrak{E}_{rb} + D_B \\
\mathfrak{K}_C &= \mathfrak{M}_{r2} - \mathfrak{M}_{l3} + \mathfrak{E}_{rc} + D_C \\
\mathfrak{K}_D &= \mathfrak{M}_{r3} - \mathfrak{M}_{l4} + \mathfrak{E}_{rd} + D_D \\
\mathfrak{K}_E &= \mathfrak{M}_{r4} - \mathfrak{E}_{l5} + \mathfrak{E}_{re} + D_E .
\end{aligned} \qquad (406)$$

auxiliary moments:

	$\mathfrak{K}_B$	$\mathfrak{K}_C$	$\mathfrak{K}_D$	$\mathfrak{K}_E$
$Y_B =$	$+ u_{bb}$	$- u_{bc}$	$+ u_{bd}$	$- u_{be}$
$Y_C =$	$- u_{cb}$	$+ u_{cc}$	$- u_{cd}$	$+ u_{ce}$
$Y_D =$	$+ u_{db}$	$- u_{dc}$	$+ u_{dd}$	$- u_{de}$
$Y_E =$	$- u_{eb}$	$+ u_{ec}$	$- u_{ed}$	$+ u_{ee}$

(407)

Beam moments:

$$M_{A1} = -(\mathfrak{L}_1 + M_{B1})\,\varepsilon_1' \qquad M_{B1} = \mathfrak{E}_{r1} - Y_B \cdot 3 S_1$$
$$M_{B2} = \mathfrak{M}_{l2} + (2 Y_B + Y_C) K_2 \qquad M_{C2} = \mathfrak{M}_{r2} - (Y_B + 2 Y_C) K_2$$
$$M_{C3} = \mathfrak{M}_{l3} + (2 Y_C + Y_D) K_3 \qquad M_{D3} = \mathfrak{M}_{r3} - (Y_C + 2 Y_D) K_3 \quad (408)$$
$$M_{D4} = \mathfrak{M}_{l4} + (2 Y_D + Y_E) K_4 \qquad M_{E4} = \mathfrak{M}_{r4} - (Y_D + 2 Y_E) K_4$$
$$M_{E5} = \mathfrak{E}_{l5} + Y_E \cdot 3 S_5 \qquad M_{F5} = -(\mathfrak{R}_5 + M_{E5})\,\varepsilon_5.$$

Column moments:

at top: $\qquad M_{Nn} = \mathfrak{E}_{rn} - Y_N \cdot 3 S_n;$ \hfill (409)

at base: $\qquad M_n = -(\mathfrak{L}_n + M_{Nn})\,\varepsilon_n.$ \hfill (410)

Check relationships:

$$\boldsymbol{D}_B + M_{B1} + M_{Bb} - M_{B2} = 0 \qquad \boldsymbol{D}_D + M_{D3} + M_{Dd} - M_{D4} = 0$$
$$\boldsymbol{D}_C + M_{C2} + M_{Cc} - M_{C3} = 0 \qquad \boldsymbol{D}_E + M_{E4} + M_{Ee} - M_{E5} = 0. \quad (411)$$

Horizontal thrusts and column shears for all Loading Conditions: as formulas (36) and (38) for the collective subscripts $n = b, c, d, e$ and $N = B, C, D, E$.

Beam restraining force:

without direct column loads (cases 1 and 3):

$$H = -(H_b + H_c + H_d + H_e); \tag{412}$$

with direct column loads (cases 2 and 4):

$$H = -(Q_{rb} + Q_{rc} + Q_{rd} + Q_{re}). \tag{413}$$

Frame Shape 11v

(Frame Shape 11 with laterally unrestrained horizontal member)

Procedure

just as for Frame Shape 1v, but for the collective subscripts:
$$n = b, c, d, e \text{ and } N = B, C, D, E$$
Specific bending moments m, according to Method I, from the formulas (407) – (410).

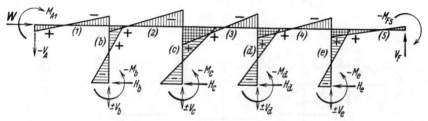

Fig. 99
Loading Condition 5 with diagram of bending moments and reactions at supports

---

Frame Shape 12v

(Frame Shape 12 with laterally unrestrained horizontal member)

Procedure

just as for Frame Shape 1v, but for the collective subscripts:
$$n = b, c, d \text{ and } N = B, C, D$$
Specific bending moments m, according to Method I, from the formulas (431) – (433).

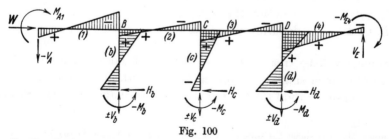

Fig. 100
Loading Condition 5 with diagram of bending moments and reactions at supports

Frame Shape 12

Single-story four-bay frame with laterally restrained horizontal member, elastically restrained at each end, and three elastically restrained columns

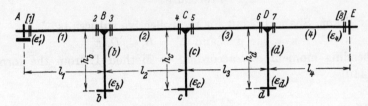

Fig. 101
Frame shape, dimensions and symbols

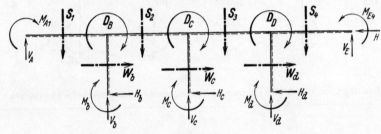

Fig. 102
Definition of positive direction for all external loads on joints and members, for all reactions at supports, and for the lateral restraining force

---

Important

All the formula sequences and formula matrices given for Frame Shape 12 can readily be extended to suit frames of similar type having any number of bays.

Note: The *collective subscripts* have the following meaning:

$\nu = 1, 2, 3, 4,$ denoting the bays (or beam spans)
$n = b, c, d,$ denoting the columns
$N = B, C, D,$ denoting the joints

The problem will be treated by the Method of Forces (Method I)* only. For a treatment of similar continuous frames with any number of bays by the Deformation Method (Method II) see Frame Shape 11, pages 89-92.

---

* Compare also the somewhat different treatment given for Frame Shape 13, in which the moments of the two elastically restrained end spans likewise appear as values X (For Frame Shape 12 these would be the beam moments X_1 and X_9).

Coefficients

Flexibility coefficients for all individual members:

$$k_\nu = \frac{J_k}{J_\nu} \cdot \frac{l_\nu}{l_k} = \frac{1}{K_\nu}; \qquad k_n = \frac{J_k}{J_n} \cdot \frac{h_n}{l_k} = \frac{1}{K_n}. \tag{414}$$

Flexibilities of the elastically restrained columns and of the end spans of the horizontal member: The values ε are assumed to be known:

$$b_1' = k_1(2 - \varepsilon_1'); \qquad b_n = k_n(2 - \varepsilon_n); \qquad b_4 = k_4(2 - \varepsilon_4). \tag{415}$$

Diagonal coefficients:

$$\left.\begin{array}{lll} O_2 = b_1' + b_b & O_4 = 2k_2 + b_c & O_6 = 2k_3 + b_d \\ O_3 = b_b + 2k_2 & O_5 = b_c + 2k_3 & O_7 = b_d + b_4. \end{array}\right\} \tag{416}$$

Moment carry-over factors and auxiliary coefficients:

$$\left.\begin{array}{ll} \text{backward reduction:} & \text{forward reduction:} \\[4pt] \begin{aligned} \sigma_d &= \frac{b_d}{r_7} \\ \iota_3 &= \frac{k_3}{r_6} \\ \sigma_c &= \frac{b_c}{r_5} \\ \iota_2 &= \frac{k_2}{r_4} \\ \sigma_b &= \frac{b_b}{r_3} \end{aligned} \quad \begin{aligned} r_7 &= O_7 \\ r_6 &= O_6 - b_d \cdot \sigma_d \\ r_5 &= O_5 - k_3 \cdot \iota_3 \\ r_4 &= O_4 - b_c \cdot \sigma_c \\ r_3 &= O_3 - k_2 \cdot \iota_2 \\ r_2 &= O_2 - b_b \cdot \sigma_b. \end{aligned} & \begin{aligned} \sigma_b' &= \frac{b_b}{v_2} \\ \iota_2' &= \frac{k_2}{v_3} \\ \sigma_c' &= \frac{b_c}{v_4} \\ \iota_3' &= \frac{k_3}{v_5} \\ \sigma_d' &= \frac{b_d}{v_6} \end{aligned} \quad \begin{aligned} v_2 &= O_2 \\ v_3 &= O_3 - b_b \cdot \sigma_b' \\ v_4 &= O_4 - k_2 \cdot \iota_2' \\ v_5 &= O_5 - b_c \cdot \sigma_c' \\ v_6 &= O_6 - k_3 \cdot \iota_3' \\ v_7 &= O_7 - b_d \cdot \sigma_d'. \end{aligned} \end{array}\right\} \begin{array}{l} (417) \\ \text{and} \\ (418) \end{array}$$

Principal influence coefficients by recursion:

$$\left.\begin{array}{ll} \text{forward:} & \text{backward:} \\[4pt] \begin{aligned} n_{22} &= 1/r_2 \\ n_{33} &= 1/r_3 + n_{22} \cdot \sigma_b^2 \\ n_{44} &= 1/r_4 + n_{33} \cdot \iota_2^2 \\ n_{55} &= 1/r_5 + n_{44} \cdot \sigma_c^2 \\ n_{66} &= 1/r_6 + n_{55} \cdot \iota_3^2 \\ n_{77} &= 1/r_7 + n_{66} \cdot \sigma_d^2. \end{aligned} & \begin{aligned} n_{77} &= 1/v_7 \\ n_{66} &= 1/v_6 + n_{77} \cdot \sigma_d'^2 \\ n_{55} &= 1/v_5 + n_{66} \cdot \iota_3'^2 \\ n_{44} &= 1/v_4 + n_{55} \cdot \sigma_c'^2 \\ n_{33} &= 1/v_3 + n_{44} \cdot \iota_2'^2 \\ n_{22} &= 1/v_2 + n_{33} \cdot \sigma_b'^2. \end{aligned} \end{array}\right\} \begin{array}{l} (419) \\ \text{and} \\ (420) \end{array}$$

Principal influence coefficients direct

$$n_{22} = \frac{1}{r_2} \qquad n_{33} = \frac{1}{r_3 + v_3 - O_3} \qquad n_{44} = \frac{1}{r_4 + v_4 - O_4} \\ n_{55} = \frac{1}{r_5 + v_5 - O_5} \qquad n_{66} = \frac{1}{r_6 + v_6 - O_6} \qquad n_{77} = \frac{1}{v_7}. \quad \Bigg\} \quad (421)$$

Computation scheme for the symmetrical influence coefficient matrix

$$\begin{array}{c|cccccc|c}
\cdot \sigma_b & (n_{22}) & n_{23} & n_{24} & n_{25} & n_{26} & n_{27} & \cdot \sigma_b' \\
 & \downarrow & \uparrow & \uparrow & \uparrow & \uparrow & \uparrow & \\
\cdot \iota_2 & n_{32} & (n_{33}) & n_{34} & n_{35} & n_{36} & n_{37} & \cdot \iota_2' \\
 & \downarrow & \downarrow & \uparrow & \uparrow & \uparrow & \cdot \uparrow & \\
\cdot \sigma_c & n_{42} & n_{43} & (n_{44}) & n_{45} & n_{46} & n_{47} & \cdot \sigma_c' \\
 & \downarrow & \downarrow & \downarrow & \uparrow & \uparrow & \uparrow & \\
\cdot \iota_3 & n_{52} & n_{53} & n_{54} & (n_{55}) & n_{56} & n_{57} & \cdot \iota_3' \\
 & \downarrow & \downarrow & \downarrow & \downarrow & \uparrow & \uparrow & \\
\cdot \sigma_d & n_{62} & n_{63} & n_{64} & n_{65} & (n_{66}) & n_{67} & \cdot \sigma_d' \\
 & \downarrow & \downarrow & \downarrow & \downarrow & \downarrow & \uparrow & \\
 & n_{72} & n_{73} & n_{74} & n_{75} & n_{76} & (n_{77}) & \\
\end{array} \quad (422)$$

Composite influence coefficients:

$$\begin{aligned}
s_{22} &= +n_{23} - n_{24} & s_{23} &= +n_{25} - n_{26} \\
s_{32} &= +n_{33} - n_{34} & s_{33} &= +n_{35} - n_{36} \\
 & & s_{43} &= +n_{45} - n_{46} \\
s_{42} &= -n_{43} + n_{44} & & \\
 & & s_{53} &= +n_{55} - n_{56} \\
s_{52} &= -n_{53} + n_{54} & & \\
s_{62} &= -n_{63} + n_{64} & s_{63} &= -n_{65} + n_{66} \\
s_{72} &= -n_{73} + n_{74} & s_{73} &= -n_{75} + n_{76};
\end{aligned} \quad \Bigg\} \quad (423)$$

$$\begin{aligned}
w_{2b} &= +n_{22} - n_{23} & w_{2c} &= +n_{24} - n_{25} & w_{2d} &= +n_{26} - n_{27} \\
 & & w_{3c} &= +n_{34} - n_{35} & w_{3d} &= +n_{36} - n_{37} \\
w_{3b} &= -n_{32} + n_{33} & w_{4c} &= +n_{44} - n_{45} & w_{4d} &= +n_{46} - n_{47} \\
w_{4b} &= -n_{42} + n_{43} & & & w_{5d} &= +n_{56} - n_{57} \\
w_{5b} &= -n_{52} + n_{53} & w_{5c} &= -n_{54} + n_{55} & w_{6d} &= +n_{66} - n_{67} \\
w_{6b} &= -n_{62} + n_{63} & w_{6c} &= -n_{64} + n_{65} & & \\
w_{7b} &= -n_{72} + n_{73} & w_{7c} &= -n_{74} + n_{75} & w_{7d} &= -n_{76} + n_{77}.
\end{aligned} \quad \Bigg\} \quad (424)$$

Loading Condition 1

All the beam spans carrying arbitrary vertical loading

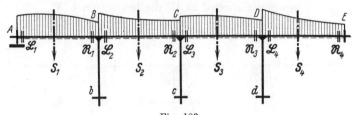

Fig. 103
Beam loading diagram with load terms $\mathfrak{L}$ and $\mathfrak{R}$

Beam load terms:

$$L_\nu = \mathfrak{L}_\nu k_\nu \qquad R_\nu = \mathfrak{R}_\nu k_\nu; \qquad \mathfrak{B}_1 = R_1 - L_1 \cdot \varepsilon_1' \qquad \mathfrak{B}_4 = L_4 - R_4 \cdot \varepsilon_4. \qquad (425)$$

Beam moments:

$$\left.\begin{aligned}
&\begin{array}{c|cccccc}
 & \mathfrak{B}_1 & L_2 & R_2 & L_3 & R_3 & \mathfrak{B}_4 \\
\hline
X_2 = M_{B1} = & -n_{22} & -n_{23} & +n_{24} & +n_{25} & -n_{26} & -n_{27} \\
X_3 = M_{B2} = & -n_{32} & -n_{33} & +n_{34} & +n_{35} & -n_{36} & -n_{37} \\
X_4 = M_{C2} = & +n_{42} & +n_{43} & -n_{44} & -n_{45} & +n_{46} & +n_{47} \\
X_5 = M_{C3} = & +n_{52} & +n_{53} & -n_{54} & -n_{55} & +n_{56} & +n_{57} \\
X_6 = M_{D3} = & -n_{62} & -n_{63} & +n_{64} & +n_{65} & -n_{66} & -n_{67} \\
X_7 = M_{D4} = & -n_{72} & -n_{73} & +n_{74} & +n_{75} & -n_{76} & -n_{77};
\end{array}\\
&M_{A1} = -(\mathfrak{L}_1 + M_{B1})\,\varepsilon_1' \qquad M_{E4} = -(\mathfrak{R}_4 + M_{D4})\,\varepsilon_4.
\end{aligned}\right\} \quad (426)$$

Column moments:

at top: $\qquad M_{Bb} = -X_2 + X_3 \qquad M_{Cc} = -X_4 + X_5 \qquad M_{Dd} = -X_6 + X_7;$ (427)

at base: $\qquad M_n = -M_{Nn} \cdot \varepsilon_n.$ (428)

Special case: All the beam spans symmetrically loaded $(R_\nu = L_\nu)$:

$$\left.\begin{aligned}
X_2 = M_{B1} &= -\mathfrak{B}_1 \cdot n_{22} - L_2 \cdot s_{22} + L_3 \cdot s_{23} - \mathfrak{B}_4 \cdot n_{27} \\
X_3 = M_{B2} &= -\mathfrak{B}_1 \cdot n_{32} - L_2 \cdot s_{32} + L_3 \cdot s_{33} - \mathfrak{B}_4 \cdot n_{37} \\
X_4 = M_{C2} &= +\mathfrak{B}_1 \cdot n_{42} - L_2 \cdot s_{42} - L_3 \cdot s_{43} + \mathfrak{B}_4 \cdot n_{47} \\
X_5 = M_{C3} &= +\mathfrak{B}_1 \cdot n_{52} - L_2 \cdot s_{52} - L_3 \cdot s_{53} + \mathfrak{B}_4 \cdot n_{57} \\
X_6 = M_{D3} &= -\mathfrak{B}_1 \cdot n_{62} + L_2 \cdot s_{62} - L_3 \cdot s_{63} - \mathfrak{B}_4 \cdot n_{67} \\
X_7 = M_{D4} &= -\mathfrak{B}_1 \cdot n_{72} + L_2 \cdot s_{72} - L_3 \cdot s_{73} - \mathfrak{B}_4 \cdot n_{77}.
\end{aligned}\right\} \quad (429)$$

Loading Conditions 2 and 3

All the columns carrying arbitrary horizontal loading (2); external rotational moments applied at all the beam joints (3)

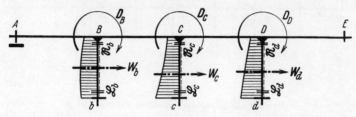

Fig. 104

Column loading diagram with load terms $\mathfrak{L}$ and $\mathfrak{R}$ for the columns and external rotational moments D

Joint load terms ($L_n = \mathfrak{L}_n k_n$ and $R_n = \mathfrak{R}_n k_n$):

$$\mathfrak{B}_N = D_N \cdot b_n + L_n \cdot \varepsilon_n - R_n. \tag{430}$$

Beam moments:

$$\left.\begin{aligned}
X_2 &= M_{B1} = -\mathfrak{B}_B \cdot w_{2b} + \mathfrak{B}_C \cdot w_{2c} - \mathfrak{B}_D \cdot w_{2d} \\
X_3 &= M_{B2} = +\mathfrak{B}_B \cdot w_{3b} + \mathfrak{B}_C \cdot w_{3c} - \mathfrak{B}_D \cdot w_{3d} \\
X_4 &= M_{C2} = -\mathfrak{B}_B \cdot w_{4b} - \mathfrak{B}_C \cdot w_{4c} + \mathfrak{B}_D \cdot w_{4d} \\
X_5 &= M_{C3} = -\mathfrak{B}_B \cdot w_{5b} + \mathfrak{B}_C \cdot w_{5c} + \mathfrak{B}_D \cdot w_{5d} \\
X_6 &= M_{D3} = +\mathfrak{B}_B \cdot w_{6b} - \mathfrak{B}_C \cdot w_{6c} - \mathfrak{B}_D \cdot w_{6d} \\
X_7 &= M_{D4} = +\mathfrak{B}_B \cdot w_{7b} - \mathfrak{B}_C \cdot w_{7c} + \mathfrak{B}_D \cdot w_{7d}; \\
M_{A1} &= -M_{B1} \cdot \varepsilon'_1 \qquad M_{E4} = -M_{D4} \cdot \varepsilon_4.
\end{aligned}\right\} \tag{431}$$

Column moments:

at top:
$$\left.\begin{aligned}
M_{Bb} &= -D_B - X_2 + X_3 \\
M_{Cc} &= -D_C - X_4 + X_5 \\
M_{Dd} &= -D_D - X_6 + X_7
\end{aligned}\right\} \tag{432};$$

at base:
$$\left.\begin{aligned}
M_b &= -(\mathfrak{L}_b + M_{Bb})\,\varepsilon_b \\
M_c &= -(\mathfrak{L}_c + M_{Cc})\,\varepsilon_c \\
M_d &= -(\mathfrak{L}_d + M_{Dd})\,\varepsilon_d.
\end{aligned}\right\} \tag{433}$$

Horizontal thrusts, column shears and lateral restraining force:

for all Loading Conditions: as formulas (36)—(39) for the collective subscripts $N = B, C, D$ and $n = b, c, d$.—In equation (37) delete H_a, and in equation (39) delete Q_{ra}.

Frame Shape 13

Single-story three-bay frame with laterally restrained horizontal member, elastically restrained at each end, and two elastically restrained columns

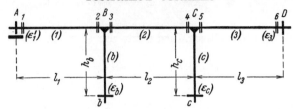

Fig. 105
Frame shape, dimensions and symbols

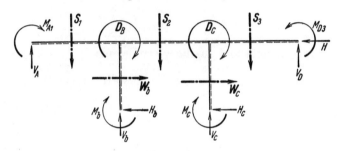

Fig. 106
Definition of positive direction for all external loads on joints and members, for all reactions at supports, and for the lateral restraining force

---
Important

All the formula sequences and formula matrices given for Frame Shape 13 can readily be extended to suit frames of similar type having any number of bays*.

Note: The *collective subscripts* have the following meaning:

$\nu = 1, 2, 3,$ denoting the bays (or beam spans)
$n = b, c,$ denoting the columns
$N = B, C,$ denoting the joints

The problem will be treated by the Method of Forces (Method I) only.** For a treatment of similar continuous frames with any number of bays by the Deformation Method (Method II) see Frame Shape 11, pages 89 to 92.

---

* The dotted lines in the sets of formulas enclose those portions of the formulas which are applicable to the two-bay Frame Shape 14ab.
** Compare also the somewhat different treatment given for Frame Shape 12, in which the moments of the two elastically restrained end spans—in the case of Frame Shape 13 these would be the moments X_1 and X_6—are excluded from the set of formulas for X.

Coefficients*

Flexibility coefficients for all individual members:

$$k_\nu = \frac{J_k}{J_\nu} \cdot \frac{l_\nu}{l_k} = \frac{1}{K_\nu}; \qquad k_n = \frac{J_k}{J_n} \cdot \frac{h_n}{l_k} = \frac{1}{K_n}. \tag{434}$$

Flexibilities of the elastically restrained columns:

$$b_b = k_b(2 - \varepsilon_b) \mid b_c = k_c(2 - \varepsilon_c). \tag{435}$$

Diagonal coefficients All the values ε are assumed to be known:

$$O_1 = \frac{k_1}{\varepsilon_1'} \quad \begin{matrix} O_2 = 2k_1 + b_b \\ O_3 = b_b + 2k_2 \end{matrix} \quad \begin{matrix} O_4 = 2k_2 + b_c \\ O_5 = b_c + 2k_3 \end{matrix} \quad O_6 = \frac{k_3}{\varepsilon_3}. \tag{436}$$

Moment carry-over factors and auxiliary coefficients:

backward reduction: | forward reduction:

$$\sigma_c = \frac{b_c}{r_5} \quad \begin{matrix} r_5 = O_5 - k_3 \cdot \varepsilon_3 \\ r_4 = O_4 - b_c \cdot \sigma_c \end{matrix} \qquad \sigma_b' = \frac{b_b}{v_2} \quad \begin{matrix} v_2 = O_2 - k_1 \cdot \varepsilon_1' \\ v_3 = O_3 - b_b \cdot \sigma_b' \end{matrix}$$

$$\iota_2 = \frac{k_2}{r_4} \quad r_3 = O_3 - k_2 \cdot \iota_2 \qquad \iota_2' = \frac{k_2}{v_3} \quad v_4 = O_4 - k_2 \cdot \iota_2' \tag{437}$$
$$\sigma_b = \frac{b_b}{r_3} \quad r_2 = O_2 - b_b \cdot \sigma_b \qquad \sigma_c' = \frac{b_c}{v_4} \quad v_5 = O_5 - b_c \cdot \sigma_c' \tag{and (438)}$$

$$\iota_1 = \frac{k_1}{r_2} \quad r_1 = O_1 - k_1 \cdot \iota_1. \qquad \iota_3' = \frac{k_3}{v_5} \quad v_6 = O_6 - k_3 \cdot \iota_3'.$$

Principal influence coefficients by recursion:

forward: | backward:

$$n_{11} = \frac{1}{r_1} = \frac{1}{k_1(1/\varepsilon_1' - \iota_1)} \qquad n_{66} = \frac{1}{v_6} = \frac{1}{k_3(1/\varepsilon_3 - \iota_3')}$$

$$n_{22} = 1/r_2 + n_{11} \cdot \iota_1^2 \qquad n_{55} = 1/v_5 + n_{66} \cdot \iota_3'^2$$

$$n_{33} = 1/r_3 + n_{22} \cdot \sigma_b^2 \qquad n_{44} = 1/v_4 + n_{55} \cdot \sigma_c'^2 \tag{439}$$

$$n_{44} = 1/r_4 + n_{33} \cdot \iota_2^2 \qquad n_{33} = 1/v_3 + n_{44} \cdot \iota_2'^2 \tag{and (440)}$$

$$n_{55} = 1/r_5 + n_{44} \cdot \sigma_c^2 \qquad n_{22} = 1/v_2 + n_{33} \cdot \sigma_b'^2$$

$$n_{66} = 1/O_6 + n_{55} \cdot \varepsilon_3^2. \qquad n_{11} = 1/O_1 + n_{22} \cdot \varepsilon_1'^2.$$

---

* For the two-bay Frame Shape 14 the symbol ι_2 should in all cases be replaced by ε_2. For Frame Shape 14 furthermore: $O_4 = k_2/\varepsilon_2 = r_4$ and $n_{44} = 1/v_4$. See also page 105. The corresponding adjustments for ν-bay frames ($\nu = 4, 5 \ldots$) should be made in similar fashion.

Principal influence coefficients direct

$$n_{11} = \frac{1}{r_1} \qquad n_{22} = \frac{1}{r_2 + v_2 - O_2} \qquad n_{33} = \frac{1}{r_3 + v_3 - O_3}$$
$$n_{44} = \frac{1}{r_4 + v_4 - O_4} \qquad n_{55} = \frac{1}{r_5 + v_5 - O_5} \qquad n_{66} = \frac{1}{v_6}.$$
(441)

Computation scheme for the symmetrical influence coefficient matrix

$$\begin{array}{l}
\cdot \iota_1 \\
\cdot \sigma_b \\
\cdot \iota_2 \\
\cdot \sigma_c \\
\cdot \varepsilon_3 \\
\end{array}
\left|
\begin{array}{cccccc}
(n_{11}) & n_{12} & n_{13} & n_{14} & n_{15} & n_{16} \\
\downarrow & \uparrow & \uparrow & \uparrow & \uparrow & \uparrow \\
n_{21} & (n_{22}) & n_{23} & n_{24} & n_{25} & n_{26} \\
\downarrow & \downarrow & \uparrow & \uparrow & \uparrow & \uparrow \\
n_{31} & n_{32} & (n_{33}) & n_{34} & n_{35} & n_{36} \\
\downarrow & \downarrow & \downarrow & \uparrow & \uparrow & \uparrow \\
n_{41} & n_{42} & n_{43} & (n_{44}) & n_{45} & n_{46} \\
\downarrow & \downarrow & \downarrow & \downarrow & \uparrow & \uparrow \\
n_{51} & n_{52} & n_{53} & n_{54} & (n_{55}) & n_{56} \\
\downarrow & \downarrow & \downarrow & \downarrow & \downarrow & \uparrow \\
n_{61} & n_{62} & n_{63} & n_{64} & n_{65} & (n_{66}) \\
\end{array}
\right|
\begin{array}{l}
\cdot \varepsilon_1' \\
\cdot \sigma_b' \\
\cdot \iota_2' \\
\cdot \sigma_c' \\
\cdot \iota_3' \\
\end{array}$$
(442)

Composite influence coefficients:

$$s_{11} = + n_{11} - n_{12} \qquad s_{12} = + n_{13} - n_{14} \qquad s_{13} = + n_{15} - n_{16}$$
$$s_{21} = - n_{21} + n_{22} \qquad s_{22} = + n_{23} - n_{24} \qquad s_{23} = + n_{25} - n_{26}$$
$$s_{31} = - n_{31} + n_{32} \qquad s_{32} = + n_{33} - n_{34} \qquad s_{33} = + n_{35} - n_{36}$$
$$s_{41} = - n_{41} + n_{42} \qquad s_{42} = - n_{43} + n_{44} \qquad s_{43} = + n_{45} - n_{46}$$
$$s_{51} = - n_{51} + n_{52} \qquad s_{52} = - n_{53} + n_{54} \qquad s_{53} = + n_{55} - n_{56}$$
$$s_{61} = - n_{61} + n_{62} \qquad s_{62} = - n_{63} + n_{64} \qquad s_{63} = - n_{65} + n_{66};$$
(443)

$$w_{1b} = + n_{12} - n_{13} \qquad w_{1c} = + n_{14} - n_{15}$$
$$w_{2b} = + n_{22} - n_{23} \qquad w_{2c} = + n_{24} - n_{25}$$
$$w_{3b} = - n_{32} + n_{33} \qquad w_{3c} = + n_{34} - n_{35}$$
$$w_{4b} = - n_{42} + n_{43} \qquad w_{4c} = + n_{44} - n_{45}$$
$$w_{5b} = - n_{52} + n_{53} \qquad w_{5c} = - n_{54} + n_{55}$$
$$w_{6b} = - n_{62} + n_{63} \qquad w_{6c} = - n_{64} + n_{65}.$$
(444)

Loading Condition 1

All the beam spans carrying arbitrary vertical loading

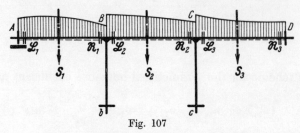

Fig. 107

Beam loading diagram with load terms $\mathfrak{L}$ and $\mathfrak{R}$

Beam moments ($L_\nu = \mathfrak{L}_\nu k_\nu$ and $R_\nu = \mathfrak{R}_\nu k_\nu$):

	L_1	R_1	L_2	R_2	L_3	R_3
$X_1 = M_{A1} =$	$-n_{11}$	$+n_{12}$	$+n_{13}$	$-n_{14}$	$-n_{15}$	$+n_{16}$
$X_2 = M_{B1} =$	$+n_{21}$	$-n_{22}$	$-n_{23}$	$+n_{24}$	$+n_{25}$	$-n_{26}$
$X_3 = M_{B2} =$	$+n_{31}$	$-n_{32}$	$-n_{33}$	$+n_{34}$	$+n_{35}$	$-n_{36}$
$X_4 = M_{C2} =$	$-n_{41}$	$+n_{42}$	$+n_{43}$	$-n_{44}$	$-n_{45}$	$+n_{46}$
$X_5 = M_{C3} =$	$-n_{51}$	$+n_{52}$	$+n_{53}$	$-n_{54}$	$-n_{55}$	$+n_{56}$
$X_6 = M_{D3} =$	$+n_{61}$	$-n_{62}$	$-n_{63}$	$+n_{64}$	$+n_{65}$	$-n_{66}$

(445)

Column moments:

at top: $\quad M_{Bb} = -X_2 + X_3 \quad | \quad M_{Cc} = -X_4 + X_5;$ (446)

at base: $\quad M_b = -M_{Bb} \cdot \varepsilon_b . \quad | \quad M_c = -M_{Cc} \cdot \varepsilon_c .$ (447)

Special case: All the beam spans symmetrically loaded ($R_\nu = L_\nu$):

$$\begin{aligned}
X_1 = M_{A1} &= -L_1 \cdot s_{11} + L_2 \cdot s_{12} \;\; -L_3 \cdot s_{13} \\
X_2 = M_{B1} &= -L_1 \cdot s_{21} - L_2 \cdot s_{22} \;\; +L_3 \cdot s_{23} \\
X_3 = M_{B2} &= -L_1 \cdot s_{31} - L_2 \cdot s_{32} \;\; +L_3 \cdot s_{33} \\
X_4 = M_{C2} &= +L_1 \cdot s_{41} - L_2 \cdot s_{42} \;\; -L_3 \cdot s_{43} \\
X_5 = M_{C3} &= +L_1 \cdot s_{51} - L_2 \cdot s_{52} \;\; -L_3 \cdot s_{53} \\
X_6 = M_{D3} &= -L_1 \cdot s_{61} + L_2 \cdot s_{62} \;\; -L_3 \cdot s_{63} .
\end{aligned}$$

(448)

Loading Conditions 2 and 3

All the columns carrying arbitrary horizontal loading (2); external rotational moments applied at both beam joints (3).

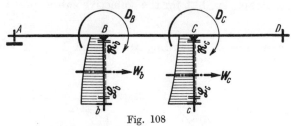

Fig. 108

Column loading diagram with load terms $\mathfrak{L}$ and $\mathfrak{R}$ for the columns and external rotational moments D

Joint load terms ($L_n = \mathfrak{L}_n k_n$ and $R_n = \mathfrak{R}_n k_n$):

$$\mathfrak{B}_B = D_B \cdot b_b + L_b \cdot \varepsilon_b - R_b \quad | \quad \mathfrak{B}_C = D_C \cdot b_c + L_c \cdot \varepsilon_c - R_c. \qquad (449)$$

Beam moments:

$$\begin{aligned}
X_1 &= M_{A1} = +\mathfrak{B}_B \cdot w_{1b} & -\mathfrak{B}_C \cdot w_{1c} \\
X_2 &= M_{B1} = -\mathfrak{B}_B \cdot w_{2b} & +\mathfrak{B}_C \cdot w_{2c} \\
X_3 &= M_{B2} = +\mathfrak{B}_B \cdot w_{3b} & +\mathfrak{B}_C \cdot w_{3c} \\
X_4 &= M_{C2} = -\mathfrak{B}_B \cdot w_{4b} & -\mathfrak{B}_C \cdot w_{4c} \\
X_5 &= M_{C3} = -\mathfrak{B}_B \cdot w_{5b} & +\mathfrak{B}_C \cdot w_{5c} \\
X_6 &= M_{D3} = +\mathfrak{B}_B \cdot w_{6b} & -\mathfrak{B}_C \cdot w_{6c}.
\end{aligned} \qquad (450)$$

Column moments:

at top: \qquad\qquad at base: \qquad (451)

$$M_{Bb} = -D_B - X_2 + X_3 \qquad M_b = -(\mathfrak{L}_b + M_{Bb})\varepsilon_b \qquad \text{and}$$

$$M_{Cc} = -D_C - X_4 + X_5; \qquad M_c = -(\mathfrak{L}_c + M_{Cc})\varepsilon_c. \qquad (452)$$

Horizontal thrusts, column shears and lateral restraining force

for all Loading Conditions: as formulas (36) – (39), for the collective subscripts $N = B, C$ and $n = b, c$.

Equation (37) is simplified to $H = -(H_b + H_c)$, and equation (39) is simplified to $H = -(Q_{rb} + Q_{rc})$.

Frame Shape 13v
(Frame Shape 13 with laterally unrestrained horizontal member)

Procedure

same as for Frame Shape 1v, but for the collective subscripts: $n = b, c$ and $N = B, C$

Specific bending moments m, according to Method I, from the formulas (450) – (452).

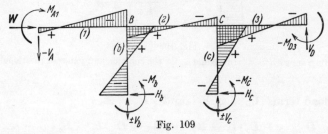

Fig. 109

Loading Condition 5 with diagram of bending moments and reactions at supports

---

Frame Shape 14
Single-story two-bay frame with laterally restrained horizontal member, elastically restrained at each end, and one elastically restrained column

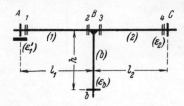 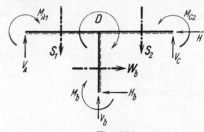

Fig. 110
Frame shape, dimensions and symbols

Fig. 111
Definition of positive direction for all external loads on joints and members, for all reactions at supports, and for the lateral restraining force

---

Important

All the formulas required for the treatment of Frame Shape 14 by Method I (Method of Forces) are already comprised in the sets of formulas for Frame Shape 13 and are defined by dotted lines. The formulas for Method II (Deformation Method) could similarly be obtained from those for Frame Shape 88, but will nevertheless, for the sake of convenience, again be given as a self-contained whole.

I. Treatment by the method of forces

Coefficients*

Flexibility coefficients of the three individual members:

$$k_1 = \frac{J_k}{J_1} \cdot \frac{l_1}{l_k} \qquad k_2 = \frac{J_k}{J_2} \cdot \frac{l_2}{l_k}; \qquad k_b = \frac{J_k}{J_b} \cdot \frac{h}{l_k}. \tag{453}$$

Flexibilities of the elastically restrained columns

$$b_b = k_b(2 - \varepsilon_b). \tag{454}$$

Diagonal coefficients:

$$O_1 = \frac{k_1}{\varepsilon_1'} \qquad O_2 = 2k_1 + b_b \qquad O_3 = b_b + 2k_2 \qquad O_4 = \frac{k_2}{\varepsilon_2}. \tag{455}$$

Moment carry-over factors and auxiliary coefficients:

backward reduction:

$$\sigma_b = \frac{b_b}{r_3}$$
$$\iota_1 = \frac{k_1}{r_2}$$
$$r_3 = O_3 - k_2 \cdot \varepsilon_2$$
$$r_2 = O_2 - b_b \cdot \sigma_b$$
$$r_1 = O_1 - k_1 \cdot \iota_1.$$

forward reduction:

$$\sigma_b' = \frac{b_b}{v_2}$$
$$\iota_2' = \frac{k_2}{v_3}$$
$$v_2 = O_2 - k_1 \cdot \varepsilon_1'$$
$$v_3 = O_3 - b_b \cdot \sigma_b'$$
$$v_4 = O_4 - k_2 \cdot \iota_2'.$$

(456) and (457)

Principal influence coefficients by recursion:

forward:

$$n_{11} = \frac{1}{r_1} = \frac{1}{k_1(1/\varepsilon_1' - \iota_1)}$$
$$n_{22} = 1/r_2 + n_{11} \cdot \iota_1^2$$
$$n_{33} = 1/r_3 + n_{22} \cdot \sigma_b^2$$
$$n_{44} = 1/O_4 + n_{33} \cdot \varepsilon_2^2.$$

backward:

$$n_{44} = \frac{1}{v_4} = \frac{1}{k_2(1/\varepsilon_2 - \iota_2')}$$
$$n_{33} = 1/v_3 + n_{44} \cdot \iota_2'^2$$
$$n_{22} = 1/v_2 + n_{33} \cdot \sigma_b'^2$$
$$n_{11} = 1/O_1 + n_{22} \cdot \varepsilon_1'^2.$$

(458) and (459)

In the 4×4 *matrix* (442) the symbol ι_2 should be replaced by ε_2. The *values s* (443) and the *values w* (444) for Frame Shape 14 are enclosed within the dotted lines.

Loading Conditions 1, 2 and 3

as on page 102 and 103, as defined by the dotted lines.

---

* Although the coefficients for Frame Shape 14 can be obtained from the corresponding coefficients for Frame Shape 13, see pages 100-101, they will nevertheless be given in full as far as the recursion formulas. In the case of Frame Shape 14 it is advisable to adopt $J_k = J_b$ and $l_k = h$, so that we obtain $k_b = 1$.

Treatment by the deformation method
Coefficients*

Stiffness coefficients of the three individual members:

$$K_1 = \frac{J_1}{J_k} \cdot \frac{l_k}{l_1} = \frac{1}{k_1} \qquad K_2 = \frac{J_2}{J_k} \cdot \frac{l_k}{l_2} = \frac{1}{k_2}; \qquad K_b = \frac{J_b}{J_k} \cdot \frac{l_k}{h} = \frac{1}{k_b}. \tag{460}$$

Stiffnesses of the three elastically restrained members:

$$S_1 = \frac{K_1}{2 - \varepsilon'_1} \qquad S_2 = \frac{K_2}{2 - \varepsilon_2}; \qquad S_b = \frac{K_b}{2 - \varepsilon_b} = \frac{1}{b_b}. \tag{461}$$

Joint stiffness

$$S_B = S_1 + S_2 + S_b. \tag{462}$$

Moment distribution factors:

$$\mu_1 = \frac{S_1}{S_B} \qquad \mu_2 = \frac{S_2}{S_B} \qquad \mu_b = \frac{S_b}{S_B}; \qquad (\mu_1 + \mu_2 + \mu_b = 1). \tag{463}$$

Loading Condition 4 — Overall loading condition
(superposition of Loading Conditions 1, 2 and 3)

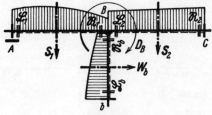

Fig. 112

Loading diagram with load terms $\mathfrak{L}$ and $\mathfrak{R}$ for all three members, and external rotational moment D at the joint.

Restraining moments of the individual members at the joints of the frame:

$$\mathfrak{C}_{r1} = -\frac{\mathfrak{R}_1 - \mathfrak{L}_1 \cdot \varepsilon'_1}{2 - \varepsilon'_1} \qquad \mathfrak{C}_{l2} = -\frac{\mathfrak{L}_2 - \mathfrak{R}_2 \cdot \varepsilon_2}{2 - \varepsilon_2} \qquad \mathfrak{C}_{rb} = -\frac{\mathfrak{R}_b - \mathfrak{L}_b \cdot \varepsilon_b}{2 - \varepsilon_b}. \tag{464}$$

Joint load term:

$$\mathfrak{R}_B = \mathfrak{C}_{r1} - \mathfrak{C}_{l2} + \mathfrak{C}_{rb} + D_B. \tag{465}$$

---

* For $J_k = J_b$ and $l_k = h$ we obtain $K_b = 1$.

Beam moments:

$$M_{B1} = \mathfrak{C}_{r1} - \mathfrak{K}_B \cdot \mu_1 \qquad M_{A1} = -(\mathfrak{L}_1 + M_{B1})\,\varepsilon_1'$$
$$M_{B2} = \mathfrak{C}_{l2} + \mathfrak{K}_B \cdot \mu_2; \qquad M_{C2} = -(\mathfrak{R}_2 + M_{B2})\,\varepsilon_2. \qquad (466)$$

Column moments:

$$M_{Bb} = \mathfrak{C}_{rb} - \mathfrak{K}_B \cdot \mu_b; \qquad M_b = -(\mathfrak{L}_b + M_{Bb})\,\varepsilon_b. \qquad (467)$$

Check relationship:

$$D_B + M_{B1} - M_{B2} + M_{Bb} = 0. \qquad (468)$$

Horizontal thrust, column shears and lateral restraining force for Loading Condition 4;

where
$$H_b = +Q_{lb} = \mathfrak{S}_{rb}/h + T_b \qquad Q_{rb} = -H = -\mathfrak{S}_{lb}/h + T_b,$$
$$T_b = (M_{Bb} - M_b)/h. \qquad (469)$$

Frame Shape 14v

(Frame Shape 14 with laterally unrestrained horizontal member)

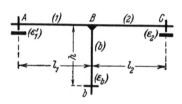

Fig. 113
Frame shape, dimensions and symbols

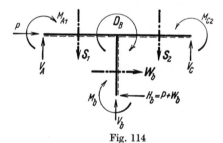

Fig. 114
Definition of positive direction for all external loads and for the reactions at the supports.

Important

In principle the procedure for analyzing the laterally unrestrained Frame Shape 14v is similar to that for Frame Shape 1v (for the collective subscripts $\nu = 1, 2;\ n = b;\ N = B$), with due reference to the formulas for the laterally restrained Frame Shape 14. However, in view of the simplicity and the importance of Frame Shape 14v, the formulas for dealing with it will nevertheless be given direct.

I. Treatment by the Method of Forces

Coefficients*

Flexibility coefficients of the three individual members:

$$k_1 = \frac{J_k}{J_1} \cdot \frac{l_1}{l_k} \qquad k_2 = \frac{J_k}{J_2} \cdot \frac{l_2}{l_k}; \qquad k_b = \frac{J_k}{J_b} \cdot \frac{h}{l_k}. \tag{470}$$

Flexibilities of the elastically restrained beam spans and of the column (the values ϵ are assumed to be known):

$$b_1 = k_1(2 - \varepsilon_1') \qquad b_2 = k_2(2 - \varepsilon_2); \qquad m_b = k_b(1/\varepsilon_b + 4). \tag{471}$$

Denominator:

$$N = b_1 b_2 + (b_1 + b_2) m_b. \tag{472}$$

Loading Condition 1

Both beam spans carrying arbitrary vertical loading

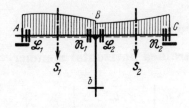
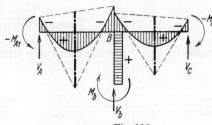

Fig. 115 — Beam loading diagram with load terms $\mathfrak{L}$ and $\mathfrak{R}$

Fig. 116 — Diagram of bending moments and reactions at supports

beam load terms:

$$\mathfrak{B}_1 = (\mathfrak{R}_1 - \mathfrak{L}_1 \cdot \varepsilon_1') k_1 \qquad \mathfrak{B}_2 = (\mathfrak{L}_2 - \mathfrak{R}_2 \cdot \varepsilon_2) k_2. \tag{473}$$

Beam moments:

$$\begin{aligned}
M_{B1} &= -\frac{\mathfrak{B}_1(m_b + b_2) + \mathfrak{B}_2 \cdot m_b}{N} & M_{A1} &= -(\mathfrak{L}_1 + M_{B1}) \varepsilon_1' \\
M_{B2} &= -\frac{\mathfrak{B}_1 \cdot m_b + \mathfrak{B}_2(b_1 + m_b)}{N} & M_{C2} &= -(\mathfrak{R}_2 + M_{B2}) \varepsilon_2.
\end{aligned} \tag{474}$$

Column moments:

$$M_{Bb} = M_b = -M_{B1} + M_{B2} = \frac{\mathfrak{B}_1 \cdot b_2 - \mathfrak{B}_2 \cdot b_1}{N}. \tag{475}$$

---

* For this Frame Shape it is advisable to adopt $J_k = J_b$ and $l_k = h$, so that $k_b = 1$.

Loading Condition 2
Column carrying arbitrary horizontal loading

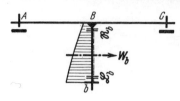 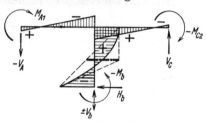

Fig. 117
Column loading diagram with load terms $\mathfrak{L}$ and $\mathfrak{R}$.

Fig. 118
Diagram of bending moments and reactions at supports

Joint load term:
$$\mathfrak{V}_B^v = [\mathfrak{S}_{lb}(1/\varepsilon_b + 1) - (\mathfrak{L}_b + \mathfrak{R}_b)] k_b. \tag{476}$$

Beam moments:
$$\begin{aligned} M_{B1} &= -\mathfrak{V}_B^v \cdot b_2/N & M_{A1} &= -M_{B1} \cdot \varepsilon_1' \\ M_{B2} &= +\mathfrak{V}_B^v \cdot b_1/N & M_{C2} &= -M_{B2} \cdot \varepsilon_2. \end{aligned} \tag{477}$$

Column moments:
$$M_{Bb} = -M_{B1} + M_{B2} = +\mathfrak{V}_B^v \cdot \frac{b_1+b_2}{N} \qquad M_b = -\mathfrak{S}_{lb} + M_{Bb}. \tag{478}$$

Loading Condition 3
External rotational moment at joint

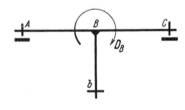 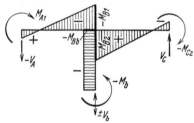

Fig. 119
External rotational moment D at joint B

Fig. 120
Diagram of bending moments and reactions at supports

Beam moments:
$$\begin{aligned} M_{B1} &= -\boldsymbol{D}_B \cdot m_b b_2/N & M_{A1} &= -M_{B1} \cdot \varepsilon_1' \\ M_{B2} &= +\boldsymbol{D}_B \cdot m_b b_1/N & M_{C2} &= -M_{B2} \cdot \varepsilon_2. \end{aligned} \tag{479}$$

Column moments:
$$M_{Bb} = M_b = -\boldsymbol{D}_B - M_{B1} + M_{B2} = -\boldsymbol{D}_B \cdot \frac{b_1 b_2}{N}. \tag{480}$$

Loading Condition 5

Horizontal concentrated load acting at level of horizontal member

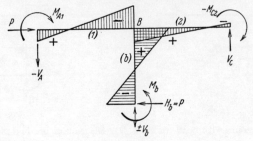

Fig. 121
Loading diagram and diagram of bending moments and reactions at supports

Beam moments:

$$M_{B1} = -Ph(1/\varepsilon_b + 1)k_b \cdot \frac{b_2}{N} \qquad M_{A1} = -M_{B1} \cdot \varepsilon_1' \\ M_{B2} = +Ph(1/\varepsilon_b + 1)k_b \cdot \frac{b_1}{N} \qquad M_{C2} = -M_{B2} \cdot \varepsilon_2. \quad \Bigg\} \quad (481)$$

Column moments:

$$M_{Bb} = -M_{B1} + M_{B2} \qquad M_b = -Ph + M_{Bb}. \qquad (482)$$

II. Treatment by the Deformation Method
Coefficients*

Stiffness coefficients of the three individual members:

$$K_1 = \frac{J_1}{J_k} \cdot \frac{l_k}{l_1} = \frac{1}{k_1} \qquad K_2 = \frac{J_2}{J_k} \cdot \frac{l_k}{l_2} = \frac{1}{k_2}; \qquad K_b = \frac{J_b}{J_k} \cdot \frac{l_k}{h} = \frac{1}{k_b}. \quad (483)$$

Stiffnesses of the elastically restrained beam spans and of the column (the values ε are assumed to be known):

$$S_1 = \frac{K_1}{2-\varepsilon_1'} = \frac{1}{b_1} \qquad S_2 = \frac{K_2}{2-\varepsilon_2} = \frac{1}{b_2}; \qquad F_b = \frac{K_b}{1/\varepsilon_b + 4} = \frac{1}{m_b}. \quad (484)$$

Joint stiffness $\qquad S_B = S_1 + S_2 + F_b.$ $\qquad\qquad\qquad\qquad$ (485)

Moment distribution factors:

$$\eta_1 = \frac{S_1}{S_B} \qquad \eta_2 = \frac{S_2}{S_B} \qquad \eta_b = \frac{F_b}{S_B}; \qquad (\eta_1 + \eta_2 + \eta_b = 1). \quad (486)$$

---

* For $J_k = J_b$ and $l_k = h$ we obtain $K_b = 1$.

Loading Condition 4 — Overall loading condition

(superposition of Loading Conditions 1, 2, 3 and 5)

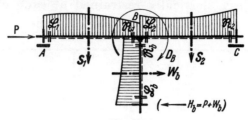

Fig. 122

Loading diagram with load terms $\mathfrak{L}$ and $\mathfrak{R}$ for all three members, external rotational moment D at joint, and horizontal concentrated load P at level of horizontal member

Restraining moments of the beam spans at the joint:

$$\mathfrak{C}_{r1} = -\frac{\mathfrak{R}_1 - \mathfrak{L}_1 \cdot \varepsilon_1'}{2 - \varepsilon_1'} \qquad \mathfrak{C}_{l2} = -\frac{\mathfrak{L}_2 - \mathfrak{R}_2 \cdot \varepsilon_2}{2 - \varepsilon_2}. \tag{487}$$

Restraining moment at column top capable of side-sway:

$$\mathfrak{F}_{rb} = \frac{(\mathfrak{S}_{lb} + Ph)(1/\varepsilon_b + 1) - (\mathfrak{L}_b + \mathfrak{R}_b)}{1/\varepsilon_b + 4}. \tag{488}$$

Joint load term:

$$\mathfrak{K}_B^v = \mathfrak{C}_{r1} - \mathfrak{C}_{l2} + \mathfrak{F}_{rb} + D_B. \tag{489}$$

Beam moments:

$$\left. \begin{aligned} M_{B1} &= \mathfrak{C}_{r1} - \mathfrak{K}_B^v \cdot \eta_1 & M_{A1} &= -(\mathfrak{L}_1 + M_{B1})\varepsilon_1' \\ M_{B2} &= \mathfrak{C}_{l2} + \mathfrak{K}_B^v \cdot \eta_2 & M_{C2} &= -(\mathfrak{R}_2 + M_{B2})\varepsilon_2 \end{aligned} \right\} \tag{490}$$

Column moments:

$$M_{Bb} = \mathfrak{F}_{rb} - \mathfrak{K}_B^v \cdot \eta_b \qquad M_b = -\mathfrak{S}_{lb} - Ph + M_{Bb}. \tag{491}$$

Check relationship:

$$M_{B1} - M_{B2} + M_{Bb} + D_B = 0. \tag{492}$$

Vertical reactions at supports (for all Loading Conditions):

$$\left. \begin{aligned} T_1 &= \frac{M_{A1} - M_{B1}}{l_1}, & V_{A1} &= \frac{\mathfrak{S}_{r1}}{l_1} - T_1 & V_{B1} &= \frac{\mathfrak{S}_{l1}}{l_1} + T_1; \\ T_2 &= \frac{M_{B2} - M_{C2}}{l_2}, & V_{B2} &= \frac{\mathfrak{S}_{r2}}{l_2} - T_2 & V_{C2} &= \frac{\mathfrak{S}_{l2}}{l_2} + T_2. \end{aligned} \right\} \tag{493}$$

Frame Shape 15
Symmetrical single-story eight-bay frame with laterally restrained horizontal member, elastically restrained at each end, and seven elastically restrained columns

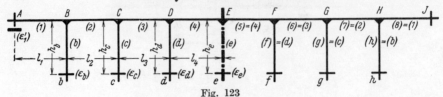

Fig. 123
Frame shape, dimensions and symbols

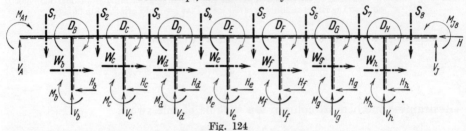

Fig. 124
Definition of positive direction for all external loads on joints and members, for all reactions at supports, and for the lateral restraining force

> All the dimensions and coefficients for the right-hand half of the frame are the same as those for the left-hand half. In the case of the columns the broken line is always placed on the right-hand side of each column despite the symmetry of the frame.

Note: For Frame Shape 15 the *coefficients and the Loading Condition* 4 for *Frame Shape* 11, enlarged by the addition of three bays, may be employed. On account of symmetry, the reduction sequences (398)/(399) and the recursion sequences (400)/(401) respectively become equal. Furthermore, the matrix of influence coefficients (403)—which in the case of Frame Shape 15 would have 7×7 elements—becomes bisymmetrical, i.e., symmetrical with respect to the principal diagonal and the secondary diagonal. If the "*load transposition method*" is employed, it is also possible to start from the following symmetrical and anti-symmetrical loading condition, which in the present case is given for the Deformation Method (Method II).*

> **Important**
> All the instructions, formula sequences and formula matrices given for Frame Shape 15 can readily be extended to suit symmetrical frames of similar type having any even number of bays.

---

* For the treatment of a similar frame by the Method of Forces (Method I) see Frame Shape 17.

a) Coefficients for symmetrical load arrangements

Stiffness coefficients of all the members, according to formulas (395), for $\nu = 1, 2, 3, 4$ and $n = b, c, d$.

Stiffnesses of the first beam span and of the columns (the values ϵ are assumed to be known):
$$S_1 = K_1/(2 - \varepsilon_1'); \qquad S_n = K_n/(2 - \varepsilon_n), \qquad (n = b, c, d). \tag{494}$$

Joint coefficients:
$$K_B = 3S_1 + 3S_b + 2K_2; \qquad \begin{aligned} K_C &= 2K_2 + 3S_c + 2K_3 \\ K_D &= 2K_3 + 3S_d + 2K_4. \end{aligned} \tag{495}$$

Rotation carry-over factors and auxiliary coefficients:

backward reduction: | forward reduction:

$$j_{3s} = \frac{K_3}{r_d'} \qquad \begin{aligned} r_d' &= K_D \\ r_c' &= K_C - K_3 \cdot j_{3s} \end{aligned} \qquad \Big| \qquad j_2'' = \frac{K_2}{v_b} \qquad \begin{aligned} v_b &= K_B \\ v_c &= K_C - K_2 \cdot j_2'' \end{aligned} \qquad \begin{matrix} (496) \\ \text{and} \\ (497) \end{matrix}$$
$$j_{2s} = \frac{K_2}{r_c'} \qquad r_b' = K_B - K_2 \cdot j_{2s}. \qquad \Big| \qquad j_3'' = \frac{K_3}{v_c} \qquad v_d = K_D - K_3 \cdot j_3''.$$

Principal influence coefficients by recursion:

forward: | backward:

$$u_{bb}' = 1/r_b' \qquad \Big| \qquad u_{dd}' = 1/v_d \qquad\qquad (498)$$
$$u_{cc}' = 1/r_c' + u_{bb}' \cdot j_{2s}^2 \qquad \Big| \qquad u_{cc}' = 1/v_c + u_{dd}' \cdot j_3''^2 \qquad \begin{matrix}\text{and}\\(499)\end{matrix}$$
$$u_{dd}' = 1/r_d' + u_{cc}' \cdot j_{3s}^2. \qquad \Big| \qquad u_{bb}' = 1/v_b + u_{cc}' \cdot j_2''^2.$$

Principal influence coefficients direct
$$u_{bb}' = 1/r_b' \qquad u_{cc}' = 1/(r_c' + v_c - K_C) \qquad u_{dd}' = 1/v_d. \tag{500}$$

Computation scheme for the symmetrical influence coefficient matrix*

$$\begin{array}{c|ccc|c} & (u_{bb}') & u_{bc}' & u_{bd}' & \cdot j_2'' \\ \cdot j_{2s} & \downarrow & \uparrow & \uparrow & \\ & u_{cb}' & (u_{cc}') & u_{cd}' & \cdot j_3'' \\ \cdot j_{3s} & \downarrow & \downarrow & \uparrow & \\ & u_{ab}' & u_{dc}' & (u_{dd}') & \end{array} \qquad (501)$$

---

* Between the elements of this 3×3 matrix for the left-hand half of the frame and the elements of the 7×7 matrix for the whole frame, as indicated on page 112, there exist, according to the "load transposition method", the following relationships:
$$u_{bb}' = u_{bb} - u_{bh} \text{ etc., up to } u_{dd}' = u_{dd} - u_{df}.$$

(The values u' with the subscripts d for the center column vanish)

b) Coefficients for anti-symmetrical load arrangements

Stiffness coefficients of all the members, according to formulas (395), for $\nu = 1, 2, 3, 4$ and $n = b, c, d$.

Stiffnesses of the first beam span and of the columns (the values ϵ are assumed to be known):

$$S_1 = \frac{K_1}{2 - \varepsilon_1'}; \quad S_n = \frac{K_n}{2 - \varepsilon_n}, \quad (n = b, c, d, e). \tag{502}$$

Joint coefficients:

$$\begin{aligned} K_B &= 3S_1 + 3S_b + 2K_2 & K_D &= 2K_3 + 3S_d + 2K_4 \\ K_C &= 2K_2 + 3S_c + 2K_3 & K_E'' &= 2K_4 + 3S_e/2. \end{aligned} \right\} \tag{503}$$

Rotation carry-over factors and auxiliary coefficients:

backward reduction:

$$j_{4t} = \frac{K_4}{r_e''} \qquad r_e'' = K_E''$$
$$j_{3t} = \frac{K_3}{r_d''} \qquad r_d'' = K_D - K_4 \cdot j_{4t}$$
$$j_{2t} = \frac{K_2}{r_c''} \qquad r_c'' = K_C - K_3 \cdot j_{3t}$$
$$\qquad\qquad r_b'' = K_B - K_2 \cdot j_{2t}.$$

forward reduction:

$$j_2' = \frac{K_2}{v_b} \qquad v_b = K_B$$
$$j_3' = \frac{K_3}{v_c} \qquad v_c = K_C - K_2 \cdot j_2'$$
$$j_4' = \frac{K_4}{v_d} \qquad v_d = K_D - K_3 \cdot j_3'$$
$$\qquad\qquad v_e'' = K_E'' - K_4 \cdot j_4'.$$

$$\left. \begin{matrix} (504) \\ \text{and} \\ (505) \end{matrix} \right.$$

Principal influence coefficients by recursion:

forward:

$$u_{bb}'' = 1/r_b''$$
$$u_{cc}'' = 1/r_c'' + u_{bb}'' \cdot j_{2t}^2$$
$$u_{dd}'' = 1/r_d'' + u_{cc}'' \cdot j_{3t}^2$$
$$u_{ee}'' = 1/r_e'' + u_{dd}'' \cdot j_{4t}^2.$$

backward:

$$u_{ee}'' = 1/v_e''$$
$$u_{dd}'' = 1/v_d + u_{ee}'' \cdot j_4'^2$$
$$u_{cc}'' = 1/v_c + u_{dd}'' \cdot j_3'^2$$
$$u_{bb}'' = 1/v_b + u_{cc}'' \cdot j_2'^2.$$

$$\left. \begin{matrix} (506) \\ \text{and} \\ (507) \end{matrix} \right.$$

Principal influence coefficients direct

$$u_{bb}'' = \frac{1}{r_b''} \quad u_{cc}'' = \frac{1}{r_c'' + v_c - K_C} \quad u_{dd}'' = \frac{1}{r_d'' + v_d - K_D} \quad u_{ee}'' = \frac{1}{v_e''}. \tag{508}$$

---

* Between the elements of the 4×4 matrix (509), page 115, for the left-hand half of the frame and the elements of the 7×7 matrix for the whole frame, as indicated on page 112, there exist, according to the "load transposition method", the following relationships:

$$u_{bb}'' = u_{bb} + u_{bh} \text{ etc., up to } u_{dd}'' = u_{dd} + u_{df}, \quad u_{ee}'' = 2u_{ee}.$$

(The values u'' of the column e of (509) here become twice as large as the values u of the 7×7 matrix).

— 115 —

Computation scheme for the symmetrical influence coefficient matrix*

$$\begin{array}{c} \cdot j_{2t} \\ \cdot j_{3t} \\ \cdot j_{4t} \end{array} \begin{vmatrix} (u''_{bb}) & u''_{bc} & u''_{bd} & u''_{be} \\ \downarrow & \uparrow & \uparrow & \uparrow \\ u''_{cb} & (u''_{cc}) & u''_{cd} & u''_{ce} \\ \downarrow & \downarrow & \uparrow & \uparrow \\ u''_{db} & u''_{dc} & (u''_{dd}) & u''_{de} \\ \downarrow & \downarrow & \downarrow & \uparrow \\ u''_{eb} & u''_{ec} & u''_{ed} & (u''_{ee}) \end{vmatrix} \begin{array}{c} \cdot j'_2 \\ \cdot j'_3 \\ \cdot j'_4 \end{array} \Bigg\} \quad (509)$$

Loading Condition 4a

Symmetrical overall loading condition

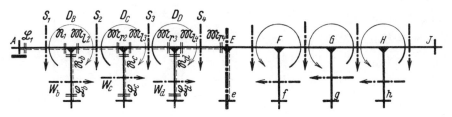

Fig. 125

Loading diagram, with fixed-end moments $\mathfrak{M}_l$ and $\mathfrak{M}_r$ of the internal spans, load terms $\mathfrak{L}$ and $\mathfrak{R}$ for the first span and the columns, and external rotational moments D, for left-hand half of frame. (Symmetrical loading of central column e and central joint E not shown).

Restraining moments of the first beam span and of the columns:

$$\mathfrak{C}_{r1} = -\frac{\mathfrak{R}_1 - \mathfrak{L}_1 \cdot \varepsilon'_1}{2 - \varepsilon'_1}; \qquad \mathfrak{C}_{rn} = -\frac{\mathfrak{R}_n - \mathfrak{L}_n \cdot \varepsilon_n}{2 - \varepsilon_n}, \quad (n = b, c, d). \qquad (510)$$

Joint load terms:

$$\mathfrak{K}_C = \mathfrak{M}_{r2} - \mathfrak{M}_{l3} + \mathfrak{C}_{rc} + \boldsymbol{D}_C$$
$$\mathfrak{K}_B = \mathfrak{C}_{r1} - \mathfrak{M}_{l2} + \mathfrak{C}_{rb} + \boldsymbol{D}_B; \qquad \mathfrak{K}_D = \mathfrak{M}_{r3} - \mathfrak{M}_{l4} + \mathfrak{C}_{rd} + \boldsymbol{D}_D. \qquad (511)$$

auxiliary moments:

$$\left.\begin{array}{l} Y'_B = +\mathfrak{K}_B \cdot u'_{bb} - \mathfrak{K}_C \cdot u'_{bc} + \mathfrak{K}_D \cdot u'_{bd} \\ Y'_C = -\mathfrak{K}_B \cdot u'_{cb} + \mathfrak{K}_C \cdot u'_{cc} - \mathfrak{K}_D \cdot u'_{cd} \\ Y'_D = +\mathfrak{K}_B \cdot u'_{db} - \mathfrak{K}_C \cdot u'_{dc} + \mathfrak{K}_D \cdot u'_{dd} \end{array}\right\} \qquad (512)$$

---

* See footnotes on page 114.

Beam moments:

$$M_{A1} = M_{J8} = -(\mathfrak{L}_1 + M_{B1})\,\varepsilon'_1;$$
$$M_{B2} = M_{H7} = \mathfrak{M}_{l2} + (2Y'_B + Y'_C)\,K_2$$
$$M_{C3} = M_{G6} = \mathfrak{M}_{l3} + (2Y'_C + Y'_D)\,K_3$$
$$M_{D4} = M_{F5} = \mathfrak{M}_{l4} + 2Y'_D \cdot K_4;$$
$$M_{B1} = M_{H8} = \mathfrak{E}_{r1} - Y'_B \cdot 3S_1$$
$$M_{C2} = M_{G7} = \mathfrak{M}_{r2} - (Y'_B + 2Y'_C)\,K_2$$
$$M_{D3} = M_{F6} = \mathfrak{M}_{r3} - (Y'_C + 2Y'_D)\,K_3$$
$$M_{E4} = M_{E5} = \mathfrak{M}_{r4} - Y'_D \cdot K_4.$$
(513)

Column moments:

at top:

$$M_{Bb} = -M_{Hh} = \mathfrak{E}_{rb} - Y'_B \cdot 3S_b$$
$$M_{Cc} = -M_{Gg} = \mathfrak{E}_{rc} - Y'_C \cdot 3S_c$$
$$M_{Dd} = -M_{Ff} = \mathfrak{E}_{rd} - Y'_D \cdot 3S_d$$
$$M_{Ee} = 0;$$

at base:

$$M_b = -M_h = -(\mathfrak{L}_b + M_{Bb})\,\varepsilon_b$$
$$M_c = -M_g = -(\mathfrak{L}_c + M_{Cc})\,\varepsilon_c$$
$$M_d = -M_f = -(\mathfrak{L}_d + M_{Dd})\,\varepsilon_d$$
$$M_e = 0.$$
(514)

Check relationships as (519), but without joint E.

Loading Condition 4b

Anti-symmetrical overall loading condition

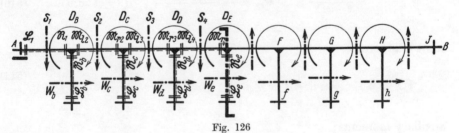

Fig. 126

Loading diagram, with fixed-end moments $\mathfrak{M}_l$ and $\mathfrak{M}_r$ of the internal spans, load terms $\mathfrak{L}$ and $\mathfrak{R}$ for the first span and the columns, and external rotational moments D_r for left-hand half of frame
Left-hand half of frame, including central column e and central joint E.

Restraining moments of the first beam span and of the columns: as formulas (510), but for $n = b, c, d, e$.

Joint load terms:

$$\left.\begin{aligned}\mathfrak{K}_B &= \mathfrak{C}_{r1} - \mathfrak{M}_{l2} + \mathfrak{C}_{rb} + \boldsymbol{D}_B & \mathfrak{K}_D &= \mathfrak{M}_{r3} - \mathfrak{M}_{l4} + \mathfrak{C}_{rd} + \boldsymbol{D}_D \\ \mathfrak{K}_C &= \mathfrak{M}_{r2} - \mathfrak{M}_{l3} + \mathfrak{C}_{rc} + \boldsymbol{D}_C & \mathfrak{K}''_E &= \mathfrak{M}_{r4} + \mathfrak{C}_{re}/2 + \boldsymbol{D}_E/2. \end{aligned}\right\} \quad (515)$$

auxiliary moments:

	$\mathfrak{K}_B$	$\mathfrak{K}_C$	$\mathfrak{K}_D$	$\mathfrak{K}''_E$
$Y''_B =$	$+u''_{bb}$	$-u''_{bc}$	$+u''_{bd}$	$-u''_{be}$
$Y''_C =$	$-u''_{cb}$	$+u''_{cc}$	$-u''_{cd}$	$+u''_{ce}$
$Y''_D =$	$+u''_{db}$	$-u''_{dc}$	$+u''_{dd}$	$-u''_{de}$
$Y''_E =$	$-u''_{eb}$	$+u''_{ec}$	$-u''_{ed}$	$+u''_{ee}.$

$$(516)$$

Beam moments:

$$\left.\begin{aligned} M_{A1} &= -M_{J8} = -(\mathfrak{L}_1 + M_{B1})\,\varepsilon'_1; \\ M_{B2} &= -M_{H7} = \mathfrak{M}_{l2} + (2Y''_B + Y''_C)\,K_2 \\ M_{C3} &= -M_{G6} = \mathfrak{M}_{l3} + (2Y''_C + Y''_D)\,K_3 \\ M_{D4} &= -M_{F5} = \mathfrak{M}_{l4} + (2Y''_D + Y''_E)\,K_4; \\ M_{B1} &= -M_{H8} = \mathfrak{C}_{r1} - Y''_B \cdot 3S_1 \\ M_{C2} &= -M_{G7} = \mathfrak{M}_{r2} - (Y''_B + 2Y''_C)\,K_2 \\ M_{D3} &= -M_{F6} = \mathfrak{M}_{r3} - (Y''_C + 2Y''_D)\,K_3 \\ M_{E4} &= -M_{E5} = \mathfrak{M}_{r4} - (Y''_D + 2Y''_E)\,K_4. \end{aligned}\right\} \quad (517)$$

Column moments:

$$\left.\begin{aligned} &\text{at top:} & &\text{at base:} \\ M_{Bb} = M_{Hh} &= \mathfrak{C}_{rb} - Y''_B \cdot 3S_b & M_b = M_h &= -(\mathfrak{L}_b + M_{Bb})\,\varepsilon_b \\ M_{Cc} = M_{Gg} &= \mathfrak{C}_{rc} - Y''_C \cdot 3S_c & M_c = M_g &= -(\mathfrak{L}_c + M_{Cc})\,\varepsilon_c \\ M_{Dd} = M_{Ff} &= \mathfrak{C}_{rd} - Y''_D \cdot 3S_d & M_d = M_f &= -(\mathfrak{L}_d + M_{Dd})\,\varepsilon_d \\ M_{Ee} &= \mathfrak{C}_{re} - Y''_E \cdot 3S_e; & M_e &= -(\mathfrak{L}_e + M_{Ee})\,\varepsilon_e. \end{aligned}\right\} \quad (518)$$

Check relationships:

$$\left.\begin{aligned} \boldsymbol{D}_B + M_{Bb} + M_{B1} - M_{B2} &= 0 & \boldsymbol{D}_D + M_{Dd} + M_{D3} - M_{D4} &= 0 \\ \boldsymbol{D}_C + M_{Cc} + M_{C2} - M_{C3} &= 0 & \boldsymbol{D}_E + M_{Ee} + 2M_{E4} &= 0. \end{aligned}\right\} \quad (519)$$

(Frame Shape 15—cont.)

Horizontal thrusts, column shears and lateral restraining force for all loading conditions

For asymmetrical loading conditions, according to formulas (179)/(180) or (181)/(182), for $n = b, c, d, e, f, g, h$ and $N = B, C, D, E, F, G, H$.

For symmetrical loading conditions, according to formulas (183)/(184) or (185)/(186), for $n = b, c, d$ and $N = B, C, D$.

For anti-symmetrical loading conditions, according to formulas (187) or (189), for $n = b, c, d, e$ and $N = B, C, D, E$. Furthermore:

$$H = -2(H_b + H_c + H_d) - H_e \quad \text{for} \quad H = -2(Q_{rb} + Q_{rc} + Q_{rd}) - Q_{re}. \quad (520)$$

Frame Shape 15v

(Frame Shape 15 with laterally unrestrained horizontal member)

Procedure

either similar to procedure for Frame Shape 1v, see pages 9-10, for the collective subscripts $\nu = 1 - 8$, $n = b - h$ and $N = B - H$, having due regard to anti-symmetry properties or, alternatively, similar to procedure for Frame Shape 8v, see pages 70-71 for the collective subscripts $\nu = 1, 2, 3, 4$; $n = b, c, d, e$ and $N = B, C, D, E$.

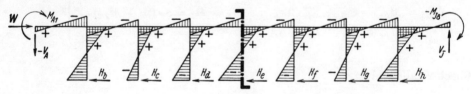

Fig. 127

Loading Condition 5 with diagram of bending moments and reactions at supports. (For convenience only the horizontal reactions H_n at the feet of the columns are shown).

Frame Shape 16
Symmetrical single-story seven-bay frame with laterally restrained horizontal member, elastically restrained at each end, and six elastically restrained columns

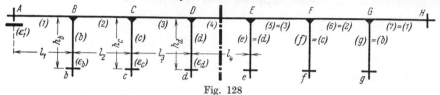

Fig. 128
Frame shape, dimensions and symbols

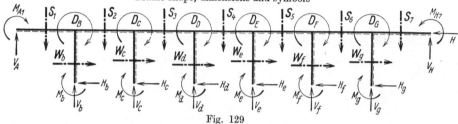

Fig. 129
Definition of positive direction for all external loads on joints and members, for all reactions at supports, and for the lateral restraining force

> All the dimensions and coefficients for the right-hand half of the frame are the same as those for the left-hand half. In the case of the columns the broken line is always placed on the right-hand side of each column, despite the symmetry of the frame.

Note: For Frame Shape 16 the *coefficients and the Loading Condition* 4 for *Frame Shape* 11, enlarged by the addition of two bays, may be employed. On account of symmetry, the reduction sequences (398)/(399) and the recursion sequences (400)/(401) respectively become equal. Furthermore, the matrix of influence coefficients (403)—which in the case of Frame Shape 16 would have 6 × 6 elements—becomes bisymmetrical, i.e., symmetrical with respect to the principal diagonal and the secondary diagonal. If the *"load transposition method"* is employed, it is also possible to start from the following symmetrical and anti-symmetrical loading condition, which in the present case is given for the Deformation Method (Method II).*

> ### Important
> All the instructions, formula sequences and formula matrices given for Frame Shape 16 can readily be extended to suit symmetrical frames of similar type having any odd number of bays.

---

* For the treatment of a similar frame by the Method of Forces (Method I) see Frame Shape 18.

a) Coefficients for symmetrical load arrangements

Stiffness coefficients of all the members, according to formulas (395), for $\nu = 1, 2, 3, 4$ and $n = b, c, d$.

Stiffnesses of the first beam span and of the columns (the values ϵ are assumed to be known):
$$S_1 = \frac{K_1}{2 - \varepsilon_1'}; \qquad S_n = \frac{K_n}{2 - \varepsilon_n}. \tag{521}$$

Joint coefficients:
$$K_B = 3S_1 + 3S_b + 2K_2; \qquad \begin{aligned} K_C &= 2K_2 + 3S_c + 2K_3 \\ K_D' &= 2K_3 + 3S_d + \mathbf{K_4} \end{aligned} \tag{522}$$

Rotation carry-over factors and auxiliary coefficients:

backward reduction: $\qquad\qquad$ forward reduction:

$$\begin{aligned} j_{3s} &= \frac{K_3}{r_d'} & r_d' &= K_D' & j_2' &= \frac{K_2}{v_b} & v_b &= K_B \\ & & r_c' &= K_C - K_3 \cdot j_{3s} & & & v_c &= K_C - K_2 \cdot j_2' \\ j_{2s} &= \frac{K_2}{r_c'} & r_b' &= K_B - K_2 \cdot j_{2s}. & j_3' &= \frac{K_3}{v_c} & v_d' &= K_D' - K_3 \cdot j_3'. \end{aligned} \qquad \begin{matrix}(523)\\ \text{and} \\ (524)\end{matrix}$$

Principal influence coefficients by recursion:

forward: $\qquad\qquad$ backward:

$$\begin{aligned} u_{bb}' &= 1/r_b' & u_{dd}' &= 1/v_d' \\ u_{cc}' &= 1/r_c' + u_{bb}' \cdot j_{2s}^2 & u_{cc}' &= 1/v_c + u_{dd}' \cdot j_3'^2 \\ u_{dd}' &= 1/r_d' + u_{cc}' \cdot j_{3s}^2. & u_{bb}' &= 1/v_b + u_{cc}' \cdot j_2'^2. \end{aligned} \qquad \begin{matrix}(525)\\ \text{and} \\ (526)\end{matrix}$$

Principal influence coefficients direct
$$u_{bb}' = \frac{1}{r_b'} \qquad u_{cc}' = \frac{1}{r_c' + v_c - K_C} \qquad u_{dd}' = \frac{1}{v_d'}. \tag{527}$$

Computation scheme for the symmetrical influence coefficient matrix*

$$\begin{array}{c|ccc|c} \cdot j_{2s} & (u_{bb}') & u_{bc}' & u_{bd}' & \cdot j_2' \\ & \downarrow & \uparrow & \uparrow & \\ & u_{cb}' & (u_{cc}') & u_{cd}' & \\ \cdot j_{3s} & \downarrow & \downarrow & \uparrow & \cdot j_3' \\ & u_{db}' & u_{dc}' & (u_{dd}') & \end{array} \tag{528}$$

---

* Between the elements of this 3×3 matrix for the left-hand half of the frame and the elements of the 6×6 matrix for the whole frame, as indicated on page 119, there exist, according to the "load transposition method", the following relationships:

$$u_{bb}' = u_{bb} + u_{bg} \quad \text{etc., up to} \quad u_{dd}' = u_{dd} + u_{de}.$$

b) Coefficients for anti-symmetrical load arrangements

Stiffness coefficients of all the members, according to formulas (395), for $\nu = 1, 2, 3, 4$ and $n = b, c, d$.

Stiffnesses of the first beam span and of the columns (the values ϵ are assumed to be known):

$$S_1 = \frac{K_1}{2 - \varepsilon_1'} \qquad S_n = \frac{K_n}{2 - \varepsilon_n}. \tag{529}$$

Joint coefficients:

$$K_B = 3S_1 + 3S_b + 2K_2 \qquad K_C = 2K_2 + 3S_c + 2K_3 \\ K_D'' = 2K_3 + 3S_d + 3K_4. \tag{530}$$

Rotation carry-over factors and auxiliary coefficients:

backward reduction: | forward reduction:

$$j_{3t} = \frac{K_3}{r_d''} \qquad r_d'' = K_D''$$
$$\qquad\qquad r_c'' = K_C - K_3 \cdot j_{3t}$$
$$j_{2t} = \frac{K_2}{r_c''} \qquad r_b'' = K_B - K_2 \cdot j_{2t}.$$

$$j_2' = \frac{K_2}{v_b} \qquad v_b = K_B$$
$$\qquad\qquad v_c = K_C - K_2 \cdot j_2'$$
$$j_3' = \frac{K_3}{v_c} \qquad v_d'' = K_D'' - K_3 \cdot j_3'.$$

(531) and (532)

Principal influence coefficients by recursion:

forward: | backward:

$$u_{bb}'' = 1/r_b''$$
$$u_{cc}'' = 1/r_c'' + u_{bb}'' \cdot j_{2t}^2$$
$$u_{dd}'' = 1/r_d'' + u_{cc}'' \cdot j_{3t}^2.$$

$$u_{dd}'' = 1/v_d''$$
$$u_{cc}'' = 1/v_c + u_{dd}'' \cdot j_3'^2$$
$$u_{bb}'' = 1/v_b + u_{cc}'' \cdot j_2'^2.$$

(533) and (534)

Principal influence coefficients direct

$$u_{bb}'' = \frac{1}{r_b''} \qquad u_{cc}'' = \frac{1}{r_c'' + v_c - K_C} \qquad u_{dd}'' = \frac{1}{v_d''}. \tag{535}$$

Computation scheme for the symmetrical influence coefficient matrix*

$$\begin{array}{c|ccc|c} & (u_{bb}'') & u_{bc}'' & u_{bd}'' & \\ \cdot j_{2t} & \downarrow & \uparrow & \uparrow & \cdot j_2' \\ & u_{cb}'' & (u_{cc}'') & u_{cd}'' & \\ \cdot j_{3t} & \downarrow & \downarrow & \uparrow & \cdot j_3' \\ & u_{db}'' & u_{dc}'' & (u_{dd}'') & \end{array} \tag{536}$$

---

* Between the elements of this 3×3 matrix for the left-hand half of the frame and the elements of the 6×6 matrix for the whole frame, as indicated on page 119, there exist, according to the "load transposition method", the following relationships:

$$u_{bb}'' = u_{bb} - u_{bg} \text{ etc., up to } u_{dd}'' = u_{dd} - u_{de}.$$

Loading Condition 4a

Symmetrical overall loading condition

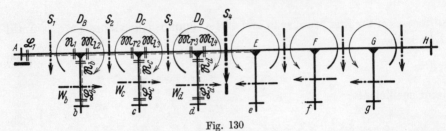

Fig. 130

Loading diagram, with fixed-end moments $\mathfrak{M}_l$ and $\mathfrak{M}_r$ of the internal spans, load terms $\mathfrak{L}$ and $\mathfrak{R}$ for the first span and the columns, and external rotational moments D, for left-hand half of frame

Restraining moments of the first beam span and of the columns:

$$\mathfrak{E}_{r1} = -\frac{\mathfrak{R}_1 - \mathfrak{L}_1 \cdot \varepsilon_1'}{2 - \varepsilon_1'}; \qquad \mathfrak{E}_{rn} = -\frac{\mathfrak{R}_n - \mathfrak{L}_n \cdot \varepsilon_n}{2 - \varepsilon_n}, \quad (n = b, c, d). \tag{537}$$

Joint load terms:

$$\left. \begin{aligned} \mathfrak{K}_B &= \mathfrak{E}_{r1} - \mathfrak{M}_{l2} + \mathfrak{E}_{rb} + D_B \\ \mathfrak{K}_C &= \mathfrak{M}_{r2} - \mathfrak{M}_{l3} + \mathfrak{E}_{rc} + D_C \\ \mathfrak{K}_D &= \mathfrak{M}_{r3} - \mathfrak{M}_{l4} + \mathfrak{E}_{rd} + D_D. \end{aligned} \right\} \tag{538}$$

auxiliary moments:

$$\left. \begin{aligned} Y_B' &= +\mathfrak{K}_B \cdot u_{bb}' - \mathfrak{K}_C \cdot u_{bc}' + \mathfrak{K}_D \cdot u_{bd}' \\ Y_C' &= -\mathfrak{K}_B \cdot u_{cb}' + \mathfrak{K}_C \cdot u_{cc}' - \mathfrak{K}_D \cdot u_{cd}' \\ Y_D' &= +\mathfrak{K}_B \cdot u_{db}' - \mathfrak{K}_C \cdot u_{dc}' + \mathfrak{K}_D \cdot u_{dd}'. \end{aligned} \right\} \tag{539}$$

Beam moments:

$$\left. \begin{aligned} M_{A1} &= M_{H7} = -(\mathfrak{L}_1 + M_{B1})\varepsilon_1' \\ M_{B2} &= M_{G6} = \mathfrak{M}_{l2} + (2Y_B' + Y_C')K_2 \\ M_{C3} &= M_{F5} = \mathfrak{M}_{l3} + (2Y_C' + Y_D')K_3 \\ M_{D4} &= M_{E4} = \mathfrak{M}_{l4} + Y_D' \cdot K_4; \\ M_{B1} &= M_{G7} = \mathfrak{E}_{r1} - Y_B' \cdot 3S_1 \\ M_{C2} &= M_{F6} = \mathfrak{M}_{r2} - (Y_B' + 2Y_C')K_2 \\ M_{D3} &= M_{E5} = \mathfrak{M}_{r3} - (Y_C' + 2Y_D')K_3. \end{aligned} \right\} \tag{540}$$

Column moments:

at top:
$$M_{Bb} = -M_{Gg} = \mathfrak{C}_{rb} - Y'_B \cdot 3 S_b$$
$$M_{Cc} = -M_{Ff} = \mathfrak{C}_{rc} - Y'_C \cdot 3 S_c$$
$$M_{Dd} = -M_{Ee} = \mathfrak{C}_{rd} - Y'_D \cdot 3 S_d;$$

at base:
$$M_b = -M_g = -(\mathfrak{L}_b + M_{Bb})\varepsilon_b$$
$$M_c = -M_f = -(\mathfrak{L}_c + M_{Cc})\varepsilon_c$$
$$M_d = -M_e = -(\mathfrak{L}_d + M_{Dd})\varepsilon_d.$$

$$(541)$$

Check relationships:

$$\boldsymbol{D}_B + M_{Bb} + M_{B1} - M_{B2}$$
$$\boldsymbol{D}_C + M_{Cc} + M_{C2} - M_{C3}$$
$$\boldsymbol{D}_D + M_{Dd} + M_{D3} - M_{D4}.$$

$$(542)$$

Loading Condition 4b

Anti-symmetrical overall loading condition

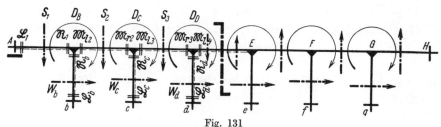

Fig. 131

Loading diagram, with fixed-end moments $\mathfrak{M}_l$ and $\mathfrak{M}_r$ of the internal spans, load terms $\mathfrak{L}$ and $\mathfrak{R}$ for the first span and the columns, and external rotational moments D_r for left-hand half of frame.

Restraining moments and joint load terms: exactly as formulas (537) and (538).

auxiliary moments:

$$Y''_B = +\mathfrak{K}_B \cdot u''_{bb} - \mathfrak{K}_C \cdot u''_{bc} + \mathfrak{K}_D \cdot u''_{bd}$$
$$Y''_C = -\mathfrak{K}_B \cdot u''_{cb} + \mathfrak{K}_C \cdot u''_{cc} - \mathfrak{K}_D \cdot u''_{cd}$$
$$Y''_D = +\mathfrak{K}_B \cdot u''_{db} - \mathfrak{K}_C \cdot u''_{dc} + \mathfrak{K}_D \cdot u''_{dd}.$$

$$(543)$$

Beam moments:

$$M_{A1} = -M_{H7} = -(\mathfrak{L}_1 + M_{B1})\varepsilon'_1$$
$$M_{B2} = -M_{G6} = \mathfrak{M}_{l2} + (2 Y''_B + Y''_C) K_2$$
$$M_{C3} = -M_{F5} = \mathfrak{M}_{l3} + (2 Y''_C + Y''_D) K_3$$
$$M_{D4} = -M_{E4} = \mathfrak{M}_{l4} + Y''_D \cdot 3 K_4;$$

$$(544)$$

— 124 —

Beam moments (cont.)

$$M_{B1} = -M_{G7} = \mathfrak{C}_{r1} - Y_B'' \cdot 3 S_1$$
$$M_{C2} = -M_{F6} = \mathfrak{M}_{r2} - (Y_B'' + 2 Y_C'') K_2 \qquad \text{(544 cont.)}$$
$$M_{D3} = -M_{E5} = \mathfrak{M}_{r3} - (Y_C'' + 2 Y_D'') K_3.$$

Column moments:

at top: at base:

$$M_{Bb} = M_{Gg} = \mathfrak{C}_{rb} - Y_B'' \cdot 3 S_b \qquad M_b = M_g = -(\mathfrak{L}_b + M_{Bb})\varepsilon_b$$
$$M_{Cc} = M_{Ff} = \mathfrak{C}_{rc} - Y_C'' \cdot 3 S_c \qquad M_c = M_f = -(\mathfrak{L}_c + M_{Cc})\varepsilon_c \qquad (545)$$
$$M_{Dd} = M_{Ee} = \mathfrak{C}_{rd} - Y_D'' \cdot 3 S_d; \qquad M_d = M_e = -(\mathfrak{L}_d + M_{Dd})\varepsilon_d.$$

Check relationships: as formulas (542).

Horizontal thrusts, column shears and lateral restraining force for all loading conditions

similar to formulas (179) – (190) for the collective subscripts $n = b - g$, $N = B - G$, or for $n = b, c, d$ and $N = B, C, D$.

Frame Shape 16v

(Frame Shape 16 with laterally unrestrained horizontal member)

Procedure

either similar to procedure for Frame Shape 1v, see pages 9-10, for the collective subscripts $\nu = 1 - 7$, $n = b - g$ and $N = B - G$, having due regard to anti-symmetry properties—or, alternatively, similar to procedure for Frame Shape 5v, see pages 42-43 for the collective subscripts $n = b, c, d$ and $N = B, C, D$.

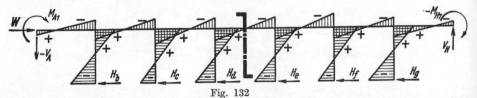

Fig. 132

Loading Condition 5 with diagram of bending moments and reactions at supports. (For convenience only the horizontal reactions H_n at the feet of the columns are shown).

Frame Shape 17

Symmetrical single-story six-bay frame with laterally restrained horizontal member, elastically restrained at each end, and five elastically restrained columns

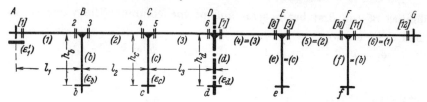

Fig. 133 Frame shape, dimensions and symbols

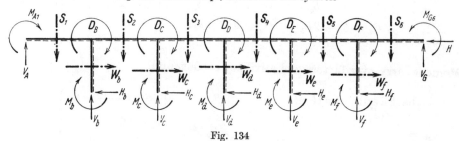

Fig. 134
Definition of positive direction for all external loads on joints and members, for all reactions at supports, and for the lateral restraining force

> All the dimensions and coefficients for the right-hand half of the frame are the same as those for the left-hand half. In the case of the columns the broken line is always placed on the right-hand side of each column, despite the symmetry of the frame.

Note: For Frame Shape 17 the *coefficients and the Loading Conditions* 1–3 for *Frame Shape* 12, enlarged by the addition of two bays, may be employed. On account of symmetry, the reduction sequences (417)/(418) and the recursion sequences (419)/(420) respectively become equal. Furthermore, the matrix of influence coefficients (422)—which in the case of Frame Shape 17 would have 10×10 elements—becomes bisymmetrical, i.e., symmetrical with respect to the principal diagonal and the secondary diagonal. If the *"load transposition method"* is employed, it is also possible to start from the following symmetrical and anti-symmetrical loading condition, which in the present case is given for the Method of Forces (Method I).*

> **Important**
>
> All the instructions, formula sequences and formula matrices given for Frame Shape 17 can readily be extended to suit symmetrical frames of similar type having any even number of bays.

---

* For the treatment of a similar frame by the Deformation Method (Method II) see Frame Shape 15.

a) Symmetrical load arrangements

Coefficients

Flexibility coefficients of all the members, according to formulas (414), for $\nu = 1, 2, 3$ and $n = b, c$.

Flexibilities of the first beam span and of the columns (the values ϵ are assumed to be known)

$$b_1 = k_1(2 - \varepsilon'_1); \qquad b_b = k_b(2 - \varepsilon_b) \qquad b_c = k_c(2 - \varepsilon_c). \qquad (546)$$

Diagonal coefficients:

$$\left. \begin{array}{lll} O_2 = b_1 + b_b & O_4 = 2k_2 + b_c & (O'_6 = 2k_3). \\ O_3 = b_b + 2k_2 & O_5 = b_c + 2k_3; & \end{array} \right\} \quad (547)$$

Moment carry-over factors and auxiliary coefficients:

$$\left. \begin{array}{ll} \text{backward reduction:} & \text{forward reduction:} \\[4pt] \begin{array}{ll} \iota_{3s} = \dfrac{1}{2} & \\ & r'_5 = O_5 - k_3 \cdot \dfrac{1}{2} \\ \sigma_{cs} = \dfrac{b_c}{r'_5} & \\ & r'_4 = O_4 - b_c \cdot \sigma_{cs} \\ \iota_{2s} = \dfrac{k_2}{r'_4} & \\ & r'_3 = O_3 - k_2 \cdot \iota_{2s} \\ \sigma_{bs} = \dfrac{b_b}{r'_3} & \\ & r'_2 = O_2 - b_b \cdot \sigma_{bs}. \end{array} & \begin{array}{l} \sigma'_b = \dfrac{b_b}{v_2} \qquad v_2 = O_2 \\[4pt] \qquad\qquad v_3 = O_3 - b_b \cdot \sigma'_b \\ \iota'_2 = \dfrac{k_2}{v_3} \\ \qquad\qquad v_4 = O_4 - k_2 \cdot \iota'_2 \\ \sigma'_c = \dfrac{b_c}{v_4} \\ \qquad\qquad v_5 = O_5 - b_c \cdot \sigma'_c \\ \iota'_3 = \dfrac{k_3}{v_5}. \end{array} \end{array} \right\} \begin{array}{c} (548) \\ \text{and} \\ (549) \end{array}$$

Principal influence coefficients by recursion:

$$\left. \begin{array}{ll} \text{forward:} & \text{backward:} \\[4pt] \begin{array}{l} n'_{22} = \dfrac{1}{r'_2} \\[4pt] n'_{33} = \dfrac{1}{r'_3} + n'_{22} \cdot \sigma^2_{bs} \\[4pt] n'_{44} = \dfrac{1}{r'_4} + n'_{33} \cdot \iota^2_{2s} \\[4pt] n'_{55} = \dfrac{1}{r'_5} + n'_{44} \cdot \sigma^2_{cs} \\[4pt] n'_{66} = \dfrac{1}{2k_3} + n'_{55} \cdot \dfrac{1}{4}. \end{array} & \begin{array}{l} n'_{66} = \dfrac{1}{k_3(2 - \iota'_3)} \\[4pt] n'_{55} = \dfrac{1}{v_5} + n'_{66} \cdot \iota'^2_3 \\[4pt] n'_{44} = \dfrac{1}{v_4} + n'_{55} \cdot \sigma'^2_c \\[4pt] n'_{33} = \dfrac{1}{v_3} + n'_{44} \cdot \iota'^2_2. \\[4pt] n'_{22} = \dfrac{1}{v_2} + n'_{33} \cdot \sigma'^2_b. \end{array} \end{array} \right\} \begin{array}{c} (550) \\ \text{and} \\ (551) \end{array}$$

Principal influence coefficients direct

$$n'_{22} = \frac{1}{r'_2} \qquad n'_{33} = \frac{1}{r'_3 + v_3 - O_3} \qquad n'_{44} = \frac{1}{r'_4 + v_4 - O_4} \qquad \left.\begin{array}{c} \\ \\ \\ \end{array}\right\} \text{(552)}$$
$$n'_{55} = \frac{1}{r'_5 + v_5 - O_5} \qquad n'_{66} = \frac{1}{k_3(2 - \iota'_3)}.$$

Computation scheme for the symmetrical influence coefficient matrix*

$$
\begin{array}{c}
\cdot \sigma_{bs} \\
\cdot \iota_{2s} \\
\cdot \sigma_{cs} \\
\cdot \frac{1}{2}
\end{array}
\left|
\begin{array}{ccccc}
(n'_{22}) & n'_{23} & n'_{24} & n'_{25} & n'_{26} \\
\downarrow & \uparrow & \uparrow & \uparrow & \uparrow \\
n'_{32} & (n'_{33}) & n'_{34} & n'_{35} & n'_{36} \\
\downarrow & \downarrow & \uparrow & \uparrow & \uparrow \\
n'_{42} & n'_{43} & (n'_{44}) & n'_{45} & n'_{46} \\
\downarrow & \downarrow & \downarrow & \uparrow & \uparrow \\
n'_{52} & n'_{53} & n'_{54} & (n'_{55}) & n'_{56} \\
\downarrow & \downarrow & \downarrow & \downarrow & \uparrow \\
n'_{62} & n'_{63} & n'_{64} & n'_{65} & (n'_{66})
\end{array}
\right|
\begin{array}{c}
\cdot \sigma'_b \\
\cdot \iota'_2 \\
\cdot \sigma'_c \\
\cdot \iota'_3
\end{array}
\quad \text{(553)}
$$

Composite influence coefficients**

$$
\begin{aligned}
s'_{22} &= +n'_{23} - n'_{24} = (s_{22} - s_{25}) & s'_{23} &= & n'_{26} &= (s_{23} - s_{24}) \\
s'_{32} &= +n'_{33} - n'_{34} = (s_{32} - s_{35}) & s'_{33} &= & n'_{36} &= (s_{33} - s_{34}) \\
s'_{42} &= -n'_{43} + n'_{44} = (s_{42} + s_{45}) & s'_{43} &= & n'_{46} &= (s_{43} - s_{44}) \\
s'_{52} &= -n'_{53} + n'_{54} = (s_{52} + s_{55}) & s'_{53} &= & n'_{56} &= (s_{53} - s_{54}) \\
s'_{62} &= -n'_{63} + n'_{64} = (s_{62} + s_{65}) & s'_{63} &= -n'_{65} + n'_{66} &= (s_{63} + s_{64});
\end{aligned}
\quad \text{(554)}
$$

$$
\begin{aligned}
w'_{2b} &= +n'_{22} - n'_{23} = (w_{2b} - w_{2f}) & w'_{2c} &= +n'_{24} - n'_{25} = (w_{2c} - w_{2e}) \\
w'_{3b} &= -n'_{32} + n'_{33} = (w_{3b} + w_{3f}) & w'_{3c} &= +n'_{34} - n'_{35} = (w_{3c} - w_{3e}) \\
w'_{4b} &= -n'_{42} + n'_{43} = (w_{4b} + w_{4f}) & w'_{4c} &= +n'_{44} - n'_{45} = (w_{4c} - w_{4e}) \\
w'_{5b} &= -n'_{52} + n'_{53} = (w_{5b} + w_{5f}) & w'_{5c} &= -n'_{54} + n'_{55} = (w_{5c} + w_{5e}) \\
w'_{6b} &= -n'_{62} + n'_{63} = (w_{6b} + w_{6f}) & w'_{6c} &= -n'_{64} + n'_{65} = (w_{6c} + w_{6e}).
\end{aligned}
\quad \text{(555)}
$$

---

* Between the elements of the matrix (553) for the left-hand half of the frame and the elements of the 10 × 10 matrix for the whole frame, as indicated on page 125, there exist, according to the "load transposition method", the following relationships:

$$n'_{22} = n_{22} + n_{2,11} \text{ etc., up to } n'_{66} = n_{66} + n_{67}.$$

** The members enclosed in parentheses have been formed with the values s and w, as indicated in formulas (423) and (424) appropriately extended to suit Frame Shape 17.

Loading Condition 1a

Horizontal member carrying vertical loading of arbitrary magnitude symmetrical about center of frame

$$(\mathfrak{R}_{7-\nu} = \mathfrak{L}_\nu, \quad \mathfrak{L}_{7-\nu} = \mathfrak{R}_\nu, \quad \nu = 1, 2, 3)$$

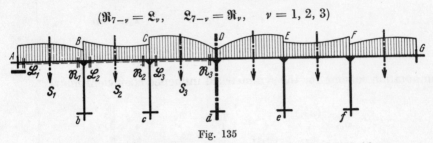

Fig. 135

Beam loading diagram with load terms $\mathfrak{L}$ and $\mathfrak{R}$ for left-hand half of frame

beam load terms:

$$L_\nu = \mathfrak{L}_\nu k_\nu, \quad R_\nu = \mathfrak{R}_\nu k_\nu, \quad \nu = 1, 2, 3; \quad \mathfrak{B}_1 = R_1 - L_1 \cdot \varepsilon_1'. \tag{556}$$

Beam moments:

	$\mathfrak{B}_1$	L_2	R_2	L_3	R_3
$X_2' = M_{B1} = M_{F6} =$	$-n_{22}'$	$-n_{23}'$	$+n_{24}'$	$+n_{25}'$	$-n_{26}'$
$X_3' = M_{B2} = M_{F5} =$	$-n_{32}'$	$-n_{33}'$	$+n_{34}'$	$+n_{35}'$	$-n_{36}'$
$X_4' = M_{C2} = M_{E5} =$	$+n_{42}'$	$+n_{43}'$	$-n_{44}'$	$-n_{45}'$	$+n_{46}'$
$X_5' = M_{C3} = M_{E4} =$	$+n_{52}'$	$+n_{53}'$	$-n_{54}'$	$-n_{55}'$	$+n_{56}'$
$X_6' = M_{D3} = M_{D4} =$	$-n_{62}'$	$-n_{63}'$	$+n_{64}'$	$+n_{65}'$	$-n_{66}';$

$$M_{A1} = M_{G6} = -(\mathfrak{L}_1 + M_{B1})\varepsilon_1'. \tag{557}$$

Column moments:

at top: at base:

$$M_{Bb} = -M_{Ff} = -X_2' + X_3' \qquad M_b = -M_f = -M_{Bb} \cdot \varepsilon_b$$

$$M_{Cc} = -M_{Ee} = -X_4' + X_5' \qquad M_c = -M_e = -M_{Cc} \cdot \varepsilon_c$$

$$M_{Dd} = 0; \qquad\qquad\qquad\qquad M_d = 0.$$

(558) and (559)

Special case: symmetrical bay loading $R_\nu = L\nu$):

$$\begin{aligned}
X_2' = M_{B1} = M_{F6} &= -\mathfrak{B}_1 \cdot n_{22}' - L_2 \cdot s_{22}' + L_3 \cdot s_{23}' \\
X_3' = M_{B2} = M_{F5} &= -\mathfrak{B}_1 \cdot n_{32}' - L_2 \cdot s_{32}' + L_3 \cdot s_{33}' \\
X_4' = M_{C2} = M_{E5} &= +\mathfrak{B}_1 \cdot n_{42}' - L_2 \cdot s_{42}' - L_3 \cdot s_{43}' \\
X_5' = M_{C3} = M_{E4} &= +\mathfrak{B}_1 \cdot n_{52}' - L_2 \cdot s_{52}' - L_3 \cdot s_{53}' \\
X_6' = M_{D3} = M_{D4} &= -\mathfrak{B}_1 \cdot n_{62}' + L_2 \cdot s_{62}' - L_3 \cdot s_{63}'.
\end{aligned} \tag{560}$$

Loading Conditions 2a and 3a

All the columns carrying arbitrary horizontal loading (2a) and external rotational moments applied at all the beam joints (3a), but load arrangement as a whole is symmetrical about center of frame

$(\mathfrak{L}_f = -\mathfrak{L}_b, \quad \mathfrak{R}_f = -\mathfrak{R}_b; \quad \mathfrak{L}_e = -\mathfrak{L}_c, \quad \mathfrak{R}_e = -\mathfrak{R}_c; \quad D_F = -D_B, \quad D_E = -D_C)$

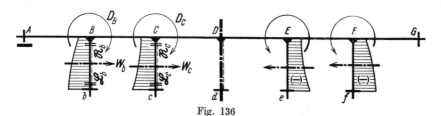

Fig. 136

Column loading diagram with load terms $\mathfrak{L}$ and $\mathfrak{R}$ for the columns, and external rotational moments D, for left-hand half of frame

(Symmetrical loading of central column d and central joint D not shown)

Joint load terms $(L_n = \mathfrak{L}_n k_n$ and $R_n = \mathfrak{R}_n k_n$ for $n = b, c)$:

$$\mathfrak{B}_B = D_B \cdot b_b + L_b \cdot \varepsilon_b - R_b \qquad \mathfrak{B}_C = D_C \cdot b_c + L_c \cdot \varepsilon_c - R_c. \qquad (561)$$

Beam moments:

$$\begin{aligned}
X'_2 &= M_{B1} = M_{F6} = -\mathfrak{B}_B \cdot w'_{2b} + \mathfrak{B}_C \cdot w'_{2c} \\
X'_3 &= M_{B2} = M_{F5} = +\mathfrak{B}_B \cdot w'_{3b} + \mathfrak{B}_C \cdot w'_{3c} \\
X'_4 &= M_{C2} = M_{E5} = -\mathfrak{B}_B \cdot w'_{4b} - \mathfrak{B}_C \cdot w'_{4c} \\
X'_5 &= M_{C3} = M_{E4} = -\mathfrak{B}_B \cdot w'_{5b} + \mathfrak{B}_C \cdot w'_{5c} \\
X'_6 &= M_{D3} = M_{D4} = +\mathfrak{B}_B \cdot w'_{6b} - \mathfrak{B}_C \cdot w'_{6c}; \\
M_{A1} &= M_{G6} = -M_{B1} \cdot \varepsilon'_1.
\end{aligned} \qquad (562)$$

Column moments:

at top:
$$\begin{aligned}
M_{Eb} &= -M_{Ff} = -D_B - X'_2 + X'_3 \\
M_{Cc} &= -M_{Ee} = -D_C - X'_4 + X'_5 \\
M_{Dd} &= 0;
\end{aligned} \qquad (563)$$

at base:
$$\begin{aligned}
M_b &= -M_f = -(\mathfrak{L}_b + M_{Bb})\,\varepsilon_b \\
M_c &= -M_e = -(\mathfrak{L}_c + M_{Cc})\,\varepsilon_c \\
M_d &= 0.
\end{aligned} \qquad (564)$$

b) Anti-symmetrical load arrangements

Coefficients

Flexibility coefficients of all the members, according to formulas (414), for $v = 1, 2, 3$ and $n = b, c, d$.

Flexibilities of the first beam span and of the columns (the values ϵ are assumed to be known)

$$b_1 = k_1(2 - \varepsilon_1'); \qquad b_n = k_n(2 - \varepsilon_n) \quad \text{for} \quad n = b, c, d. \tag{565}$$

Diagonal coefficients:

$$\left. \begin{array}{lll} O_2 = b_1 + b_b & O_4 = 2k_2 + b_c & O_6'' = 2k_3 + 2b_d \\ O_3 = b_b + 2k_2 & O_5 = b_c + 2k_3 & \end{array} \right\} \tag{566}$$

Moment carry-over factors and auxiliary coefficients:

backward reduction:	forward reduction:

$$\begin{aligned} \iota_{3t} &= \frac{k_3}{r_6''} & r_6'' &= O_6'' \\ \sigma_{ct} &= \frac{b_c}{r_5''} & r_5'' &= O_5 - k_3 \cdot \iota_{3t} \\ \iota_{2t} &= \frac{k_2}{r_4''} & r_4'' &= O_4 - b_c \cdot \sigma_{ct} \\ \sigma_{bt} &= \frac{b_b}{r_3''} & r_3'' &= O_3 - k_2 \cdot \iota_{2t} \\ & & r_2'' &= O_2 - b_b \cdot \sigma_{bt}. \end{aligned} \qquad \begin{aligned} & & v_2 &= O_2 \\ \sigma_b' &= \frac{b_b}{v_2} & v_3 &= O_3 - b_b \cdot \sigma_b' \\ \iota_2' &= \frac{k_2}{v_3} & v_4 &= O_4 - k_2 \cdot \iota_2' \\ \sigma_c' &= \frac{b_c}{v_4} & v_5 &= O_5 - b_c \cdot \sigma_c' \\ \iota_3' &= \frac{k_3}{v_5} & v_6'' &= O_6'' - k_3 \cdot \iota_3'. \end{aligned} \tag{567}$$
and (568)

Principal influence coefficients by recursion:

forward:	backward:

$$\begin{aligned} n_{22}'' &= \frac{1}{r_2''} \\ n_{33}'' &= \frac{1}{r_3''} + n_{22}'' \cdot \sigma_{bt}^2 \\ n_{44}'' &= \frac{1}{r_4''} + n_{33}'' \cdot \iota_{2t}^2 \\ n_{55}'' &= \frac{1}{r_5''} + n_{44}'' \cdot \sigma_{ct}^2 \\ n_{66}'' &= \frac{1}{r_6''} + n_{55}'' \cdot \iota_{3t}^2. \end{aligned} \qquad \begin{aligned} n_{66}'' &= \frac{1}{v_6''} \\ n_{55}'' &= \frac{1}{v_5} + n_{66}'' \cdot \iota_3'^2 \\ n_{44}'' &= \frac{1}{v_4} + n_{55}'' \cdot \sigma_c'^2 \\ n_{33}'' &= \frac{1}{v_3} + n_{44}'' \cdot \iota_2'^2 \\ n_{22}'' &= \frac{1}{v_2} + n_{33}'' \cdot \sigma_b'^2. \end{aligned} \tag{569}$$
and (570)

Principal influence coefficients direct

$$n''_{22} = \frac{1}{r'_2} \quad n''_{33} = \frac{1}{r'_3 + v_3 - O_3} \quad n''_{44} = \frac{1}{r'_4 + v_4 - O_4} \quad \Biggr\} \quad (571)$$
$$n''_{55} = \frac{1}{r'_5 + v_5 - O_5} \quad n''_{66} = \frac{1}{v''_6}.$$

Computation scheme for the symmetrical influence coefficient matrix*

$$\begin{array}{c|ccccc|c}
 & (n''_{22}) & n''_{23} & n''_{24} & n''_{25} & n''_{26} & \\
\cdot \sigma_{bt} & \downarrow & \uparrow & \uparrow & \uparrow & \uparrow & \cdot \sigma'_b \\
 & n''_{32} & (n''_{33}) & n''_{34} & n''_{35} & n''_{36} & \\
\cdot l_{2t} & \downarrow & \downarrow & \uparrow & \uparrow & \uparrow & \cdot l'_2 \\
 & n''_{42} & n''_{43} & (n''_{44}) & n''_{45} & n''_{46} & \\
\cdot \sigma_{ct} & \downarrow & \downarrow & \downarrow & \uparrow & \uparrow & \cdot \sigma'_c \\
 & n''_{52} & n''_{53} & n''_{54} & (n''_{55}) & n''_{56} & \\
\cdot l_{3t} & \downarrow & \downarrow & \downarrow & \downarrow & \uparrow & \cdot l'_3 \\
 & n''_{62} & n''_{63} & n''_{64} & n''_{65} & (n''_{66}) & \\
\end{array} \quad (572)$$

Composite influence coefficients**:

$$\begin{aligned}
s''_{22} &= + n''_{23} - n''_{24} = (s_{22} + s_{25}) & s''_{23} &= + n''_{25} - n''_{26} = (s_{23} + s_{24}) \\
s''_{32} &= + n''_{33} - n''_{34} = (s_{32} + s_{35}) & s''_{33} &= + n''_{35} - n''_{36} = (s_{33} + s_{34}) \\
 & & s''_{43} &= + n''_{45} - n''_{46} = (s_{43} + s_{44}) \\
s''_{42} &= - n''_{43} + n''_{44} = (s_{42} - s_{45}) & s''_{53} &= + n''_{55} - n''_{56} = (s_{53} + s_{54}) \\
s''_{52} &= - n''_{53} + n''_{54} = (s_{52} - s_{55}) & & \\
s''_{62} &= - n''_{63} + n''_{64} = (s_{62} - s_{65}) & s''_{63} &= - n''_{65} + n''_{66} = (s_{63} - s_{64});
\end{aligned} \quad (573)$$

$$\begin{aligned}
w''_{2b} &= + n''_{22} - n''_{23} = (w_{2b} + w_{2f}) & w''_{2c} &= + n''_{24} - n''_{25} = (w_{2c} + w_{2e}) \\
 & & w''_{3c} &= + n''_{34} - n''_{35} = (w_{3c} + w_{3e}) \\
w''_{3b} &= - n''_{32} + n''_{33} = (w_{3b} - w_{3f}) & w''_{4c} &= + n''_{44} - n''_{45} = (w_{4c} + w_{4e}) \\
w''_{4b} &= - n''_{42} + n''_{43} = (w_{4b} - w_{4f}) & & \\
w''_{5b} &= - n''_{52} + n''_{53} = (w_{5b} - w_{5f}) & w''_{5c} &= - n''_{54} + n''_{55} = (w_{5c} - w_{5e}) \\
w''_{6b} &= - n''_{62} + n''_{63} = (w_{6b} - w_{6f}) & w''_{6c} &= - n''_{64} + n''_{65} = (w_{6c} - w_{6e}).
\end{aligned} \quad (574)$$

* Between the elements of the matrix (572) for the left-hand half of the frame and the elements of the 10 × 10 matrix for the whole frame, as indicated on page 125, there exist, according to the "load transposition method", the following relationships:

$$n''_{22} = n_{22} - n_{2,11} \text{ etc., up to } n''_{66} = n_{66} - n_{67}.$$

** The members enclosed in parentheses have been formed with the values s and w, as indicated in formulas (423) and (424) appropriately extended to suit Frame Shape 17.

Loading Condition 1b

Horizontal member carrying vertical loading of arbitrary magnitude, but anti-symmetrical about center of frame

$$(\mathfrak{R}_{7-\nu} = -\mathfrak{L}_\nu, \quad \mathfrak{L}_{7-\nu} = -\mathfrak{R}_\nu; \quad \nu = 1, 2, 3)$$

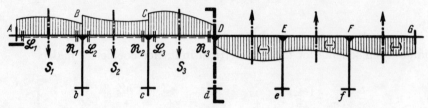

Fig. 137
Beam loading diagram with load terms $\mathfrak{L}$ and $\mathfrak{R}$ for left-hand half of frame

Beam load terms:

$$L_\nu = \mathfrak{L}_\nu k_\nu, \quad R_\nu = \mathfrak{R}_\nu k_\nu, \quad \nu = 1, 2, 3; \quad \boldsymbol{B}_1 = \boldsymbol{R}_1 - \boldsymbol{L}_1 \cdot \varepsilon_1'. \tag{575}$$

Beam moments:

	$\boldsymbol{B}_1$	L_2	R_2	L_3	R_3
$X_2'' = M_{B1} = -M_{F6} =$	$-n_{22}''$	$-n_{23}''$	$+n_{24}''$	$+n_{25}''$	$-n_{26}''$
$X_3'' = M_{B2} = -M_{F5} =$	$-n_{32}''$	$-n_{33}''$	$+n_{34}''$	$+n_{35}''$	$-n_{36}''$
$X_4'' = M_{C2} = -M_{E5} =$	$+n_{42}''$	$+n_{43}''$	$-n_{44}''$	$-n_{45}''$	$+n_{46}''$
$X_5'' = M_{C3} = -M_{E4} =$	$+n_{52}''$	$+n_{53}''$	$-n_{54}''$	$-n_{55}''$	$+n_{56}''$
$X_6'' = M_{D3} = -M_{D4} =$	$-n_{62}''$	$-n_{63}''$	$+n_{64}''$	$+n_{65}''$	$-n_{66}'';$

$$M_{A1} = -M_{G6} = -(\mathfrak{L}_1 + M_{B1})\varepsilon_1'. \tag{576}$$

Column moments:

at top: at base:

$$M_{Bb} = M_{Ff} = -X_2'' + X_3'' \qquad M_b = M_f = -M_{Bb} \cdot \varepsilon_b$$
$$M_{Cc} = M_{Ee} = -X_4'' + X_5'' \qquad M_c = M_e = -M_{Cc} \cdot \varepsilon_c \tag{577}$$
$$M_{Dd} = -2X_6''; \qquad M_d = -M_{Dd} \cdot \varepsilon_d. \tag{578}$$

Special case: symmetrical bay loading $(R_\nu = L_\nu)$:

$$\begin{aligned}
X_2'' = M_{B1} = -M_{F6} &= -\boldsymbol{B}_1 \cdot n_{22}'' - \boldsymbol{L}_2 \cdot s_{22}'' + \boldsymbol{L}_3 \cdot s_{23}'' \\
X_3'' = M_{B2} = -M_{F5} &= -\boldsymbol{B}_1 \cdot n_{32}'' - \boldsymbol{L}_2 \cdot s_{32}'' + \boldsymbol{L}_3 \cdot s_{33}'' \\
X_4'' = M_{C2} = -M_{E5} &= +\boldsymbol{B}_1 \cdot n_{42}'' - \boldsymbol{L}_2 \cdot s_{42}'' - \boldsymbol{L}_3 \cdot s_{43}'' \\
X_5'' = M_{C3} = -M_{E4} &= +\boldsymbol{B}_1 \cdot n_{52}'' - \boldsymbol{L}_2 \cdot s_{52}'' - \boldsymbol{L}_3 \cdot s_{53}'' \\
X_6'' = M_{D3} = -M_{D4} &= -\boldsymbol{B}_1 \cdot n_{62}'' + \boldsymbol{L}_2 \cdot s_{62}'' - \boldsymbol{L}_3 \cdot s_{63}''.
\end{aligned} \tag{579}$$

Loading Conditions 2b and 3b

All the columns carrying arbitrary horizontal loading (2b) and external rotational moments applied to all the beam joints (3b), but load arrangement as a whole is anti-symmetrical about center of frame

$(\mathfrak{L}_f = \mathfrak{L}_b,\ \mathfrak{R}_f = \mathfrak{R}_b;\quad \mathfrak{L}_e = \mathfrak{L}_c,\ \mathfrak{R}_e = \mathfrak{R}_c;\quad D_F = D_B,\ D_E = D_C)$

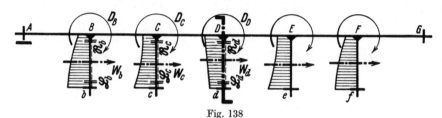

Fig. 138

Column loading diagram with load terms $\mathfrak{L}$ and $\mathfrak{R}$ for the columns, and external rotational moments D, for left-hand half of frame
including the central column and central joint)

Joint load terms $(L_n = \mathfrak{L}_n k_n\ \text{and}\ R_n = \mathfrak{R}_n k_n)$:

$$\mathfrak{B}_N = D_N \cdot b_n + L_n \cdot \varepsilon_n - R_n;\qquad \mathfrak{B}_D'' = \mathfrak{B}_D/2;\quad \text{for}\ N = B, C, D\ \text{and}\ n = b, c, d. \tag{580}$$

Beam moments:

$$\begin{aligned}
X_2'' = M_{B1} = -M_{F6} &= -\mathfrak{B}_B \cdot w_{2b}'' + \mathfrak{B}_C \cdot w_{2c}'' - \mathfrak{B}_D'' \cdot n_{26}'' \\
X_3'' = M_{B2} = -M_{F5} &= +\mathfrak{B}_B \cdot w_{3b}'' + \mathfrak{B}_C \cdot w_{3c}'' - \mathfrak{B}_D'' \cdot n_{36}'' \\
X_4'' = M_{C2} = -M_{E5} &= -\mathfrak{B}_B \cdot w_{4b}'' - \mathfrak{B}_C \cdot w_{4c}'' + \mathfrak{B}_D'' \cdot n_{46}'' \\
X_5'' = M_{C3} = -M_{E4} &= -\mathfrak{B}_B \cdot w_{5b}'' + \mathfrak{B}_C \cdot w_{5c}'' + \mathfrak{B}_D'' \cdot n_{56}'' \\
X_6'' = M_{D3} = -M_{D4} &= +\mathfrak{B}_B \cdot w_{6b}'' - \mathfrak{B}_C \cdot w_{6c}'' - \mathfrak{B}_D'' \cdot n_{66}''; \\
M_{A1} = -M_{G6} &= -M_{B1} \cdot \varepsilon_1'.
\end{aligned} \tag{581}$$

Column moments:

at top:
$$\begin{aligned}
M_{Bb} = M_{Ff} &= -D_B - X_2'' + X_3'' \\
M_{Cc} = M_{Ee} &= -D_C - X_4'' + X_5'' \\
M_{Dd} &= -D_D - 2 X_6'';
\end{aligned} \tag{582}$$

at base:
$$\begin{aligned}
M_b = M_f &= -(\mathfrak{L}_b + M_{Bb})\varepsilon_b \\
M_c = M_e &= -(\mathfrak{L}_c + M_{Cc})\varepsilon_c \\
M_d &= -(\mathfrak{L}_d + M_{Dd})\varepsilon_d.
\end{aligned} \tag{583}$$

(Frame Shape 17—cont.)

Horizontal thrusts, column shears and lateral restraining force for all loading conditions

For asymmetrical loading conditions, according to formulas (179)/(180) or (181)/(182), for $n = b, c, d, e, f$ and $N = B, C, D, E, F$.

For symmetrical loading conditions, according to formulas (183)/(184) or (185)/(186), for $n = b, c$ and $N = B, C$.

For anti-symmetrical loading conditions, according to formulas (187) or (189), for $n = b, c, d$ and $N = B, C, D$. Furthermore:

$$H = -2(H_b + H_c) - H_d \quad \text{or} \quad H = -2(Q_{rb} + Q_{rc}) - Q_{rd}. \qquad (584)$$

Frame Shape 17v

(Frame Shape 17 with laterally unrestrained horizontal member)

Procedure

either similar to procedure for Frame Shape 1v, see pages 9-10, for the collective subscripts $v = 1 - 6$, $n = b - f$ and $N = B - F$, having due regard to anti-symmetry properties or, alternatively, similar to procedure for Frame Shape 8v, see pages 70-71, for the collective subscripts $v = 1, 2, 3$; $n = b, c, d$ and $N = B, C, D$.

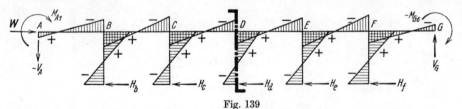

Fig. 139

Loading Condition 5 with diagram of anti-symmetrical bending moments and reactions at supports. (For convenience only the horizontal reactions H_n at the feet of the columns are shown.)

Frame Shape 18

Symmetrical single-story five-bay frame with laterally restrained horizontal member, elastically restrained at each end, and four elastically restrained columns

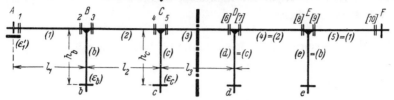

Fig. 140   Frame shape, dimensions and symbols

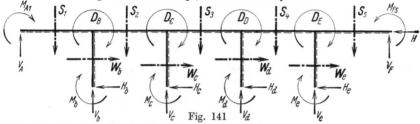

Definition of positive direction for all external loads on joints and members, for all reactions at supports, and for the lateral restraining force

> All the dimensions and coefficients for the right-hand half of the frame are the same as those for the left-hand half. In the case of the columns the broken line is always placed on the right-hand side of each column, despite the symmetry of the frame.

Note: For Frame Shape 18 the *coefficients and the Loading Conditions* 1—3 for *Frame Shape* 13, enlarged by the addition of two bays, may be employed. On account of symmetry, the reduction sequences (437)/(438) and the recursion sequences (439)/(440) respectively become equal. Furthermore, the matrix of influence coefficients (442)—which in the case of Frame Shape 18 would have 10×10 elements—becomes bisymmetrical, i.e., symmetrical with respect to the principal diagonal and the secondary diagonal. If the *"load transposition method"* is employed, it is also possible to start from the following symmetrical and anti-symmetrical loading condition, which in the present case is given for the Method of Forces (Method I)*

> **Important**
>
> All the instructions, formula sequences and formula matrices given for Frame Shape 18 can readily be extended to suit symmetrical frames of similar type having any odd number of bays.

---

* For the treatment of a similar frame by the Deformation Method (Method II) see Frame Shape 16.

a) Symmetrical load arrangements

Coefficients

Flexibility coefficients of all the members, according to formulas (434), for $v = 1, 2, 3$ and $n = b, c$.

Flexibilities of the semi-rigidly fixed columns: just as formulas (435).

Diagonal coefficients:

$$\left. \begin{array}{ll} O_1 = \dfrac{k_1}{\varepsilon_1'} & \begin{array}{lll} O_2 = 2k_1 + b_b & O_4 = 2k_2 + b_c \\ O_3 = b_b + 2k_2 & O_5' = b_c + 3k_3. \end{array} \end{array} \right\} \quad (585)$$

Moment carry-over factors and auxiliary coefficients:

$$\left. \begin{array}{l} \text{backward reduction:} \\ \sigma_{cs} = \dfrac{b_c}{O_5'} \\ \qquad\qquad r_4' = O_4 - b_c \cdot \sigma_{cs} \\ \iota_{2s} = \dfrac{k_2}{r_4'} \\ \qquad\qquad r_3' = O_3 - k_2 \cdot \iota_{2s} \\ \sigma_{bs} = \dfrac{b_b}{r_3'} \\ \qquad\qquad r_2' = O_2 - b_b \cdot \sigma_{bs} \\ \iota_{1s} = \dfrac{k_1}{r_2'}. \end{array} \middle| \begin{array}{l} \text{forward reduction:} \\ \\ \sigma_b' = \dfrac{b_b}{v_2} \qquad \begin{array}{l} v_2 = O_2 - k_1 \cdot \varepsilon_1' \\ v_3 = O_3 - b_b \cdot \sigma_b' \end{array} \\ \iota_2' = \dfrac{k_2}{v_3} \\ \qquad\qquad v_4 = O_4 - k_2 \cdot \iota_2' \\ \sigma_c' = \dfrac{b_c}{v_4} \\ \qquad\qquad v_5' = O_5' - b_c \cdot \sigma_c'. \end{array} \right\} \begin{array}{c} (586) \\ \text{and} \\ (587) \end{array}$$

Principal influence coefficients by recursion:

$$\left. \begin{array}{l} \text{forward:} \\ n_{11}' = \dfrac{1}{k_1(1/\varepsilon_1' - \iota_{1s})} \\ n_{22}' = \dfrac{1}{r_2'} + n_{11}' \cdot \iota_{1s}^2 \\ n_{33}' = \dfrac{1}{r_3'} + n_{22}' \cdot \sigma_{bs}^2 \\ n_{44}' = \dfrac{1}{r_4'} + n_{33}' \cdot \iota_{2s}^2 \\ n_{55}' = \dfrac{1}{O_5'} + n_{44}' \cdot \sigma_{cs}^2. \end{array} \middle| \begin{array}{l} \text{backward:} \\ n_{55}' = \dfrac{1}{v_5'} \\ n_{44}' = \dfrac{1}{v_4} + n_{55}' \cdot \sigma_c'^2 \\ n_{33}' = \dfrac{1}{v_3} + n_{44}' \cdot \iota_2'^2 \\ n_{22}' = \dfrac{1}{v_2} + n_{33}' \cdot \sigma_b'^2 \\ n_{11}' = \dfrac{1}{O_1} + n_{22}' \cdot \varepsilon_1'^2. \end{array} \right\} \begin{array}{c} (588) \\ \text{and} \\ (589) \end{array}$$

Principal influence coefficients direct

$$n'_{11} = \frac{1}{k_1(1/\varepsilon'_1 - \iota_{1s})} \qquad n'_{22} = \frac{1}{r'_2 + v_2 - O_2}$$

$$n'_{33} = \frac{1}{r'_3 + v_3 - O_3} \qquad n'_{44} = \frac{1}{r'_4 + v_4 - O_4} \qquad n'_{55} = \frac{1}{v'_5} \,. \quad \Bigg\} \quad (590)$$

Computation scheme for the symmetrical influence coefficient matrix*

$$
\begin{array}{c|ccccc|c}
 & (n'_{11}) & n'_{12} & n'_{13} & n'_{14} & n'_{15} & \\
\cdot\, \iota_{1s} & \downarrow & \uparrow & \uparrow & \uparrow & \uparrow & \cdot\, \varepsilon'_1 \\
 & n'_{21} & (n'_{22}) & n'_{23} & n'_{24} & n'_{25} & \\
\cdot\, \sigma_{bs} & \downarrow & \downarrow & \uparrow & \uparrow & \uparrow & \cdot\, \sigma'_b \\
 & n'_{31} & n'_{32} & (n'_{33}) & n'_{34} & n'_{35} & \\
\cdot\, \iota_{2s} & \downarrow & \downarrow & \downarrow & \uparrow & \uparrow & \cdot\, \iota'_2 \\
 & n'_{41} & n'_{42} & n'_{43} & (n'_{44}) & n'_{45} & \\
\cdot\, \sigma_{cs} & \downarrow & \downarrow & \downarrow & \downarrow & \uparrow & \cdot\, \sigma'_c \\
 & n'_{51} & n'_{52} & n'_{53} & n'_{54} & (n'_{55}) & \\
\end{array}
\quad \Bigg\} \quad (591)
$$

Composite influence coefficients**

$$
\begin{aligned}
s'_{11} &= +n'_{11} - n'_{12} = (s_{11} + s_{15}) & s'_{12} &= +n'_{13} - n'_{14} = (s_{12} + s_{14}) \\
 & & s'_{22} &= +n'_{23} - n'_{24} = (s_{22} + s_{24}) \\
s'_{21} &= -n'_{21} + n'_{22} = (s_{21} - s_{25}) & s'_{32} &= +n'_{33} - n'_{34} = (s_{32} + s_{34}) \\
s'_{31} &= -n'_{31} + n'_{32} = (s_{31} - s_{35}) & & \\
s'_{41} &= -n'_{41} + n'_{42} = (s_{41} - s_{45}) & s'_{42} &= -n'_{43} + n'_{44} = (s_{42} - s_{44}) \\
s'_{51} &= -n'_{51} + n'_{52} = (s_{51} - s_{55}) & s'_{52} &= -n'_{53} + n'_{54} = (s_{52} - s_{54});
\end{aligned}
\quad \Bigg\} \quad (592)
$$

$$
\begin{aligned}
w'_{1b} &= +n'_{12} - n'_{13} = (w_{1b} + w_{1e}) & w'_{1c} &= +n'_{14} - n'_{15} = (w_{1c} + w_{1d}) \\
w'_{2b} &= +n'_{22} - n'_{23} = (w_{2b} + w_{2e}) & w'_{2c} &= +n'_{24} - n'_{25} = (w_{2c} + w_{2d}) \\
w'_{3b} &= -n'_{32} + n'_{33} = (w_{3b} - w_{3e}) & w'_{3c} &= +n'_{34} - n'_{35} = (w_{3c} + w_{3d}) \\
w'_{4b} &= -n'_{42} + n'_{43} = (w_{4b} - w_{4e}) & w'_{4c} &= +n'_{44} - n'_{45} = (w_{4c} + w_{4d}) \\
w'_{5b} &= -n'_{52} + n'_{53} = (w_{5b} - w_{5e}) & w'_{5c} &= -n'_{54} + n'_{55} = (w_{5c} - w_{5d}).
\end{aligned}
\quad \Bigg\} \quad (593)
$$

---

* Between the elements of the matrix (591) for the left-hand half of the frame and the elements of the 10 × 10 matrix for the whole frame, as indicated on page 135, there exist, according to the "load transposition method", the following relationships:

$$n'_{11} = n_{11} - n_{1,10} \text{ etc., up to } n'_{55} = n_{55} - n_{56}.$$

** The members enclosed in parentheses have been formed with the values s and w, as indicated in formulas (443) and (444) appropriately extended to suit Frame Shape 18.

Loading Condition 1a

Horizontal member carrying vertical loading of arbitrary magnitude symmetrical about center of frame

$$(\mathfrak{R}_5 = \mathfrak{L}_1, \quad \mathfrak{L}_5 = \mathfrak{R}_1; \quad \mathfrak{R}_4 = \mathfrak{L}_2, \quad \mathfrak{L}_4 = \mathfrak{R}_2; \quad \mathfrak{R}_3 = \mathfrak{L}_3)$$

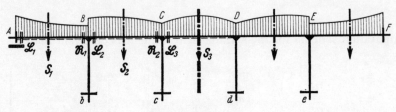

Fig. 142

Beam loading diagram with load terms $\mathfrak{L}$ and $\mathfrak{R}$ for left-hand half of frame

Beam moments ($L_\nu = \mathfrak{L}_\nu k_\nu$ and $R_\nu = \mathfrak{R}_\nu k_\nu$):

	L_1	R_1	L_2	R_2	L_3
$X'_1 = M_{A1} = M_{F5} =$	$-n'_{11}$	$+n'_{12}$	$+n'_{13}$	$-n'_{14}$	$-n'_{15}$
$X'_2 = M_{B1} = M_{E5} =$	$+n'_{21}$	$-n'_{22}$	$-n'_{23}$	$+n'_{24}$	$+n'_{25}$
$X'_3 = M_{B2} = M_{E4} =$	$+n'_{31}$	$-n'_{32}$	$-n'_{33}$	$+n'_{34}$	$+n'_{35}$
$X'_4 = M_{C2} = M_{D4} =$	$-n'_{41}$	$+n'_{42}$	$+n'_{43}$	$-n'_{44}$	$-n'_{45}$
$X'_5 = M_{C3} = M_{D3} =$	$-n'_{51}$	$+n'_{52}$	$+n'_{53}$	$-n'_{54}$	$-n'_{55}.$

(594)

Column moments:

at top:

$$M_{Bb} = -M_{Ee} = -X'_2 + X'_3$$
$$M_{Cc} = -M_{Dd} = -X'_4 + X'_5;$$

at base:

$$M_b = -M_e = -M_{Bb} \cdot \varepsilon_b$$
$$M_c = -M_d = -M_{Cc} \cdot \varepsilon_c.$$

(595) and (596)

Special case: symmetrical bay loading ($R_1 = L_1$ and $R_2 = L_2$):

$$\begin{aligned}
X'_1 &= M_{A1} = M_{F5} = -L_1 \cdot s'_{11} + L_2 \cdot s'_{12} - L_3 \cdot n'_{15} \\
X'_2 &= M_{B1} = M_{E5} = -L_1 \cdot s'_{21} - L_2 \cdot s'_{22} + L_3 \cdot n'_{25} \\
X'_3 &= M_{B2} = M_{E4} = -L_1 \cdot s'_{31} - L_2 \cdot s'_{32} + L_3 \cdot n'_{35} \\
X'_4 &= M_{C2} = M_{D4} = +L_1 \cdot s'_{41} - L_2 \cdot s'_{42} - L_3 \cdot n'_{45} \\
X'_5 &= M_{C3} = M_{D3} = +L_1 \cdot s'_{51} - L_2 \cdot s'_{52} - L_3 \cdot n'_{55}.
\end{aligned}$$

(597)

Loading Conditions 2a and 3a

All the columns carrying arbitrary horizontal loading (2a) and external rotational moments applied at all the beam joints (3a), but load arrangement as a whole is symmetrical about center of frame

$$(\mathfrak{L}_e = -\mathfrak{L}_b, \quad \mathfrak{R}_e = -\mathfrak{R}_b; \quad \mathfrak{L}_d = -\mathfrak{L}_c, \quad \mathfrak{R}_d = -\mathfrak{R}_c; \quad D_E = -D_B, \quad D_D = -D_C)$$

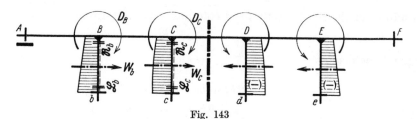

Fig. 143

Column loading diagram with load terms $\mathfrak{L}$ and $\mathfrak{R}$ for the columns, and external rotational moments D, for left-hand half of frame

Joint load terms:

$$(L_n = \mathfrak{L}_n k_n \quad \text{and} \quad R_n = \mathfrak{R}_n k_n \quad \text{for} \quad n = b, c);$$

$$\left. \begin{array}{l} \mathfrak{B}_B = D_B \cdot b_b + L_b \cdot \varepsilon_b - R_b \\ \mathfrak{B}_C = D_C \cdot b_c + L_c \cdot \varepsilon_c - R_c . \end{array} \right\} \quad (598)$$

Beam moments:

$$\left. \begin{array}{l} X'_1 = M_{A1} = M_{F5} = +\mathfrak{B}_B \cdot w'_{1b} - \mathfrak{B}_C \cdot w'_{1c} \\ X'_2 = M_{B1} = M_{E5} = -\mathfrak{B}_B \cdot w'_{2b} + \mathfrak{B}_C \cdot w'_{2c} \\ X'_3 = M_{B2} = M_{E4} = +\mathfrak{B}_B \cdot w'_{3b} + \mathfrak{B}_C \cdot w'_{3c} \\ X'_4 = M_{C2} = M_{D4} = -\mathfrak{B}_B \cdot w'_{4b} - \mathfrak{B}_C \cdot w'_{4c} \\ X'_5 = M_{C3} = M_{D3} = -\mathfrak{B}_B \cdot w'_{5b} + \mathfrak{B}_C \cdot w'_{5c} . \end{array} \right\} \quad (599)$$

Column moments:

at top:
$$\left. \begin{array}{l} M_{Bb} = -M_{Ee} = -D_B - X'_2 + X'_3 \\ M_{Cc} = -M_{Dd} = -D_C - X'_4 + X'_5; \end{array} \right\} \quad (600)$$

at base:
$$\left. \begin{array}{l} M_b = -M_e = -(\mathfrak{L}_b + M_{Bb})\,\varepsilon_b \\ M_c = -M_d = -(\mathfrak{L}_c + M_{Cc})\,\varepsilon_c . \end{array} \right\} \quad (601)$$

b) Anti-symmetrical load arrangements.

Coefficients

Flexibility coefficients of all the members, according to formulas (434), for $v = 1, 2, 3$ and $n = b, c$.

Flexibilities of the semi-rigidly fixed columns: just as formulas (435).

Diagonal coefficients:

$$\left. \begin{array}{lll} O_1 = \dfrac{k_1}{\varepsilon_1'} & O_2 = 2k_1 + b_b & O_4 = 2k_2 + b_c \\ & O_3 = b_b + 2k_2 & O_5'' = b_c + k_3 . \end{array} \right\} \quad (602)$$

Moment carry-over factors and auxiliary coefficients:

$$\left. \begin{array}{l|l} \text{backward reduction:} & \text{forward reduction:} \\[4pt] \begin{array}{ll} \sigma_{ct} = \dfrac{b_c}{O_5''} & \\ & r_4'' = O_4 - b_c \cdot \sigma_{ct} \\ \iota_{2t} = \dfrac{k_2}{r_4''} & \\ & r_3'' = O_3 - k_2 \cdot \iota_{2t} \\ \sigma_{bt} = \dfrac{b_b}{r_3''} & \\ & r_2'' = O_2 - b_b \cdot \sigma_{bt} \\ \iota_{1t} = \dfrac{k_1}{r_2''} . & \end{array} & \begin{array}{ll} & v_2 = O_2 - k_1 \cdot \varepsilon_1' \\ \sigma_b' = \dfrac{b_b}{v_2} & \\ & v_3 = O_3 - b_b \cdot \sigma_b' \\ \iota_2' = \dfrac{k_2}{v_3} & \\ & v_4 = O_4 - k_2 \cdot \iota_2' \\ \sigma_c' = \dfrac{b_c}{v_4} & \\ & v_5'' = O_5'' - b_c \cdot \sigma_c' . \end{array} \end{array} \right\} \quad \begin{array}{c} (603) \\ \text{and} \\ (604) \end{array}$$

Principal influence coefficients by recursion:

$$\left. \begin{array}{l|l} \text{forward:} & \text{backward:} \\[4pt] n_{11}'' = \dfrac{1}{k_1(1/\varepsilon_1' - \iota_{1t})} & n_{55}'' = \dfrac{1}{v_5''} \\[4pt] n_{22}'' = \dfrac{1}{r_2''} + n_{11}'' \cdot \iota_{1t}^2 & n_{44}'' = \dfrac{1}{v_4} + n_{55}'' \cdot \sigma_c'^2 \\[4pt] n_{33}'' = \dfrac{1}{r_3''} + n_{22}'' \cdot \sigma_{bt}^2 & n_{33}'' = \dfrac{1}{v_3} + n_{44}'' \cdot \iota_2'^2 \\[4pt] n_{44}'' = \dfrac{1}{r_4''} + n_{33}'' \cdot \iota_{2t}^2 & n_{22}'' = \dfrac{1}{v_2} + n_{33}'' \cdot \sigma_b'^2 \\[4pt] n_{55}'' = \dfrac{1}{O_5''} + n_{44}'' \cdot \sigma_{ct}^2 . & n_{11}'' = \dfrac{1}{O_1} + n_{22}'' \cdot \varepsilon_1'^2 . \end{array} \right\} \quad \begin{array}{c} (605) \\ \text{and} \\ (606) \end{array}$$

Principal influence coefficients direct

$$n_{11}'' = \frac{1}{k_1(1/\varepsilon_1' - \iota_{1t})} \qquad n_{22}'' = \frac{1}{r_2'' + v_2 - O_2}$$

$$n_{33}'' = \frac{1}{r_3'' + v_3 - O_3} \qquad n_{44}'' = \frac{1}{r_4'' + v_4 - O_4} \qquad n_{55}'' = \frac{1}{v_5''} \cdot \qquad (607)$$

Computation scheme for the symmetrical influence coefficient matrix*

$$
\begin{array}{c|ccccc|c}
 & (n_{11}'') & n_{12}'' & n_{13}'' & n_{14}'' & n_{15}'' & \\
\cdot\,\iota_{1t} & \downarrow & \uparrow & \uparrow & \uparrow & \uparrow & \cdot\,\varepsilon_1' \\
 & n_{21}'' & (n_{22}'') & n_{23}'' & n_{24}'' & n_{25}'' & \\
\cdot\,\sigma_{bt} & \downarrow & \downarrow & \uparrow & \uparrow & \uparrow & \cdot\,\sigma_b' \\
 & n_{31}'' & n_{32}'' & (n_{33}'') & n_{34}'' & n_{35}'' & \\
\cdot\,\iota_{2t} & \downarrow & \downarrow & \downarrow & \uparrow & \uparrow & \cdot\,\iota_2' \\
 & n_{41}'' & n_{42}'' & n_{43}'' & (n_{44}'') & n_{45}'' & \\
\cdot\,\sigma_{ct} & \downarrow & \downarrow & \downarrow & \downarrow & \uparrow & \cdot\,\sigma_c' \\
 & n_{51}'' & n_{52}'' & n_{53}'' & n_{54}'' & (n_{55}'') & \\
\end{array} \qquad (608)
$$

Composite influence coefficients**

$$s_{11}'' = + n_{11}'' - n_{12}'' = (s_{11} - s_{15}) \qquad s_{12}'' = + n_{13}'' - n_{14}'' = (s_{12} - s_{14})$$

$$s_{22}'' = + n_{23}'' - n_{24}'' = (s_{22} - s_{24})$$

$$s_{21}'' = - n_{21}'' + n_{22}'' = (s_{21} + s_{25}) \qquad s_{32}'' = + n_{33}'' - n_{34}'' = (s_{32} - s_{34})$$

$$s_{31}'' = - n_{31}'' + n_{32}'' = (s_{31} + s_{35})$$

$$s_{41}'' = - n_{41}'' + n_{42}'' = (s_{41} + s_{45}) \qquad s_{42}'' = - n_{43}'' + n_{44}'' = (s_{42} + s_{44}) \qquad (609)$$

$$s_{51}'' = - n_{51}'' + n_{52}'' = (s_{51} + s_{55}) \qquad s_{52}'' = - n_{53}'' + n_{54}'' = (s_{52} + s_{54});$$

$$w_{1b}'' = + n_{12}'' - n_{13}'' = (w_{1b} - w_{1e}) \qquad w_{1c}'' = + n_{14}'' - n_{15}'' = (w_{1c} - w_{1d})$$

$$w_{2b}'' = + n_{22}'' - n_{23}'' = (w_{2b} - w_{2e}) \qquad w_{2c}'' = + n_{24}'' - n_{25}'' = (w_{2c} - w_{2d})$$

$$w_{3b}'' = - n_{32}'' + n_{33}'' = (w_{3b} + w_{3e}) \qquad w_{3c}'' = + n_{34}'' - n_{35}'' = (w_{3c} - w_{3d}) \qquad (610)$$

$$w_{4b}'' = - n_{42}'' + n_{43}'' = (w_{4b} + w_{4e}) \qquad w_{4c}'' = + n_{44}'' - n_{45}'' = (w_{4c} - w_{4d})$$

$$w_{5b}'' = - n_{52}'' + n_{53}'' = (w_{5b} + w_{5e}) \qquad w_{5c}'' = - n_{54}'' + n_{55}'' = (w_{5c} + w_{5d}).$$

---

* Between the elements of the matrix (608) for the left-hand half of the frame and the elements of the 10 × 10 matrix for the whole frame, as indicated on page 135, there exist, according to the "load transposition method", the following relationships:

$$n_{11}'' = n_{11} + n_{1,10} \text{ etc., up to } n_{55}'' = n_{55} + n_{56}.$$

** The members enclosed in parentheses have been formed with the values s and w, as indicated in formulas (443) and (444) appropriately extended to suit Frame Shape 18.

Loading Condition 1b

Horizontal member carrying vertical loading of arbitrary magnitude, but anti-symmetrical about center of frame

$$(\mathfrak{R}_5 = -\mathfrak{L}_1, \quad \mathfrak{L}_5 = -\mathfrak{R}_1; \quad \mathfrak{R}_4 = -\mathfrak{L}_2, \quad \mathfrak{L}_4 = -\mathfrak{R}_2; \quad \mathfrak{R}_3 = -\mathfrak{L}_3)$$

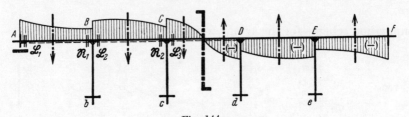

Fig. 144
Beam loading diagram with load terms $\mathfrak{L}$ and $\mathfrak{R}$ for left-hand half of frame

Beam moments ($L_\nu = \mathfrak{L}_\nu k_\nu$ and $R_\nu = \mathfrak{R}_\nu k_\nu$):

	L_1	R_1	L_2	R_2	L_3
$X_1'' = M_{A1} = -M_{F5} =$	$-n_{11}''$	$+n_{12}''$	$+n_{13}''$	$-n_{14}''$	$-n_{15}''$
$X_2'' = M_{B1} = -M_{E5} =$	$+n_{21}''$	$-n_{22}''$	$-n_{23}''$	$+n_{24}''$	$+n_{25}''$
$X_3'' = M_{B2} = -M_{E4} =$	$+n_{31}''$	$-n_{32}''$	$-n_{33}''$	$+n_{34}''$	$+n_{35}''$
$X_4'' = M_{C2} = -M_{D4} =$	$-n_{41}''$	$+n_{42}''$	$+n_{43}''$	$-n_{44}''$	$-n_{45}''$
$X_5'' = M_{C3} = -M_{D3} =$	$-n_{51}''$	$+n_{52}''$	$+n_{53}''$	$-n_{54}''$	$-n_{55}''.$

(611)

Column moments:

at top: $\qquad\qquad$ at base:

$$M_{Bb} = M_{Ee} = -X_2'' + X_3'' \qquad M_b = M_e = -M_{Bb} \cdot \varepsilon_b$$
$$M_{Cc} = M_{Dd} = -X_4'' + X_5''; \qquad M_c = M_d = -M_{Cc} \cdot \varepsilon_c.$$

(612) and (613)

Special case: symmetrical bay loading ($R_1 = L_1$ and $R_2 = L_2$):

$$\begin{aligned}
X_1'' = M_{A1} &= -M_{F5} = -L_1 \cdot s_{11}'' + L_2 \cdot s_{12}'' \\
X_2'' = M_{B1} &= -M_{E5} = -L_1 \cdot s_{21}'' - L_2 \cdot s_{22}'' \\
X_3'' = M_{B2} &= -M_{E4} = -L_1 \cdot s_{31}'' - L_2 \cdot s_{32}'' \\
X_4'' = M_{C2} &= -M_{D4} = +L_1 \cdot s_{41}'' - L_2 \cdot s_{42}'' \\
X_5'' = M_{C3} &= -M_{D3} = +L_1 \cdot s_{51}'' - L_2 \cdot s_{52}''.
\end{aligned}$$

(614)

Loading Conditions 2b and 3b

All the columns carrying arbitrary horizontal loading (2b) and external rotational moments applied to all the beam joints (3b), but load arrangement as a whole is anti-symmetrical about center of frame

$$(\mathfrak{L}_e = \mathfrak{L}_b, \quad \mathfrak{R}_e = \mathfrak{R}_b; \quad \mathfrak{L}_d = \mathfrak{L}_c, \quad \mathfrak{R}_d = \mathfrak{R}_c; \quad D_E = D_B, \quad D_D = D_C)$$

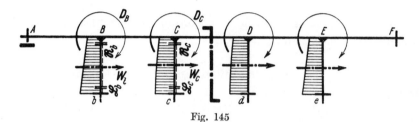

Fig. 145

Column loading diagram with load terms $\mathfrak{L}$ and $\mathfrak{R}$ for the columns, and external rotational moments D, for left-hand half of frame

Joint load terms:

$$(L_n = \mathfrak{L}_n k_n \quad \text{and} \quad R_n = \mathfrak{R}_n k_n \quad \text{for} \quad n = b, c);$$

$$\left. \begin{array}{l} \mathfrak{B}_B = D_B \cdot b_b + L_b \cdot \varepsilon_b - R_b \\ \mathfrak{B}_C = D_C \cdot b_c + L_c \cdot \varepsilon_c - R_c \end{array} \right\} \quad (615)$$

Beam moments:

$$\left. \begin{array}{l} X_1'' = M_{A1} = -M_{F5} = +\mathfrak{B}_B \cdot w_{1b}'' - \mathfrak{B}_C \cdot w_{1c}'' \\ X_2'' = M_{B1} = -M_{E5} = -\mathfrak{B}_B \cdot w_{2b}'' + \mathfrak{B}_C \cdot w_{2c}'' \\ X_3'' = M_{B2} = -M_{E4} = +\mathfrak{B}_B \cdot w_{3b}'' + \mathfrak{B}_C \cdot w_{3c}'' \\ X_4'' = M_{C2} = -M_{D4} = -\mathfrak{B}_B \cdot w_{4b}'' - \mathfrak{B}_C \cdot w_{4c}'' \\ X_5'' = M_{C3} = -M_{D3} = -\mathfrak{B}_B \cdot w_{5b}'' + \mathfrak{B}_C \cdot w_{5c}'' \end{array} \right\} \quad (616)$$

Column moments:

at top:
$$\left. \begin{array}{l} M_{Lb} = M_{Ee} = -D_B - X_2'' + X_3'' \\ M_{Cc} = M_{Dd} = -D_C - X_4'' + X_5'' ; \end{array} \right\} \quad (617)$$

at base:
$$\left. \begin{array}{l} M_b = M_e = -(\mathfrak{L}_b + M_{Eb}) \varepsilon_b \\ M_c = M_d = -(\mathfrak{L}_c + M_{Cc}) \varepsilon_c . \end{array} \right\} \quad (618)$$

(Frame Shape 18—cont.)

Horizontal thrusts, column shears and lateral restraining force for all loading conditions

For asymmetrical loading conditions, according to formulas (179)/(180) or (181)/(182), for $n = b, c, d, e$ and $N = B, C, D, E$.

For symmetrical loading conditions, according to formulas (183)/(184) or (185)/(186), for $n = b, c$ and $N = B, C$.

For anti-symmetrical loading conditions, according to formulas (187)/(188) or (189)/(190) for $n = b, c$ and $N = B, C$.

Frame Shape 18v

(Frame Shape 18 with laterally unrestrained horizontal member)

Procedure

either similar to procedure for Frame Shape 1v, see page 9/10, for the collective subscripts $\nu = 1 - 5$, $n = b - e$ and $N = B - E$, having due regard to anti-symmetry properties or, alternatively, similar to procedure for Frame Shape 5v, see pages 42-43 for the collective subscripts $\nu = 1, 2, 3$; $n = b, c$ and $N = B, C$.

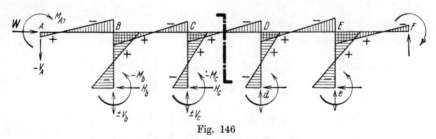

Fig. 146

Loading Condition 5 with diagram of bending moments and reactions at supports.

Frame Shape 19

Symmetrical single-story four-bay frame with laterally restrained horizontal member, elastically restrained at each end, and three elastically restrained columns

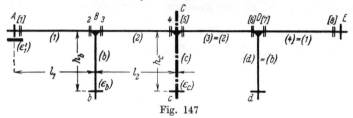

Fig. 147
Frame shape, dimensions and symbols

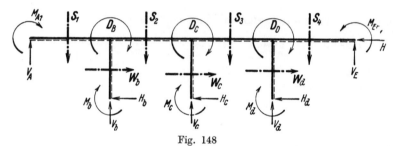

Fig. 148
Definition of positive direction for all external loads on joints and members, for all reactions at supports, and for the lateral restraining force

> All the dimensions and coefficients for the right-hand half of the frame are the same as those for the left-hand half. In the case of the columns the broken line is always placed on the right-hand side of each column, despite the symmetry of the frame.

Note: For Frame Shape 19 the formulas for Frame Shape 12, or alternatively the formulas for Frame Shape 13 enlarged by the addition of one bay, or alternatively the formulas for Frame Shape 11 reduced by the removal of one bay, may be employed. On account of symmetry, the backward and forward reduction become identical, as also do the forward and backward recursion. Furthermore, the square matrix of influence coefficients (with 6×6 elements in the case of Frame Shape 12, 8×8 elements in the case of Frame Shape 13, and 3×3 elements in the case of Frame Shape 11) becomes bisymmetrical, i.e., symmetrical with respect to the principal diagonal and the secondary diagonal. If the *"load transposition method"* is employed, it is also possible to start from the following symmetrical and anti-symmetrical loading conditions.

I. Treatment by the Method of Forces

a) Symmetrical load arrangements

Coefficients

Flexibility coefficients of all the members, according to formulas (414), for $\nu = 1, 2$ and $n = b, c$.

Flexibilities of the elastically restrained members:

$$b_1 = k_1(2 - \varepsilon_1'); \qquad b_b = k_b(2 - \varepsilon_b) \qquad b_c = k_c(2 - \varepsilon_c). \tag{619}$$

Diagonal coefficients:

$$O_2 = b_1 + b_b \qquad O_3 = b_b + 2k_2; \qquad (O_4' = 2k_2). \tag{620}$$

Moment carry-over factors and auxiliary coefficients:

$$
\begin{array}{l|l}
\text{backward reduction:} & \text{forward reduction:} \\
\iota_{2s} = \dfrac{1}{2} & v_2 = O_2 \\
\qquad r_3' = O_3 - k_2 \cdot \dfrac{1}{2} & \sigma_b' = \dfrac{b_b}{v_2} \\
\sigma_{bs} = \dfrac{b_b}{r_3'} & \qquad v_3 = O_3 - b_b \cdot \sigma_b' \\
\qquad r_2' = O_2 - b_b \cdot \sigma_{bs}. & \iota_2' = \dfrac{k_2}{v_3}.
\end{array}
\qquad
\begin{array}{l}
(621) \\
\text{and} \\
(622)
\end{array}
$$

Principal influence coefficients by recursion:

$$
\begin{array}{l|l}
\text{forward:} & \text{backward:} \\
n_{22}' = 1/r_2' & n_{44}' = 1/k_2(2 - \iota_2') \\
n_{33}' = 1/r_3' + n_{22}' \cdot \sigma_{os}^2 & n_{33}' = 1/v_3 + n_{44}' \cdot \iota_2'^2 \\
n_{44}' = 1/O_4' + n_{33}' \cdot 1/4. & n_{22}' = 1/v_2 + n_{33}' \cdot \sigma_b'^2.
\end{array}
\qquad
\begin{array}{l}
(623) \\
\text{and} \\
(624)
\end{array}
$$

Principal influence coefficients direct

$$n_{22}' = \frac{1}{r_2'} \qquad n_{33}' = \frac{1}{r_3' + v_3 - O_3} \qquad n_{44}' = \frac{1}{k_2(2 - \iota_2')}. \tag{625}$$

Computation scheme for the symmetrical influence coefficient matrix[*]

$$
\begin{array}{c}
\cdot \sigma_{bs} \\
\cdot \dfrac{1}{2}
\end{array}
\left|
\begin{array}{ccc}
(n_{22}') & n_{23}' & n_{24}' \\
\downarrow & \uparrow & \uparrow \\
n_{32}' & (n_{33}') & n_{34}' \\
\downarrow & \downarrow & \uparrow \\
n_{42}' & n_{43}' & (n_{44}')
\end{array}
\right|
\begin{array}{c}
\cdot \sigma_b' \\
\cdot \iota_2'
\end{array}
\qquad (626)
$$

---

[*] Between the elements of the matrix (626) for the left-hand half of the frame and the elements of the matrix (422) for the whole frame, as indicated on page 96, there exist according to the "load transposition method", the following relationships:

$$n_{22}' = n_{22} + n_{27}, \quad n_{23}' = n_{23} + n_{26}, \quad \text{etc., up to} \quad n_{44}' = n_{44} + n_{45}.$$

Loading Condition 1a

Horizontal member carrying vertical loading of arbitrary magnitude, but symmetrical about center of frame

$$(\mathfrak{R}_4 = \mathfrak{L}_1, \quad \mathfrak{L}_4 = \mathfrak{R}_1; \quad \mathfrak{R}_3 = \mathfrak{L}_2, \quad \mathfrak{L}_3 = \mathfrak{R}_2)$$

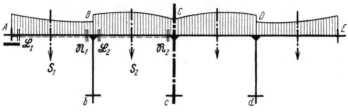

Fig. 149

Beam loading diagram with load terms $\mathfrak{L}$ and $\mathfrak{R}$ for left-hand half of frame

beam load terms:

$$L_\nu = \mathfrak{L}_\nu k_\nu \quad \text{and} \quad R_\nu = \mathfrak{R}_\nu k_\nu \quad \text{for} \quad \nu = 1, 2; \qquad \mathfrak{B}_1 = R_1 - L_1 \cdot \varepsilon_1'. \qquad (627)$$

Beam moments:

$$\left. \begin{array}{l} X_2' = M_{B1} = M_{D4} = -\mathfrak{B}_1 \cdot n_{22}' - L_2 \cdot n_{23}' + R_2 \cdot n_{24}' \\ X_3' = M_{B2} = M_{D3} = -\mathfrak{B}_1 \cdot n_{32}' - L_2 \cdot n_{33}' + R_2 \cdot n_{34}' \\ X_4' = M_{C2} = M_{C3} = +\mathfrak{B}_1 \cdot n_{42}' + L_2 \cdot n_{43}' - R_2 \cdot n_{44}'; \\ M_{A1} = M_{E4} = -(\mathfrak{L}_1 + M_{B1})\varepsilon_1'. \end{array} \right\} \qquad (628)$$

Column moments:

at top: $\qquad M_{Bb} = -M_{Dd} = -X_2' + X_3' \qquad M_{Cc} = 0; \qquad (629)$

at base: $\qquad M_b \;\; = -M_d \;\; = -M_{Eb} \cdot \varepsilon_b \qquad M_c \;\; = 0. \qquad (630)$

Special case: symmetrical bay loading ($R_\nu = L_\nu$):

$$\left. \begin{array}{l} X_2' = M_{B1} = M_{D4} = -L_1(1 - \varepsilon_1') \cdot n_{22}' - L_2 \cdot s_{22}' \\ X_3' = M_{B2} = M_{D3} = -L_1(1 - \varepsilon_1') \cdot n_{32}' - L_2 \cdot s_{32}' \\ X_4' = M_{C2} = M_{C3} = +L_1(1 - \varepsilon_1') \cdot n_{42}' - L_2 \cdot s_{42}'; \end{array} \right\} \qquad (631)$$

where*

$$\left. \begin{array}{ll} s_{22}' = & n_{24}' = (s_{22} - s_{23}) \\ s_{32}' = & n_{34}' = (s_{32} - s_{33}) \\ s_{42}' = -n_{43}' + n_{44}' = (s_{42} + s_{43}). \end{array} \right\} \qquad (632)$$

---

* The members enclosed in parentheses in (632) have been formed with the values s as indicated in (423).

Loading Conditions 2a and 3a

All the columns carrying arbitrary horizontal loading (2a) and external rotational moments applied at all the beam joints (3a), but load arrangement as a whole is symmetrical about center of frame

$$(\mathfrak{L}_d = -\mathfrak{L}_b, \quad \mathfrak{R}_d = -\mathfrak{R}_b; \quad \boldsymbol{D}_D = -\boldsymbol{D}_B)$$

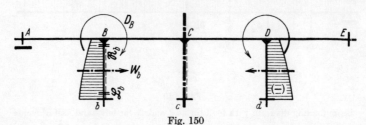

Fig. 150

Column loading diagram with load terms $\mathfrak{L}$ and $\mathfrak{R}$ for the columns, and external rotational moments D, for left-hand half of frame. (Symmetrical loading of central column c and central joint C not shown.)

Joint load terms:

$$\mathfrak{B}_B = \boldsymbol{D}_B \cdot b_b + (\mathfrak{L}_b \cdot \varepsilon_b - \mathfrak{R}_b) k_b. \tag{633}$$

Beam moments:

$$\left.\begin{aligned} X'_2 &= M_{B1} = M_{D4} = -\mathfrak{B}_B \cdot w'_{2b} \\ X'_3 &= M_{B2} = M_{D3} = +\mathfrak{B}_B \cdot w'_{3b} \\ X'_4 &= M_{C2} = M_{C3} = -\mathfrak{B}_B \cdot w'_{4b}; \\ M_{A1} &= M_{E4} = -M_{B1} \cdot \varepsilon'_1; \end{aligned}\right\} \tag{634}$$

where *

$$\left.\begin{aligned} w'_{2b} &= +n'_{22} - n'_{23} = (w_{2b} - w_{2d}) \\ w'_{3b} &= -n'_{32} + n'_{33} = (w_{3b} + w_{3d}) \\ w'_{4b} &= -n'_{42} + n'_{43} = (w_{4b} + w_{4d}). \end{aligned}\right\} \tag{635}$$

Column moments:

at top: $\qquad M_{Eb} = -M_{Dd} = -\boldsymbol{D}_B - X'_2 + X'_3 \qquad M_{Cc} = 0; \qquad (636)$

at base: $\qquad M_b = -M_d = -(\mathfrak{L}_b + M_{Bb})\varepsilon_b \qquad M_c = 0. \qquad (637)$

---

* The members enclosed in parentheses in (635) have been formed with the values w as indicated in (424).

b) Anti-symmetrical load arrangements
Coefficients

Flexibility coefficients and flexibilities: just as on page 146.

Diagonal coefficients:

$$O_2 = b_1 + b_b \qquad O_3 = b_b + 2k_2 \qquad O_4'' = 2k_2 + 2b_c. \tag{638}$$

Moment carry-over factors and auxiliary coefficients:

backward reduction:

$$\iota_{2t} = \frac{k_2}{r_4''}$$
$$\sigma_{bt} = \frac{b_b}{r_3''}$$

$$r_4'' = O_4''$$
$$r_3'' = O_3 - k_2 \cdot \iota_{2t}$$
$$r_2'' = O_2 - b_b \cdot \sigma_{bt}.$$

forward reduction:

$$v_2 = O_2$$
$$\sigma_b' = \frac{b_b}{v_2}$$
$$v_3 = O_3 - b_b \cdot \sigma_b'$$
$$\iota_2' = \frac{k_2}{v_3}$$
$$v_4'' = O_4'' - k_2 \cdot \iota_2'.$$

(639) and (640)

Principal influence coefficients by recursion:

forward:

$$n_{22}'' = 1/r_2''$$
$$n_{33}'' = 1/r_3'' + n_{22}'' \cdot \sigma_{bt}^2$$
$$n_{44}'' = 1/r_4'' + n_{33}'' \cdot \iota_{2t}^2.$$

backward:

$$n_{44}'' = 1/v_4''$$
$$n_{33}'' = 1/v_3 + n_{44}'' \cdot \iota_2'^2$$
$$n_{22}'' = 1/v_2 + n_{33}'' \cdot \sigma_b'^2.$$

(641) and (642)

Principal influence coefficients direct

$$n_{22}'' = \frac{1}{r_2''} \qquad n_{33}'' = \frac{1}{r_3'' + v_3 - O_3} \qquad n_{44}'' = \frac{1}{v_4''}. \tag{643}$$

Computation scheme for the symmetrical influence coefficient matrix*

$$\begin{array}{c} \\ \cdot \sigma_{bt} \\ \cdot \iota_{2t} \end{array} \left| \begin{array}{ccc} (n_{22}'') & n_{23}'' & n_{24}'' \\ \downarrow & \uparrow & \uparrow \\ n_{32}'' & (n_{33}'') & n_{34}'' \\ \downarrow & \downarrow & \uparrow \\ n_{42}'' & n_{43}'' & (n_{44}'') \end{array} \right| \begin{array}{c} \\ \cdot \sigma_b' \\ \cdot \iota_2' \end{array} \right\} \tag{644}$$

---

* Between the elements of the matrix (644) for the left-hand half of the frame and the elements of the matrix (422) for the whole frame, as indicated on page 96, there exist, according to the "load transposition method", the following relationships:
$$n_{22}'' = n_{22} - n_{27}, \quad n_{23}'' = n_{23} - n_{26} \text{ etc., up to } n_{44}'' = n_{44} - n_{45}.$$

Loading Condition 1b

Horizontal member carrying vertical loading of arbitrary magnitude, but anti-symmetrical about center of frame

$$(\mathfrak{R}_4 = -\mathfrak{L}_1, \quad \mathfrak{L}_4 = -\mathfrak{R}_1; \quad \mathfrak{R}_3 = -\mathfrak{L}_2, \quad \mathfrak{L}_3 = -\mathfrak{R}_2)$$

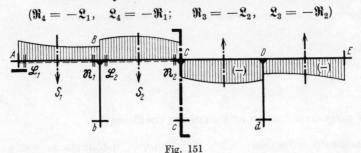

Fig. 151

Beam loading diagram with load terms $\mathfrak{L}$ and $\mathfrak{R}$ for left-hand half of frame

Beam load terms:

$$L_\nu = \mathfrak{L}_\nu k_\nu \quad \text{and} \quad R_\nu = \mathfrak{R}_\nu k_\nu \quad \text{for} \quad \nu = 1, 2; \quad \mathfrak{B}_1 = R_1 - L_1 \cdot \varepsilon_1'. \tag{645}$$

Beam moments:

$$\left.\begin{aligned}
X_2'' &= M_{B1} = -M_{D4} = -\mathfrak{B}_1 \cdot n_{22}'' - L_2 \cdot n_{23}'' + R_2 \cdot n_{24}'' \\
X_3'' &= M_{B2} = -M_{D3} = -\mathfrak{B}_1 \cdot n_{32}'' - L_2 \cdot n_{33}'' + R_2 \cdot n_{34}'' \\
X_4'' &= M_{C2} = -M_{C3} = +\mathfrak{B}_1 \cdot n_{42}'' + L_2 \cdot n_{43}'' - R_2 \cdot n_{44}''; \\
M_{A1} &= -M_{E4} = -(\mathfrak{L}_1 + M_{B1})\varepsilon_1'.
\end{aligned}\right\} \tag{646}$$

Column moments:

at top: $\qquad M_{Bb} = M_{Dd} = -X_2'' + X_3'' \qquad M_{Cc} = -2X_4''; \tag{647}$

at base: $\qquad M_b = M_d = -M_{Lb} \cdot \varepsilon_b \qquad M_c = -M_{Cc} \cdot \varepsilon_c. \tag{648}$

Special case: symmetrical bay loading $R_\nu = L_\nu$:

$$\left.\begin{aligned}
X_2'' &= M_{B1} = -M_{D4} = -L_1(1 - \varepsilon_1') \cdot n_{22}'' - L_2 \cdot s_{22}'' \\
X_3'' &= M_{B2} = -M_{D3} = -L_1(1 - \varepsilon_1') \cdot n_{32}'' - L_2 \cdot s_{32}'' \\
X_4'' &= M_{C2} = -M_{C3} = +L_1(1 - \varepsilon_1') \cdot n_{42}'' - L_2 \cdot s_{42}'';
\end{aligned}\right\} \tag{649}$$

where*

$$\left.\begin{aligned}
s_{22}'' &= +n_{23}'' - n_{24}'' = (s_{22} + s_{23}) \\
s_{32}'' &= +n_{33}'' - n_{34}'' = (s_{32} + s_{33}) \\
s_{42}'' &= -n_{43}'' + n_{44}'' = (s_{42} - s_{43}).
\end{aligned}\right\} \tag{650}$$

---

* The members enclosed in parentheses in (650) have been formed with the values s as indicated in (423).

Loading Conditions 2b and 3b

All the columns carrying arbitrary horizontal loading (2b) and external rotational moments applied to all the beam joints (3b), but load arrangement as a whole is anti-symmetrical about center of frame

$$(\mathfrak{L}_d = \mathfrak{L}_b, \quad \mathfrak{R}_d = \mathfrak{R}_b; \quad D_D = D_B)$$

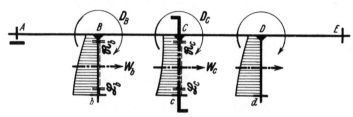

Fig. 152

Column loading diagram with load terms $\mathfrak{L}$ and $\mathfrak{R}$ for the columns, and external rotational moments D, for left-hand half of frame including the central column and central joint

Joint load terms ($L_n = \mathfrak{L}_n k_n$ and $R_n = \mathfrak{R}_n k_n$):

$$\mathfrak{B}_B = D_B \cdot b_b + L_b \cdot \varepsilon_b - R_b \qquad \mathfrak{B}_C'' = (D_C \cdot b_c + L_c \cdot \varepsilon_c - R_c)/2. \qquad (651)$$

Beam moments:

$$\left. \begin{aligned} X_2'' &= M_{B1} = -M_{D4} = -\mathfrak{B}_B \cdot w_{2b}'' + \mathfrak{B}_C'' \cdot n_{24}'' \\ X_3'' &= M_{B2} = -M_{D3} = +\mathfrak{B}_B \cdot w_{3b}'' + \mathfrak{B}_C'' \cdot n_{34}'' \\ X_4'' &= M_{C2} = -M_{C3} = -\mathfrak{B}_B \cdot w_{4b}'' - \mathfrak{B}_C'' \cdot n_{44}''; \\ M_{A1} &= -M_{E4} = -M_{B1} \cdot \varepsilon_1'; \end{aligned} \right\} \qquad (652)$$

where:*

$$\left. \begin{aligned} w_{2b}'' &= +n_{22}'' - n_{23}'' = (w_{2b} + w_{2d}) \\ w_{3b}'' &= -n_{32}'' + n_{33}'' = (w_{3b} - w_{3d}) \\ w_{4b}'' &= -n_{42}'' + n_{43}'' = (w_{4b} - w_{4d}). \end{aligned} \right\} \qquad (653)$$

Column moments:

at top: $\quad M_{Bb} = M_{Dd} = -D_B - X_2'' + X_3'' \quad M_{Cc} = -D_C - 2X_4''; \quad (654)$

at base: $\quad M_b = M_d = -(\mathfrak{L}_b + M_{Bb})\varepsilon_b \quad M_c = -(\mathfrak{L}_c + M_{Cc})\varepsilon_c. \quad (655)$

---

* The members enclosed in parentheses in (653) have been formed with the values w as indicated in (424).

II. Treatment by the Deformation Method

a) Coefficients for symmetrical load arrangements

$$K_1 = \frac{J_1}{J_k} \cdot \frac{l_k}{l_1} \qquad K_2 = \frac{J_2}{J_k} \cdot \frac{l_k}{l_2} \qquad K_b = \frac{J_b}{J_k} \cdot \frac{l_k}{h_b}. \tag{656}$$

$$S_1 = K_1/(2 - \varepsilon_1') \qquad S_b = K_b/(2 - \varepsilon_b); \qquad S_B = 3S_1 + 3S_b + 2K_2. \tag{657}$$

$$\mu_1 = \frac{3S_1}{S_B} \qquad \mu_b = \frac{3S_b}{S_B} \qquad \mu_2 = \frac{2K_2}{S_B}; \qquad (\mu_1 + \mu_b + \mu_2 = 1). \tag{658}$$

Loading Condition 4a

Symmetrical overall loading condition

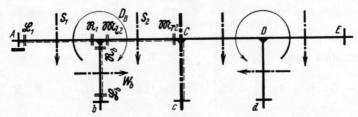

Fig. 153

Loading diagram with symbols indicated for left-hand half of frame. (Symmetrical loading of central column c and central joint C not shown)

Restraining moments of members 1 and b:

$$\mathfrak{E}_{r1} = -(\mathfrak{R}_1 - \mathfrak{L}_1 \cdot \varepsilon_1')/(2 - \varepsilon_1') \qquad \mathfrak{E}_{rb} = -(\mathfrak{R}_b - \mathfrak{L}_b \cdot \varepsilon_b)/(2 - \varepsilon_b). \tag{659}$$

Joint load term:

$$\mathfrak{K}_B = \mathfrak{E}_{r1} + \mathfrak{E}_{rb} - \mathfrak{M}_{l2} + \boldsymbol{D}_B. \tag{660}$$

Beam moments:

$$\left. \begin{array}{ll} M_{B1} = M_{D4} = \mathfrak{E}_{r1} - \mathfrak{K}_B \cdot \mu_1 & M_{A1} = M_{E4} = -(\mathfrak{L}_1 + M_{B1})\varepsilon_1' \\ M_{B2} = M_{D3} = \mathfrak{M}_{l2} + \mathfrak{K}_B \cdot \mu_2 & M_{C2} = M_{C3} = \mathfrak{M}_{r2} - \mathfrak{K}_B \cdot \mu_2/2. \end{array} \right\} \tag{661}$$

Column moments:

$$\left. \begin{array}{lll} \text{at head} & M_{Eb} = -M_{Dd} = \mathfrak{E}_{rb} - \mathfrak{K}_B \cdot \mu_b & M_{Cc} = 0; \\ \text{at foot} & M_b = -M_d = -(\mathfrak{L}_b + M_{Bb})\varepsilon_b & M_c = 0. \end{array} \right\} \tag{662}$$

Check relationship:

$$\boldsymbol{D}_B + M_{Eb} + M_{B1} - M_{B2} = 0.$$

Loading Condition 4b

Anti-symmetrical overall loading condition

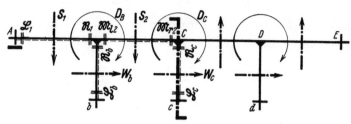

Fig. 154

Loading diagram with symbols indicated for left-hand half of frame, including central column and central joint

Restraining moments of the elastically restrained members:

$$\mathfrak{E}_{r1} = -\frac{\mathfrak{R}_1 - \mathfrak{L}_1 \cdot \varepsilon_1'}{2 - \varepsilon_1'}; \qquad \mathfrak{E}_{rb} = -\frac{\mathfrak{R}_b - \mathfrak{L}_b \cdot \varepsilon_b}{2 - \varepsilon_b} \qquad \mathfrak{E}_{rc} = -\frac{\mathfrak{R}_c - \mathfrak{L}_c \cdot \varepsilon_c}{2 - \varepsilon_c}. \qquad (664)$$

Joint load terms:

$$\mathfrak{K}_B = \mathfrak{E}_{r1} + \mathfrak{E}_{rb} - \mathfrak{M}_{l2} + D_B \qquad \mathfrak{K}_C'' = \mathfrak{M}_{r2} + \mathfrak{E}_{rc}/2 + D_C/2. \qquad (665)$$

auxiliary moments:

$$\left. \begin{aligned} Y_B'' &= + \mathfrak{K}_B \cdot u_{bb}'' - \mathfrak{K}_C'' \cdot u_{bc}'' \\ Y_C'' &= - \mathfrak{K}_B \cdot u_{cb}'' + \mathfrak{K}_C'' \cdot u_{cc}''. \end{aligned} \right\} \qquad (666)$$

Beam moments:

$$\left. \begin{aligned} M_{B1} &= -M_{D4} = \mathfrak{E}_{r1} - Y_B'' \cdot 3 S_1 \\ M_{C2} &= -M_{C3} = \mathfrak{M}_{r2} - (Y_B'' + 2 Y_C'') K_2; \\ M_{A1} &= -M_{E4} = -(\mathfrak{L}_1 + M_{B1}) \varepsilon_1' \\ M_{B2} &= -M_{D3} = \mathfrak{M}_{l2} + (2 Y_B'' + Y_C'') K_2. \end{aligned} \right\} \qquad (667)$$

Column moments:

at top: at base:

$$\left. \begin{aligned} M_{Bb} = M_{Dd} &= \mathfrak{E}_{rb} - Y_B'' \cdot 3 S_b & M_b = M_d &= -(\mathfrak{L}_b + M_{Bb}) \varepsilon_b \\ M_{Cc} &= \mathfrak{E}_{rc} - Y_C'' \cdot 3 S_c; & M_c &= -(\mathfrak{L}_c + M_{Cc}) \varepsilon_c. \end{aligned} \right\} \qquad (668)$$

Check relationships:

$$D_B + M_{Bb} + M_{B1} - M_{B2} = 0 \qquad D_C + M_{Cc} + 2 M_{C2} = 0. \qquad (669)$$

Note: For coefficients relating to Loading Condition 4b see page 154.

b) Coefficients for anti-symmetrical load arrangements

$$K_1 = \frac{J_1}{J_k} \cdot \frac{l_k}{l_1} \qquad K_2 = \frac{J_2}{J_k} \cdot \frac{l_k}{l_2} \qquad K_b = \frac{J_b}{J_k} \cdot \frac{l_k}{h_b} \qquad K_c = \frac{J_c}{J_k} \cdot \frac{l_k}{h_c}. \qquad (670)$$

$$\left. \begin{array}{l} S_1 = K_1/(2-\varepsilon_1') \qquad S_b = K_b/(2-\varepsilon_b) \qquad S_c = K_c/(2-\varepsilon_c); \\ K_B = 3S_1 + 3S_b + 2K_2 \qquad K_C'' = 2K_2 + 3S_c/2. \end{array} \right\} \qquad (671)$$

$$\left. \begin{array}{l} j_{2t} = \dfrac{K_2}{K_C''} \qquad j_2' = \dfrac{K_2}{K_B}; \qquad u_{bb}'' = \dfrac{1}{K_B - K_2 \cdot j_{2t}} \qquad u_{cc}'' = \dfrac{1}{K_C'' - K_2 \cdot j_2'}; \\ u_{bc}'' = u_{cb}'' = u_{bb}'' \cdot j_{2t} = u_{cc}'' \cdot j_2'. \end{array} \right\} \qquad (672)$$

Horizontal thrusts, column shears and lateral restraining force for all loading conditions

Formulas (179) – (187) and (189) are applicable, for the collective subscripts $n = b, c, d$ and $N = B, C, D$, having due regard to symmetry and anti-symmetry of loading. The formulas (188) and (190) are simplified to:

$$H = -2H_b - H_c \quad \text{and} \quad H = -2Q_{rb} - Q_{rc}. \qquad (673)$$

Frame Shape 19v

(Frame Shape 19 with laterally unrestrained horizontal member)

Procedure

either similar to procedure for Frame Shape 1v, see pages 9-10, for the collective subscripts $v = 1, 2, 3, 4$; $n = b, c, d$ and $N = B, C, D$, having due regard to anti-symmetry properties, or, alternatively, similar to procedure for Frame Shape 8v, see pages 70-71, for the collective subscripts $v = 1, 2$; $n = b, c$ and $N = B, C$.

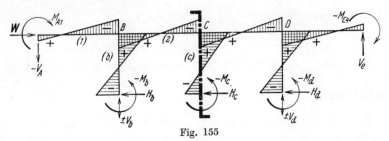

Fig. 155

Loading Condition 5 with diagram of bending moments and reactions at supports.

Frame Shape 20

Symmetrical single-story three-bay frame with laterally restrained horizontal member, elastically restrained at each end, and two elastically restrained columns

(Symmetrical form of Frame Shape 13)

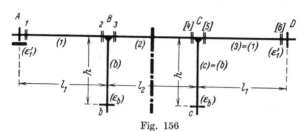

Fig. 156

Frame shape, dimensions and symbols

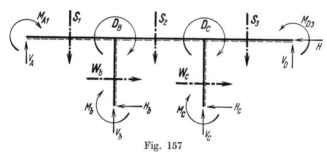

Fig. 157

Definition of positive direction for all external loads on joints and members, for all reactions at supports, and for the lateral restraining force

> All the dimensions and coefficients for the right-hand half of the frame are the same as those for the left-hand half. In the case of the columns the broken line is always placed on the right-hand side of each column, despite the symmetry of the frame.

Note: For Frame Shape 20 the *coefficients and the Loading Conditions 1 – 3 for Frame Shape* 13 may be employed. On account of symmetry the reduction sequences (437)/(438) and the recursion sequences (439)/(440) respectively become equal. Furthermore, the square matrix (442) becomes bisymmetrical, i.e., symmetrical with respect to the principal diagonal and the secondary diagonal. If the *"load transposition method"* is employed, it is also possible to start from the following symmetrical and anti-symmetrical loading conditions.

I. Treatment by the Method of Forces

a) Coefficients for symmetrical load arrangements

Flexibility coefficients of all the horizontal members, and flexibility of columns:

$$k_1 = \frac{J_k}{J_1} \cdot \frac{l_1}{l_k} \qquad k_2 = \frac{J_k}{J_2} \cdot \frac{l_2}{l_k} \qquad k_b = \frac{J_k}{J_b} \cdot \frac{h}{l_k} \; ; \qquad b_b = k_b(2 - \varepsilon_b). \tag{674}$$

Diagonal coefficients:

$$O_1 = k_1/\varepsilon_1' \qquad O_2 = 2k_1 + b_b \qquad O_3' = b_b + 3k_2. \tag{675}$$

Moment carry-over factors and auxiliary coefficients:

backward reduction: | forward reduction:

$$\sigma_{bs} = \frac{b_b}{r_3'} \qquad \begin{aligned} r_3' &= O_3' \\ r_2' &= O_2 - b_b \cdot \sigma_{bs} \end{aligned} \qquad \sigma_b' = \frac{b_b}{v_2} \qquad \begin{aligned} v_2 &= O_2 - k_1 \cdot \varepsilon_1' \\ v_3' &= O_3' - b_b \cdot \sigma_b'. \end{aligned} \tag{676, 677}$$

$$\iota_{1s} = \frac{k_1}{r_2'} \qquad r_1' = O_1 - k_1 \cdot \iota_{1s}.$$

Principal influence coefficients by recursion:

forward: | backward:

$$n_{11}' = \frac{1}{r_1'} = \frac{1}{k_1(1/\varepsilon_1' - \iota_{1s})} \qquad n_{33}' = \frac{1}{v_3'}$$

$$n_{22}' = \frac{1}{r_2'} + n_{11}' \cdot \iota_{1s}^2 \qquad n_{22}' = \frac{1}{v_2} + n_{33}' \cdot \sigma_b'^2 \tag{678, 679}$$

$$n_{33}' = \frac{1}{r_3'} + n_{22}' \cdot \sigma_{bs}^2. \qquad n_{11}' = \frac{1}{O_1} + n_{22}' \cdot \varepsilon_1'^2.$$

Principal influence coefficients direct

$$n_{11}' = 1/r_1' \qquad n_{22}' = 1/(r_2' + v_2 - O_2) \qquad n_{33}' = 1/v_3'. \tag{680}$$

Computation scheme for the symmetrical influence coefficient matrix*

$$\begin{array}{c} \cdot \iota_{1s} \\ \cdot \sigma_{bs} \end{array} \left| \begin{array}{ccc} (n_{11}') & n_{12}' & n_{13}' \\ \downarrow & \uparrow & \uparrow \\ n_{21}' & (n_{22}') & n_{23}' \\ \downarrow & \downarrow & \uparrow \\ n_{31}' & n_{32}' & (n_{33}') \end{array} \right| \begin{array}{c} \cdot \varepsilon_1' \\ \cdot \sigma_b' \end{array} \tag{681}$$

---

* Between the elements of the matrix (681) for the left-hand half of the frame and the elements of the matrix (442) for the whole frame, as indicated on page 101, there exist according to the "load transposition method", the following relationships:

$$n_{11}' = n_{11} - n_{16}, \quad n_{12}' = n_{12} - n_{15} \text{ etc., up to } n_{33}' = n_{33} - n_{34}.$$

b) Coefficients for anti-symmetrical load arrangements

Flexibility coefficients of all the horizontal members, and flexibility of columns:

$$k_1 = \frac{J_k}{J_1} \cdot \frac{l_1}{l_k} \qquad k_2 = \frac{J_k}{J_2} \cdot \frac{l_2}{l_k} \qquad k_b = \frac{J_k}{J_b} \cdot \frac{h}{l_k}; \qquad b_b = k_b(2 - \varepsilon_b), \qquad (682)$$

Diagonal coefficients:

$$O_1 = k_1/\varepsilon_1' \qquad O_2 = 2k_1 + b_b \qquad O_3'' = b_b + k_2. \qquad (683)$$

Moment carry-over factors and auxiliary coefficients:

backward reduction:

$$\sigma_{bt} = \frac{b_b}{r_3''}$$

$$\iota_{1t} = \frac{k_1}{r_2''}$$

$$r_3'' = O_3''$$

$$r_2'' = O_2 - b_b \cdot \sigma_{bt}$$

$$r_1'' = O_1 - k_1 \cdot \iota_{1t}.$$

forward reduction:

$$\sigma_b' = \frac{b_b}{v_2}$$

$$v_2 = O_2 - k_1 \cdot \varepsilon_1'$$

$$v_3'' = O_3'' - b_b \cdot \sigma_b'.$$

(684) and (685)

Principal influence coefficients by recursion:

forward:

$$n_{11}'' = \frac{1}{r_1''} = \frac{1}{k_1(1/\varepsilon_1' - \iota_{1t})}$$

$$n_{22}'' = \frac{1}{r_2''} + n_{11}'' \cdot \iota_{1t}^2$$

$$n_{33}'' = \frac{1}{r_3''} + n_{22}'' \cdot \sigma_{bt}^2.$$

backward:

$$n_{33}'' = \frac{1}{v_3''}$$

$$n_{22}'' = \frac{1}{v_2} + n_{33}'' \cdot \sigma_b'^2$$

$$n_{11}'' = \frac{1}{O_1} + n_{22}'' \cdot \varepsilon_1'^2.$$

(686) and (687)

Principal influence coefficients direct

$$n_{11}'' = 1/r_1'' \qquad n_{22}'' = 1/(r_2'' + v_2 - O_2) \qquad n_{33}'' = 1/v_3''. \qquad (688)$$

Computation scheme for the symmetrical influence coefficient matrix*

$$\begin{array}{c|ccc|c}
& (n_{11}'') & n_{12}'' & n_{13}'' & \\
\cdot \iota_{1t} & \downarrow & \uparrow & \uparrow & \cdot \varepsilon_1' \\
& n_{21}'' & (n_{22}'') & n_{23}'' & \\
\cdot \sigma_{bt} & \downarrow & \downarrow & \uparrow & \cdot \sigma_b' \\
& n_{31}'' & n_{32}'' & (n_{33}'') &
\end{array} \qquad (689)$$

---

* Between the elements of the matrix (689) for the left-hand half of the frame and the elements of the matrix (442) for the whole frame, as indicated on page 101, there exist, according to the "load transposition method", the following relationships:

$$n_{11}'' = n_{11} + n_{16}, \quad n_{12}'' = n_{12} + n_{15} \text{ etc., up to } n_{33}'' = n_{33} + n_{34}.$$

Loading Condition 1a

Horizontal member carrying vertical loading of arbitrary magnitude, but symmetrical about center of frame

$$(\mathfrak{R}_3 = \mathfrak{L}_1, \quad \mathfrak{L}_3 = \mathfrak{R}_1; \quad \mathfrak{R}_2 = \mathfrak{L}_2)$$

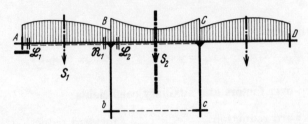

Fig. 158
Beam loading diagram with load terms $\mathfrak{L}$ and $\mathfrak{R}$ for left-hand half of frame

Beam moments $(L_1 = \mathfrak{L}_1 k_1, \quad R_1 = \mathfrak{R}_1 k_1; \quad L_2 = \mathfrak{L}_2 k_2)$:

$$\left. \begin{array}{l} X'_1 = M_{A1} = M_{D3} = \\ X'_2 = M_{B1} = M_{C3} = \\ X'_3 = M_{B2} = M_{C2} = \end{array} \right| \begin{array}{ccc} L_1 & R_1 & L_2 \\ \hline -n'_{11} & +n'_{12} & +n'_{13} \\ +n'_{21} & -n'_{22} & -n'_{23} \\ +n'_{31} & -n'_{32} & -n'_{33}. \end{array} \quad (690)$$

Column moments:

at top: at base:

$$M_{Bb} = -M_{Cc} = -X'_2 + X'_3 \qquad M_b = -M_c = -M_{Bb} \cdot \varepsilon_b. \qquad (691)$$

Special case: symmetrical bay loading $R_1 = L_1$):

$$\left. \begin{array}{l} X'_1 = M_{A1} = M_{D3} = -L_1 \cdot s'_{11} + L_2 \cdot n'_{13} \\ X'_2 = M_{B1} = M_{C3} = -L_1 \cdot s'_{21} - L_2 \cdot n'_{23} \\ X'_3 = M_{B2} = M_{C2} = -L_1 \cdot s'_{31} - L_2 \cdot n'_{33}; \end{array} \right\} \qquad (692)$$

where*   belongs to page 160*

$$\begin{array}{l} s'_{11} = +n'_{11} - n'_{12} = (s_{11} + s_{13}) \\ s'_{21} = -n'_{21} + n'_{22} = (s_{21} - s_{23}) \\ s'_{31} = -n'_{31} + n'_{32} = (s_{31} - s_{33}). \end{array} \quad \begin{array}{l} w'_{1b} = +n'_{12} - n'_{13} = (w_{1b} + w_{1c}) \\ w'_{2b} = +n'_{22} - n'_{23} = (w_{2b} + w_{2c}) \\ w'_{3b} = -n'_{32} + n'_{33} = (w_{3b} - w_{3c}). \end{array} \quad \begin{array}{l} (693) \\ \text{and} \\ (694) \end{array}$$

* The members enclosed in parentheses in (693) and (694) have been formed with the values s in (443) and w in (444).

Loading Condition 1b

Horizontal member carrying vertical loading of arbitrary magnitude, antisymmetrical about center of frame

$$(\mathfrak{R}_3 = -\mathfrak{L}_1, \quad \mathfrak{L}_3 = -\mathfrak{R}_1; \quad \mathfrak{R}_2 = -\mathfrak{L}_2)$$

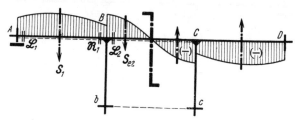

Fig. 159
Beam loading diagram with load terms $\mathfrak{L}$ and $\mathfrak{R}$ for left-hand half of frame

Beam moments ($L_1 = \mathfrak{L}_1 k_1, \quad R_1 = \mathfrak{R}_1 k_1; \quad L_2 = \mathfrak{L}_2 k_2$):

	L_1	R_1	L_2
$X_1'' = M_{A1} = -M_{D3} =$	$-n_{11}''$	$+n_{12}''$	$+n_{13}''$
$X_2'' = M_{B1} = -M_{C3} =$	$+n_{21}''$	$-n_{22}''$	$-n_{23}''$
$X_3'' = M_{B2} = -M_{C2} =$	$+n_{31}''$	$-n_{32}''$	$-n_{33}''$.

(695)

Column moments:

at top: $\qquad$ at base:

$$M_{Bb} = M_{Cc} = -X_2'' + X_3'' \qquad M_b = M_c = -M_{Bb} \cdot \varepsilon_b. \qquad (696)$$

Special case: symmetrical loading of the End Spans ($R_1 = L_1$):

$$\begin{aligned} X_1'' &= M_{A1} = -M_{D3} = -L_1 \cdot s_{11}'' \\ X_2'' &= M_{B1} = -M_{C3} = -L_1 \cdot s_{21}'' \\ X_3'' &= M_{B2} = -M_{C2} = -L_1 \cdot s_{31}''; \end{aligned} \qquad (697)$$

where*: $\qquad\qquad$ belongs to page 160*

$$s_{11}'' = +n_{11}'' - n_{12}'' = (s_{11} - s_{13}) \qquad w_{1b}'' = +n_{12}'' - n_{13}'' = (w_{1b} - w_{1c})$$
$$s_{21}'' = -n_{21}'' + n_{22}'' = (s_{21} + s_{23}) \qquad w_{2b}'' = +n_{22}'' - n_{23}'' = (w_{2b} - w_{2c})$$
$$s_{31}'' = -n_{31}'' + n_{32}'' = (s_{31} + s_{33}). \qquad w_{3b}'' = -n_{32}'' + n_{33}'' = (w_{3b} + w_{3c}).$$

(698) and (699)

---

* The members enclosed in parentheses in (698) and (699) have been formed with the values s in (443) and w in (444).

Loading Conditions 2a and 3a

Symmetrical load arrangement for columns (2a) and joints (3a)

$$(\mathfrak{L}_c = -\mathfrak{L}_b, \quad \mathfrak{R}_c = -\mathfrak{R}_b; \quad D_C = -D_B)$$

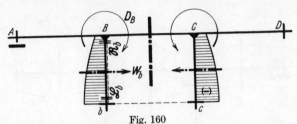

Fig. 160

Loading diagram with symbols indicated for left-hand half of frame

Joint load term: $\qquad \mathfrak{B}_B = D_B \cdot b_b + (\mathfrak{L}_b \cdot \varepsilon_b - \mathfrak{R}_b) k_b$. $\qquad$ (700)

Beam moments: $\quad \begin{aligned} X'_1 &= M_{A1} = M_{D3} = + \mathfrak{B}_B \cdot w'_{1b} \\ X'_2 &= M_{B1} = M_{C3} = - \mathfrak{B}_B \cdot w'_{2b} \\ X'_3 &= M_{B2} = M_{C2} = + \mathfrak{B}_B \cdot w'_{3b}.* \end{aligned} \right\}$ (701)

Column moments:
$$M_{Bb} = -M_{Cc} = -D_B - X'_2 + X'_3 \qquad M_b = -M_c = -(\mathfrak{L}_b + M_{Bb})\varepsilon_b. \quad (702)$$

Loading Conditions 2b and 3b

Anti-symmetrical load arrangement for columns (2b) and joints (3b)

$$(\mathfrak{L}_c = \mathfrak{L}_b, \quad \mathfrak{R}_c = \mathfrak{R}_b; \quad D_C = D_B)$$

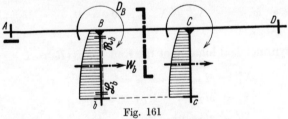

Fig. 161

Loading diagram with symbols indicated for left-hand half of frame

Beam moments: $\quad X''_1 = M_{A1} = -M_{D3} = +\mathfrak{B}_B \cdot w''_{1b}$

($\mathfrak{B}_B$ as above) $\quad X''_2 = M_{B1} = -M_{C3} = -\mathfrak{B}_B \cdot w''_{2b} \qquad$ (703)

$\qquad \qquad \qquad X''_3 = M_{B2} = -M_{C2} = +\mathfrak{B}_B \cdot w''_{3b}.*$

Column moments:
$$M_{Bb} = M_{Cc} = -D_B - X''_2 + X''_3 \qquad M_b = M_c = -(\mathfrak{L}_b + M_{Bb})\varepsilon_b. \quad (704)$$

---

* For the values w' see page 158, ; for the values w'' see page 159. (This arrangement has been adopted in order to save space).

II. Treatment by the Deformation Method

a) Coefficients for symmetrical load arrangements

$$K_1 = \frac{J_1}{J_k} \cdot \frac{l_k}{l_1} = \frac{1}{k_1} \qquad K_2 = \frac{J_2}{J_k} \cdot \frac{l_k}{l_2} = \frac{1}{k_2} \qquad K_b = \frac{J_b}{J_k} \cdot \frac{l_k}{h} = \frac{1}{k_b}. \qquad (705)$$

$$S_1 = K_1/(2 - \varepsilon_1') \qquad S_b = K_b/(2 - \varepsilon_b); \qquad K_B' = 3S_1 + 3S_b + \mathbf{K_2}. \qquad (706)$$

$$\mu_1' = \frac{3S_1}{K_B'} \qquad \mu_2' = \frac{K_2}{K_B'} \qquad \mu_b' = \frac{3S_b}{K_B'}; \qquad (\mu_1' + \mu_2' + \mu_b' = 1). \qquad (707)$$

Loading Condition 4a

Symmetrical overall loading condition

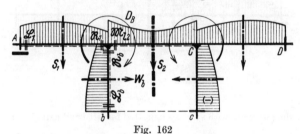

Fig. 162

Loading diagram with symbols indicated for left-hand half of frame

Restraining moments of the elastically restrained members:

$$\mathfrak{C}_{r1} = -\frac{\mathfrak{R}_1 - \mathfrak{L}_1 \cdot \varepsilon_1'}{2 - \varepsilon_1'} \qquad \mathfrak{C}_{rb} = -\frac{\mathfrak{R}_b - \mathfrak{L}_b \cdot \varepsilon_b}{2 - \varepsilon_b}. \qquad (708)$$

Joint load term:

$$\mathfrak{R}_B = \mathfrak{C}_{r1} + \mathfrak{C}_{rb} - \mathfrak{M}_{l2} + \mathbf{D}_B. \qquad (709)$$

Beam moments:

$$\left.\begin{array}{c} M_{B1} = M_{C3} = \mathfrak{C}_{r1} - \mathfrak{R}_B \cdot \mu_1' \qquad M_{B2} = M_{C2} = \mathfrak{M}_{l2} + \mathfrak{R}_B \cdot \mu_2' \\ M_{A1} = M_{D3} = -(\mathfrak{L}_1 + M_{B1})\varepsilon_1'. \end{array}\right\} \qquad (710)$$

Column moments:

$$\text{at top:} \qquad\qquad\qquad \text{at base:}$$

$$M_{Bb} = -M_{Cc} = \mathfrak{C}_{rb} - \mathfrak{R}_B \cdot \mu_b' \qquad M_b = -M_c = -(\mathfrak{L}_b + M_{Bb})\varepsilon_b. \qquad (711)$$

Check relationship:

$$\mathbf{D}_B + M_{B1} - M_{B2} + M_{Bb} = 0. \qquad (712)$$

b) Coefficients for anti-symmetrical load arrangements

K_1, K_2, K_b as equation (705); S_1, S_b as equation (706).

Joint stiffness:

$$S_B = S_1 + S_b + K_2. \tag{713}$$

$$\mu_1'' = \frac{S_1}{S_B} \quad \mu_2'' = \frac{K_2}{S_B} \quad \mu_b'' = \frac{S_b}{S_B}; \quad (\mu_1'' + \mu_2'' + \mu_b'' = 1). \tag{714}$$

Loading Condition 4b
Anti-symmetrical overall loading condition

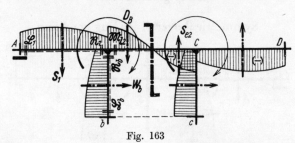

Fig. 163
Loading diagram with symbols indicated for left-hand half of frame

Restraining moments of the elastically restrained members:

Beam moments:

$$M_{B1} = -M_{C3} = \mathfrak{E}_{r1} - \mathfrak{R}_B \cdot \mu_1'' \quad M_{B2} = -M_{C2} = \mathfrak{M}_{l2} + \mathfrak{R}_B \cdot \mu_2'';$$
$$M_{A1} = -M_{D3} = -(\mathfrak{L}_1 + M_{B1})\varepsilon_1'. \tag{715}$$

Column moments:

at top: at base:
$$M_{Bb} = M_{Cc} = \mathfrak{E}_{rb} - \mathfrak{R}_B \cdot \mu_b'' \quad M_b = M_c = -(\mathfrak{L}_b + M_{Bb})\varepsilon_b. \tag{716}$$

Check relationship: as equation (712).

Horizontal thrusts, column shears and lateral restraining force for all loading conditions

Formulas (179) – (182) are applicable, for the collective subscripts $n = b, c$ and $N = B, C$ or, for symmetrical or anti-symmetrical load arrangements, formulas (183) – (190) are applicable, for the collective subscripts $n = b$ and $N = B$.

Frame Shape 20v

(Frame Shape 20 with laterally unrestrained horizontal member)

Note: Horizontal displacement ("side-sway") of the horizontal member will occur only with asymmetrical and anti-symmetrical loading conditions (for all symmetrical loading conditions of Frame Shape 20: $H = 0$). Since any asymmetrical loading condition can, by means of the "load transposition method," be split up into a symmetrical and an anti-symmetrical loading condition, only the anti-symmetrical loading conditions will be considered here. The Deformation Method (Method II) is employed.

Coefficients for anti-symmetrical load arrangements

Stiffness coefficients of the individual members:

$$K_1 = \frac{J_1}{J_k} \cdot \frac{l_k}{l_1} \qquad K_b = \frac{J_b}{J_k} \cdot \frac{l_k}{h} \qquad K_2 = \frac{J_2}{J_k} \cdot \frac{l_k}{l_2}. \tag{717}$$

Stiffnesses of the members and joint stiffness:

$$S_1 = K_1/(2 - \varepsilon_1') \qquad F_b = K_b/(1/\varepsilon_b + 4); \qquad S_B = S_1 + F_b + K_2. \tag{718}$$

Moment distribution factors:

$$\eta_1 = \frac{S_1}{S_B} \qquad \eta_b = \frac{F_b}{S_B} \qquad \eta_2 = \frac{K_2}{S_B}; \qquad (\eta_1 + \eta_b + \eta_2 = 1). \tag{719}$$

Loading Condition 5

Horizontal concentrated load P at level of horizontal member

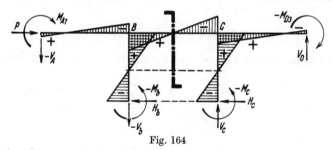

Fig. 164
Loading diagram, with diagram of bending moments and reactions at supports

Beam moments:

$$\left. \begin{array}{c} M_{A1} = -M_{D3} = -M_{B1} \cdot \varepsilon_1'; \\[4pt] M_{C3} = -M_{B1} = Ph \cdot \dfrac{1/\varepsilon_b + 1}{1/\varepsilon_b + 4} \cdot \eta_1 \qquad M_{B2} = -M_{C2} = Ph \cdot \dfrac{1/\varepsilon_b + 1}{1/\varepsilon_b + 4} \cdot \eta_2. \end{array} \right\} \tag{720}$$

Column moments:

$$M_{Bb} = M_{Cc} = M_{B2} - M_{B1} \qquad M_b = M_c = -Ph + M_{Bb}. \tag{721}$$

Loading Conditions 1b and 2b

Horizontal member carrying arbitrary vertical loading, columns carrying arbitrary horizontal loading, but loading as a whole is anti-symmetrical about center of frame

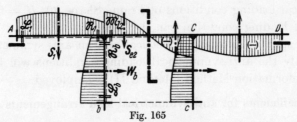

Fig. 165

Loading diagram with symbols $\mathfrak{L}$, $\mathfrak{R}$ and $\mathfrak{M}_l$ indicated for left-hand half of frame

Restraining moments of the elastically restrained members:

$$\mathfrak{E}_{r1} = -\frac{\mathfrak{R}_1 - \mathfrak{L}_1 \cdot \varepsilon_1'}{2 - \varepsilon_1'} \qquad \mathfrak{F}_{rb} = \frac{\mathfrak{S}_{lb}(1/\varepsilon_b + 1) - (\mathfrak{L}_b + \mathfrak{R}_b)}{1/\varepsilon_b + 4}. \tag{722}$$

Joint load term:

$$\mathfrak{K}_B^v = \mathfrak{E}_{r1} + \mathfrak{F}_{rb} - \mathfrak{M}_{l2}. \tag{723}$$

Beam moments:

$$\left. \begin{aligned} M_{A1} &= -M_{D3} = -(\mathfrak{L}_1 + M_{B1})\varepsilon_1'; \\ M_{B1} = -M_{C3} &= \mathfrak{E}_{r1} - \mathfrak{K}_B^v \cdot \eta_1 \qquad M_{B2} = -M_{C2} = \mathfrak{M}_{l2} + \mathfrak{K}_B^v \cdot \eta_2. \end{aligned} \right\} \tag{724}$$

Column moments:

$$(M_{B1} + M_{Bb} - M_{B2} = 0).$$

$$M_{Bb} = M_{Cc} = \mathfrak{F}_{rb} - \mathfrak{K}_B^v \cdot \eta_b \qquad M_b = M_c = -\mathfrak{S}_{lb} + M_{Bb}. \tag{725}$$

Loading Condition 3b

Anti-symmetrical pair of rotational moments D at the joints

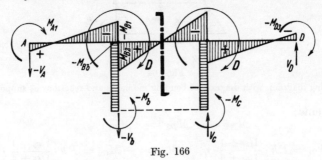

Fig. 166

Loading diagram, with diagram of bending moments and reactions at supports

Beam and column moments:

$$\left. \begin{aligned} M_{B1} = -M_{C3} &= -D \cdot \eta_1 \qquad & M_{B2} = -M_{C2} &= +D \cdot \eta_2 \\ M_{Bb} = M_{Cc} = M_b = M_c &= -D \cdot \eta_b \qquad & M_{A1} = -M_{D3} &= -M_{B1} \cdot \varepsilon_1'. \end{aligned} \right\} \tag{726}$$

Frame Shape 21

Symmetrical single-story two-bay frame with laterally restrained horizontal member, elastically restrained at each end, and one elastically restrained central column

(Symmetrical form of Frame Shape 14)

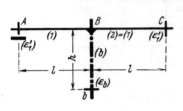
Fig. 167
Frame shape, dimensions and symbols

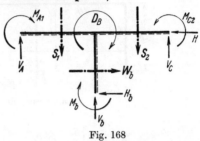
Fig. 168
Definition of positive direction for all external loads, for the column and beam reactions, and for the lateral restraining force

Note: The formulas for Frame Shape 14, as treated by the Deformation Method (Method II)—see page 104 —are applicable to Frame Shape 21, subject to introduction of the following simplifications into the coefficients indicated there:

$$\varepsilon_2 = \varepsilon_1' \qquad K_2 = K_1 \qquad S_2 = S_1 \qquad \mu_2 = \mu_1.$$

Frame Shape 21v

(Frame Shape 21 with laterally unrestrained horizontal member—symmetrical form of Frame Shape 14v)

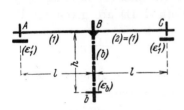
Fig. 169
Frame shape, dimensions and symbols

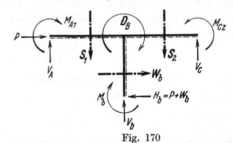
Fig. 170
Definition of positive direction for all external loads, and for the column and beam reactions

Note: The formulas for Frame Shape 14v, as treated by the Deformation Method (Method II)—see page 107 —are applicable to Frame Shape 21v, subject to introduction of the following simplifications into the coefficients indicated there:

$$\varepsilon_2 = \varepsilon_1' \qquad K_2 = K_1 \qquad S_2 = S_1 \qquad \eta_2 = \eta_1.$$

Frame Shape 22

Single-story three-bay frame with laterally restrained horizontal member, elastically restrained at its right-hand end, and three elastically restrained columns

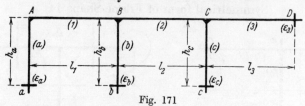

Fig. 171

Frame shape, dimensions and symbols

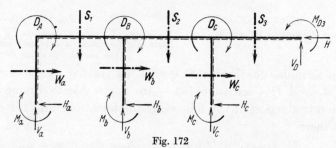

Fig. 172

Definition of positive direction for all external loads on joints and members, for all column and beam reactions, and for the lateral restraining force

Note: Frame Shape 22 is constructed at its left-hand end like Frame Shape 1 or 3, and at its right-hand end like Frame Shape 11, 12 or 13. Hence all the necessary formulas and formula matrices for Frame Shape 22 can readily be derived from those applicable to these other Frame Shapes, having due regard to the relevant subscripts. Under the present heading only the formulas for the Deformation Method (Method II) are given for Frame Shape 22.*

Important

All the formula sequences and formula matrices given for Frame Shape 22, as treated by the Deformation Method, can readily be extended to suit frames of similar type having any number of bays

---

Note: The *collective subscripts* have the following meaning:

$\nu = 1, 2, 3$ denoting the bays (or beam spans)
$n = a, b, c$ denoting the columns
$N = A, B, C$ denoting the joints

---

* For the treatment of a similar frame by the Method of Forces (Method I) see, for example, Frame Shape 23.

Coefficients

Stiffness coefficients of all the individual members:

$$K_\nu = \frac{J_\nu}{J_k} \cdot \frac{l_k}{l_\nu} = \frac{1}{k_\nu} \; ; \qquad K_n = \frac{J_n}{J_k} \cdot \frac{l_k}{h_n} = \frac{1}{k_n} \; . \tag{727}$$

Stiffnesses of the elastically restrained columns and of the right-hand end span of the horizontal member (the values are assumed to be known):

$$S_n = \frac{K_n}{2 - \varepsilon_n} \; ; \qquad S_3 = \frac{K_3}{2 - \varepsilon_3} \; . \tag{728}$$

Joint coefficients:

$$K_A = 3S_a + 2K_1 \qquad K_B = 2K_1 + 3S_b + 2K_2 \qquad K_C = 2K_2 + 3S_c + 3S_3 . \tag{729}$$

Rotation carry-over factors and auxiliary coefficients:

backward reduction: | forward reduction:

$$\begin{aligned}
j_2 &= \frac{K_2}{r_c} & r_c &= K_C \\
& & r_b &= K_B - K_2 \cdot j_2 \\
j_1 &= \frac{K_1}{r_b} & & \\
& & r_a &= K_A - K_1 \cdot j_1 .
\end{aligned}
\qquad
\begin{aligned}
& & v_a &= K_A \\
j_1' &= \frac{K_1}{v_a} & & \\
& & v_b &= K_B - K_1 \cdot j_1' \\
j_2' &= \frac{K_2}{v_b} & & \\
& & v_c &= K_C - K_2 \cdot j_2' .
\end{aligned}
\qquad
\begin{matrix}(730) \\ \text{and} \\ (731)\end{matrix}$$

Principal influence coefficients by recursion:

forward: | backward:

$$\begin{aligned}
u_{aa} &= 1/r_a \\
u_{bb} &= 1/r_b + u_{aa} \cdot j_1^2 \\
u_{cc} &= 1/r_c + u_{bb} \cdot j_2^2 .
\end{aligned}
\qquad
\begin{aligned}
u_{cc} &= 1/v_c \\
u_{bb} &= 1/v_b + u_{cc} \cdot j_2'^2 \\
u_{aa} &= 1/v_a + u_{bb} \cdot j_1'^2 .
\end{aligned}
\qquad
\begin{matrix}(732) \\ \text{and} \\ (733)\end{matrix}$$

Principal influence coefficients direct

$$u_{aa} = \frac{1}{r_a} \qquad u_{bb} = \frac{1}{r_b + v_b - K_B} \qquad u_{cc} = \frac{1}{v_c} . \tag{734}$$

Computation scheme for the symmetrical influence coefficient matrix

$$
\begin{array}{c|ccc|c}
 & (u_{aa}) & u_{ab} & u_{ac} & \\
\cdot j_1 & \downarrow & \uparrow & \uparrow & \cdot j_1' \\
 & u_{ba} & (u_{bb}) & u_{bc} & \\
\cdot j_2 & \downarrow & \downarrow & \uparrow & \cdot j_2' \\
 & u_{ca} & u_{cb} & (u_{cc}) & \\
\end{array}
\tag{735}
$$

Loading Condition 4

Overall loading condition*

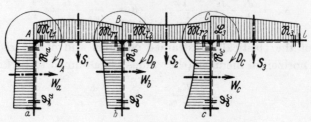

Fig. 173

Loading diagram, with fixed-end moments $\mathfrak{M}_l$ and $\mathfrak{M}_r$ of beam spans 1 and 2, load terms $\mathfrak{L}$ and $\mathfrak{R}$ for the columns and beam span 3, and external rotational moments D at the joints

Restraining moments of the elastically restrained members:

$$\mathfrak{C}_{rn} = -\frac{\mathfrak{R}_n - \mathfrak{L}_n \cdot \varepsilon_n}{2 - \varepsilon_n} \; ; \qquad \mathfrak{C}_{l3} = -\frac{\mathfrak{L}_3 - \mathfrak{R}_3 \cdot \varepsilon_3}{2 - \varepsilon_3}. \tag{736}$$

Joint load terms:

$$\left. \begin{aligned} \mathfrak{K}_A &= \phantom{\mathfrak{M}_{r1} -{}} -\mathfrak{M}_{l1} + \mathfrak{C}_{ra} + D_A \\ \mathfrak{K}_B &= \mathfrak{M}_{r1} - \mathfrak{M}_{l2} + \mathfrak{C}_{rb} + D_B \\ \mathfrak{K}_C &= \mathfrak{M}_{r2} - \mathfrak{C}_{l3} + \mathfrak{C}_{rc} + D_C. \end{aligned} \right\} \tag{737}$$

auxiliary moments:

$$\left. \begin{array}{c|ccc} & \mathfrak{K}_A & \mathfrak{K}_B & \mathfrak{K}_C \\ \hline Y_A = & +u_{aa} & -u_{ab} & +u_{ac} \\ Y_B = & -u_{ba} & +u_{bb} & -u_{bc} \\ Y_C = & +u_{ca} & -u_{cb} & +u_{cc}. \end{array} \right\} \tag{738}$$

Beam moments:

$$\left. \begin{aligned} M_{A1} &= \mathfrak{M}_{l1} + (2Y_A + Y_B)K_1 & M_{B1} &= \mathfrak{M}_{r1} - (Y_A + 2Y_B)K_1 \\ M_{B2} &= \mathfrak{M}_{l2} + (2Y_B + Y_C)K_2 & M_{C2} &= \mathfrak{M}_{r2} - (Y_B + 2Y_C)K_2 \\ M_{C3} &= \mathfrak{C}_{l3} + Y_C \cdot 3S_3; & M_{D3} &= -(\mathfrak{R}_3 + M_{C3})\varepsilon_3. \end{aligned} \right\} \tag{739}$$

---

* Superposition of the Loading Conditions: 1—arbitrary beam loading; 2—arbitrary column loading; 3—rotational moments at the joints.

Column moments:

at top:
$$M_{Aa} = \mathfrak{C}_{ra} - Y_A \cdot 3S_a$$
$$M_{Bb} = \mathfrak{C}_{rb} - Y_B \cdot 3S_b$$
$$M_{Cc} = \mathfrak{C}_{rc} - Y_C \cdot 3S_c;$$

at base:
$$M_a = -(\mathfrak{L}'_a + M_{Aa})\varepsilon_a$$
$$M_b = -(\mathfrak{L}_b + M_{Bb})\varepsilon_b$$
$$M_c = -(\mathfrak{L}_c + M_{Cc})\varepsilon_c.$$

(740) and (741)

Check relationships:

$$D_A + M_{Aa} - M_{A1} = 0; \quad D_B + M_{Bb} + M_{B1} - M_{B2} = 0$$
$$D_C + M_{Cc} + M_{C2} - M_{C3} = 0.$$

(742)

Horizontal thrusts and column shears: according to formulas (36) and (38) for the collective subscripts $n = a, b, c$ and $N = A, B, C$.

Lateral restraining force:

$$H = -(H_a + H_b + H_c) \quad \text{or} \quad H = -(Q_{ra} + Q_{rb} + Q_{rc}). \quad (743)$$

Frame Shape 22v
(Frame Shape 22 with laterally unrestrained horizontal member)

Procedure

as for Frame Shape 1v, but for collective subscripts:
$$n = a, b, c \text{ and } N = A, B, C$$

Specific bending moments m: according to Method II, from formulas (738) – (741).

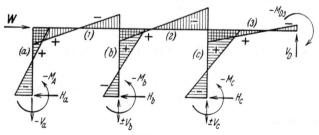

Fig. 174
Loading Condition 5 with diagram of bending moments and reactions at supports

Frame Shape 23

Single-story two-bay frame with laterally restrained horizontal member, elastically restrained at its right-hand end, and two elastically restrained columns

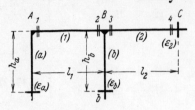

Fig. 175
Frame shape, dimensions and symbols

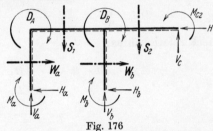

Fig. 176
Definition of positive direction for all external loads on joints and members, for all column and beam reactions, and for the lateral restraining force

Note: Frame Shape 23 is constructed at its left-hand end like Frame Shape 1 or 3, and at its right-hand end like Frame Shape 11, 12 or 13. Hence all the necessary formulas and formula matrices for Frame Shape 23 can readily be derived from those applicable to these other Frame Shapes, having due regard to the relevant subscripts. Under the present heading only the formulas for the Method of Forces (Method I) are given for Frame Shape 23.*

Important

All the formula sequences and formula matrices given for Frame Shape 23, as treated by the Method of Forces, can readily be extended to suit frames of similar type having any number of bays

---

Note: The *collective subscripts* have the following meaning:

$\nu = 1, 2$ denoting the bays (or beam spans)
$n = a, b$ denoting the columns
$N = A, B$ denoting the joints

Coefficients

Flexibility coefficients of all the horizontal members, and flexibilities of columns:

$$k_\nu = \frac{J_k}{J_\nu} \cdot \frac{l_\nu}{l_k} \qquad k_n = \frac{J_k}{J_n} \cdot \frac{h_n}{l_k}; \qquad b_n = k_n(2 - \varepsilon_n). \qquad (744)/(745)$$

Diagonal coefficients:

$$O_1 = b_a + 2k_1 \qquad O_2 = 2k_1 + b_b \qquad O_3 = b_b + 2k_2 \qquad O_4 = k_2/\varepsilon_2. \qquad (746)$$

---

* For the treatment of a similar frame by the Deformation Method (Method II) see, for example, Frame Shape 22.

Moment carry-over factors and auxiliary coefficients:

backward reduction:

$$\sigma_b = \frac{b_b}{r_3}$$

$$\iota_1 = \frac{k_1}{r_2}$$

$$r_3 = O_3 - k_2 \cdot \varepsilon_2$$

$$r_2 = O_2 - b_b \cdot \sigma_b$$

$$r_1 = O_1 - k_1 \cdot \iota_1.$$

forward reduction:

$$\iota_1' = \frac{k_1}{v_1}$$

$$\sigma_b' = \frac{b_b}{v_2}$$

$$\iota_2' = \frac{k_2}{v_3}$$

$$v_1 = O_1$$

$$v_2 = O_2 - k_1 \cdot \iota_1'$$

$$v_3 = O_3 - b_b \cdot \sigma_b'$$

$$v_4 = O_4 - k_2 \cdot \iota_2'.$$

(747) and (748)

Principal influence coefficients by recursion:

forward:

$$n_{11} = \frac{1}{r_1}$$

$$n_{22} = \frac{1}{r_2} + n_{11} \cdot \iota_1^2$$

$$n_{33} = \frac{1}{r_3} + n_{22} \cdot \sigma_b^2$$

$$n_{44} = \frac{1}{O_4} + n_{33} \cdot \varepsilon_2^2.$$

backward:

$$n_{44} = \frac{1}{v_4} = \frac{1}{k_2(1/\varepsilon_2 - \iota_2')}$$

$$n_{33} = \frac{1}{v_3} + n_{44} \cdot \iota_2'^2$$

$$n_{22} = \frac{1}{v_2} + n_{33} \cdot \sigma_b'^2$$

$$n_{11} = \frac{1}{v_1} + n_{22} \cdot \iota_1'^2.$$

(749) and (750)

Principal influence coefficients direct

$$n_{11} = \frac{1}{r_1} \quad n_{22} = \frac{1}{r_2 + v_2 - O_2} \quad n_{33} = \frac{1}{r_3 + v_3 - O_3} \quad n_{44} = \frac{1}{v_4}. \tag{751}$$

Computation scheme for the symmetrical influence coefficient matrix

$$\begin{array}{c|cccc|c}
 & (n_{11}) & n_{12} & n_{13} & n_{14} & \\
\cdot \iota_1 & \downarrow & \uparrow & \uparrow & \uparrow & \cdot \iota_1' \\
 & n_{21} & (n_{22}) & n_{23} & n_{24} & \\
\cdot \sigma_b & \downarrow & \downarrow & \uparrow & \uparrow & \cdot \sigma_b' \\
 & n_{31} & n_{32} & (n_{33}) & n_{34} & \\
\cdot \varepsilon_2 & \downarrow & \downarrow & \downarrow & \uparrow & \cdot \iota_2' \\
 & n_{41} & n_{42} & n_{43} & (n_{44}) & \\
\end{array} \tag{752}$$

Loading Condition 1

Both beam spans carrying arbitrary vertical loading

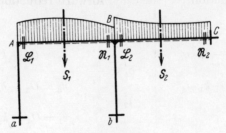

Fig. 177
Loading diagram with load terms $\mathfrak{L}$ and $\mathfrak{R}$

Beam moments ($L_\nu = \mathfrak{L}_\nu k_\nu$ and $R_\nu = \mathfrak{R}_\nu k_\nu$):

$$\begin{array}{r|cccc}
 & L_1 & R_1 & L_2 & R_2 \\
\hline
X_1 = M_{A1} = & -n_{11} & +n_{12} & +n_{13} & -n_{14} \\
X_2 = M_{B1} = & +n_{21} & -n_{22} & -n_{23} & +n_{24} \\
X_3 = M_{B2} = & +n_{31} & -n_{32} & -n_{33} & +n_{34} \\
X_4 = M_{C2} = & -n_{41} & +n_{42} & +n_{43} & -n_{44} \,.
\end{array} \qquad (753)$$

Column moments:

at top: $\qquad M_{Aa} = +X_1 \qquad\qquad M_{Bb} = -X_2 + X_3;$ (754)

at base: $\qquad M_a = -M_{Aa} \cdot \varepsilon_a \qquad M_b = -M_{Bb} \cdot \varepsilon_b\,.$ (755)

Special case: symmetrical bay loading ($R_\nu = L_\nu$):

$$\left.\begin{aligned}
X_1 = M_{A1} &= -L_1 \cdot s_{11} + L_2 \cdot s_{12} \\
X_2 = M_{B1} &= -L_1 \cdot s_{21} - L_2 \cdot s_{22} \\
X_3 = M_{B2} &= -L_1 \cdot s_{31} - L_2 \cdot s_{32} \\
X_4 = M_{C2} &= +L_1 \cdot s_{41} - L_2 \cdot s_{42};
\end{aligned}\right\} \quad (756)$$

where:

$$\left.\begin{aligned}
s_{11} &= +n_{11} - n_{12} & s_{12} &= +n_{13} - n_{14} \\
 & & s_{22} &= +n_{23} - n_{24} \\
s_{21} &= -n_{21} + n_{22} & s_{32} &= +n_{33} - n_{34} \\
s_{31} &= -n_{31} + n_{32} & & \\
s_{41} &= -n_{41} + n_{42}; & s_{42} &= -n_{43} + n_{44}\,.
\end{aligned}\right\} \quad (757)$$

Loading Conditions 2 and 3

Both columns carrying arbitrary horizontal loading (2), external rotational moments at both joints (3)

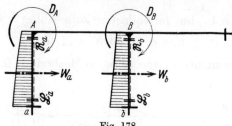

Fig. 178

Loading diagram with load terms $\mathfrak{L}$ and $\mathfrak{R}$ of the columns and with external rotational moments D

Joint load terms:

$$(L_n = \mathfrak{L}_n k_n \quad \text{and} \quad R_n = \mathfrak{R}_n k_n);$$

$$\mathfrak{B}_A = D_A \cdot b_a + L_a \cdot \varepsilon_a - R_a \qquad \mathfrak{B}_B = D_b \cdot b_b + L_b \cdot \varepsilon_b - R_b. \qquad (758)$$

Beam moments*

$$\begin{aligned}
X_1 = M_{A1} &= +\mathfrak{B}_A \cdot n_{11} + \mathfrak{B}_B \cdot w_{1b} \\
X_2 = M_{B1} &= -\mathfrak{B}_A \cdot n_{21} - \mathfrak{B}_B \cdot w_{2b} \\
X_3 = M_{B2} &= -\mathfrak{B}_A \cdot n_{31} + \mathfrak{B}_B \cdot w_{3b} \\
X_4 = M_{C2} &= +\mathfrak{B}_A \cdot n_{41} - \mathfrak{B}_B \cdot w_{4b}:
\end{aligned} \right\} \qquad (759)$$

where

$$\left. \begin{aligned} w_{1b} &= +n_{12} - n_{13} & w_{3b} &= -n_{32} + n_{33} \\ w_{2b} &= +n_{22} - n_{23} & w_{4b} &= -n_{42} + n_{43} \end{aligned} \right\} \qquad (760)$$

Column moments:

at top: $\quad M_{Aa} = -D_A + X_1 \qquad M_{Bb} = -D_B - X_2 + X_3; \qquad (761)$

at base: $\quad M_a = -(\mathfrak{L}_a + M_{Aa})\varepsilon_a \qquad M_b = -(\mathfrak{L}_b + M_{Bb})\varepsilon_b. \qquad (762)$

Horizontal thrusts and column shears for all loading conditions: according to equations (36) and (38), for the collective subscripts $n = a, b$ and $N = A, B$.

Lateral restraining force:

$$H = -H_a - H_b \quad \text{or} \quad H = -Q_{ra} - Q_{rb}. \qquad (763)$$

---

* The beam moments for a continuous frame of similar type, but with four bays, are given by the equations (91), but omitting the $\mathfrak{B}_E$ column.

Frame Shape 23v
(Frame Shape 23 with laterally unrestrained horizontal member)
Procedure

as for Frame Shape 1v, but for collective subscripts:

$$n = a, b \text{ and } N = A, B$$

Specific bending moments m: according to Method I, from formulas (759)–(762).

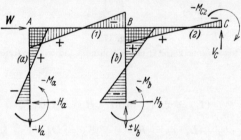

Fig. 179
Loading Condition 5 with diagram of bending moments and reactions at supports

---

Frame Shape 24
Single-story single-bay frame with laterally restrained horizontal member, elastically restrained at its right-hand end, and one elastically restrained column

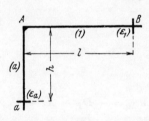 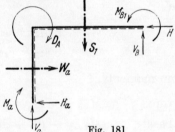

Fig. 180
Frame shape, dimensions and symbols

Fig. 181
Definition of positive direction of loads on members, of rotational moment, and of all reactions at supports

I. Treatment by the Method of Forces

Coefficients for $J_k = J_1$ and $l_k = l$, hence $k_1 = 1$:

$$k_a = \frac{J_1}{J_a} \cdot \frac{h}{l}; \qquad b_a = k_a(2 - \varepsilon_a) \qquad b_1 = (2 - \varepsilon_1); \qquad (764)$$

$$\iota'_1 = 1/(b_a + 2); \qquad N_1 = b_a + b_1 \qquad N_2 = 1/\varepsilon_1 - \iota'_1. \qquad (765)$$

Loading Condition 1
Horizontal member carrying arbitrary vertical loading

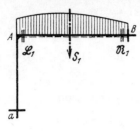

Fig. 182
Loading diagram with load terms $\mathfrak{L}$ and $\mathfrak{R}$

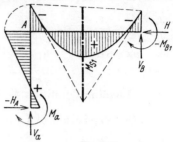

Fig. 183
Diagram of bending moments and reactions at supports

Beam and column moments ($L_1 = \mathfrak{L}_1$, $R_1 = \mathfrak{R}_1$, since $k_1 = 1$):

$$M_{A1} = M_{Aa} = -(L_1 - R_1 \cdot \varepsilon_1)/N_1 \qquad M_{B1} = -(R_1 - L_1 \cdot \iota'_1)/N_2$$
$$M_a = -M_{Aa} \cdot \varepsilon_a. \tag{766}$$

Special case: symmetrical load on beam span ($R_1 = L_1$):

$$M_{A1} = M_{Aa} = -L_1(1 - \varepsilon_1)/N_1 \qquad M_{B1} = -L_1(1 - \iota'_1)/N_2. \tag{767}$$

Loading Conditions 2 and 3
Column carrying arbitrary horizontal loading, and external rotational moment at corner of frame

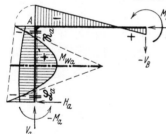
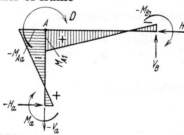

Fig. 184/185
Loading diagrams with load terms $\mathfrak{L}$ and $\mathfrak{R}$ for the column (Loading Condition 2), and with rotational moment D (Loading Condition 3), bending moment diagrams and reactions at supports

Joint load term:

$$(L_a = \mathfrak{L}_a k_a, \quad R_a = \mathfrak{R}_a k_a); \qquad \mathfrak{B}_A = D_A \cdot b_a + L_a \cdot \varepsilon_a - R_a. \tag{768}$$

Beam and column moments:

$$M_{A1} = \mathfrak{B}_A/N_1 \qquad M_{B1} = -M_{A1} \cdot \varepsilon_1 \qquad M_{Aa} = -D_A + M_{A1}$$
$$M_a = -(\mathfrak{L}_a + M_{Aa})\varepsilon_a. \tag{769}$$

II. Treatment by the Deformation Method

Coefficients for $J_k = J_1$ and $l_k = l$, hence $K_1 = 1$

$$K_a = \frac{J_a}{J_1} \cdot \frac{l}{h} = \frac{1}{k_a}; \quad S_a = \frac{K_a}{2-\varepsilon_a} \quad S_1 = \frac{1}{2-\varepsilon_1}; \quad S_A = S_a + S_1; \quad (770)$$

$$\mu_a = S_a/S_A \quad \mu_1 = S_1/S_A; \quad (\mu_a + \mu_1 = 1). \tag{771}$$

Loading Condition 4—Overall loading condition

(Superposition of Loading Conditions 1, 2 and 3)

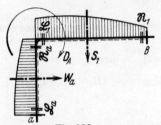

Restraining moments:

$$\mathfrak{C}_{l1} = -\frac{\mathfrak{L}_1 - \mathfrak{R}_1 \cdot \varepsilon_1}{2 - \varepsilon_1};$$

$$\mathfrak{C}_{ra} = -\frac{\mathfrak{R}_a - \mathfrak{L}_a \cdot \varepsilon_a}{2 - \varepsilon_a}. \tag{772}$$

Fig. 186
Loading diagram with load terms $\mathfrak{L}$ and $\mathfrak{R}$ and rotational moment D.

Joint load term:

$$\mathfrak{R}_A = \mathfrak{C}_{ra} - \mathfrak{C}_{l1} + D_A. \tag{773}$$

Beam and column moments:

$$M_{A1} = \mathfrak{C}_{l1} + \mathfrak{R}_A \cdot \mu_1 \quad M_{B1} = -(\mathfrak{R}_1 + M_{A1})\varepsilon_1;$$

$$M_{Aa} = \mathfrak{C}_{ra} - \mathfrak{R}_A \cdot \mu_a \quad M_a = -(\mathfrak{L}_a + M_{Aa})\varepsilon_a. \tag{774}$$

Check relationship:

$$D_A + M_{Aa} - M_{A1} = 0. \tag{775}$$

Support reactions for all loading conditions:

$$T_1 = \frac{M_{B1} - M_{A1}}{l}; \quad V_a = \frac{\mathfrak{S}_{r1}}{l} + T_1 \quad V_B = \frac{\mathfrak{S}_{l1}}{l} - T_1;$$

$$T_a = \frac{M_{Aa} - M_a}{h}; \quad H_a = \frac{\mathfrak{S}_{ra}}{h} + T_a \quad H = \frac{\mathfrak{S}_{la}}{h} - T_a. \tag{776}$$

Special case: symmetrical span loads ($\mathfrak{R}_1 = \mathfrak{L}_1; \quad \mathfrak{R}_a = \mathfrak{L}_a$):

$$\mathfrak{C}_{l1} = -\mathfrak{L}_1(1-\varepsilon_1)/(2-\varepsilon_1) \quad \mathfrak{C}_{ra} = -\mathfrak{L}_a(1-\varepsilon_a)/(2-\varepsilon_a); \tag{777}$$

$$V_a = \frac{S_1}{2} + T_1 \quad V_b = \frac{S_1}{2} - T_1; \quad H_a = \frac{W_a}{2} + T_a \quad H = \frac{W_a}{2} - T_a. \tag{778}$$

Frame Shape 24v

(Frame Shape 24 with laterally unrestrained horizontal member)

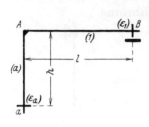

Fig. 187
Frame shape, dimensions and symbols

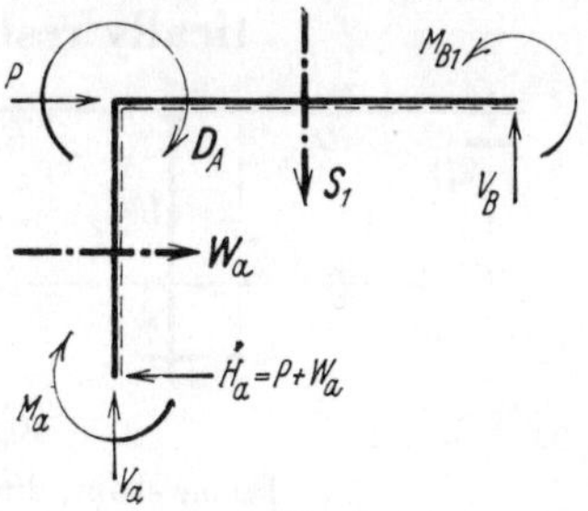

Fig. 188
Definition of positive direction for all external loads and for reactions at supports

Note: On account of the simplicity of Frame Shape 24v, direct formulas will be given for the overall loading condition, according to the Deformation Method (Method II).*

Coefficients for $J_k = J_1$ and $l_k = l$, hence $K_1 = 1$:

$$K_a = \frac{J_a}{J_1} \cdot \frac{l}{h} \qquad F_a = \frac{K_a}{1/\varepsilon_a + 4} \qquad S_1 = \frac{1}{2 - \varepsilon_1}; \qquad S_A = F_a + S_1. \qquad (779)$$

$$\eta_a = F_a/S_A \qquad \eta_1 = S_1/S_A; \qquad (\eta_a + \eta_1 = 1). \qquad (780)$$

Loading Conditions 1, 2, 3 and 5

Restraining moments at corner A:

$$\mathfrak{F}_{ra} = \frac{(\mathfrak{S}_{la} + Ph)(1/\varepsilon_a + 1) - (\mathfrak{L}_a + \mathfrak{R}_a)}{1/\varepsilon_a + 4} \; ; \qquad \mathfrak{C}_{l1} = - \frac{\mathfrak{L}_1 - \mathfrak{R}_1 \cdot \varepsilon_1}{2 - \varepsilon_1}. \qquad (781)$$

Joint load term:

$$\mathfrak{R}_A^v = \mathfrak{F}_{ra} - \mathfrak{C}_{l1} + D_A. \qquad (782)$$

Beam and column moments:

$$\begin{aligned} M_{A1} &= \mathfrak{C}_{l1} + \mathfrak{R}_A^v \cdot \eta_1 & M_{B1} &= -(\mathfrak{R}_1 + M_{A1})\varepsilon_1; \\ M_{Aa} &= \mathfrak{F}_{ra} - \mathfrak{R}_A^v \cdot \eta_a & M_a &= -\mathfrak{S}_{la} - Ph + M_{Aa}. \end{aligned} \right\} \qquad (783)$$

Check relationship:

$$D_A + M_{Aa} - M_{A1} = 0. \qquad (784)$$

Vertical support reactions:

$$T_1 = \frac{M_{B1} - M_{A1}}{l}; \qquad V_a = \frac{\mathfrak{S}_{r1}}{l} + T_1 \qquad V_B = \frac{\mathfrak{S}_{l1}}{l} - T_1. \qquad (785)$$

---

* For comparison see the treatment by the Method of Forces (Method I), for the individual loading conditions, in the case of the mirror-symmetrical Frame Shape 27v.

Frame Shape 25

Single-story three-bay frame with laterally restrained horizontal member, elastically restrained at its left-hand end, and three elastically restrained columns

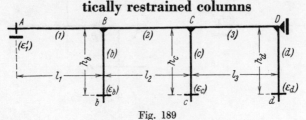

Fig. 189

Frame shape, dimensions and symbols

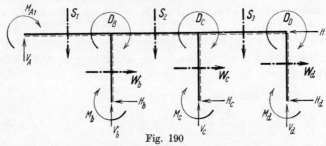

Fig. 190

Definition of positive direction for all external loads on joints and members, for all column and beam reactions, and for the lateral restraining force

Note: Frame Shape 25 is constructed at its left-hand end like Frame Shape 11, 12 or 13, and at its right-hand end like Frame Shape 1. Hence all the necessary formulas and formula matrices for Frame Shape 25 can readily be derived from those applicable to these other Frame Shapes, having due regard to the relevant subscripts. Under the present heading only the formulas for the Deformation Method (Method II) are given for Frame Shape 25.*

Important

All the formula sequences and formula matrices given for Frame Shape 25, as treated by the Deformation Method, can readily be extended to suit frames of similar type having any number of bays

Note: The *collective subscripts* have the following meaning:

ν = 1, 2, 3   denoting the bays (or beam spans)
n = b, c, d denoting the columns
N = B, C, D denoting the joints

---

* For the treatment of a similar frame by the Method of Forces (Method I) see, for example, Frame Shape 26.

Coefficients

Stiffness coefficients of all the individual members:

$$K_\nu = \frac{J_\nu}{J_k} \cdot \frac{l_k}{l_\nu} = \frac{1}{k_\nu}; \qquad K_n = \frac{J_n}{J_k} \cdot \frac{l_k}{h_n} = \frac{1}{k_n}. \qquad (786)$$

Stiffnesses of the elastically restrained columns and of the right-hand end span of the horizontal member (the values ϵ are assumed to be known).

$$S_1 = \frac{K_1}{2 - \varepsilon_1'}; \qquad S_n = \frac{K_n}{2 - \varepsilon_n}. \qquad (787)$$

Joint coefficients:

$$K_B = 3S_1 + 3S_b + 2K_2 \qquad K_C = 2K_2 + 3S_c + 2K_3 \qquad K_D = 2K_3 + 3S_d. \qquad (788)$$

Rotation carry-over factors and auxiliary coefficients:

$$\begin{array}{l|l}
\text{backward reduction:} & \text{forward reduction:} \\
\end{array}$$

$$\left.\begin{array}{ll}
j_3 = \dfrac{K_3}{r_d} & \quad r_d = K_D \\
 & \quad r_c = K_C - K_3 \cdot j_3 \\
j_2 = \dfrac{K_2}{r_c} & \quad r_b = K_B - K_2 \cdot j_2.
\end{array} \;\middle|\; \begin{array}{ll}
j_2' = \dfrac{K_2}{v_b} & \quad v_b = K_B \\
 & \quad v_c = K_C - K_2 \cdot j_2' \\
j_3' = \dfrac{K_3}{v_c} & \quad v_d = K_D - K_3 \cdot j_3'.
\end{array}\right\} \begin{array}{c}(789)\\ \text{and}\\ (790)\end{array}$$

Principal influence coefficients by recursion:

$$\left.\begin{array}{l|l}
\text{forward:} & \text{backward:} \\
u_{bb} = 1/r_b & u_{dd} = 1/v_d \\
u_{cc} = 1/r_c + u_{bb} \cdot j_2^2 & u_{cc} = 1/v_c + u_{dd} \cdot j_3'^2 \\
u_{dd} = 1/r_d + u_{cc} \cdot j_3^2. & u_{bb} = 1/v_b + u_{cc} \cdot j_2'^2.
\end{array}\right\} \begin{array}{c}(791)\\ \text{and}\\ (792)\end{array}$$

Principal influence coefficients direct

$$u_{bb} = \frac{1}{r_b} \qquad u_{cc} = \frac{1}{r_c + v_c - K_C} \qquad u_{dd} = \frac{1}{v_d}. \qquad (793)$$

Computation scheme for the symmetrical influence coefficient matrix

$$\left.\begin{array}{c}
 \\
\cdot j_2 \\
\cdot j_3
\end{array} \;\middle|\; \begin{array}{ccc}
(u_{bb}) & u_{bc} & u_{bd} \\
\downarrow & \uparrow & \uparrow \\
u_{cb} & (u_{cc}) & u_{cd} \\
\downarrow & \downarrow & \uparrow \\
u_{db} & u_{dc} & (u_{dd})
\end{array} \;\middle|\; \begin{array}{c}
 \\
\cdot j_2' \\
\cdot j_3'
\end{array}\right\} \qquad (794)$$

Loading Condition 4

Overall loading condition*

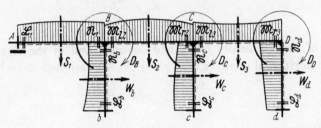

Fig. 191

Loading diagram, with fixed-end moments $\mathfrak{M}_l$ and $\mathfrak{M}_r$ of beam spans 1 and 2, load terms $\mathfrak{L}$ and $\mathfrak{R}$ for the columns and beam span 3, and external rotational moments D at the joints

Restraining moments of the elastically restrained members:

$$\mathfrak{C}_{r1} = -\frac{\mathfrak{R}_1 - \mathfrak{L}_1 \cdot \varepsilon_1'}{2 - \varepsilon_1'} \qquad \mathfrak{C}_{rn} = -\frac{\mathfrak{R}_n - \mathfrak{L}_n \cdot \varepsilon_n}{2 - \varepsilon_n}. \tag{795}$$

Joint load terms:

$$\left. \begin{aligned} \mathfrak{K}_B &= \mathfrak{C}_{r1} - \mathfrak{M}_{l2} + \mathfrak{C}_{rb} + D_B \\ \mathfrak{K}_C &= \mathfrak{M}_{r2} - \mathfrak{M}_{l3} + \mathfrak{C}_{rc} + D_C \\ \mathfrak{K}_D &= \mathfrak{M}_{r3} \phantom{- \mathfrak{M}_{l3}} + \mathfrak{C}_{rd} + D_D \,. \end{aligned} \right\} \tag{796}$$

auxiliary moments:

$$\left. \begin{array}{c|ccc} & \mathfrak{K}_B & \mathfrak{K}_C & \mathfrak{K}_D \\ \hline Y_B = & +u_{bb} & -u_{bc} & +u_{bd} \\ Y_C = & -u_{cb} & +u_{cc} & -u_{cd} \\ Y_D = & +u_{db} & -u_{dc} & +u_{dd} \,. \end{array} \right\} \tag{797}$$

Beam moments:

$$\left. \begin{aligned} M_{A1} &= -(\mathfrak{L}_1 + M_{B1})\varepsilon_1' & M_{B1} &= \mathfrak{C}_{r1} - Y_B \cdot 3S_1 \\ M_{B2} &= \mathfrak{M}_{l2} + (2Y_B + Y_C)K_2 & M_{C2} &= \mathfrak{M}_{r2} - (Y_B + 2Y_C)K_2 \\ M_{C3} &= \mathfrak{M}_{l3} + (2Y_C + Y_D)K_3; & M_{D3} &= \mathfrak{M}_{r3} - (Y_C + 2Y_D)K_3 \,. \end{aligned} \right\} \tag{798}$$

---

* Superposition of the Loading Conditions: 1—arbitrary beam loading; 2—arbitrary loading; 3—rotational moments at the joints. — Note: in Fig. 191 at corner D the indication of the horizontally restrained bearing is missing (see Fig. 189).

Column moments:

$$\begin{array}{ll}
\text{at top:} & \text{at base:} \\
M_{Bb} = \mathfrak{C}_{rb} - Y_B \cdot 3S_b & M_b = -(\mathfrak{L}_b + M_{Bb})\varepsilon_b \\
M_{Cc} = \mathfrak{C}_{rc} - Y_C \cdot 3S_c & M_c = -(\mathfrak{L}_c + M_{Cc})\varepsilon_c \\
M_{Dd} = \mathfrak{C}_{rd} - Y_D \cdot 3S_d; & M_d = -(\mathfrak{L}_d + M_{Dd})\varepsilon_d.
\end{array} \quad \begin{array}{c} (799) \\ \text{and} \\ (800) \end{array}$$

Check relationships:

$$\left. \begin{array}{l}
D_B + M_{Bb} + M_{B1} - M_{B2} = 0 \\
D_C + M_{Cc} + M_{C2} - M_{C3} = 0 \\
D_D + M_{Dd} + M_{D3} = 0.
\end{array} \right\} \quad (801)$$

Horizontal thrusts and column shears: according to formulas (36) and (38) for the collective subscripts $n = b, c, d$ and $N = B, C, D$.

Lateral restraining force:

$$H = -(H_b + H_c + H_d) \quad \text{or} \quad H = -(Q_{rb} + Q_{rc} + Q_{rd}). \quad (802)$$

Frame Shape 25v

(Frame Shape 25 with laterally unrestrained horizontal member)

Procedure

as for Frame Shape 1v, but for collective subscripts:
$$n = b, c, d \text{ and } N = B, C, D.$$

Specific bending moments m: according to Method II, from formulas (797) to (800), page 180/181.

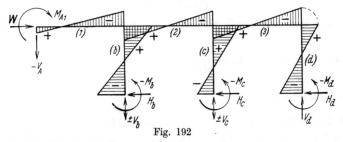

Fig. 192

Loading Condition 5 with diagram of bending moments and reactions at supports

Frame Shape 26

Single-story two-bay frame with laterally restrained horizontal member, elastically restrained at its left-hand end, and two elastically restrained columns

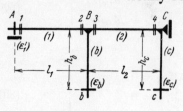

Fig. 193
Frame shape, dimensions and symbols

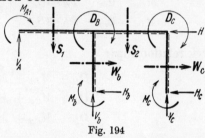

Fig. 194
Definition of positive direction for all external loads on joints and members, for all column and beam reactions, and for the lateral restraining force

Note: Frame Shape 26 is constructed at its left-hand end like Frame Shape 11, 12 or 13, and at its right-hand end like Frame Shape 1. Hence all the necessary formulas and formula matrices for Frame Shape 26 can readily be derived from those applicable to these other Frame Shapes, having due regard to the relevant subscripts. Under the present heading only the formulas for the Method of Forces (Method I) are given for Frame Shape 26.*

Important

All the formula sequences and formula matrices given for Frame Shape 26, as treated by the Method of Forces, can readily be extended to suit frames of similar type having any number of bays.

Note: The *collective subscripts* have the following meaning:

$\nu = 1, 2$ denoting the bays (or beam spans)
$n = b, c$ denoting the columns
$N = B, C$ denoting the joints

Coefficients

Flexibility coefficients of all the horizontal members, and flexibilities of columns:

$$k_\nu = \frac{J_k}{J_\nu} \cdot \frac{l_\nu}{l_k} \qquad k_n = \frac{J_k}{J_n} \cdot \frac{h_n}{l_k} \ ; \qquad b_n = k_n(2 - \varepsilon_n) \cdot \qquad (803)/(804)$$

Diagonal coefficients:

$$O_1 = k_1/\varepsilon_1' \qquad O_2 = 2k_1 + b_b \qquad O_3 = b_b + 2k_2 \qquad O_4 = 2k_2 + b_c. \qquad (805)$$

---

* For the treatment of a similar frame by the Deformation Method (Method II) see, for example, Frame Shape 25.

Moment carry-over factors and auxiliary coefficients:

backward reduction:

$$\iota_2 = \frac{k_2}{r_4}$$
$$\sigma_b = \frac{b_b}{r_3}$$
$$\iota_1 = \frac{k_1}{r_2}$$

$$r_4 = O_4$$
$$r_3 = O_3 - k_2 \cdot \iota_2$$
$$r_2 = O_2 - b_b \cdot \sigma_b$$
$$r_1 = O_1 - k_1 \cdot \iota_1.$$

forward reduction:

$$\sigma_b' = \frac{b_b}{v_2}$$
$$\iota_2' = \frac{k_2}{v_3}$$

$$v_2 = O_2 - k_1 \cdot \varepsilon_1'$$
$$v_3 = O_3 - b_b \cdot \sigma_b'$$
$$v_4 = O_4 - k_2 \cdot \iota_2'.$$

(806) and (807)

Principal influence coefficients by recursion:

forward:

$$n_{11} = \frac{1}{r_1} = \frac{1}{k_1(1/\varepsilon_1' - \iota_1)}$$
$$n_{22} = \frac{1}{r_2} + n_{11} \cdot \iota_1^2$$
$$n_{33} = \frac{1}{r_3} + n_{22} \cdot \sigma_b^2$$
$$n_{44} = \frac{1}{r_4} + n_{33} \cdot \iota_2^2.$$

backward:

$$n_{44} = \frac{1}{v_4}$$
$$n_{33} = \frac{1}{v_3} + n_{44} \cdot \iota_2'^2$$
$$n_{22} = \frac{1}{v_2} + n_{33} \cdot \sigma_b'^2$$
$$n_{11} = \frac{1}{O_1} + n_{22} \cdot \varepsilon_1'^2.$$

(808) and (809)

Principal influence coefficients direct

$$n_{11} = \frac{1}{r_1} \qquad n_{22} = \frac{1}{r_2 + v_2 - O_2} \qquad n_{33} = \frac{1}{r_3 + v_3 - O_3} \qquad n_{44} = \frac{1}{v_4}. \qquad (810)$$

Computation scheme for the symmetrical influence coefficient matrix

$$
\begin{array}{c|cccc|c}
 & (n_{11}) & n_{12} & n_{13} & n_{14} & \\
\cdot \iota_1 & \downarrow & \uparrow & \uparrow & \uparrow & \cdot \varepsilon_1' \\
 & n_{21} & (n_{22}) & n_{23} & n_{24} & \\
\cdot \sigma_b & \downarrow & \downarrow & \uparrow & \uparrow & \cdot \sigma_b' \\
 & n_{31} & n_{32} & (n_{33}) & n_{34} & \\
\cdot \iota_2 & \downarrow & \downarrow & \downarrow & \uparrow & \cdot \iota_2' \\
 & n_{41} & n_{42} & n_{43} & (n_{44}) & \\
\end{array}
\qquad (811)
$$

Loading Condition 1

Both beam spans carrying arbitrary vertical loading

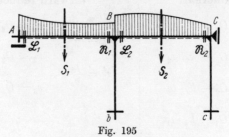

Fig. 195
Loading diagram with load terms $\mathfrak{L}$ and $\mathfrak{R}$

Beam moments ($L_\nu = \mathfrak{L}_\nu k_\nu$ and $R_\nu = \mathfrak{R}_\nu k_\nu$):

$$
\begin{array}{c|cccc}
 & L_1 & R_1 & L_2 & R_2 \\
\hline
X_1 = M_{A1} = & -n_{11} & +n_{12} & +n_{13} & -n_{14} \\
X_2 = M_{B1} = & +n_{21} & -n_{22} & -n_{23} & +n_{24} \\
X_3 = M_{B2} = & +n_{31} & -n_{32} & -n_{33} & +n_{34} \\
X_4 = M_{C2} = & -n_{41} & +n_{42} & +n_{43} & -n_{44}
\end{array}
\qquad (812)
$$

Column moments:

at top: $\qquad M_{Bb} = -X_2 + X_3 \qquad M_{Cc} = -X_4;$ \qquad (813)

at base: $\qquad M_b = -M_{Bb}\cdot \varepsilon_b \qquad M_c = -M_{Cc}\cdot \varepsilon_c.$ \qquad (814)

Special case: symmetrical bay loading $(R_\nu = L_\nu)$:

$$
\begin{aligned}
X_1 = M_{A1} &= -L_1 \cdot s_{11} + L_2 \cdot s_{12} \\
X_2 = M_{B1} &= -L_1 \cdot s_{21} - L_2 \cdot s_{22} \\
X_3 = M_{B2} &= -L_1 \cdot s_{31} - L_2 \cdot s_{32} \\
X_4 = M_{C2} &= +L_1 \cdot s_{41} - L_2 \cdot s_{42};
\end{aligned}
\qquad (815)
$$

where

$$
\begin{aligned}
s_{11} &= +n_{11} - n_{12} & s_{12} &= +n_{13} - n_{14} \\
& & s_{22} &= +n_{23} - n_{24} \\
s_{21} &= -n_{21} + n_{22} & s_{32} &= +n_{33} - n_{34} \\
s_{31} &= -n_{31} + n_{32} & & \\
s_{41} &= -n_{41} + n_{42}; & s_{42} &= -n_{43} + n_{44}.
\end{aligned}
\qquad (816)
$$

Loading Conditions 2 and 3

Both columns carrying arbitrary horizontal loading (2), external rotational moments at both joints (3)

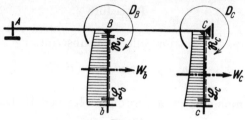

Fig. 196

Loading diagram with load terms $\mathfrak{L}$ and $\mathfrak{R}$ of the columns and with external rotational moments D

Joint load terms:

$$(L_n = \mathfrak{L}_n k_n \quad \text{and} \quad R_n = \mathfrak{R}_n k_n);$$

$$\mathfrak{B}_B = \boldsymbol{D}_B \cdot b_b + \boldsymbol{L}_b \cdot \varepsilon_b - \boldsymbol{R}_b \qquad \mathfrak{B}_C = \boldsymbol{D}_C \cdot b_c + \boldsymbol{L}_c \cdot \varepsilon_c - \boldsymbol{R}_c. \qquad (817)$$

Beam moments*:

$$\left.\begin{aligned}
X_1 &= M_{A1} = +\mathfrak{B}_B \cdot w_{1b} - \mathfrak{B}_C \cdot n_{14} \\
X_2 &= M_{B1} = -\mathfrak{B}_B \cdot w_{2b} + \mathfrak{B}_C \cdot n_{24} \\
X_3 &= M_{B2} = +\mathfrak{B}_B \cdot w_{3b} + \mathfrak{B}_C \cdot n_{34} \\
X_4 &= M_{C2} = -\mathfrak{B}_B \cdot w_{4b} - \mathfrak{B}_C \cdot n_{44};
\end{aligned}\right\} \qquad (818)$$

where:

$$\left.\begin{aligned}
w_{1b} &= +n_{12} - n_{13} & w_{3b} &= -n_{32} + n_{33} \\
w_{2b} &= +n_{22} - n_{23} & w_{4b} &= -n_{42} + n_{43}.
\end{aligned}\right\} \qquad (819)$$

Column moments:

at top: $\qquad M_{Bb} = -\boldsymbol{D}_B - X_2 + X_3 \qquad M_{Cc} = -\boldsymbol{D}_C - X_4; \qquad (820)$

at base: $\qquad M_b = -(\mathfrak{L}_b + M_{Bb})\varepsilon_b \qquad M_c = -(\mathfrak{L}_c + M_{Cc})\varepsilon_c. \qquad (821)$

Horizontal thrusts and column shears for all loading conditions: according to equations (36) and (38), for the collective subscripts $n = a, b$ and $N = A, B$.

Lateral restraining force:

$$H = -H_b - H_c \quad \text{or} \quad H = -Q_{rb} - Q_{rc}. \qquad (822)$$

---

* The beam moments for a continuous frame of similar type, but with four bays, are given by the equations (91), but omitting the $\mathfrak{B}_A$ column.

Frame Shape 26v

(Frame Shape 26 with laterally unrestrained horizontal member)

Procedure

as for Frame Shape 1v, but for collective subscripts:

$$n = b, c \text{ and } N = B, C$$

Specific bending moments m: according to Method II, from formulas (818)–(821).

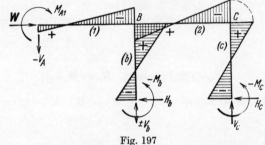

Fig. 197
Loading Condition 5 with diagram of bending moments and reactions at supports

Frame Shape 27

Single-story single-bay frame with laterally restrained horizontal member, elastically restrained at its left-hand end, and one elastically restrained column

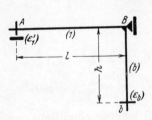

Fig. 198
Frame shape, dimensions and symbols

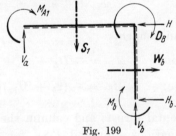

Fig. 199
Definition of positive direction of loads on members, of rotational moment, and of all reactions at supports

Note: The analysis will be treated for the overall loading condition by the Deformation Method (Method II) only. For a treatment of a frame of similar type by the Method of Forces (Method I) see the mirror image Frame Shape 24.

Coefficients for $J_k = J_1$ and $l_k = l$, hence $K_1 = 1$:

$$K_b = \frac{J_b}{J_1} \cdot \frac{l}{h}; \qquad S_1 = \frac{1}{2-\varepsilon_1'} \qquad S_b = \frac{K_b}{2-\varepsilon_b}; \qquad S_B = S_1 + S_b. \qquad (823)$$

$$\mu_1 = S_1/S_B \qquad \mu_b = S_b/S_B; \qquad (\mu_1 + \mu_b = 1). \qquad (824)$$

Loading Condition 4—Overall loading condition

(Superposition of Loading Conditions 1, 2 and 3)

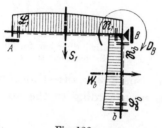

Fig. 186
Loading diagram with load terms $\mathfrak{L}$ and $\mathfrak{R}$ and rotational moment D

Restraining moments:

$$\left. \begin{aligned} \mathfrak{E}_{r1} &= -\frac{\mathfrak{R}_1 - \mathfrak{L}_1 \cdot \varepsilon_1'}{2-\varepsilon_1'}; \\ \mathfrak{E}_{rb} &= -\frac{\mathfrak{R}_b - \mathfrak{L}_b \cdot \varepsilon_b}{2-\varepsilon_b}. \end{aligned} \right\} \qquad (825)$$

Joint load term:

$$\mathfrak{R}_B = \mathfrak{E}_{r1} + \mathfrak{E}_{rb} + D_B. \qquad (826)$$

Beam and column moments:

$$\left. \begin{aligned} M_{B1} &= \mathfrak{E}_{r1} - \mathfrak{R}_B \cdot \mu_1 & M_{A1} &= -(\mathfrak{L}_1 + M_{B1})\varepsilon_1'; \\ M_{Bb} &= \mathfrak{E}_{rb} - \mathfrak{R}_B \cdot \mu_b & M_b &= -(\mathfrak{L}_b + M_{Bb})\varepsilon_b. \end{aligned} \right\} \qquad (827)$$

Check relationship:

$$D_B + M_{B1} + M_{Bb} = 0. \qquad (828)$$

Support reactions for all loading conditions:

$$\left. \begin{aligned} T_1 &= \frac{M_{B1} - M_{A1}}{l}; & V_a &= \frac{\mathfrak{E}_{r1}}{l} + T_1 & V_b &= \frac{\mathfrak{E}_{l1}}{l} - T_1; \\ T_b &= \frac{M_{Bb} - M_b}{h}; & H_b &= \frac{\mathfrak{E}_{rb}}{h} + T_b & H &= \frac{\mathfrak{E}_{lb}}{h} - T_b. \end{aligned} \right\} \qquad (829)$$

Special case: symmetrical span loads $(\mathfrak{R}_1 = \mathfrak{L}_1; \;\; \mathfrak{R}_b = \mathfrak{L}_b)$:

$$\mathfrak{E}_{r1} = -\mathfrak{L}_1 \cdot \frac{1-\varepsilon_1'}{2-\varepsilon_1'} \qquad \mathfrak{E}_{rb} = -\mathfrak{L}_b \cdot \frac{1-\varepsilon_b}{2-\varepsilon_b}; \qquad (830)$$

$$V_a = S/2 + T_1 \quad V_b = S/2 - T_1 \quad H_b = W/2 + T_b \quad H = W/2 - T_b. \quad (831)$$

Frame Shape 27v

(Frame Shape 27 with laterally unrestrained horizontal member)

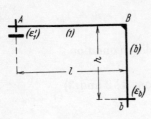

Fig. 201
Frame shape, dimensions and symbols

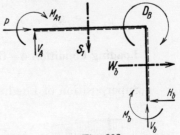

Fig. 202
Definition of positive direction for all external loads and for reactions at supports

Note: On account of the simplicity of Frame Shape 27v, direct formulas will be given for all the individual loading cases, according to the Method of Forces (Method I).*

Coefficients for $J_k = J_1$ and $l_k = l$, hence $k_1 = 1$:

$$k_b = \frac{J_1}{J_b} \cdot \frac{h}{l}; \qquad b_1 = (2 - \varepsilon_1') \qquad m_b = k_b(1/\varepsilon_b + 4); \qquad (832)$$
$$N = b_1 + m_b.$$

Loading Condition 1—Horizontal member carrying arbitrary vertical loading

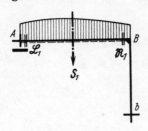

Fig. 203
Beam loading diagram with load terms $\mathfrak{L}$ and $\mathfrak{R}$

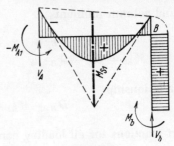

Fig. 204
Diagram of bending moments and reactions at supports

Beam and column moments:

$$M_{Bb} = M_b = -M_{B1} = \frac{\mathfrak{R}_1 - \mathfrak{L}_1 \cdot \varepsilon_1'}{N} \qquad M_{A1} = -(\mathfrak{L}_1 + M_{B1})\varepsilon_1'. \qquad (833)$$

Vertical support reactions:

$$T_1 = \frac{M_{B1} - M_{A1}}{l}; \qquad V_A = \frac{\mathfrak{S}_{r1}}{l} + T_1 \qquad V_b = \frac{\mathfrak{S}_{l1}}{l} - T_1. \qquad (834)$$

---

* For comparison see the treatment by the Deformation Method (Method II), for the overall loading condition, in the case of the mirror-symmetrical Frame Shape 24v.

Loading Condition 2—Column carrying arbitrary horizontal loading

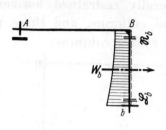
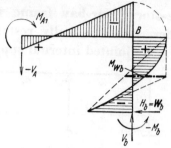

Fig. 205
Column loading diagram with load terms $\mathfrak{L}$ and $\mathfrak{R}$

Fig. 206
Diagram of bending moments and reactions at supports

$$M_{Bb} = -M_{B1} = [\mathfrak{S}_{lb}(1/\varepsilon_b + 1) - (\mathfrak{L}_b + \mathfrak{R}_b)]k_b/N$$
$$M_{A1} = -M_{B1} \cdot \varepsilon_1' \qquad M_b = -\mathfrak{S}_{lb} + M_{Bb}. \tag{835}$$

Loading Condition 3—
Rotational moment at corner B.

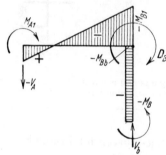

Fig. 207
Loading diagram, diagram of bending moments and reactions at the supports

$$M_{B1} = -D_B \cdot m_b/N \qquad M_{A1} = -M_{B1} \cdot \varepsilon_1'$$
$$M_{Bb} = M_b = -D_B - M_{B1} = -D_B \cdot b_1/N. \tag{836}$$

Loading Condition 5—
Horizontal point load at level of horizontal member

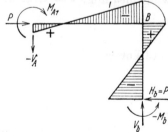

Fig. 208
Loading diagram, diagram of bending moments and reactions at the supports

$$M_{Bb} = -M_{B1} = Ph \cdot (1/\varepsilon_b + 1)k_b/N \qquad M_{A1} = -M_{B1} \cdot \varepsilon_1'$$
$$M_b = -Ph + M_{Bb} = -Ph \cdot (b_1 + 3k_b)/N. \tag{837}$$

Formula for the vertical support reactions for Loading Conditions 2, 3 and 5:
$$V_b = -V_A = -T_1 = -M_{B1}(1 + \varepsilon_1')/l. \tag{838}$$

Frame Shape 28

Single-story four-bay frame with laterally restrained horizontal member, two elastically restrained end columns, and three pin-jointed internal supports (or rocker columns)

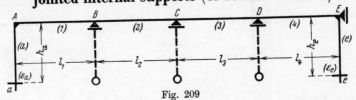

Fig. 209
Frame shape, dimensions and symbols

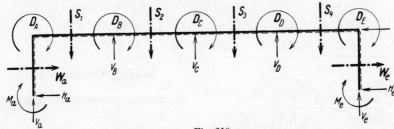

Fig. 210
Definition of positive direction for all external loads on joints and members, for all reactions at the supports, and for the lateral restraining force

---
Important

All the formula sequences and formula matrices given for Frame Shape 28 can readily be extended to suit frames of similar type having any even number of bays.*

Note: The *collective subscripts* have the following meaning: $\nu = 1, 2, 3, 4$ (column numbers $n = a, e$)

I. Treatment by the Method of Forces

Coefficients

Flexibility coefficients of all horizontal members:

$$k_a = \frac{J_k}{J_a} \cdot \frac{h_a}{l_k} \qquad k_\nu = \frac{J_k}{J_\nu} \cdot \frac{l_\nu}{l_k} \qquad k_e = \frac{J_k}{J_e} \cdot \frac{h_e}{l_k}. \qquad (839)$$

Flexibilities of the elastically restrained columns:

$$b_a = k_a(2 - \varepsilon_a) \qquad b_e = k_e(2 - \varepsilon_e). \qquad (840)$$

---
* The dotted lines in the sets of formulas relate to Frame Shape 29.

Diagonal coefficients:

$$k_A = b_a + 2k_1$$
$$k_B = 2k_1 + 2k_2$$
$$k_C = 2k_2 + 2k_3$$
$$k_D = 2k_3 + 2k_4$$
$$k_E = 2k_4 + b_e.$$
\hfill (841)

Moment carry-over factors and auxiliary coefficients:

backward reduction:

$$\iota_4 = \frac{k_4}{r_e} \qquad r_e = k_E$$
$$\iota_3 = \frac{k_3}{r_d} \qquad r_d = k_D - k_4 \cdot \iota_4$$
$$\iota_2 = \frac{k_2}{r_c} \qquad r_c = k_C - k_3 \cdot \iota_3$$
$$\iota_1 = \frac{k_1}{r_b} \qquad r_b = k_B - k_2 \cdot \iota_2$$
$$\qquad\qquad r_a = k_A - k_1 \cdot \iota_1.$$

forward reduction:

$$\iota'_1 = \frac{k_1}{v_a} \qquad v_a = k_A$$
$$\iota'_2 = \frac{k_2}{v_b} \qquad v_b = k_B - k_1 \cdot \iota'_1$$
$$\iota'_3 = \frac{k_3}{v_c} \qquad v_c = k_C - k_2 \cdot \iota'_2$$
$$\iota'_4 = \frac{k_4}{v_d} \qquad v_d = k_D - k_3 \cdot \iota'_3$$
$$\qquad\qquad v_e = k_E - k_4 \cdot \iota'_4.$$

(842) and (843)

Principal influence coefficients by recursion:

forward:

$$n_{aa} = 1/r_a$$
$$n_{bb} = 1/r_b + n_{aa} \cdot \iota_1^2$$
$$n_{cc} = 1/r_c + n_{bb} \cdot \iota_2^2$$
$$n_{dd} = 1/r_d + n_{cc} \cdot \iota_3^2$$
$$n_{ee} = 1/r_e + n_{dd} \cdot \iota_4^2.$$

backward:

$$n_{ee} = 1/v_e$$
$$n_{dd} = 1/v_d + n_{ee} \cdot \iota'^2_4$$
$$n_{cc} = 1/v_c + n_{dd} \cdot \iota'^2_3$$
$$n_{bb} = 1/v_b + n_{cc} \cdot \iota'^2_2$$
$$n_{aa} = 1/v_a + n_{bb} \cdot \iota'^2_1.$$

(844) and (845)

Computation scheme for the symmetrical influence coefficient matrix*

$$
\begin{array}{c|ccccc|c}
\cdot \iota_1 & (n_{aa}) & n_{ab} & n_{ac} & n_{ad} & n_{ae} & \\
 & \downarrow & \uparrow & \uparrow & \uparrow & \uparrow & \cdot \iota'_1 \\
\cdot \iota_2 & n_{ba} & (n_{bb}) & n_{bc} & n_{bd} & n_{be} & \\
 & \downarrow & \downarrow & \downarrow & \uparrow & \uparrow & \cdot \iota'_2 \\
\cdot \iota_3 & n_{ca} & n_{cb} & (n_{cc}) & n_{cd} & n_{ce} & \\
 & \downarrow & \downarrow & \downarrow & \uparrow & \uparrow & \cdot \iota'_3 \\
\cdot \iota_4 & n_{da} & n_{db} & n_{dc} & (n_{dd}) & n_{de} & \\
 & \downarrow & \downarrow & \downarrow & \downarrow & \uparrow & \cdot \iota'_4 \\
 & n_{ea} & n_{eb} & n_{ec} & n_{ed} & (n_{ee}) & \\
\end{array}
$$
\hfill (846)

---

* Principal influence coefficients direct: see p. 192.

Principal influence coefficients direct:

$$n_{aa} = 1/r_a \qquad n_{bb} = 1/(r_b + v_b - k_B) \qquad n_{cc} = 1/(r_c + v_c - k_C) \atop n_{dd} = 1/(r_d + v_d - k_D) \qquad n_{ee} = 1/v_e. \qquad (847)$$

Loading Condition 1
All the beam spans carrying arbitrary vertical loading

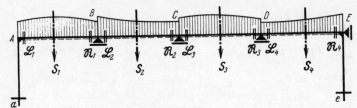

Fig. 211
Beam loading diagram with load terms $\mathfrak{L}$ and $\mathfrak{R}$

Corner and beam moments ($L_\nu = \mathfrak{L}_\nu k_\nu$ and $R_\nu = \mathfrak{R}_\nu k_\nu$):

	L_1	$(R_1 + L_2)$	$(R_2 + L_3)$	$(R_3 + L_4)$	R_4
$+ M_{Aa} = M_{A1} =$	$- n_{aa}$	$+ n_{ab}$	$- n_{ac}$	$+ n_{ad}$	$- n_{ae}$
$M_B =$	$+ n_{ba}$	$- n_{bb}$	$+ n_{bc}$	$- n_{bd}$	$+ n_{be}$
$M_C =$	$- n_{ca}$	$+ n_{cb}$	$- n_{cc}$	$+ n_{cd}$	$- n_{ce}$
$M_D =$	$+ n_{da}$	$- n_{db}$	$+ n_{dc}$	$- n_{dd}$	$+ n_{de}$
$- M_{Ee} = M_{E4} =$	$- n_{ea}$	$+ n_{eb}$	$- n_{ec}$	$+ n_{ed}$	$- n_{ee}$

(848)

Column base moments:

$$M_a = - M_{Aa} \cdot \varepsilon_a \qquad M_e = - M_{Ee} \cdot \varepsilon_e. \qquad (849)$$

Special case: symmetrical bay loading ($R_\nu = L_\nu$):

	L_1	L_2	L_3	L_4
$+ M_{Aa} = M_{A1} =$	$- s_{a1}$	$+ s_{a2}$	$- s_{a3}$	$+ s_{a4}$
$M_B =$	$- s_{b1}$	$- s_{b2}$	$+ s_{b3}$	$- s_{b4}$
$M_C =$	$+ s_{c1}$	$- s_{c2}$	$- s_{c3}$	$+ s_{c4}$
$M_D =$	$- s_{d1}$	$+ s_{d2}$	$- s_{d3}$	$- s_{d4}$
$- M_{Ee} = M_{E4} =$	$+ s_{e1}$	$- s_{e2}$	$+ s_{e3}$	$- s_{e4}$

(850)

Note: values s are given on next page.

Composite influence coefficients for Loading Condition 1:

$$s_{a1} = + n_{aa} - n_{ab} \qquad s_{a2} = + n_{ab} - n_{ac}$$
$$s_{b2} = + n_{bb} - n_{bc}$$
$$s_{b1} = - n_{ba} + n_{bb}$$
$$s_{c1} = - n_{ca} + n_{cb} \qquad s_{c2} = - n_{cb} + n_{cc}$$
$$s_{d1} = - n_{da} + n_{db} \qquad s_{d2} = - n_{db} + n_{dc}$$
$$s_{e1} = - n_{ea} + n_{eb} \qquad s_{e2} = - n_{eb} + n_{ec}$$

$$s_{a3} = + n_{ac} - n_{ad} \qquad s_{a4} = + n_{ad} - n_{ae}$$
$$s_{b3} = + n_{bc} - n_{bd} \qquad s_{b4} = + n_{bd} - n_{be}$$
$$s_{c3} = + n_{cc} - n_{cd} \qquad s_{c4} = + n_{cd} - n_{ce}$$
$$s_{d4} = + n_{dd} - n_{de}$$
$$s_{d3} = - n_{dc} + n_{dd}$$
$$s_{e3} = - n_{ec} + n_{ed} \qquad s_{e4} = - n_{ed} + n_{ee} .$$

(851)

Loading Conditions 2 and 3

Both columns carrying arbitrary horizontal loading (2) external rotational moments applied at the corners of the frame (3)

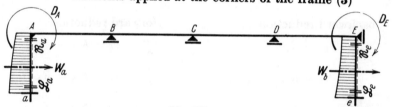

Fig. 212
Diagram of loading on columns, with load terms $\mathfrak{L}$ and $\mathfrak{R}$, and rotational moments D at corners

Joint load terms $(L_n = \mathfrak{L}_n k_n \text{ and } R_n = \mathfrak{R}_n k_n)$:

$$\mathfrak{B}_A = D_A \cdot b_a + L_a \cdot \varepsilon_a - R_a \qquad \mathfrak{B}_E = D_E \cdot b_e + L_e \cdot \varepsilon_e - R_e . \qquad (852)$$

Beam moments: | **Column moments:**

(853) at top:

$$M_{A1} = + \mathfrak{B}_A \cdot n_{aa} - \mathfrak{B}_E \cdot n_{ae} \qquad M_{Aa} = - D_A + M_{A1}$$
$$M_B = - \mathfrak{B}_A \cdot n_{ba} + \mathfrak{B}_E \cdot n_{be} \qquad M_{Ee} = - D_E - M_{E4} ;$$

(854)

$$M_C = + \mathfrak{B}_A \cdot n_{ca} - \mathfrak{B}_E \cdot n_{ce}$$

at base:

$$M_D = - \mathfrak{B}_A \cdot n_{da} + \mathfrak{B}_E \cdot n_{de} \qquad M_a = - (\mathfrak{L}_a + M_{Aa}) \varepsilon_a$$
$$M_{E4} = + \mathfrak{B}_A \cdot n_{ea} - \mathfrak{B}_E \cdot n_{ee} . \qquad M_e = - (\mathfrak{L}_e + M_{Ee}) \varepsilon_e .$$

(855)

II. Treatment by the Deformation Method

Coefficients

Stiffness coefficients for all individual members:

$$K_a = \frac{J_a}{J_k} \cdot \frac{l_k}{h_a} = \frac{1}{k_a} \; ; \quad K_v = \frac{J_v}{J_k} \cdot \frac{l_k}{l_v} = \frac{1}{k_v} \; ; \quad K_e = \frac{J_e}{J_k} \cdot \frac{l_k}{h_e} = \frac{1}{k_e} \, . \tag{856}$$

Stiffnesses of the elastically restrained columns:

$$S_a = \frac{K_a}{2 - \varepsilon_a} = \frac{1}{b_a} \qquad S_e = \frac{K_e}{2 - \varepsilon_e} = \frac{1}{b_e} \, . \tag{857}$$

Joint coefficients:

$$\left. \begin{array}{ll} K_A = 3 S_a + 2 K_1 & \\ K_B = 2 K_1 + 2 K_2 & \end{array} \; K_C = 2 K_2 + 2 K_3 \; \begin{array}{l} K_D = 2 K_3 + 2 K_4 \\ K_E = 2 K_4 + 3 S_e \, . \end{array} \right\} \tag{858}$$

Rotation carry-over factors

$$\left. \begin{array}{ll} \text{backward reduction:} & \text{forward reduction:} \\[4pt] j_4 = \dfrac{K_4}{r_e} \quad\; r_e = K_E & j'_1 = \dfrac{K_1}{v_a} \quad\; v_a = K_A \\[6pt] j_3 = \dfrac{K_3}{r_d} \quad\; r_d = K_D - K_4 \cdot j_4 & j'_2 = \dfrac{K_2}{v_b} \quad\; v_b = K_B - K_1 \cdot j'_1 \\[6pt] j_2 = \dfrac{K_2}{r_c} \quad\; r_c = K_C - K_3 \cdot j_3 & j'_3 = \dfrac{K_3}{v_c} \quad\; v_c = K_C - K_2 \cdot j'_2 \\[6pt] j_1 = \dfrac{K_1}{r_b} \quad\; r_b = K_B - K_2 \cdot j_2 & j'_4 = \dfrac{K_4}{v_d} \quad\; v_d = K_D - K_3 \cdot j'_3 \\[6pt] & r_a = K_A - K_1 \cdot j_1 . \quad v_e = K_E - K_4 \cdot j'_4 . \end{array} \right\} \begin{array}{c} (859) \\ \text{and} \\ (860) \end{array}$$

Principal influence coefficients by recursion:

$$\left. \begin{array}{ll} \text{forward:} & \text{backward:} \\[4pt] u_{aa} = 1/r_a & u_{ee} = 1/v_e \\ u_{bb} = 1/r_b + u_{aa} \cdot j_1^2 & u_{dd} = 1/v_d + u_{ee} \cdot j'^2_4 \\ u_{cc} = 1/r_c + u_{bb} \cdot j_2^2 & u_{cc} = 1/v_c + u_{dd} \cdot j'^2_3 \\ u_{dd} = 1/r_d + u_{cc} \cdot j_3^2 & u_{bb} = 1/v_b + u_{cc} \cdot j'^2_2 \\ u_{ee} = 1/r_e + u_{dd} \cdot j_4^2 . & u_{aa} = 1/v_a + u_{bb} \cdot j'^2_1 . \end{array} \right\} \begin{array}{c} (861) \\ \text{and} \\ (862) \end{array}$$

Principal influence coefficients direct

$$u_{aa} = 1/r_a \quad u_{bb} = 1/(r_b + v_b - K_B) \quad u_{cc} = 1/(r_c + v_c - K_C) \\ u_{dd} = 1/(r_d + v_d - K_D) \quad u_{ee} = 1/v_e. \tag{863}$$

Computation scheme for the symmetrical influence coefficient matrix

$$\begin{array}{c|ccc|cc|c}
\cdot j_1 & (u_{aa}) & u_{ab} & u_{ac} & u_{ad} & u_{ae} & \cdot j_1' \\
 & \downarrow & \uparrow & \uparrow & \uparrow & \uparrow & \\
\cdot j_2 & u_{ba} & (u_{bb}) & u_{bc} & u_{bd} & u_{be} & \cdot j_2' \\
 & \downarrow & \downarrow & \uparrow & \uparrow & \uparrow & \\
\cdot j_3 & u_{ca} & u_{cb} & (u_{cc}) & u_{cd} & u_{ce} & \cdot j_3' \\
 & \downarrow & \downarrow & \downarrow & \uparrow & \uparrow & \\
\cdot j_4 & u_{da} & u_{db} & u_{dc} & (u_{dd}) & u_{de} & \cdot j_4' \\
 & \downarrow & \downarrow & \downarrow & \downarrow & \uparrow & \\
 & u_{ea} & u_{eb} & u_{ec} & u_{ed} & (u_{ee}) &
\end{array} \tag{864}$$

Loading Condition 4—Overall loading condition

(Superposition of Loading Conditions 1, 2 and 3)

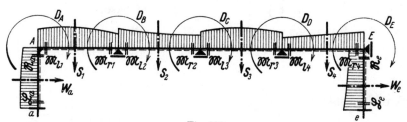

Fig. 213

Loading diagram with fixed-end moments $\mathfrak{M}_l$ and $\mathfrak{M}_r$ for the beam spans, load terms $\mathfrak{L}$ and $\mathfrak{R}$ for the columns, and external rotational moments D at the corners and supports

Column restraining moments at the corner joints:

$$\mathfrak{C}_{ra} = -\frac{\mathfrak{R}_a - \mathfrak{L}_a \cdot \varepsilon_a}{2 - \varepsilon_a} \qquad \mathfrak{C}_{re} = -\frac{\mathfrak{R}_e - \mathfrak{L}_e \cdot \varepsilon_e}{2 - \varepsilon_e}. \tag{865}$$

Joint load terms:

$$\begin{aligned}
\mathfrak{K}_A &= D_A + \mathfrak{C}_{ra} - \mathfrak{M}_{l1}; & \mathfrak{K}_B &= D_B + \mathfrak{M}_{r1} - \mathfrak{M}_{l2} \\
 & & \mathfrak{K}_C &= D_C + \mathfrak{M}_{r2} - \mathfrak{M}_{l3} \\
\mathfrak{K}_E &= D_E + \mathfrak{M}_{r4} + \mathfrak{C}_{re}. & \mathfrak{K}_D &= D_D + \mathfrak{M}_{r3} - \mathfrak{M}_{l4};
\end{aligned} \tag{866}$$

auxiliary moments:

	$\mathfrak{K}_A$	$\mathfrak{K}_B$	$\mathfrak{K}_C$	$\mathfrak{K}_D$	$\mathfrak{K}_E$
$Y_A =$	$+u_{aa}$	$-u_{ab}$	$+u_{ac}$	$-u_{ad}$	$+u_{ae}$
$Y_B =$	$-u_{ba}$	$+u_{bb}$	$-u_{bc}$	$+u_{bd}$	$-u_{be}$
$Y_C =$	$+u_{ca}$	$-u_{cb}$	$+u_{cc}$	$-u_{cd}$	$+u_{ce}$
$Y_D =$	$-u_{da}$	$+u_{db}$	$-u_{dc}$	$+u_{dd}$	$-u_{de}$
$Y_E =$	$+u_{ea}$	$-u_{eb}$	$+u_{ec}$	$-u_{ed}$	$+u_{ee}.$

(867)

Beam moments:

$$M_{A1} = \mathfrak{M}_{l1} + (2Y_A + Y_B)K_1 \qquad M_{B1} = \mathfrak{M}_{r1} - (Y_A + 2Y_B)K_1$$
$$M_{B2} = \mathfrak{M}_{l2} + (2Y_B + Y_C)K_2 \qquad M_{C2} = \mathfrak{M}_{r2} - (Y_B + 2Y_C)K_2$$
$$M_{C3} = \mathfrak{M}_{l3} + (2Y_C + Y_D)K_3 \qquad M_{D3} = \mathfrak{M}_{r3} - (Y_C + 2Y_D)K_3$$
$$M_{D4} = \mathfrak{M}_{l4} + (2Y_D + Y_E)K_4; \qquad M_{E4} = \mathfrak{M}_{r4} - (Y_D + 2Y_E)K_4.$$

(868)

Column moments:

$$\text{at top:} \qquad\qquad \text{at base:}$$
$$M_{Aa} = \mathfrak{C}_{ra} - Y_A \cdot 3S_a \qquad M_a = -(\mathfrak{L}_a + M_{Aa})\varepsilon_a$$
$$M_{Ee} = \mathfrak{C}_{re} - Y_E \cdot 3S_e; \qquad M_e = -(\mathfrak{L}_e + M_{Ee})\varepsilon_e.$$

(869) and (870)

Check relationships:

$$\boldsymbol{D}_A + M_{Aa} - M_{A1} = 0; \qquad \boldsymbol{D}_B + M_{B1} - M_{B2} = 0$$
$$\boldsymbol{D}_C + M_{C2} - M_{C3} = 0$$
$$\boldsymbol{D}_E + M_{E4} + M_{Ee} = 0. \qquad \boldsymbol{D}_D + M_{D3} - M_{D4} = 0;$$

(871)

Horizontal thrusts and column shears for all loading conditions: according to equations (36) and (38), for the collective subscripts $n = a, e$ and $N = A, E$.

Lateral restraining force for Loading Conditions 1 and 3, and 2 and 4 respectively:

$$H = -H_a - H_e \quad \text{or} \quad H = -Q_{ra} - Q_{re}. \tag{872}$$

Frame Shape 28v

(Frame Shape 28 with laterally unrestrained horizontal member)

Procedure

as for Frame Shape 1v, but for collective subscripts:

$$n = a, e \text{ and } N = A, E$$

Specific bending moments m: according to Method I, from formulas (853)–(855) or according to Method II, from formulas (867)–(870).

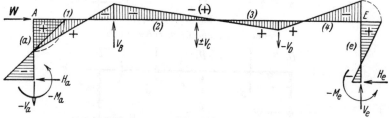

Fig. 214

Loading Condition 5 with diagram of bending moments and reactions at supports.

Frame Shape 29v

(Frame Shape 29 with laterally unrestrained horizontal member)

Procedure

as for Frame Shape 1v, but for collective subscripts:

$$n = a, c \text{ and } N = A, C$$

Specific bending moments m: according to Method I, from formulas (853)–(855) or according to Method II, from formulas (867)–(870), having due regard to the dotted lines defining the scope of the formulas for Frame Shape 29.

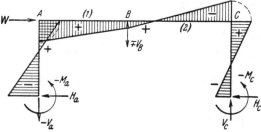

Fig. 215

Loading Condition 5 with diagram of bending moments and reactions at supports.

Frame Shape 29

Single-story two-bay frame with laterally restrained horizontal member, two elastically restrained end columns, and one pin-jointed internal column (or rocker column)

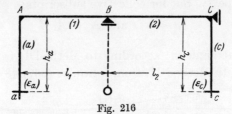

Fig. 216
Frame shape, dimensions and symbols

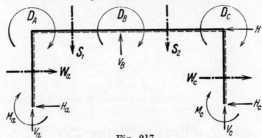

Fig. 217
Definition of positive direction for all external loads on joints and members, for all reactions at the supports, and for the lateral restraining force

Important

All the formulas required for Frame Shape 29 are already contained in the formulas for Frame Shape 28 and are there indicated by dotted boundary lines. The relevant portions of the formulas comprise the collective subscripts $\nu = 1, 2$, $N = A, C$ and $n = a, c$. All the formulas without collective subscripts are in each case valid up to the dotted lines; in the case of backward reduction and recursion they are, however, valid from those lines onward.

In particular, the modifications applicable to the formulas for Frame Shape 28 to obtain those for Frame Shape 29 are as follows:

The subscripts e and E must be replaced by c and C;

in (841) put $k_C = 2k_2 + b_c$;
in (848) and (850) we get $-M_{Cc} = M_{C2}$;
in (854) we get $-D_C - M_{C2}$;
in (858) put $K_C = 2K_2 + 3S_c$;
in (866) put $\mathfrak{K}_C = D_C + \mathfrak{M}_{r2} + \mathfrak{E}_{rc}$;
in (871) we get $D_C + M_{C2} + M_{Cc} = 0$.

Frame Shape 30

Single-story three-bay frame with laterally restrained horizontal member, two elastically restrained end columns, and two pin-jointed internal supports (or rocker columns)

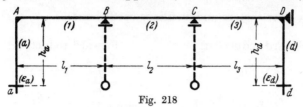

Fig. 218
Frame shape, dimensions and symbols

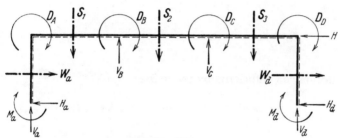

Fig. 219
Definition of positive direction for all external loads on joints and members, for all reactions at the supports, and for the lateral restraining force

Important

All the formula sequences and formula matrices given for Frame Shape 30 can readily be extended to suit frames of similar type having any odd number of bays.

Note: The *collective subscripts* have the following meaning: $\nu = 1, 2, 3$ (column numbers $n = a, d$)

I. Treatment by the Method of Forces

Coefficients

Flexibility coefficients of all horizontal members:

$$k_a = \frac{J_k}{J_a} \cdot \frac{h_a}{l_k} \qquad k_\nu = \frac{J_k}{J_\nu} \cdot \frac{l_\nu}{l_k} \qquad k_d = \frac{J_k}{J_d} \cdot \frac{h_d}{l_k}. \tag{873}$$

Flexibilities of the elastically restrained columns:

$$b_a = k_a(2 - \varepsilon_a) \qquad b_d = k_d(2 - \varepsilon_d). \tag{874}$$

Diagonal coefficients:

$$\left.\begin{aligned} k_A &= b_a + 2k_1 & k_C &= 2k_2 + 2k_3 \\ k_B &= 2k_1 + 2k_2 & k_D &= 2k_3 + b_d \end{aligned}\right\} \quad (875)$$

Moment carry-over factors and auxiliary coefficients:

$$\left.\begin{array}{l|l} \text{backward reduction:} & \text{forward reduction:} \\[4pt] \begin{aligned} \iota_3 &= \frac{k_3}{r_d} & r_d &= k_D \\ \iota_2 &= \frac{k_2}{r_c} & r_c &= k_C - k_3 \cdot \iota_3 \\ \iota_1 &= \frac{k_1}{r_b} & r_b &= k_B - k_2 \cdot \iota_2 \\ & & r_a &= k_A - k_1 \cdot \iota_1. \end{aligned} & \begin{aligned} \iota_1' &= \frac{k_1}{v_a} & v_a &= k_A \\ \iota_2' &= \frac{k_2}{v_b} & v_b &= k_B - k_1 \cdot \iota_1' \\ \iota_3' &= \frac{k_3}{v_c} & v_c &= k_C - k_2 \cdot \iota_2' \\ & & v_d &= k_D - k_3 \cdot \iota_3'. \end{aligned} \end{array}\right\} \begin{array}{l} (876) \\ \text{and} \\ (877) \end{array}$$

Principal influence coefficients by recursion:

$$\left.\begin{array}{l|l} \text{forward:} & \text{backward:} \\[4pt] \begin{aligned} n_{aa} &= \frac{1}{r_a} \\ n_{bb} &= \frac{1}{r_b} + n_{aa} \cdot \iota_1^2 \\ n_{cc} &= \frac{1}{r_c} + n_{bb} \cdot \iota_2^2 \\ n_{dd} &= \frac{1}{r_d} + n_{cc} \cdot \iota_3^2. \end{aligned} & \begin{aligned} n_{dd} &= \frac{1}{v_d} \\ n_{cc} &= \frac{1}{v_c} + n_{dd} \cdot \iota_3'^2 \\ n_{bb} &= \frac{1}{v_b} + n_{cc} \cdot \iota_2'^2 \\ n_{aa} &= \frac{1}{v_a} + n_{bb} \cdot \iota_1'^2. \end{aligned} \end{array}\right\} \begin{array}{l} (878) \\ \text{and} \\ (879) \end{array}$$

Principal influence coefficients direct

$$n_{aa} = \frac{1}{r_a} \quad n_{bb} = \frac{1}{r_b + v_b - k_B} \quad n_{cc} = \frac{1}{r_c + v_c - k_C} \quad n_{dd} = \frac{1}{v_d}. \quad (880)$$

Computation scheme for the symmetrical influence coefficient matrix

$$\left.\begin{array}{r|cccc|l} & (n_{aa}) & n_{ab} & n_{ac} & n_{ad} & \\ \cdot \iota_1 & \downarrow & \uparrow & \uparrow & \uparrow & \cdot \iota_1' \\ & n_{ba} & (n_{bb}) & n_{bc} & n_{bd} & \\ \cdot \iota_2 & \downarrow & \downarrow & \uparrow & \uparrow & \cdot \iota_2' \\ & n_{ca} & n_{cb} & (n_{cc}) & n_{cd} & \\ \cdot \iota_3 & \downarrow & \downarrow & \downarrow & \uparrow & \cdot \iota_3' \\ & n_{da} & n_{db} & n_{dc} & (n_{dd}) & \end{array}\right\} \quad (881)$$

Loading Condition 1
All the beam spans carrying arbitrary vertical loading

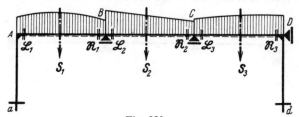

Fig. 220
Beam loading diagram with load terms
$\mathfrak{L}$ and $\mathfrak{R}$

Corner and beam moments $(L_\nu = \mathfrak{L}_\nu k_\nu$ and $R_\nu = \mathfrak{R}_\nu k_\nu)$:

$$\begin{array}{r|cccc}
 & L_1 & (R_1 + L_2) & (R_2 + L_3) & R_3 \\
+M_{Aa} = M_{A1} = & -n_{aa} & +n_{ab} & -n_{ac} & +n_{ad} \\
M_B = & +n_{ba} & -n_{bb} & +n_{bc} & -n_{bd} \\
M_C = & -n_{ca} & +n_{cb} & -n_{cc} & +n_{cd} \\
-M_{Dd} = M_{D3} = & +n_{da} & -n_{db} & +n_{dc} & -n_{dd}
\end{array} \qquad (882)$$

Column base moments:

$$M_a = -M_{Aa} \cdot \varepsilon_a \qquad M_d = -M_{Dd} \cdot \varepsilon_d. \qquad (883)$$

Special case: symmetrical bay loading $(R_\nu = L_\nu)$:

$$\begin{aligned}
+M_{Aa} = M_{A1} &= -L_1 \cdot s_{a1} + L_2 \cdot s_{a2} - L_3 \cdot s_{a3} \\
M_B &= -L_1 \cdot s_{b1} - L_2 \cdot s_{b2} + L_3 \cdot s_{b3} \\
M_C &= +L_1 \cdot s_{c1} - L_2 \cdot s_{c2} - L_3 \cdot s_{c3} \\
-M_{Dd} = M_{D3} &= -L_1 \cdot s_{d1} + L_2 \cdot s_{d2} - L_3 \cdot s_{d3};
\end{aligned} \qquad (884)$$

where:

$$\begin{aligned}
s_{a1} &= +n_{aa} - n_{ab} & s_{a2} &= +n_{ab} - n_{ac} & s_{a3} &= +n_{ac} - n_{ad} \\
s_{b1} &= -n_{ba} + n_{bb} & s_{b2} &= +n_{bb} - n_{bc} & s_{b3} &= +n_{bc} - n_{bd} \\
s_{c1} &= -n_{ca} + n_{cb} & s_{c2} &= -n_{cb} + n_{cc} & s_{c3} &= +n_{cc} - n_{cd} \\
s_{d1} &= -n_{da} + n_{db} & s_{d2} &= -n_{db} + n_{dc} & s_{d3} &= -n_{dc} + n_{dd}.
\end{aligned} \qquad (885)$$

Loading Conditions 2 and 3

Both columns carrying arbitrary horizontal loading (2) external rotational moments applied at the corners of the frame (3)

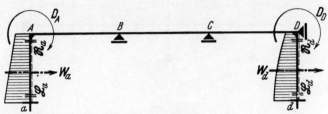

Fig. 221

Diagram of loading on columns, with load terms $\mathfrak{L}$ and $\mathfrak{R}$, and rotational moments D at corners

Joint load terms ($L_n = \mathfrak{L}_n k_n$ and $R_n = \mathfrak{R}_n k_n$):

$$\mathfrak{B}_A = D_A \cdot b_a + L_a \cdot \varepsilon_a - R_a \qquad \mathfrak{B}_D = D_D \cdot b_d + L_d \cdot \varepsilon_d - R_d. \qquad (886)$$

Beam moments:

$$\left. \begin{aligned} M_{A1} &= +\mathfrak{B}_A \cdot n_{aa} + \mathfrak{B}_D \cdot n_{ad} \\ M_B &= -\mathfrak{B}_A \cdot n_{ba} - \mathfrak{B}_D \cdot n_{bd} \\ M_C &= +\mathfrak{B}_A \cdot n_{ca} + \mathfrak{B}_D \cdot n_{cd} \\ M_{D3} &= -\mathfrak{B}_A \cdot n_{da} - \mathfrak{B}_D \cdot n_{dd} \,. \end{aligned} \right\} \qquad (887)$$

Column moments:

$$\left. \begin{array}{ll} \text{at top:} & \text{at base:} \\ M_{Aa} = -D_A + M_{A1} & M_a = -(\mathfrak{L}_a + M_{Aa})\varepsilon_a \\ M_{Dd} = -D_D - M_{D3}; & M_d = -(\mathfrak{L}_d + M_{Dd})\varepsilon_d \,. \end{array} \right\} \begin{array}{l} (888) \\ \text{and} \\ (889) \end{array}$$

II. Treatment by the Deformation Method

Coefficients

Stiffness coefficients for all individual members:

$$K_a = \frac{J_a}{J_k} \cdot \frac{l_k}{h_a} = \frac{1}{k_a} \qquad K_v = \frac{J_v}{J_k} \cdot \frac{l_k}{l_v} = \frac{1}{k_v} \qquad K_d = \frac{J_d}{J_k} \cdot \frac{l_k}{h_d} = \frac{1}{k_d}. \qquad (890)$$

Stiffnesses of the elastically restrained columns:

$$S_a = \frac{K_a}{2 - \varepsilon_a} = \frac{1}{b_a} \qquad S_d = \frac{K_d}{2 - \varepsilon_d} = \frac{1}{b_d} \qquad (891)$$

Joint coefficients:

$$K_A = 3S_a + 2K_1 \qquad K_C = 2K_2 + 2K_3$$
$$K_B = 2K_1 + 2K_2 \qquad K_D = 2K_3 + 3S_d . \tag{892}$$

Moment carry-over factors and auxiliary coefficients:

backward reduction:

$$j_3 = \frac{K_3}{r_d} \qquad r_d = K_D$$
$$j_2 = \frac{K_2}{r_c} \qquad r_c = K_C - K_3 \cdot j_3$$
$$j_1 = \frac{K_1}{r_b} \qquad r_b = K_B - K_2 \cdot j_2$$
$$\qquad r_a = K_A - K_1 \cdot j_1 .$$

forward reduction:

$$\qquad v_a = K_A$$
$$j_1' = \frac{K_1}{v_a} \qquad v_b = K_B - K_1 \cdot j_1'$$
$$j_2' = \frac{K_2}{v_b} \qquad v_c = K_C - K_2 \cdot j_2'$$
$$j_3' = \frac{K_3}{v_c} \qquad v_d = K_D - K_3 \cdot j_3' .$$

$$\tag{893}$$
$$\text{and}$$
$$\tag{894}$$

Principal influence coefficients by recursion:

forward:

$$u_{aa} = \frac{1}{r_a}$$
$$u_{bb} = \frac{1}{r_b} + u_{aa} \cdot j_1^2$$
$$u_{cc} = \frac{1}{r_c} + u_{bb} \cdot j_2^2$$
$$u_{dd} = \frac{1}{r_d} + u_{cc} \cdot j_3^2 .$$

backward:

$$u_{dd} = \frac{1}{v_d}$$
$$u_{cc} = \frac{1}{v_c} + u_{dd} \cdot j_3'^2$$
$$u_{bb} = \frac{1}{v_b} + u_{cc} \cdot j_2'^2$$
$$u_{aa} = \frac{1}{v_a} + u_{bb} \cdot j_1'^2 .$$

$$\tag{895}$$
$$\text{and}$$
$$\tag{896}$$

Principal influence coefficients direct

$$u_{aa} = \frac{1}{r_a} \qquad u_{bb} = \frac{1}{r_b + v_b - K_B} \qquad u_{cc} = \frac{1}{r_c + v_c - K_C} \qquad u_{dd} = \frac{1}{v_d} . \tag{897}$$

Computation scheme for the symmetrical influence coefficient matrix

$$\begin{array}{c|cccc|c}
 & (u_{aa}) & u_{ab} & u_{ac} & u_{ad} & \\
\cdot j_1 & \downarrow & \uparrow & \uparrow & \uparrow & \cdot j_1' \\
 & u_{ba} & (u_{bb}) & u_{bc} & u_{bd} & \\
\cdot j_2 & \downarrow & \downarrow & \uparrow & \uparrow & \cdot j_2' \\
 & u_{ca} & u_{cb} & (u_{cc}) & u_{cd} & \\
\cdot j_3 & \downarrow & \downarrow & \downarrow & \uparrow & \cdot j_3' \\
 & u_{da} & u_{db} & u_{dc} & (u_{dd}) &
\end{array} \tag{898}$$

Loading Condition 4—Overall loading condition

(Superposition of Loading Conditions 1, 2 and 3)

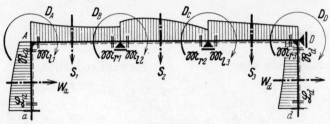

Fig. 222

Loading diagram with fixed-end moments $\mathfrak{M}_l$ and $\mathfrak{M}_r$ for the beam spans, load terms $\mathfrak{L}$ and $\mathfrak{R}$ for the columns, and external rotational moments D at the corners and supports

Column restraining moments at the corner joints:

$$\mathfrak{C}_{ra} = -\frac{\mathfrak{R}_a - \mathfrak{L}_a \cdot \varepsilon_a}{2 - \varepsilon_a} \qquad \mathfrak{C}_{rd} = -\frac{\mathfrak{R}_d - \mathfrak{L}_d \cdot \varepsilon_d}{2 - \varepsilon_d}. \tag{899}$$

Joint load terms:

$$\left. \begin{array}{ll} \mathfrak{K}_A = D_A + \mathfrak{C}_{ra} - \mathfrak{M}_{l1} & \mathfrak{K}_C = D_C + \mathfrak{M}_{r2} - \mathfrak{M}_{l3} \\ \mathfrak{K}_B = D_B + \mathfrak{M}_{r1} - \mathfrak{M}_{l2} & \mathfrak{K}_D = D_D + \mathfrak{M}_{r3} + \mathfrak{C}_{rd} \end{array} \right\} \tag{900}$$

auxiliary moments:

	$\mathfrak{K}_A$	$\mathfrak{K}_B$	$\mathfrak{K}_C$	$\mathfrak{K}_D$
$Y_A =$	$+u_{aa}$	$-u_{ab}$	$+u_{ac}$	$-u_{ad}$
$Y_B =$	$-u_{ba}$	$+u_{bb}$	$-u_{bc}$	$+u_{bd}$
$Y_C =$	$+u_{ca}$	$-u_{cb}$	$+u_{cc}$	$-u_{cd}$
$Y_D =$	$-u_{da}$	$+u_{db}$	$-u_{dc}$	$+u_{dd}$

(901)

Beam moments:

$$\left. \begin{array}{ll} M_{A1} = \mathfrak{M}_{l1} + (2Y_A + Y_B)K_1 & M_{B1} = \mathfrak{M}_{r1} - (Y_A + 2Y_B)K_1 \\ M_{B2} = \mathfrak{M}_{l2} + (2Y_B + Y_C)K_2 & M_{C2} = \mathfrak{M}_{r2} - (Y_B + 2Y_C)K_2 \\ M_{C3} = \mathfrak{M}_{l3} + (2Y_C + Y_D)K_3; & M_{D3} = \mathfrak{M}_{r3} - (Y_C + 2Y_D)K_3. \end{array} \right\} \tag{902}$$

Column moments:

at top: at base:

$$M_{Aa} = \mathfrak{C}_{ra} - Y_A \cdot 3S_a \qquad M_a = -(\mathfrak{L}_a + M_{Aa})\varepsilon_a$$
$$M_{Dd} = \mathfrak{C}_{rd} - Y_D \cdot 3S_d; \qquad M_d = -(\mathfrak{L}_d + M_{Dd})\varepsilon_d.$$

(903) and (904)

Check relationships:

$$D_A + M_{Aa} - M_{A1} = 0 \qquad D_C + M_{C2} - M_{C3} = 0$$
$$D_B + M_{B1} - M_{B2} = 0 \qquad D_D + M_{D3} + M_{Dd} = 0.$$

(905)

Horizontal thrusts and column shears for all loading conditions according to equations (36) and (38), for the collective subscripts $n = a, d$ and $N = A, D$.

Lateral restraining force for Loading Conditions 1 and 3, and 2 and 4 respectively: $H = -H_a - H_d$ or $H = -Q_{ra} - Q_{rd}$.   (906)

Frame Shape 30v

(Frame Shape 30 with laterally unrestrained horizontal member)

Procedure

as for Frame Shape 1v, but for collective subscripts:

$$n = a, d \text{ and } N = A, D.$$

Specific bending moments m according to Method I from formulas (887)-(889), page 202, or according to Method II from formulas (901)-(904), pages 204-205.

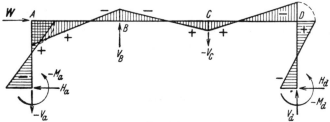

Fig. 223
Loading Condition 5 with diagram of bending moments and reactions at supports

Frame Shape 31

Symmetrical single-story six-bay frame with laterally restrained horizontal member, two elastically restrained end columns, and five pin-jointed internal supports (or rocker columns)

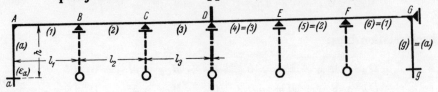

Fig. 224

Frame shape, dimensions and symbols

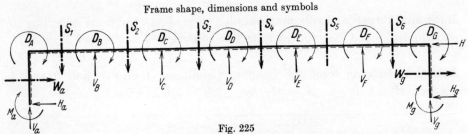

Fig. 225

Definition of positive direction for all external loads on joints and members, for all support reactions, and for the lateral restraining force

All the dimensions and coefficients for the right-hand half of the frame are the same as those for the left-hand half. In the case of the columns the broken line is always placed on the right-hand side of each column, despite the symmetry of the frame.

Note: For Frame Shape 31 the *coefficients* and the *Loading Conditions* 1–4 for *Frame Shape* 28, enlarged by the addition of two bays, may be employed. On account of symmetry, the forward reduction becomes the same as the backward reduction, and the backward recursion becomes the same as the forward recursion, both in Method I and Method II. Furthermore, the *square matrices of influence coefficients* (each with 7×7 elements) are bisymmetrical, i.e., symmetrical about the principal diagonal and the secondary diagonal. If the *"load transposition method"* is used, however, it is also possible to start from the following symmetrical and anti-symmetrical loading conditions.

Important

All the instructions, formula sequences and formula matrices given for Frame Shape 31 can readily be extended to suit symmetrical frames of similar type having any even number of bays.

I. Treatment by the Method of Forces

a) Coefficients for symmetrical load arrangements

Flexibility coefficients of all the horizontal members and column flexibility:

$$k_\nu = \frac{J_k}{J_\nu} \cdot \frac{l_\nu}{l_k} \quad \text{for} \quad \nu = 1, 2, 3; \qquad k_a = \frac{J_k}{J_a} \cdot \frac{h}{l_k}; \qquad b_a = k_a(2 - \varepsilon_a). \tag{907}$$

Diagonal coefficients:

$$k_A = b_a + 2k_1 \qquad k_B = 2k_1 + 2k_2 \qquad k_C = 2k_2 + 2k_3; \qquad (k'_D = 2k_3). \tag{908}$$

Moment carry-over factors and auxiliary coefficients:

backward reduction:

$$\iota_{3s} = \frac{1}{2}$$
$$\iota_{2s} = \frac{k_2}{r'_c}$$
$$\iota_{1s} = \frac{k_1}{r'_b}$$

$$r'_c = k_C - k_3 \cdot \frac{1}{2}$$
$$r'_b = k_B - k_2 \cdot \iota_{2s}$$
$$r'_a = k_A - k_1 \cdot \iota_{1s}.$$

forward reduction:

$$\iota'_1 = \frac{k_1}{v_a}$$
$$\iota'_2 = \frac{k_2}{v_b}$$
$$\iota'_3 = \frac{k_3}{v_c}.$$

$$v_a = k_A$$
$$v_b = k_B - k_1 \cdot \iota'_1$$
$$v_c = k_C - k_2 \cdot \iota'_2$$

(909) and (910)

Principal influence coefficients by recursion:

forward:

$$n'_{aa} = 1/r'_a$$
$$n'_{bb} = 1/r'_b + n'_{aa} \cdot \iota^2_{1s}$$
$$n'_{cc} = 1/r'_c + n'_{bb} \cdot \iota^2_{2s}$$
$$n'_{dd} = 1/2k_3 + n'_{cc} \cdot 1/4.$$

backward:

$$n'_{dd} = 1/k_3(2 - \iota'_3)$$
$$n'_{cc} = 1/v_c + n'_{dd} \cdot \iota'^2_3$$
$$n'_{bb} = 1/v_b + n'_{cc} \cdot \iota'^2_2$$
$$n'_{aa} = 1/v_a + n'_{bb} \cdot \iota'^2_1.$$

(911) and (912)

Computation scheme for the symmetrical influence coefficient matrix

$$\begin{array}{c|cccc|c}
 & (n'_{aa}) & n'_{ab} & n'_{ac} & n'_{ad} & \\
\cdot \iota_{1s} & \downarrow & \uparrow & \uparrow & \uparrow & \cdot \iota'_1 \\
 & n'_{ba} & (n'_{bb}) & n'_{bc} & n'_{bd} & \\
\cdot \iota_{2s} & \downarrow & \downarrow & \uparrow & \uparrow & \cdot \iota'_2 \\
 & n'_{ca} & n'_{cb} & (n'_{cc}) & n'_{cd} & \\
\cdot \frac{1}{2} & \downarrow & \downarrow & \downarrow & \uparrow & \cdot \iota'_3 \\
 & n'_{da} & n'_{db} & n'_{dc} & (n'_{dd}) & \\
\end{array} \tag{913}$$

---

* Between the elements of matrix (913) for the left-hand half of the frame and the elements of matrix (846), enlarged to 7 × 7 elements, there exist, according to the "load transposition method" the following relationships:

$$n'_{aa} = n_{aa} + n_{ag}, \quad n'_{ab} = n_{ab} + n_{af} \text{ etc., up to } n'_{dd} = 2n_{dd}.$$

(The values n' of the column d of (913) here become twice as large as the values n of the column d of the 7 × 7 matrix).

Principal influence coefficients direct:

$$n'_{aa} = \frac{1}{r'_a} \qquad n'_{bb} = \frac{1}{r'_b + v_b - k_B} \qquad n'_{cc} = \frac{1}{r'_c + v_c - k_C} \qquad n'_{dd} = \frac{1}{k_3(2 - \iota'_3)} \,. \tag{914}$$

Loading Condition 1a

Horizontal member carrying arbitrary vertical loading symmetrical about center of frame

$$(\mathfrak{R}_6 = \mathfrak{L}_1, \quad \mathfrak{L}_6 = \mathfrak{R}_1; \quad \mathfrak{R}_5 = \mathfrak{L}_2, \quad \mathfrak{L}_5 = \mathfrak{R}_2; \quad \mathfrak{R}_4 = \mathfrak{L}_3, \quad \mathfrak{L}_4 = \mathfrak{R}_3)$$

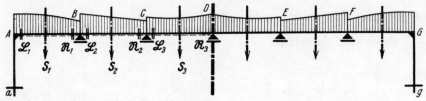

Fig. 226

Beam loading diagram with load terms $\mathfrak{L}$ and $\mathfrak{R}$ for left-hand half of frame

Beam moments $(L_\nu = \mathfrak{L}_\nu k_\nu \text{ and } R_\nu = \mathfrak{R}_\nu k_\nu)$:

$$\left. \begin{array}{l|cccc}
 & L_1 & (R_1 + L_2) & (R_2 + L_3) & R_3 \\
\hline
M_{A1} = M_{G6} = & -n'_{aa} & +n'_{ab} & -n'_{ac} & +n'_{ad} \\
M_B = M_F = & +n'_{ba} & -n'_{bb} & +n'_{bc} & -n'_{bd} \\
M_C = M_E = & -n'_{ca} & +n'_{cb} & -n'_{cc} & +n'_{cd} \\
M_D = & +n'_{da} & -n'_{db} & +n'_{dc} & -n'_{dd}\,.
\end{array} \right\} \tag{915}$$

Column moments:

$$M_{Aa} = -M_{Gg} = M_{A1} \qquad M_a = -M_g = -M_{Aa} \cdot \varepsilon_a. \tag{916}$$

Special case: symmetrical bay loading $(R_\nu = L_\nu)$:

$$\left. \begin{array}{l}
M_{A1} = M_{G6} = -L_1 \cdot s'_{a1} + L_2 \cdot s'_{a2} - L_3 \cdot s'_{a3} \\
M_B = M_F = -L_1 \cdot s'_{b1} - L_2 \cdot s'_{b2} + L_3 \cdot s'_{b3} \\
M_C = M_E = +L_1 \cdot s'_{c1} - L_2 \cdot s'_{c2} - L_3 \cdot s'_{c3} \\
M_D = -L_1 \cdot s'_{d1} + L_2 \cdot s'_{d2} - L_3 \cdot s'_{d3}\,.
\end{array} \right\} \tag{917}$$

Note: values s are given on next page.

Composite influence coefficients for Loading Condition 1a:

$$\begin{aligned}
s'_{a1} &= +n'_{aa} - n'_{ab} = (s_{a1} - s_{a6}) & s'_{a2} &= +n'_{ab} - n'_{ac} = (s_{a2} - s_{a5}) \\
& & s'_{b2} &= +n'_{bb} - n'_{bc} = (s_{b2} - s_{b5}) \\
s'_{b1} &= -n'_{ba} + n'_{bb} = (s_{b1} + s_{b6}) & & \\
s'_{c1} &= -n'_{ca} + n'_{cb} = (s_{c1} + s_{c6}) & s'_{c2} &= -n'_{cb} + n'_{cc} = (s_{c2} + s_{c5}) \\
s'_{d1} &= -n'_{da} + n'_{db} = (s_{d1} + s_{d6}) & s'_{d2} &= -n'_{db} + n'_{dc} = (s_{d2} + s_{d5}) \\
s'_{a3} &= & n'_{ad} &= (s_{a3} - s_{a4}) \\
s'_{b3} &= & n'_{bd} &= (s_{b3} - s_{b4}) \\
s'_{c3} &= & n'_{cd} &= (s_{c3} - s_{c4}) \\
s'_{d3} &= -n'_{dc} + n'_{dd} = (s_{d3} + s_{d4}).^{*} & &
\end{aligned} \quad (918)$$

Loading Conditions 2a and 3a

The two columns carrying arbitrary horizontal loading (directed towards center of frame) symmetrical about center of frame (2a), and a symmetrical pair of rotational moments at corners of frame (3a)

$$(\mathfrak{L}_g = -\mathfrak{L}_a, \quad \mathfrak{R}_g = -\mathfrak{R}_a; \quad D_G = -D_A)$$

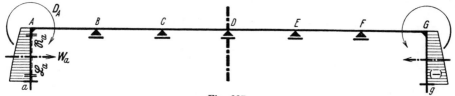

Fig. 227

Column loading diagram with load terms $\mathfrak{L}$ and $\mathfrak{R}$, and rotational moment D, for left-hand half of frame

Joint load term:

$$(L_a = \mathfrak{L}_a k_a \quad \text{and} \quad R_a = \mathfrak{R}_a k_a); \qquad \mathfrak{B}_A = D_A \cdot b_a + L_a \cdot \varepsilon_a - R_a. \quad (919)$$

Beam moments:

$$\left. \begin{aligned}
M_{A1} &= M_{G6} = +\mathfrak{B}_A \cdot n'_{aa} & M_C &= M_E = +\mathfrak{B}_A \cdot n'_{ca} \\
M_B &= M_F = -\mathfrak{B}_A \cdot n'_{ba} & M_D &= -M_C/2 = -\mathfrak{B}_A \cdot n'_{da}
\end{aligned} \right\} \quad (920)$$

Column moments:

$$M_{Aa} = -M_{Gg} = -D_A + M_{A1}; \qquad M_a = -M_g = -(\mathfrak{L}_a + M_{Aa})\,\varepsilon_a. \quad (921)$$

---

* The members enclosed in parentheses have been formed with the values s (851) appropriately extended for the six-bay frame.

b) Coefficients for anti-symmetrical load arrangements

Flexibility coefficients of all the horizontal members, and column flexibility: same as formulas (907).

Diagonal coefficients:

$$k_A = b_a + 2k_1 \qquad k_B = 2k_1 + 2k_2 \qquad k_C = 2k_2 + 2k_3. \tag{922}$$

Moment carry-over factors and auxiliary coefficients:

backward reduction: | forward reduction:

$$(\iota_{3t} = 0)$$
$$\iota_{2t} = \frac{k_2}{r_c''}$$
$$\iota_{1t} = \frac{k_1}{r_b''}$$

$$r_c'' = k_C$$
$$r_b'' = k_B - k_2 \cdot \iota_{2t}$$
$$r_a'' = k_A - k_1 \cdot \iota_{1t}.$$

$$\iota_1' = \frac{k_1}{v_a}$$
$$\iota_2' = \frac{k_2}{v_b}$$
$$\iota_3' = \frac{k_3}{v_c}.$$

$$v_a = k_A$$
$$v_b = k_B - k_1 \cdot \iota_1'$$
$$v_c = k_C - k_2 \cdot \iota_2'$$

(923) and (924)

Principal influence coefficients by recursion:

forward: | backward:

$$n_{aa}'' = 1/r_a''$$
$$n_{bb}'' = 1/r_b'' + n_{aa}'' \cdot \iota_{1t}^2$$
$$n_{cc}'' = 1/r_c'' + n_{bb}'' \cdot \iota_{2t}^2.$$

$$n_{cc}'' = 1/v_c$$
$$n_{bb}'' = 1/v_b + n_{cc}'' \cdot \iota_2'^2$$
$$n_{aa}'' = 1/v_a + n_{bb}'' \cdot \iota_1'^2.$$

(925) and (926)

Principal influence coefficients direct

$$n_{aa}'' = \frac{1}{r_a''} \qquad n_{bb}'' = \frac{1}{r_b'' + v_b - k_B} \qquad n_{cc}'' = \frac{1}{v_c}. \tag{927}$$

Computation scheme for the symmetrical influence coefficient matrix*

$$\begin{array}{c} \\ \cdot \iota_{1t} \\ \cdot \iota_{2t} \end{array} \left| \begin{array}{ccc} (n_{aa}'') & n_{ab}'' & n_{ac}'' \\ \downarrow & \uparrow & \uparrow \\ n_{ba}'' & (n_{bb}'') & n_{bc}'' \\ \downarrow & \downarrow & \uparrow \\ n_{ca}'' & n_{cb}'' & (n_{cc}'') \end{array} \right| \begin{array}{c} \cdot \iota_1' \\ \\ \cdot \iota_2' \\ \\ \end{array} \qquad (928)$$

---

* Between the elements of matrix (928) for the left-hand half of the frame and the elements of matrix (846), enlarged to 7 × 7 elements, there exist, according to the "load transposition method" the following relationships:

$$n_{aa}'' = n_{aa} - n_{ag}, \quad n_{ab}'' = n_{ab} - n_{af} \text{ etc., up to } n_{cc}'' = n_{cc} - n_{ce}.$$

(The values n'' with the subscripts d of the central support vanish).

Loading Condition 1b

Horizontal member carrying vertical loading of arbitrary magnitude, but anti-symmetrical about center of frame

$$(\mathfrak{R}_6 = -\mathfrak{L}_1, \quad \mathfrak{L}_6 = -\mathfrak{R}_1; \quad \mathfrak{R}_5 = -\mathfrak{L}_2, \quad \mathfrak{L}_5 = -\mathfrak{R}_2; \quad \mathfrak{R}_4 = -\mathfrak{L}_3)$$

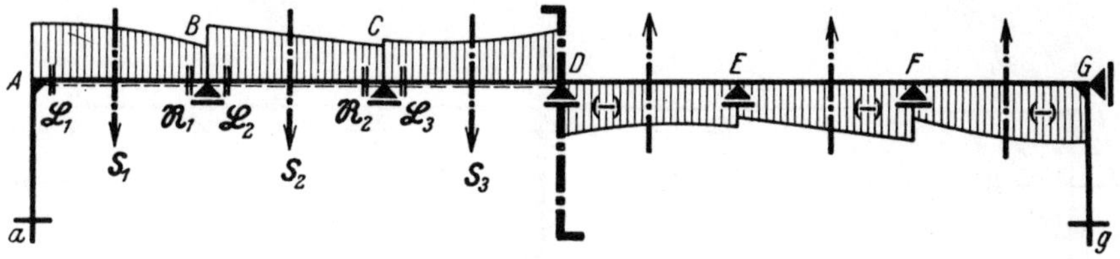

Fig. 228
Beam loading diagram with load terms $\mathfrak{L}$ and $\mathfrak{R}$ for left-hand half of frame

Beam moments $(L_\nu = \mathfrak{L}_\nu k_\nu \text{ and } R_\nu = \mathfrak{R}_\nu k_\nu)$:

	L_1	$(R_1 + L_2)$	$(R_2 + L_3)$
$M_{A1} = -M_{G6} =$	$-n''_{aa}$	$+n''_{ab}$	$-n''_{ac}$
$M_B = -M_F =$	$+n''_{ba}$	$-n''_{bb}$	$+n''_{bc}$
$M_C = -M_E =$	$-n''_{ca}$	$+n''_{cb}$	$-n''_{cc}$;
$M_D = 0$.			

(929)

Column moments:

$$M_{Aa} = M_{Gg} = M_{A1} \qquad M_a = M_g = -M_{Aa} \cdot \varepsilon_a. \tag{930}$$

Special case: symmetrical bay loading $(R_\nu = L_\nu)$:

$$\begin{aligned}
M_{A1} &= -M_{G6} = -L_1 \cdot s''_{a1} + L_2 \cdot s''_{a2} - L_3 \cdot n''_{ac} \\
M_B &= -M_F = -L_1 \cdot s''_{b1} - L_2 \cdot s''_{b2} + L_3 \cdot n''_{bc} \\
M_C &= -M_E = +L_1 \cdot s''_{c1} - L_2 \cdot s''_{c2} - L_3 \cdot n''_{cc}.
\end{aligned} \tag{931}$$

where

$$\begin{aligned}
s''_{a1} &= +n''_{aa} - n''_{ab} = (s_{a1} + s_{a6}) & s''_{a2} &= +n''_{ab} - n''_{ac} = (s_{a2} + s_{a5}) \\
s''_{b1} &= -n''_{ba} + n''_{bb} = (s_{b1} - s_{b6}) & s''_{b2} &= +n''_{bb} - n''_{bc} = (s_{b2} + s_{b5}) \\
s''_{c1} &= -n''_{ca} + n''_{cb} = (s_{c1} - s_{c6}) & s''_{c2} &= -n''_{cb} + n''_{cc} = (s_{c2} - s_{c5}).
\end{aligned} \tag{932}$$

---

* The members enclosed in parentheses have been formed with the values s (851) appropriately extended for the six-bay frame.

Loading Conditions 2b and 3b

The two columns carrying arbitrary horizontal loading (acting from left to right) anti-symmetrical about center of frame (2b), and an anti-symmetrical pair of rotational moments at corners of frame (3b)

$$(\mathfrak{L}_g = \mathfrak{L}_a, \quad \mathfrak{R}_g = \mathfrak{R}_a; \quad D_G = D_A)$$

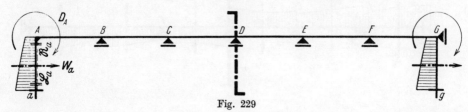

Fig. 229

Column loading diagram with load terms $\mathfrak{L}$ and $\mathfrak{R}$, and rotational moment D, for left-hand half of frame

Joint load term:

$$(\boldsymbol{L}_a = \mathfrak{L}_a k_a \quad \text{and} \quad \boldsymbol{R}_a = \mathfrak{R}_a k_a); \qquad \mathfrak{B}_A = \boldsymbol{D}_A \cdot b_a + \boldsymbol{L}_a \cdot \varepsilon_a - \boldsymbol{R}_a. \qquad (933)$$

Beam moments:

$$\left. \begin{array}{ll} M_{A1} = -M_{G6} = +\mathfrak{B}_A \cdot n''_{aa} & M_C = -M_E = +\mathfrak{B}_A \cdot n''_{ca} \\ M_B = -M_F = -\mathfrak{B}_A \cdot n''_{ba} & M_D = 0. \end{array} \right\} \qquad (934)$$

Column moments:

$$M_{Aa} = M_{Gg} = -\boldsymbol{D}_A + M_{A1}; \qquad M_a = M_g = -(\mathfrak{L}_a + M_{Aa})\varepsilon_a. \qquad (935)$$

Horizontal thrusts, column shears and lateral restraining force for all loading conditions

For asymmetrical Loading Conditions 1 and 3, and 2 and 4, respectively: according to formulas (179) and (181) respectively, for the subscripts $n = a$, g and $N = A, G$.
$$H = -H_a - H_g \quad \text{and} \quad H = -Q_{ra} - Q_{rg}. \qquad (936)$$

For symmetrical Loading Conditions 1a and 3a, and 2a and 4a, respectively: according to formulas (179) and (181) respectively, for the subscripts $n = a$ and $N = A$.
$$H = 0. \qquad (937)$$

For anti-symmetrical Loading Conditions 1b and 3b, and 2b and 4b, respectively: according to formulas (179) and (181) respectively, for the subscripts $n = a$ and $N = A$.
$$H = -2H_a \quad \text{and} \quad H = -2Q_{ra}. \qquad (938)$$

II. Treatment by the Deformation Method

a) Coefficients for symmetrical load arrangements

Stiffness coefficients of all the horizontal members and column stiffnesses:

$$K_\nu = \frac{J_\nu}{J_k} \cdot \frac{l_k}{l_\nu} \quad \text{for} \quad \nu = 1, 2, 3; \qquad K_a = \frac{J_a}{J_k} \cdot \frac{l_k}{h}; \qquad S_a = \frac{K_a}{2 - \varepsilon_a}. \tag{939}$$

Joint coefficients:

$$K_A = 3 S_a + 2 K_1 \qquad K_B = 2 K_1 + 2 K_2 \qquad K_C = 2 K_2 + 2 K_3. \tag{940}$$

Rotation carry-over factors and auxiliary coefficients:

backward reduction:

$$(j_{3s} = 0)$$
$$j_{2s} = \frac{K_2}{r'_c} \qquad r'_c = K_C$$
$$j_{1s} = \frac{K_1}{r'_b} \qquad r'_b = K_B - K_2 \cdot j_{2s}$$
$$r'_a = K_A - K_1 \cdot j_{1s}.$$

forward reduction:

$$j'_1 = \frac{K_1}{v_a} \qquad v_a = K_A$$
$$j'_2 = \frac{K_2}{v_b} \qquad v_b = K_B - K_1 \cdot j'_1$$
$$v_c = K_C - K_2 \cdot j'_2.$$

(941) and (942)

Principal influence coefficients by recursion:

forward:

$$u'_{aa} = 1/r'_a$$
$$u'_{bb} = 1/r'_b + u'_{aa} \cdot j^2_{1s}$$
$$u'_{cc} = 1/r'_c + u'_{bb} \cdot j^2_{2s}.$$

backward:

$$u'_{cc} = 1/v_c$$
$$u'_{bb} = 1/v_b + u'_{cc} \cdot j'^2_2$$
$$u'_{aa} = 1/v_a + u'_{bb} \cdot j'^2_1.$$

(943) and (944)

Principal influence coefficients direct

$$u'_{aa} = 1/r'_a \qquad u'_{bb} = 1/(r'_b + v_b - K_B) \qquad u'_{cc} = 1/v_c. \tag{945}$$

Computation scheme for the symmetrical influence coefficient matrix*

$$\begin{array}{c|ccc|c}
 & (u'_{aa}) & u'_{ab} & u'_{ac} & \\
\cdot j_{1s} & \downarrow & \uparrow & \uparrow & \cdot j'_1 \\
 & u'_{ba} & (u'_{bb}) & u'_{bc} & \\
\cdot j_{2s} & \downarrow & \downarrow & \uparrow & \cdot j'_2 \\
 & u'_{ca} & u'_{cb} & (u'_{cc}) &
\end{array} \tag{946}$$

---

* Between the elements of matrix (946) for the left-hand half of the frame and the elements of matrix (864), enlarged to 7 × 7 elements, there exist, according to the "load transposition method" the following relationships: $u'_{aa} = u_{aa} - u_{ag}$ etc., up to $u'_{cc} = u_{cc} - u_{ce}$.
(The values u'' with the subscripts d of the central support vanish).

Loading Condition 4a—Symmetrical overall loading condition

(Superposition of Loading Conditions 1a, 2a and 3a)

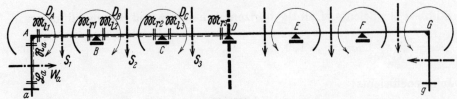

Fig. 230

Loading diagram, with fixed-end moments $\mathfrak{M}_l$ and $\mathfrak{M}_r$ of the beam spans, load terms $\mathfrak{L}$ and $\mathfrak{R}$ for the column, and external rotational moments D, for left-hand half of frame

Column restraining moment:

$$\mathfrak{C}_{ra} = -\frac{\mathfrak{R}_a - \mathfrak{L}_a \cdot \varepsilon_a}{2 - \varepsilon_a}. \qquad (947)$$

Joint load terms:

$$\mathfrak{K}_A = D_A + \mathfrak{C}_{ra} - \mathfrak{M}_{l1} \qquad \begin{aligned} \mathfrak{K}_B &= D_B + \mathfrak{M}_{r1} - \mathfrak{M}_{l2} \\ \mathfrak{K}_C &= D_C + \mathfrak{M}_{r2} - \mathfrak{M}_{l3}. \end{aligned} \qquad (948)$$

auxiliary moments:

$$\begin{aligned} Y'_A &= +\mathfrak{K}_A \cdot u'_{aa} - \mathfrak{K}_B \cdot u'_{ab} + \mathfrak{K}_C \cdot u'_{ac} \\ Y'_B &= -\mathfrak{K}_A \cdot u'_{ba} + \mathfrak{K}_B \cdot u'_{bb} - \mathfrak{K}_C \cdot u'_{bc} \\ Y'_C &= +\mathfrak{K}_A \cdot u'_{ca} - \mathfrak{K}_B \cdot u'_{cb} + \mathfrak{K}_C \cdot u'_{cc}. \end{aligned} \qquad (949)$$

Beam moments:

$$\begin{aligned} M_{A1} &= M_{G6} = \mathfrak{M}_{l1} + (2Y'_A + Y'_B)K_1 \\ M_{B2} &= M_{F5} = \mathfrak{M}_{l2} + (2Y'_B + Y'_C)K_2 \\ M_{C3} &= M_{E4} = \mathfrak{M}_{l3} + 2Y'_C \cdot K_3; \\ M_{B1} &= M_{F6} = \mathfrak{M}_{r1} - (Y'_A + 2Y'_B)K_1 \\ M_{C2} &= M_{E5} = \mathfrak{M}_{r2} - (Y'_B + 2Y'_C)K_2 \\ M_{D3} &= M_{D4} = \mathfrak{M}_{r3} - Y'_C \cdot K_3. \end{aligned} \qquad (950)$$

Column moments:

$$M_{Aa} = -M_{Gg} = \mathfrak{C}_{ra} - Y'_A \cdot 3S_a \qquad M_a = -M_g = -(\mathfrak{L}_a + M_{Aa})\varepsilon_a. \qquad (951)$$

Check relationships:

$$D_A + M_{Aa} = M_{A1} \qquad D_B + M_{B1} = M_{B2} \qquad D_C + M_{C2} = M_{C3}. \qquad (952)$$

b) Coefficients for anti-symmetrical load arrangements

Stiffness coefficients of all the members, and column stiffness: same as formulas (939)

Diagonal coefficients:

$$K_A = 3S_a + 2K_1 \qquad K_B = 2K_1 + 2K_2 \qquad K_C'' = 2K_2 + 3K_3/2. \tag{953}$$

Rotation carry-over factors and auxiliary coefficients:

$$\left. \begin{array}{l|l}
\text{backward reduction:} & \text{forward reduction:} \\
\begin{aligned} j_{2t} &= \dfrac{K_2}{r_c''} \\ j_{1t} &= \dfrac{K_1}{r_b''} \end{aligned} \quad \begin{aligned} r_c'' &= K_C'' \\ r_b'' &= K_B - K_2 \cdot j_{2t} \\ r_a'' &= K_A - K_1 \cdot j_{1t}. \end{aligned} & \begin{aligned} j_1' &= \dfrac{K_1}{v_a} \\ j_2' &= \dfrac{K_2}{v_b} \end{aligned} \quad \begin{aligned} v_a &= K_A \\ v_b &= K_B - K_1 \cdot j_1' \\ v_c'' &= K_C'' - K_2 \cdot j_2'. \end{aligned}
\end{array} \right\} \begin{array}{c} (954) \\ \text{and} \\ (955) \end{array}$$

Principal influence coefficients by recursion:

$$\left. \begin{array}{l|l}
\text{forward:} & \text{backward:} \\
\begin{aligned} u_{aa}'' &= 1/r_a'' \\ u_{bb}'' &= 1/r_b'' + u_{aa}'' \cdot j_{1t}^2 \\ u_{cc}'' &= 1/r_c'' + u_{bb}'' \cdot j_{2t}^2. \end{aligned} & \begin{aligned} u_{cc}'' &= 1/v_c'' \\ u_{bb}'' &= 1/v_b + u_{cc}'' \cdot j_2'^2 \\ u_{aa}'' &= 1/v_a + u_{bb}'' \cdot j_1'^2. \end{aligned}
\end{array} \right\} \begin{array}{c} (956) \\ \text{and} \\ (957) \end{array}$$

Principal influence coefficients direct

$$u_{aa}'' = \frac{1}{r_a''} \qquad u_{bb}'' = \frac{1}{r_b'' + v_b - K_B} \qquad u_{cc}'' = \frac{1}{v_c''}. \tag{958}$$

Computation scheme for the symmetrical influence coefficient matrix*

$$\left. \begin{array}{c|ccc|c}
& (u_{aa}'') & u_{ab}'' & u_{ac}'' & \\
\cdot j_{1t} & \downarrow & \uparrow & \uparrow & \cdot j_1' \\
& u_{ba}'' & (u_{bb}'') & u_{bc}'' & \\
\cdot j_{2t} & \downarrow & \downarrow & \uparrow & \cdot j_2' \\
& u_{ca}'' & u_{cb}'' & (u_{cc}'') & \\
\end{array} \right\} \tag{959}$$

---

* Matrix (959) which has only 3 × 3 elements cannot be directly derived according to the "load transposition method" from the 7 × 7 matrix mentioned in the footnote to page 213. Coefficient K''_c and load term $\mathfrak{R}''_C$, page 216 take into account in advance that the central support D acts as a hinge under anti-symmetrical loading.

Loading Condition 4b—Anti-symmetrical overall loading condition

(Superposition of Loading Conditions 1b, 2b and 3b)

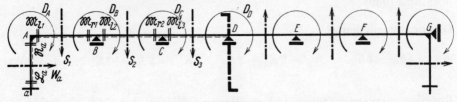

Fig. 231

Loading diagram, with fixed-end moments $\mathfrak{M}_l$ and $\mathfrak{M}_r$ of the beam spans, load terms $\mathfrak{L}$ and $\mathfrak{R}$ of the column and innermost beam span, and external rotational moments D, for left-hand half of frame

Fixed-end moments for column a and beam span 3*:

$$\mathfrak{C}_{ra} = -(\mathfrak{R}_a - \mathfrak{L}_a \cdot \varepsilon_a)/(2 - \varepsilon_a); \qquad \mathfrak{M}'_{l3} = -\mathfrak{L}_3/2 + D_D/4. \tag{960}$$

Joint load terms:

$$\left. \begin{array}{ll} \mathfrak{R}_A = D_A + \mathfrak{C}_{ra} - \mathfrak{M}_{l1} & \mathfrak{R}_B = D_B + \mathfrak{M}_{r1} - \mathfrak{M}_{l2} \\ & \mathfrak{R}''_C = D_C + \mathfrak{M}_{r2} - \mathfrak{M}'_{l3}. \end{array} \right\} \tag{961}$$

auxiliary moments:

$$\left. \begin{array}{c|ccc} & \mathfrak{R}_A & \mathfrak{R}_B & \mathfrak{R}''_C \\ \hline Y''_A = & +u''_{aa} & -u''_{ab} & +u''_{ac} \\ Y''_B = & -u''_{ba} & +u''_{bb} & -u''_{bc} \\ Y''_C = & +u''_{ca} & -u''_{cb} & +u''_{cc}. \end{array} \right\} \tag{962}$$

Beam moments:

$$\left. \begin{array}{l} M_{A1} = -M_{G6} = \mathfrak{M}_{l1} + (2Y''_A + Y''_B) K_1 \\ M_{B2} = -M_{F5} = \mathfrak{M}_{l2} + (2Y''_B + Y''_C) K_2 \\ M_{C3} = -M_{E4} = \mathfrak{M}'_{l3} + Y''_C \cdot 3K_3/2; \\ \qquad M_{B1} = -M_{F6} = \mathfrak{M}_{r1} - (Y''_A + 2Y''_B) K_1 \\ \qquad M_{C2} = -M_{E5} = \mathfrak{M}_{r2} - (Y''_B + 2Y''_C) K_2 \\ \qquad M_{D3} = -M_{D4} = -D_D/2. \end{array} \right\} \tag{963}$$

Column moments:

$$M_{Aa} = M_{Gg} = \mathfrak{C}_{ra} - Y''_A \cdot 3S_a \qquad M_a = M_g = -(\mathfrak{L}_a + M_{Aa})\varepsilon_a. \tag{964}$$

Check relationships: same as (952).

---
* See footnote on p. 215.

Frame Shape 31v
(Frame Shape 31 with laterally unrestrained horizontal member)*

Note: Lateral displacement of the horizontal member occurs only for asymmetrical and anti-symmetrical loading conditions (for all symmetrical loading cases of Frame Shape 31 we have $H = 0$). As every asymmetrical loading condition can, with the aid of the "load transposition method", be split up into a symmetrical and an anti-symmetrical loading condition, only the anti-symmetrical loading conditions for Frame Shape 31v will here be considered, while the symmetrical loading conditions can be analysed with the formulas given for Frame Shape 31.

I. Treatment by the Method of Forces
Coefficients

Flexibility coefficients of all the horizontal members and column flexibility:

$$k_\nu = \frac{J_k}{J_\nu} \cdot \frac{l_\nu}{l_k} \quad \text{for } \nu = 1, 2, 3; \quad k_a = \frac{J_k}{J_a} \cdot \frac{h}{l_k}; \quad m_a = k_a\left(\frac{1}{\varepsilon_a} + 4\right). \tag{965}$$

Diagonal coefficients:

$$k_A^v = m_a + 2k_1 \qquad k_B = 2k_1 + 2k_2 \qquad k_C = 2k_2 + 2k_3. \tag{966}$$

Moment carry-over factors and auxiliary coefficients:

backward reduction: | forward reduction:

$$\iota_2 = \frac{k_2}{r_c} \qquad r_c = k_C \qquad \qquad \iota_1' = \frac{k_1}{v_a} \qquad v_a = k_A^v$$
$$\qquad\qquad r_b = k_B - k_2 \cdot \iota_2 \qquad\qquad\qquad v_b = k_B - k_1 \cdot \iota_1'$$
$$\iota_1 = \frac{k_1}{r_b} \qquad r_a = k_A^v - k_1 \cdot \iota_1. \qquad \iota_2' = \frac{k_2}{v_b} \qquad v_c = k_C - k_2 \cdot \iota_2'.$$

$$\tag{967 and 968}$$

Principal influence coefficients by recursion:

forward: | backward:

$$n_{aa} = 1/r_a \qquad\qquad n_{cc} = 1/v_c$$
$$n_{bb} = 1/r_b + n_{aa} \cdot \iota_1^2 \qquad n_{bb} = 1/v_b + n_{cc} \cdot \iota_2'^2$$
$$n_{cc} = 1/r_c + n_{bb} \cdot \iota_2^2. \qquad n_{aa} = 1/v_a + n_{bb} \cdot \iota_1'^2.$$

$$\tag{969 and 970}$$

Principal influence coefficients direct

$$n_{aa} = \frac{1}{r_a} \qquad n_{bb} = \frac{1}{r_b + v_b - k_B} \qquad n_{cc} = \frac{1}{v_c}. \tag{971}$$

---

* See Fig. 224 and 225 on page 206: in the case of Frame Shape 31v the restraining bearing G and therefore also the lateral restraining force H are absent. The information given under the heading "Important" on that page is also applicable to Frame Shape 31v.

Computation scheme for the symmetrical influence coefficient matrix:

$$\begin{array}{c} \cdot l_1 \\ \cdot l_2 \end{array} \begin{vmatrix} (n_{aa}) & n_{ab} & n_{ac} \\ \downarrow & \uparrow & \uparrow \\ n_{ba} & (n_{bb}) & n_{bc} \\ \downarrow & \downarrow & \uparrow \\ n_{ca} & n_{cb} & (n_{cc}) \end{vmatrix} \begin{array}{c} \cdot l_1' \\ \cdot l_2' \end{array} \quad (972)$$

Loading Condition 1b

Horizontal member carrying vertical loading of arbitrary magnitude antisymmetrical about center of frame

$$(\Re_6 = -\mathfrak{L}_1, \quad \mathfrak{L}_6 = -\Re_1; \quad \Re_5 = -\mathfrak{L}_2, \quad \mathfrak{L}_5 = -\Re_2; \quad \Re_4 = -\mathfrak{L}_3)$$

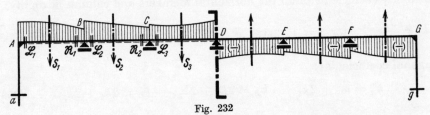

Fig. 232

Beam loading diagram with load terms $\mathfrak{L}$ and $\Re$ for left-hand half of frame

Beam moments ($L_\nu = \mathfrak{L}_\nu k_\nu$ and $R_\nu = \Re_\nu k_\nu$):

	L_1	$(R_1 + L_2)$	$(R_2 + L_3)$
$X_a = M_{A1} = -M_{G6} =$	$-n_{aa}$	$+n_{ab}$	$-n_{ac}$
$X_b = M_B = -M_F =$	$+n_{ba}$	$-n_{bb}$	$+n_{bc}$
$X_c = M_C = -M_E =$	$-n_{ca}$	$+n_{cb}$	$-n_{cc}$;
$M_D = 0.$			

(973)

Column moments: $(M_{Aa} = M_a) = (M_{Gg} = M_g) = X_a.$ \quad (974)

Special case: symmetrical bay loading ($R_\nu = L_\nu$):

$$\left.\begin{array}{l} X_a = M_{A1} = -M_{G6} = -L_1 \cdot s_{a1} + L_2 \cdot s_{a2} - L_3 \cdot n_{ac} \\ X_b = M_B = -M_F = -L_1 \cdot s_{b1} - L_2 \cdot s_{b2} + L_3 \cdot n_{bc} \\ X_c = M_C = -M_E = +L_1 \cdot s_{c1} - L_2 \cdot s_{c2} - L_3 \cdot n_{cc}. \end{array}\right\} \quad (975)$$

Note: values s are given on next page.

Composite influence coefficients for Loading Condition 1b.

$$s_{a1} = + n_{aa} - n_{ab} \qquad s_{a2} = + n_{ab} - n_{ac}$$
$$s_{b1} = - n_{ba} + n_{bb} \qquad s_{b2} = + n_{bb} - n_{bc} \qquad (976)$$
$$s_{c1} = - n_{ca} + n_{cb} \qquad s_{c2} = - n_{cb} + n_{cc}.$$

Loading Conditions 2b and 3b

The two columns carrying arbitrary horizontal loading (acting from left to right) anti-symmetrical about center of frame (2b), and an anti-symmetrical pair of rotational moments at corners of frame (3b)

$$(\mathfrak{L}_g = \mathfrak{L}_a, \quad \mathfrak{R}_g = \mathfrak{R}_a; \quad D_G = D_A)$$

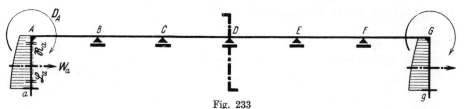

Fig. 233

Column loading diagram with load terms $\mathfrak{L}$ and $\mathfrak{R}$, and rotational moment D, for left-hand half of frame

Joint load terms:

$$\mathfrak{B}_a = \mathfrak{S}_{la}(3k_a + 2k_1) + (\mathfrak{L}_a + \mathfrak{R}_a)k_a \qquad \mathfrak{B}_1 = \mathfrak{S}_{la} \cdot k_1. \qquad (977)$$

Beam and column moments

$$X_a = M_a = + M_g = - \mathfrak{B}_a \cdot n_{aa} + \mathfrak{B}_1 \cdot n_{ab}$$
$$X_b = M_B = - M_F = + \mathfrak{B}_a \cdot n_{ba} - \mathfrak{B}_1 \cdot n_{bb} \qquad (978)$$
$$X_c = M_C = - M_E = - \mathfrak{B}_a \cdot n_{ca} + \mathfrak{B}_1 \cdot n_{cb};$$

$$M_D = 0; \qquad M_{Aa} = M_{Gg} = X_a + \mathfrak{S}_{la} \qquad M_{A1} = -M_{G6} = M_{Aa} + D_A. \quad (979)$$

Column forces:

$$(H_a = H_g) = Q_{la} = W_a \qquad Q_{ra} = 0. \qquad (980)$$

II. Treatment by the Deformation Method

Coefficients

Stiffness coefficients of all the horizontal members, and column stiffness

$$K_\nu = \frac{J_\nu}{J_k} \cdot \frac{l_k}{l_\nu} \quad \text{for} \quad \nu = 1, 2, 3; \qquad K_a = \frac{J_a}{J_k} \cdot \frac{l_k}{h}; \qquad F_a = \frac{K_a}{1/\varepsilon_a + 4}. \qquad (981)$$

Joint coefficients:

$$K_A^v = 3F_a + 2K_1 \qquad K_B = 2K_1 + 2K_2 \qquad K_C'' = 2K_2 + 3K_3/2. \tag{982}$$

Rotation carry-over factors and auxiliary coefficients:

$$
\left.\begin{array}{ll}
\text{backward reduction:} & \text{forward reduction:} \\[4pt]
j_2 = \dfrac{K_2}{r_c} \quad
\begin{aligned}
r_c &= K_C'' \\
r_b &= K_B - K_2 \cdot j_2
\end{aligned}
&
j_1' = \dfrac{K_1}{v_a} \quad
\begin{aligned}
v_a &= K_A^v \\
v_b &= K_B - K_1 \cdot j_1'
\end{aligned} \\[10pt]
j_1 = \dfrac{K_1}{r_b} \quad r_a = K_A^v - K_1 \cdot j_1. &
j_2' = \dfrac{K_2}{v_b} \quad v_c = K_C'' - K_2 \cdot j_2'.
\end{array}\right\}
\begin{array}{l}(983)\\ \text{and} \\ (984)\end{array}
$$

Principal influence coefficients by recursion:

$$
\left.\begin{array}{ll}
\text{forward:} & \text{backward:} \\[4pt]
u_{aa} = 1/r_a & u_{cc} = 1/v_c \\
u_{bb} = 1/r_b + u_{aa} \cdot j_1^2 & u_{bb} = 1/v_b + u_{cc} \cdot j_2'^2 \\
u_{cc} = 1/r_c + u_{bb} \cdot j_2^2. & u_{aa} = 1/v_a + u_{bb} \cdot j_1'^2.
\end{array}\right\}
\begin{array}{l}(985)\\ \text{and} \\ (986)\end{array}
$$

Principal influence coefficients direct

$$u_{aa} = 1/r_a \qquad u_{bb} = 1/(r_b + v_b - K_B) \qquad u_{cc} = 1/v_c. \tag{987}$$

Computation scheme for the symmetrical influence coefficient matrix

$$
\left.\begin{array}{c|ccc|c}
\cdot j_1 & (u_{aa}) & u_{ab} & u_{ac} & \cdot j_1' \\
& \downarrow & \uparrow & \uparrow & \\
& u_{ba} & (u_{bb}) & u_{bc} & \\
\cdot j_2 & \downarrow & \downarrow & \uparrow & \cdot j_2' \\
& u_{ca} & u_{cb} & (u_{cc}) &
\end{array}\right\} \tag{988}
$$

Loading Condition 4b—Anti-symmetrical overall loading condition

(Superposition of Loading Conditions 1b, 2b, 3b and 5)

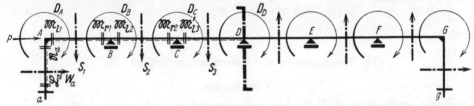

Fig. 234

Loading diagram, with fixed-end moments $\mathfrak{M}_l$ and $\mathfrak{M}_r$ of the beam spans, load terms $\mathfrak{L}$ and $\mathfrak{R}$ for the column, and external rotational moments D, for left-hand half of frame; furthermore with horizontal concentrated load P at level of horizontal member

Restraining moments for column a and beam span 3

$$\mathfrak{F}_{ra} = \frac{(Ph/2 + \mathfrak{S}_{la})(1/\varepsilon_a + 1) - (\mathfrak{L}_a + \mathfrak{R}_a)}{1/\varepsilon_a + 4} \; ; \qquad \mathfrak{M}'_{l3} = -\frac{\mathfrak{L}_3}{2} + \frac{D_D}{4}. \qquad (989)$$

Joint load terms:

$$\begin{aligned} \mathfrak{K}^v_A &= D_A + \mathfrak{F}_{ra} - \mathfrak{M}_{l1} \\ \mathfrak{K}_B &= D_B + \mathfrak{M}_{r1} - \mathfrak{M}_{l2} \\ \mathfrak{K}''_C &= D_C + \mathfrak{M}_{r2} - \mathfrak{M}'_{l3}. \end{aligned} \qquad (990)$$

auxiliary moments:

	$\mathfrak{K}^v_A$	$\mathfrak{K}_B$	$\mathfrak{K}''_C$
$Y_A =$	$+u_{aa}$	$-u_{ab}$	$+u_{ac}$
$Y_B =$	$-u_{ba}$	$+u_{bb}$	$-u_{bc}$
$Y_C =$	$+u_{ca}$	$-u_{cb}$	$+u_{cc}.$

(991)

Beam moments:

$$\begin{aligned} M_{A1} &= -M_{G6} = \mathfrak{M}_{l1} + (2Y_A + Y_B)K_1 \\ M_{B2} &= -M_{F5} = \mathfrak{M}_{l2} + (2Y_B + Y_C)K_2 \\ M_{C3} &= -M_{E4} = \mathfrak{M}'_{l3} + Y_C \cdot 3K_3/2; \\ M_{B1} &= -M_{F6} = \mathfrak{M}_{r1} - (Y_A + 2Y_B)K_1 \\ M_{C2} &= -M_{E5} = \mathfrak{M}_{r2} - (Y_B + 2Y_C)K_2 \\ M_{D3} &= -M_{D4} = -D_D/2. \end{aligned} \qquad (992)$$

Column moments:

$$M_{Aa} = M_{Gg} = \mathfrak{F}_{ra} - Y_A \cdot 3F_a \qquad M_a = M_g = -\frac{Ph}{2} - \mathfrak{S}_{la} + M_{Aa}. \qquad (993)$$

Check relationships:

$$D_A + M_{Aa} = M_{A1} \qquad D_B + M_{B1} = M_{B2} \qquad D_C + M_{C2} = M_{C3}. \qquad (994)$$

Column forces:

$$(H_a = Q_{la}) = (H_g = Q_{lg}) = W_a + P/2 \qquad (Q_{ra} = Q_{rg}) = P/2. \qquad (995)$$

Frame Shape 32

Symmetrical single-story five-bay frame with laterally restrained horizontal member, two elastically restrained end columns, and four pin-jointed internal supports (or rocker columns)

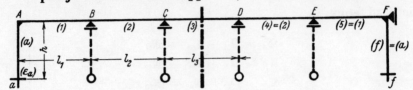

Fig. 235 Frame shape, dimensions and symbols

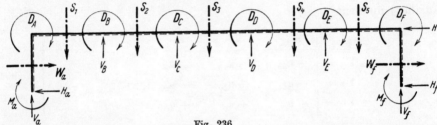

Fig. 236

Definition of positive direction for all external loads on joints and members, for all support reactions, and for the lateral restraining force

All the dimensions and coefficients for the right-hand half of the frame are the same as those for the left-hand half. In the case of the columns the broken line is always placed on the right-hand side of each column, despite the symmetry of the frame.

Note: For Frame Shape 32 the *coefficients and the Loading Conditions* 1–4 for *Frame Shape* 30, enlarged by the addition of two bays, may be employed. On account of symmetry, the forward reduction becomes the same as the backward reduction, and the backward recursion becomes the same as the forward recursion, both in Method I and Method II. Furthermore, the *square matrices of influence coefficients* (each with 6 × 6 elements) are bisymmetrical, i.e., symmetrical about the principal diagonal and the secondary diagonal. If the *"load transposition method"* is used, however, it is also possible to start from the following symmetrical and anti-symmetrical loading conditions.

Important

All the instructions, formula sequences and formula matrices given for Frame Shape 32 can readily be extended to suit symmetrical frames of similar type having any odd number of bays.

I. Treatment by the Method of Forces
a) Coefficients for symmetrical load arrangements

Flexibility coefficients of all the horizontal members and column flexibility:

$$k_\nu = \frac{J_k}{J_\nu} \cdot \frac{l_\nu}{l_k} \quad \text{for} \quad \nu = 1, 2, 3; \qquad k_a = \frac{J_k}{J_a} \cdot \frac{h}{l_k}; \qquad b_a = k_a(2 - \varepsilon_a). \qquad (996)$$

Diagonal coefficients:

$$k_A = b_a + 2k_1 \qquad k_B = 2k_1 + 2k_2 \qquad k'_C = 2k_2 + 3k_3. \qquad (997)$$

Moment carry-over factors and auxiliary coefficients:

backward reduction:

$$\iota_{2s} = \frac{k_2}{r'_c} \qquad \begin{aligned} r'_c &= k'_C \\ r'_b &= k_B - k_2 \cdot \iota_{2s} \end{aligned}$$

$$\iota_{1s} = \frac{k_1}{r'_b} \qquad r'_a = k_A - k_1 \cdot \iota_{1s}.$$

forward reduction:

$$\iota'_1 = \frac{k_1}{v_a} \qquad \begin{aligned} v_a &= k_A \\ v_b &= k_B - k_1 \cdot \iota'_1 \end{aligned}$$

$$\iota'_2 = \frac{k_2}{v_b} \qquad v'_c = k'_C - k_2 \cdot \iota'_2.$$

$$(998) \text{ and } (999)$$

Principal influence coefficients by recursion:

forward:

$$n'_{aa} = 1/r'_a$$
$$n'_{bb} = 1/r'_b + n'_{aa} \cdot \iota^2_{1s}$$
$$n'_{cc} = 1/r'_c + n'_{bb} \cdot \iota^2_{2s}.$$

backward:

$$n'_{cc} = 1/v'_c$$
$$n'_{bb} = 1/v_b + n'_{cc} \cdot \iota'^2_2$$
$$n'_{aa} = 1/v_a + n'_{bb} \cdot \iota'^2_1.$$

$$(1000) \text{ and } (1001)$$

Principal influence coefficients direct:

$$n'_{aa} = \frac{1}{r'_a} \qquad n'_{bb} = \frac{1}{r'_b + v_b - k_B} \qquad n'_{cc} = \frac{1}{v'_c}. \qquad (1002)$$

Computation scheme for the symmetrical influence coefficient matrix*

$$\begin{array}{c}
\cdot \iota_{1s} \\
\\
\cdot \iota_{2s}
\end{array}
\begin{vmatrix}
(n'_{aa}) & n'_{ab} & n'_{ac} \\
\downarrow & \uparrow & \uparrow \\
n'_{ba} & (n'_{bb}) & n'_{bc} \\
\downarrow & \downarrow & \uparrow \\
n'_{ca} & n'_{cb} & (n'_{cc})
\end{vmatrix}
\begin{array}{c}
\cdot \iota'_1 \\
\\
\cdot \iota'_2
\end{array} \qquad (1003)$$

---

* Between the elements of matrix (1003) for the left-hand half of the frame and the elements of matrix (881), page 200, enlarged to 6 × 6 elements, there exist, according to the "load transposition method", the following relationships:

$$n'_{aa} = n_{aa} - n_{af}, \quad n'_{ab} = n_{ab} - n_{ae} \text{ etc., up to } n'_{cc} = n_{cc} - n_{cd}.$$

Loading Condition 1a

Horizontal member carrying arbitrary vertical loading symmetrical about center of frame

$$(\mathfrak{R}_5 = \mathfrak{L}_1, \quad \mathfrak{L}_5 = \mathfrak{R}_1; \quad \mathfrak{R}_4 = \mathfrak{L}_2, \quad \mathfrak{L}_4 = \mathfrak{R}_2; \quad \mathfrak{R}_3 = \mathfrak{L}_3)$$

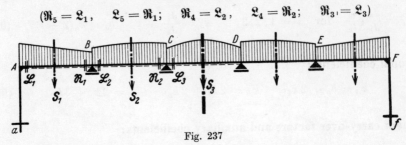

Fig. 237

Beam loading diagram with load terms $\mathfrak{L}$ and $\mathfrak{R}$ for left-hand half of frame

Beam moments ($L_\nu = \mathfrak{L}_\nu k_\nu$ and $R_\nu = \mathfrak{R}_\nu k_\nu$):

$$
\begin{array}{c|ccc}
 & L_1 & (R_1 + L_2) & (R_2 + L_3) \\
\hline
M_{A1} = M_{F5} = & -n'_{aa} & +n'_{ab} & -n'_{ac} \\
M_B = M_E = & +n'_{ba} & -n'_{bb} & +n'_{bc} \\
M_C = M_D = & -n'_{ca} & +n'_{cb} & -n'_{cc}
\end{array}
\qquad (1004)
$$

Column moments:

$$M_{Aa} = -M_{Ff} = M_{A1} \qquad M_a = -M_f = -M_{Aa} \cdot \varepsilon_a. \qquad (1005)$$

Special case: symmetrical bay loading $(R_\nu = L_\nu)$:

$$
\begin{aligned}
M_{A1} = M_{F5} &= -L_1 \cdot s'_{a1} + L_2 \cdot s'_{a2} - L_3 \cdot n'_{ac} \\
M_B = M_E &= -L_1 \cdot s'_{b1} - L_2 \cdot s'_{b2} + L_3 \cdot n'_{bc} \\
M_C = M_D &= +L_1 \cdot s'_{c1} - L_2 \cdot s'_{c2} - L_3 \cdot n'_{cc};
\end{aligned}
\qquad (1006)
$$

where

$$
\begin{aligned}
s'_{a1} &= +n'_{aa} - n'_{ab} = (s_{a1} + s_{a5}) & s'_{a2} &= +n'_{ab} - n'_{ac} = (s_{a2} + s_{a4}) \\
s'_{b1} &= -n'_{ba} + n'_{bb} = (s_{b1} - s_{b5}) & s'_{b2} &= +n'_{bb} - n'_{bc} = (s_{b2} + s_{b4}) \\
s'_{c1} &= -n'_{ca} + n'_{cb} = (s_{c1} - s_{c5}) & s'_{c2} &= -n'_{cb} + n'_{cc} = (s_{c2} - s_{c4}).
\end{aligned}
\qquad (1007)
$$

** The members enclosed in parentheses have been formed with the values s (885) appropriately extended for the five-bay frame.

Loading Conditions 2a and 3a (2b and 3b)*

The two columns carrying arbitrary horizontal loading directed towards center of frame (acting from left to right) and symmetrical (anti-symmetrical) about center of frame, and symmetrical (anti-symmetrical) pair of rotational moments at corners of frame.

Joint load term:

$$(L_a = \mathfrak{L}_a k_a \quad \text{and} \quad R_a = \mathfrak{R}_a k_a); \qquad \mathfrak{B}_A = D_A \cdot b_a + L_a \cdot \varepsilon_a - R_a. \qquad (1008)$$

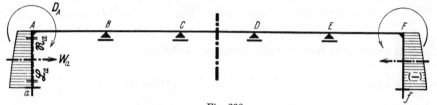

Fig. 238
Loading Conditions 2a and 3a—symmetry

$$(\mathfrak{L}_f = -\mathfrak{L}_a, \quad \mathfrak{R}_f = -\mathfrak{R}_a; \quad D_F = -D_A).$$

Column and beam moments:

$$M_{Aa} = -M_{Ff} = -D_A + M_{A1}$$
$$M_a = -M_f = -(\mathfrak{L}_a + M_{Aa})\varepsilon_a;$$

$$\left.\begin{array}{l} M_{A1} = M_{F5} = +\mathfrak{B}_A \cdot n'_{aa} \\ M_B = M_E = -\mathfrak{B}_A \cdot n'_{ba} \\ M_C = M_D = +\mathfrak{B}_A \cdot n'_{ca} \end{array}\right\} \begin{array}{c} (1009) \\ \text{and} \\ (1010) \end{array}$$

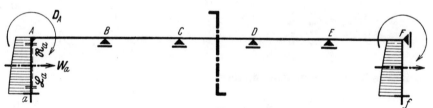

Fig. 239
Loading Conditions 2b and 3b—anti-symmetry

$$(\mathfrak{L}_f = \mathfrak{L}_a, \quad \mathfrak{R}_f = \mathfrak{R}_a; \quad D_F = D_A).$$

Column and beam moments:

$$M_{Aa} = M_{Ff} = -D_A + M_{A1}$$
$$M_a = M_f = -(\mathfrak{L}_a + M_{Aa})\varepsilon_a;$$

$$\left.\begin{array}{l} M_{A1} = -M_{F5} = +\mathfrak{B}_A \cdot n''_{aa} \\ M_B = -M_E = -\mathfrak{B}_A \cdot n''_{ba} \\ M_C = -M_D = +\mathfrak{B}_A \cdot n''_{ca} \end{array}\right\} \begin{array}{c} (1011) \\ \text{and} \\ (1012) \end{array}$$

---

* This arrangement has been adopted in order to save space. For the coefficients to the anti-symmetrical Loading Conditions 2b and 3b see page 226.

b) Coefficients for anti-symmetrical load arrangements

Flexibility coefficients of all the horizontal members, and column flexibility:

$$k_\nu = \frac{J_k}{J_\nu} \cdot \frac{l_\nu}{l_k} \quad \text{for} \quad \nu = 1, 2, 3; \qquad k_a = \frac{J_k}{J_a} \cdot \frac{h}{l_k}; \qquad b_a = k_a(2 - \varepsilon_a). \tag{1013}$$

Diagonal coefficients:

$$k_A = b_a + 2k_1 \qquad k_B = 2k_1 + 2k_2 \qquad k_C'' = 2k_2 + k_3. \tag{1014}$$

Moment carry-over factors and auxiliary coefficients:

backward reduction:

$$\iota_{2t} = \frac{k_2}{r_c''}$$
$$\iota_{1t} = \frac{k_1}{r_b''}$$

$$r_c'' = k_C''$$
$$r_b'' = k_B - k_2 \cdot \iota_{2t}$$
$$r_a'' = k_A - k_1 \cdot \iota_{1t}.$$

forward reduction:

$$\iota_1' = \frac{k_1}{v_a}$$
$$\iota_2' = \frac{k_2}{v_b}$$

$$v_a = k_A$$
$$v_b = k_B - k_1 \cdot \iota_1'$$
$$v_c'' = k_C'' - k_2 \cdot \iota_2'.$$

(1015) and (1016)

Principal influence coefficients by recursion:

forward:

$$n_{aa}'' = 1/r_a''$$
$$n_{bb}'' = 1/r_b'' + n_{aa}'' \cdot \iota_{1t}^2$$
$$n_{cc}'' = 1/r_c'' + n_{bb}'' \cdot \iota_{2t}^2.$$

backward:

$$n_{cc}'' = 1/v_c''$$
$$n_{bb}'' = 1/v_b + n_{cc}'' \cdot \iota_2'^2$$
$$n_{aa}'' = 1/v_a + n_{bb}'' \cdot \iota_1'^2.$$

(1017) and (1018)

Principal influence coefficients direct

$$n_{aa}'' = \frac{1}{r_a''} \qquad n_{bb}'' = \frac{1}{r_b'' + v_b - k_B} \qquad n_{cc}'' = \frac{1}{v_c''}. \tag{1019}$$

Computation scheme for the symmetrical influence coefficient matrix*

$$\begin{array}{c} \\ \cdot \iota_{1t} \\ \\ \cdot \iota_{2t} \\ \end{array} \left| \begin{array}{ccc} (n_{aa}'') & n_{ab}'' & n_{ac}'' \\ \downarrow & \uparrow & \uparrow \\ n_{ba}'' & (n_{bb}'') & n_{bc}'' \\ \downarrow & \downarrow & \uparrow \\ n_{ca}'' & n_{cb}'' & (n_{cc}'') \end{array} \right| \begin{array}{c} \cdot \iota_1' \\ \\ \cdot \iota_2' \\ \end{array} \quad\quad (1020)$$

---

* Between the elements of matrix (1020) for the left-hand half of the frame and the elements of matrix (881), enlarged to 6×6 elements, there exist, according to the "load transposition method", the following relationships:

$$n_{aa}'' = n_{aa} + n_{af}, \quad n_{ab}'' = n_{ab} + n_{ae} \text{ etc., up to } n_{cc}'' = n_{cc} + n_{cd}.$$

Loading Condition 1b

Horizontal member carrying vertical loading of arbitrary magnitude anti-symmetrical about center of frame

$$(\mathfrak{R}_5 = -\mathfrak{L}_1, \quad \mathfrak{L}_5 = -\mathfrak{R}_1; \quad \mathfrak{R}_4 = -\mathfrak{L}_2, \quad \mathfrak{L}_4 = -\mathfrak{R}_2; \quad \mathfrak{R}_3 = -\mathfrak{L}_3)$$

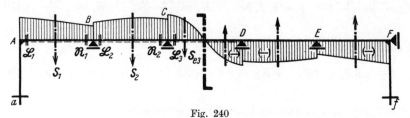

Fig. 240

Beam loading diagram with load terms $\mathfrak{L}$ and $\mathfrak{R}$ for left-hand half of frame

Beam moments ($L_\nu = \mathfrak{L}_\nu k_\nu$ and $R_\nu = \mathfrak{R}_\nu k_\nu$):

	L_1	$(R_1 + L_2)$	$(R_2 + L_3)$
$M_{A1} = -M_{F5} =$	$-n''_{aa}$	$+n''_{ab}$	$-n''_{ac}$
$M_B = -M_E =$	$+n''_{ba}$	$-n''_{bb}$	$+n''_{bc}$
$M_C = -M_D =$	$-n''_{ca}$	$+n''_{cb}$	$-n''_{cc}$.

(1021)

Column moments:

$$M_{Aa} = M_{Ff} = M_{A1} \qquad M_a = M_f = -M_{Aa} \cdot \varepsilon_a. \qquad (1022)$$

Special case: symmetrical bay loading ($R_1 = L_1$ and $R_2 = L_2$):

$$\left.\begin{aligned}
M_{A1} &= -M_{F5} = -L_1 \cdot s''_{a1} + L_2 \cdot s''_{a2} \\
M_B &= -M_E = -L_1 \cdot s''_{b1} - L_2 \cdot s''_{b2} \\
M_C &= -M_D = +L_1 \cdot s''_{c1} - L_2 \cdot s''_{c2};
\end{aligned}\right\} \qquad (1023)$$

where *

$$\left.\begin{aligned}
s''_{a1} &= +n''_{aa} - n''_{ab} = (s_{a1} - s_{a5}) & s''_{a2} &= +n''_{ab} - n''_{ac} = (s_{a2} - s_{a4}) \\
s''_{b1} &= -n''_{ba} + n''_{bb} = (s_{b1} + s_{b5}) & s''_{b2} &= +n''_{bb} - n''_{bc} = (s_{b2} - s_{b4}) \\
s''_{c1} &= -n''_{ca} + n''_{cb} = (s_{c1} + s_{c5}) & s''_{c2} &= -n''_{cb} + n''_{cc} = (s_{c2} + s_{c4}).
\end{aligned}\right\} \qquad (1024)$$

---

* The members enclosed in parentheses have been formed with the values s (885) appropriately extended for the five-bay frame.

II. Treatment by the Deformation Method

a) Coefficients for symmetrical load arrangements

Stiffness coefficients of all the horizontal members and column stiffnesses:

$$K_\nu = \frac{J_\nu}{J_k} \cdot \frac{l_k}{l_\nu} \quad \text{for} \quad \nu = 1, 2, 3; \qquad K_a = \frac{J_a}{J_k} \cdot \frac{l_k}{h}; \qquad S_a = \frac{K_a}{2 - \varepsilon_a}. \quad (1025)$$

Joint coefficients:

$$K_A = 3S_a + 2K_1 \qquad K_B = 2K_1 + 2K_2 \qquad K_C' = 2K_2 + K_3. \quad (1026)$$

Rotation carry-over factors and auxiliary coefficients:

$$\left.\begin{array}{l}
\text{backward reduction:} \\[4pt]
j_{2s} = \dfrac{K_2}{r_c'} \qquad \begin{array}{l} r_c' = K_C' \\[4pt] r_b' = K_B - K_2 \cdot j_{2s} \end{array} \\[16pt]
j_{1s} = \dfrac{K_1}{r_b'} \qquad r_a' = K_A - K_1 \cdot j_{1s}.
\end{array}\right|\left.\begin{array}{l}
\text{forward reduction:} \\[4pt]
j_1' = \dfrac{K_1}{v_a} \qquad \begin{array}{l} v_a = K_A \\[4pt] v_b = K_B - K_1 \cdot j_1' \end{array} \\[16pt]
j_2' = \dfrac{K_2}{v_b} \qquad v_c' = K_C' - K_2 \cdot j_2'.
\end{array}\right\} \begin{array}{c} (1027) \\ \text{and} \\ (1028) \end{array}$$

Principal influence coefficients by recursion:

$$\left.\begin{array}{l}
\text{forward:} \\[4pt]
u_{aa}' = 1/r_a' \\[4pt]
u_{bb}' = 1/r_b' + u_{aa}' \cdot j_{1s}^2 \\[4pt]
u_{cc}' = 1/r_c' + u_{bb}' \cdot j_{2s}^2.
\end{array}\right|\left.\begin{array}{l}
\text{backward:} \\[4pt]
u_{cc}' = 1/v_c' \\[4pt]
u_{bb}' = 1/v_b + u_{cc}' \cdot j_2'^2 \\[4pt]
u_{aa}' = 1/v_a + u_{bb}' \cdot j_1'^2.
\end{array}\right\} \begin{array}{c} (1029) \\ \text{and} \\ (1030) \end{array}$$

Principal influence coefficients direct

$$u_{aa}' = \frac{1}{r_a'} \qquad u_{bb}' = \frac{1}{r_b' + v_b - K_B} \qquad u_{cc}' = \frac{1}{v_c'}. \quad (1031)$$

Computation scheme for the symmetrical influence coefficient matrix[*]

$$\left.\begin{array}{c} \\ \cdot j_{1s} \\ \cdot j_{2s} \end{array}\right| \begin{array}{ccc} (u_{aa}') & u_{ab}' & u_{ac}' \\ \downarrow & \uparrow & \uparrow \\ u_{ba}' & (u_{bb}') & u_{bc}' \\ \downarrow & \downarrow & \uparrow \\ u_{ca}' & u_{cb}' & (u_{cc}') \end{array} \left|\begin{array}{c} \cdot j_1' \\ \\ \cdot j_2' \end{array}\right.\right\} \quad (1032)$$

---

[*] Between the elements of matrix (1032) for the left-hand half of the frame and the elements of matrix (898), enlarged to 6 × 6 elements, there exist, according to the "load transposition method", the following relationships:

$$u_{aa}' = u_{aa} + u_{af} \text{ etc., up to } u_{cc}' = u_{cc} + u_{cd}.$$

Loading Condition 4a—Symmetrical overall loading condition

(Superposition of Loading Conditions 1a, 2a and 3a)

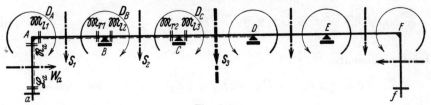

Fig. 241

Loading diagram, with fixed-end moments $\mathfrak{M}_l$ and $\mathfrak{M}_r$ of the beam spans, load terms $\mathfrak{L}$ and $\mathfrak{R}$ for the column, and external rotational moments D, for left-hand half of frame

Column restraining moment:

$$\mathfrak{C}_{ra} = -\frac{\mathfrak{R}_a - \mathfrak{L}_a \cdot \varepsilon_a}{2 - \varepsilon_a}. \qquad (1033)$$

Joint load terms:

$$\mathfrak{R}_A = D_A + \mathfrak{C}_{ra} - \mathfrak{M}_{l1} \qquad \begin{matrix} \mathfrak{R}_B = D_B + \mathfrak{M}_{r1} - \mathfrak{M}_{l2} \\ \mathfrak{R}_C = D_C + \mathfrak{M}_{r2} - \mathfrak{M}_{l3}. \end{matrix} \qquad (1034)$$

auxiliary moments:

$$\begin{aligned} Y'_A &= +\mathfrak{R}_A \cdot u'_{aa} - \mathfrak{R}_B \cdot u'_{ab} + \mathfrak{R}_C \cdot u'_{ac} \\ Y'_B &= -\mathfrak{R}_A \cdot u'_{ba} + \mathfrak{R}_B \cdot u'_{bb} - \mathfrak{R}_C \cdot u'_{bc} \\ Y'_C &= +\mathfrak{R}_A \cdot u'_{ca} - \mathfrak{R}_B \cdot u'_{cb} + \mathfrak{R}_C \cdot u'_{cc}. \end{aligned} \qquad (1035)$$

Beam moments:

$$\begin{aligned} M_{A1} &= M_{F5} = \mathfrak{M}_{l1} + (2Y'_A + Y'_B) K_1 \\ M_{B2} &= M_{E4} = \mathfrak{M}_{l2} + (2Y'_B + Y'_C) K_2 \\ M_{C3} &= M_{D3} = \mathfrak{M}_{l3} + Y'_C \cdot K_3; \\ M_{B1} &= M_{E5} = \mathfrak{M}_{r1} - (Y'_A + 2Y'_B) K_1 \\ M_{C2} &= M_{D4} = \mathfrak{M}_{r2} - (Y'_B + 2Y'_C) K_2. \end{aligned} \qquad (1036)$$

Column moments:

$$M_{Aa} = -M_{Ff} = \mathfrak{C}_{ra} - Y'_A \cdot 3S_a \qquad M_a = -M_f = -(\mathfrak{L}_a + M_{Aa}) \varepsilon_a. \qquad (1037)$$

Check relationships:

$$D_A + M_{Aa} = M_{A1} \qquad D_B + M_{B1} = M_{B2} \qquad D_C + M_{C2} = M_{C3}. \qquad (1038)$$

b) Coefficients for anti-symmetrical load arrangements

Stiffness coefficients of all the members, and column stiffness:

$$K_\nu = \frac{J_\nu}{J_k} \cdot \frac{l_k}{l_\nu} \quad \text{for} \quad \nu = 1, 2, 3; \quad K_a = \frac{J_a}{J_k} \cdot \frac{l_k}{h}; \quad S_a = \frac{K_a}{2-\varepsilon_a}. \quad (1039)$$

Joint coefficients:

$$K_A = 3S_a + 2K_1 \quad K_B = 2K_1 + 2K_2 \quad K_C'' = 2K_2 + 3\mathbf{K_3}. \quad (1040)$$

Rotation carry-over factors and auxiliary coefficients:

backward reduction:

$$j_{2t} = \frac{K_2}{r_c''} \quad \begin{array}{l} r_c'' = K_C'' \\ r_b'' = K_B - K_2 \cdot j_{2t} \end{array}$$
$$j_{1t} = \frac{K_1}{r_b''} \quad r_a'' = K_A - K_1 \cdot j_{1t}.$$

forward reduction:

$$j_1' = \frac{K_1}{v_a} \quad \begin{array}{l} v_a = K_A \\ v_b = K_B - K_1 \cdot j_1' \end{array}$$
$$j_2' = \frac{K_2}{v_b} \quad v_c'' = K_C'' - K_2 \cdot j_2'$$

$$(1041)$$
and
$$(1042)$$

Principal influence coefficients by recursion:

forward:

$$u_{aa}'' = 1/r_a''$$
$$u_{bb}'' = 1/r_b'' + u_{aa}'' \cdot j_{1t}^2$$
$$u_{cc}'' = 1/r_c'' + u_{bb}'' \cdot j_{2t}^2.$$

backward:

$$u_{cc}'' = 1/v_c''$$
$$u_{bb}'' = 1/v_b + u_{cc}'' \cdot j_2'^2$$
$$u_{aa}'' = 1/v_a + u_{bb}'' \cdot j_1'^2.$$

$$(1043)$$
and
$$(1044)$$

Principal influence coefficients direct

$$u_{aa}'' = \frac{1}{r_a''} \qquad u_{bb}'' = \frac{1}{r_b'' + v_b - K_B} \qquad u_{cc}'' = \frac{1}{v_c''}. \quad (1045)$$

Computation scheme for the symmetrical influence coefficient matrix*

$$\begin{array}{c} \\ \cdot j_{1t} \\ \cdot j_{2t} \end{array} \left| \begin{array}{ccc} (u_{aa}'') & u_{ab}'' & u_{ac}'' \\ \downarrow & \uparrow & \uparrow \\ u_{ba}'' & (u_{bb}'') & u_{bc}'' \\ \downarrow & \uparrow & \uparrow \\ u_{ca}'' & u_{cb}'' & (u_{cc}'') \end{array} \right| \begin{array}{c} \cdot j_1' \\ \\ \cdot j_2' \end{array} \right\} \quad (1046)'$$

---

* Between the elements of matrix (1046) for the left-hand half of the frame and the elements of matrix (898), enlarged to 6 × 6 elements, there exist, according to the "load transposition method", the following relationships:

$$u_{aa}'' = u_{aa} - u_{af} \text{ etc., up to } u_{cc}'' = u_{cc} - u_{cd}.$$

Loading Condition 4b—Anti-symmetrical overall loading condition

(Superposition of Loading Conditions 1b, 2b and 3b)

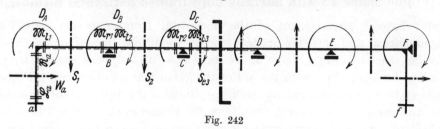

Fig. 242

Loading diagram, with fixed-end moments $\mathfrak{M}_l$ and $\mathfrak{M}_r$ of the beam spans, load terms $\mathfrak{L}$ and $\mathfrak{R}$ of the column and innermost beam span, and external rotational moments D, for left-hand half of frame

Column restraining moment, joint load terms and check relationships: same as formulas (1033), (1034) and (1038).

auxiliary moments:

$$\left.\begin{aligned}
Y_A'' &= +\mathfrak{K}_A \cdot u_{aa}'' - \mathfrak{K}_B \cdot u_{ab}'' + \mathfrak{K}_C \cdot u_{ac}'' \\
Y_B'' &= -\mathfrak{K}_A \cdot u_{ba}'' + \mathfrak{K}_B \cdot u_{bb}'' - \mathfrak{K}_C \cdot u_{bc}'' \\
Y_C'' &= +\mathfrak{K}_A \cdot u_{ca}'' - \mathfrak{K}_B \cdot u_{cb}'' + \mathfrak{K}_C \cdot u_{cc}''.
\end{aligned}\right\} \quad (1047)$$

Beam moments:

$$\left.\begin{aligned}
M_{A1} &= -M_{F5} = \mathfrak{M}_{l1} + (2\,Y_A'' + Y_B'')K_1 \\
M_{B2} &= -M_{E4} = \mathfrak{M}_{l2} + (2\,Y_B'' + Y_C'')K_2 \\
M_{C3} &= -M_{D3} = \mathfrak{M}_{l3} + (2\,Y_C'') \cdot 3\,K_3; \\
M_{B1} &= -M_{E5} = \mathfrak{M}_{r1} - (Y_A'' + 2\,Y_B'')K_1 \\
M_{C2} &= -M_{D4} = \mathfrak{M}_{r2} - (Y_B'' + 2\,Y_C'')K_2.
\end{aligned}\right\} \quad (1048)$$

Column moments:

$$M_{Aa} = M_{Ff} = \mathfrak{C}_{ra} - Y_A'' \cdot 3\,S_a \qquad M_a = M_f = -(\mathfrak{L}_a + M_{Aa})\varepsilon_a. \quad (1049)$$

Horizontal thrusts, column shears and lateral restraining force for all loading conditions

Same as at foot of page 212, except that subscript g must be replaced by f.

Lateral restraining forces:

asymmetrical case: $\quad H = -H_a - H_f \quad$ or $\quad H = -Q_{ra} - Q_{rf};$ (1050)

symmetrical case $\qquad\qquad H = 0;$ (1051)

anti-symmetrical case: $H = -2H_a \quad$ or $\quad H = -2Q_{ra}.$ (1052)

Frame Shape 32v

(Frame Shape 32 with laterally unrestrained horizontal member)*

Note: Lateral displacement of the horizontal member occurs only for asymmetrical and anti-symmetrical loading conditions (for all symmetrical loading cases of Frame Shape 32 we have H = 0). As every asymmetrical loading condition can, with the aid of the "load transposition method", be split up into a symmetrical and an anti-symmetrical loading condition, only the anti-symmetrical loading conditions for Frame Shape 32v will here be considered, while the symmetrical loading conditions can be analyzed with the formulas given for Frame Shape 32.

I. Treatment by the Method of Forces
Coefficients

Flexibility coefficients of all the horizontal members and column flexibility:

$$k_v = \frac{J_k}{J_v} \cdot \frac{l_v}{l_k} \quad \text{for} \quad v = 1, 2, 3; \quad k_a = \frac{J_k}{J_a} \cdot \frac{h}{l_k}; \quad m_a = k_a \left(\frac{1}{\varepsilon_a} + 4 \right). \tag{1053}$$

Diagonal coefficients:

$$k_A^v = m_a + 2 k_1 \qquad k_B = 2 k_1 + 2 k_2 \qquad k_C'' = 2 k_2 + k_3. \tag{1054}$$

Moment carry-over factors and auxiliary coefficients:

backward reduction:

$$\iota_2 = \frac{k_2}{r_c} \qquad r_c = k_C''$$
$$r_b = k_B - k_2 \cdot \iota_2$$
$$\iota_1 = \frac{k_1}{r_b} \qquad r_a = k_A^v - k_1 \cdot \iota_1.$$

forward reduction:

$$\iota_1' = \frac{k_1}{v_a} \qquad v_a = k_A^v$$
$$v_b = k_B - k_1 \cdot \iota_1'$$
$$\iota_2' = \frac{k_2}{v_b} \qquad v_c = k_C'' - k_2 \cdot \iota_2'.$$

$$(1055)$$
and
$$(1056)$$

Principal influence coefficients by recursion:

forward:

$$n_{aa} = 1/r_a$$
$$n_{bb} = 1/r_b + n_{aa} \cdot \iota_1^2$$
$$n_{cc} = 1/r_c + n_{bb} \cdot \iota_2^2.$$

backward:

$$n_{cc} = 1/v_c$$
$$n_{bb} = 1/v_b + n_{cc} \cdot \iota_1'^2$$
$$n_{aa} = 1/v_a + n_{bb} \cdot \iota_2'^2.$$

$$(1057)$$
and
$$(1058)$$

Principal influence coefficients direct

$$n_{aa} = \frac{1}{r_a} \qquad n_{bb} = \frac{1}{r_b + v_b - k_B} \qquad n_{cc} = \frac{1}{v_c}. \tag{1059}$$

---

* See Figs. 235 and 236 on page 222: in the case of Frame Shape 32v the restraining bearing F and therefore also the lateral restraining force H are absent. The information given under the heading "Important" on that page is also applicable to Frame Shape 32v.

Computation scheme for the symmetrical influence coefficient matrix:

$$\begin{array}{c} \cdot l_1 \\ \cdot l_2 \end{array} \left| \begin{array}{ccc} (n_{aa}) & n_{ab} & n_{ac} \\ \downarrow & \uparrow & \uparrow \\ n_{ba} & (n_{bb}) & n_{bc} \\ \downarrow & \downarrow & \uparrow \\ n_{ca} & n_{cb} & (n_{cc}) \end{array} \right| \begin{array}{c} \cdot l'_1 \\ \cdot l'_2 \end{array} \qquad (1060)$$

Loading Condition 1b

Horizontal member carrying vertical loading of arbitrary magnitude anti-symmetrical about center of frame

$(\Re_5 = -\mathfrak{L}_1, \quad \mathfrak{L}_5 = -\Re_1; \quad \Re_4 = -\mathfrak{L}_2, \quad \mathfrak{L}_4 = -\Re_2; \quad \Re_3 = -\mathfrak{L}_3)$

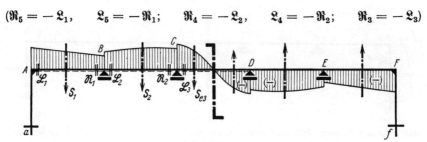

Fig. 243

Beam loading diagram with load terms $\mathfrak{L}$ and $\Re$ for left-hand half of frame

Beam moments ($L_\nu = \mathfrak{L}_\nu k_\nu$ and $R_\nu = \Re_\nu k_\nu$):

$$\left. \begin{array}{llll} & & L_1 & (R_1 + L_2) & (R_2 + L_3) \\ X_a = M_{A1} = -M_{F5} = & -n_{aa} & +n_{ab} & -n_{ac} \\ X_b = M_B \;\; = -M_E \;\; = & +n_{ba} & -n_{bb} & +n_{bc} \\ X_c = M_C \;\; = -M_D \;\; = & -n_{ca} & +n_{cb} & -n_{cc}. \end{array} \right\} \qquad (1061)$$

Column moments:

$$(M_{Aa} = M_a) = (M_{Ff} = M_f) = X_a. \qquad (1062)$$

Special case: symmetrical bay loading ($R_1 = L_1$ and $R_2 = L_2$):

$$\left. \begin{array}{l} X_a = M_{A1} = -M_{F5} = -L_1 \cdot s_{a1} + L_2 \cdot s_{a2} \\ X_b = M_B \;\; = -M_E \;\; = -L_1 \cdot s_{b1} - L_2 \cdot s_{b2} \\ X_c = M_C \;\; = -M_D \;\; = +L_1 \cdot s_{c1} - L_2 \cdot s_{c2}. \end{array} \right\} \qquad (1063)$$

Note: values s are given on next page.

Composite influence coefficients for Loading Condition 1b.

$$\begin{aligned}
s_{a1} &= +n_{aa} - n_{ab} & s_{a2} &= +n_{ab} - n_{ac} \\
& & s_{b2} &= +n_{bb} - n_{bc} \\
s_{b1} &= -n_{ba} + n_{bb} & & \\
s_{c1} &= -n_{ca} + n_{cb} & s_{c2} &= -n_{cb} + n_{cc}.
\end{aligned} \qquad (1064)$$

Loading Conditions 2b and 3b

The two columns carrying arbitrary horizontal loading (acting from left to right) anti-symmetrical about center of frame (2b), and an anti-symmetrical pair of rotational moments at corners of frame (3b)

$$(\mathfrak{L}_f = \mathfrak{L}_a, \qquad \mathfrak{R}_f = \mathfrak{R}_a; \qquad D_F = D_A)$$

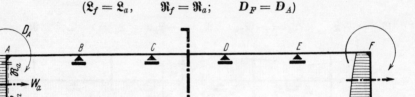

Fig. 244
Column loading diagram with load terms $\mathfrak{L}$ and $\mathfrak{R}$, and rotational moment D, for left-hand half of frame

Joint load terms:

$$\mathfrak{B}_a = \mathfrak{S}_{la}(3k_a + 2k_1) + (\mathfrak{L}_a + \mathfrak{R}_a)k_a \qquad \mathfrak{B}_1 = \mathfrak{S}_{la} \cdot k_1. \qquad (1065)$$

Beam and column moments

$$\begin{aligned}
X_a &= M_a = +M_f = -\mathfrak{B}_a \cdot n_{aa} + \mathfrak{B}_1 \cdot n_{ab} \\
X_b &= M_B = -M_E = +\mathfrak{B}_a \cdot n_{ba} - \mathfrak{B}_1 \cdot n_{bb} \\
X_c &= M_C = -M_D = -\mathfrak{B}_a \cdot n_{ca} + \mathfrak{B}_1 \cdot n_{cb};
\end{aligned} \qquad (1066)$$

$$M_{Aa} = M_{Ff} = X_a + \mathfrak{S}_{la} \qquad M_{A1} = -M_{F5} = M_{Aa} + D_A. \qquad (1067)$$

Column forces:

$$(H_a = H_f) = Q_{la} = W_a \qquad Q_{ra} = 0. \qquad (1068)$$

II. Treatment by the Deformation Method

Coefficients

Stiffness coefficients of all the horizontal members, and column stiffness

$$K_\nu = \frac{J_\nu}{J_k} \cdot \frac{l_k}{l_\nu} \quad \text{for} \quad \nu = 1, 2, 3; \quad K_a = \frac{J_a}{J_k} \cdot \frac{l_k}{h}; \quad F_a = \frac{K_a}{1/\varepsilon_a + 4}. \qquad (1069)$$

Joint coefficients:

$$K_A^v = 3F_a + 2K_1 \qquad K_B = 2K_1 + 2K_2 \qquad K_C'' = 2K_2 + 3K_3. \qquad (1070)$$

Rotation carry-over factors and auxiliary coefficients:

backward reduction: | forward reduction:

$$j_2 = \frac{K_2}{r_c} \qquad \begin{array}{l} r_c = K_C'' \\ r_b = K_B - K_2 \cdot j_2 \end{array} \qquad j_1' = \frac{K_1}{v_a} \qquad \begin{array}{l} v_a = K_A^v \\ v_b = K_B - K_1 \cdot j_1' \end{array} \qquad \begin{array}{c} (1071) \\ \text{and} \\ (1072) \end{array}$$

$$j_1 = \frac{K_1}{r_b} \qquad r_a = K_A^v - K_1 \cdot j_1. \qquad j_2' = \frac{K_2}{v_b} \qquad v_c = K_C'' - K_2 \cdot j_2'.$$

Principal influence coefficients by recursion:

forward: | backward:

$$\begin{array}{ll} u_{aa} = 1/r_a & u_{cc} = 1/v_c \\ u_{bb} = 1/r_b + u_{aa} \cdot j_1^2 & u_{bb} = 1/v_b + u_{cc} \cdot j_2'^2 \\ u_{cc} = 1/r_c + u_{bb} \cdot j_2^2. & u_{aa} = 1/v_a + u_{bb} \cdot j_1'^2. \end{array} \qquad \begin{array}{c} (1073) \\ \text{and} \\ (1074) \end{array}$$

Principal influence coefficients direct

$$u_{aa} = 1/r_a \qquad u_{bb} = 1/(r_b + v_b - K_B) \qquad v_{cc} = 1/v_c. \qquad (1075)$$

Computation scheme for the symmetrical influence coefficient matrix

$$\begin{array}{c} \cdot j_1 \\ \\ \cdot j_2 \end{array} \left| \begin{array}{ccc} (u_{aa}) & u_{ab} & u_{ac} \\ \downarrow & \uparrow & \uparrow \\ u_{ba} & (u_{bb}) & u_{bc} \\ \downarrow & \downarrow & \uparrow \\ u_{ca} & u_{cb} & (u_{cc}) \end{array} \right| \begin{array}{c} \cdot j_1' \\ \\ \cdot j_2' \end{array} \qquad (1076)$$

Loading Condition 4b—Anti-symmetrical overall loading condition

(Superposition of Loading Conditions 1b, 2b, 3b and 5)

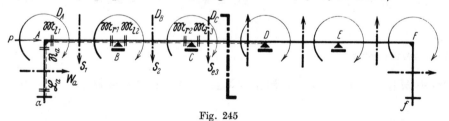

Fig. 245

Loading diagram, with fixed-end moments $\mathfrak{M}_l$ and $\mathfrak{M}_r$ of the beam spans, load terms $\mathfrak{L}$ and $\mathfrak{R}$ for the column, and external rotational moments D, for left-hand half of frame;
furthermore with horizontal concentrated load P at level of horizontal member

Restraining moment for column a:

$$\mathfrak{F}_{ra} = \frac{(Ph/2 + \mathfrak{S}_{la})(1/\varepsilon_a + 1) - (\mathfrak{L}_a + \mathfrak{R}_a)}{1/\varepsilon_a + 4}. \tag{1077}$$

Joint load terms:

$$\left. \begin{aligned} \mathfrak{R}_A^v &= D_A + \mathfrak{F}_{ra} - \mathfrak{M}_{l1} \\ \mathfrak{R}_B &= D_B + \mathfrak{M}_{r1} - \mathfrak{M}_{l2} \\ \mathfrak{R}_C &= D_C + \mathfrak{M}_{r2} - \mathfrak{M}_{l3}. \end{aligned} \right\} \tag{1078}$$

auxiliary moments:

$$\left. \begin{array}{c|ccc} & \mathfrak{R}_A^v & \mathfrak{R}_B & \mathfrak{R}_C \\ \hline Y_A = & +u_{aa} & -u_{ab} & +u_{ac} \\ Y_B = & -u_{ba} & +u_{bb} & -u_{bc} \\ Y_C = & +u_{ca} & -u_{cb} & +u_{cc}. \end{array} \right\} \tag{1079}$$

Beam moments:

$$\left. \begin{aligned} M_{A1} &= -M_{F5} = \mathfrak{M}_{l1} + (2Y_A + Y_B)K_1 \\ M_{B2} &= -M_{E4} = \mathfrak{M}_{l2} + (2Y_B + Y_C)K_2 \\ M_{C3} &= -M_{D3} = \mathfrak{M}_{l3} + Y_C \cdot 3K_3; \\ M_{B1} &= -M_{E5} = \mathfrak{M}_{r1} - (Y_A + 2Y_B)K_1 \\ M_{C2} &= -M_{D4} = \mathfrak{M}_{r2} - (Y_B + 2Y_C)K_2. \end{aligned} \right\} \tag{1080}$$

Column moments:

$$M_{Aa} = M_{Ff} = \mathfrak{F}_{ra} - Y_A \cdot 3F_a \qquad M_a = M_f = -\frac{Ph}{2} - \mathfrak{S}_{la} + M_{Aa}. \tag{1081}$$

Check relationships:

$$D_A + M_{Aa} = M_{A1} \qquad D_B + M_{B1} = M_{B2} \qquad D_C + M_{C2} = M_{C3}. \tag{1082}$$

Column forces:

$$(H_a = Q_{la}) = (H_f = Q_{lf}) = W_a + P/2 \qquad (Q_{ra} = Q_{rf}) = P/2. \tag{1083}$$

Frame Shape 33

Symmetrical single-story four-bay frame with laterally restrained horizontal member, two elastically restrained end columns, and three pin-jointed internal supports (or rocker columns)

(Symmetrical form of Frame Shape 28)

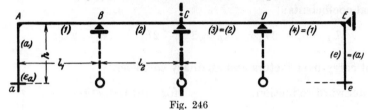

Fig. 246
Frame shape, dimensions and symbols

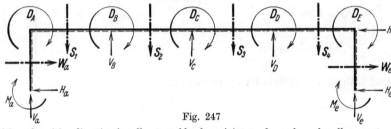

Fig. 247

Definition of positive direction for all external loads on joints and members, for all support reactions, and for the lateral restraining force

All the dimensions and coefficients for the right-hand half of the frame are the same as those for the left-hand half. In the case of the columns the broken line is always placed on the right-hand side of each column, despite the symmetry of the frame.

Note: For Frame Shape 33 the coefficients and the *Loading Conditions* 1 – 4 *for Frame Shape* 28 may be employed. On account of symmetry, the forward reduction becomes the same as the backward reduction, and the backward recursion becomes the same as the forward recursion, both in Method I and Method II; in particular, (843) becomes equal to (842), (845) to (844), (860) to (859), and (862) to (861). Furthermore, the *square matrices of influence coefficients* (846) and (864) are bisymmetrical, i.e., symmetrical about the principal diagonal and the secondary diagonal. If the "load transposition method" is used, however, it is also possible to start from the following symmetrical and anti-symmetrical loading conditions, which in the present case are given for the Method of Forces (Method I). [For the treatment by the Deformation Method (Method II) see the relevant formulas for Frame Shape 31, page 206, appropriately reducing the number of bays by two].

a) Coefficients for symmetrical load arrangements

Flexibility coefficients of all the horizontal members and column flexibility:

$$k_a = \frac{J_k}{J_a} \cdot \frac{h}{l_k}\ ; \qquad k_1 = \frac{J_k}{J_1} \cdot \frac{l_1}{l_k} \qquad k_2 = \frac{J_k}{J_2} \cdot \frac{l_2}{l_k}\ ; \qquad b_a = k_a(2 - \varepsilon_a). \quad (1084)$$

Diagonal coefficients:

$$k_A = b_a + 2k_1 \qquad k_B = 2k_1 + 2k_2 \qquad (k'_C = 2k_2). \quad (1085)$$

Moment carry-over factors and auxiliary coefficients:

backward reduction:

$$\iota_{2s} = \frac{1}{2}$$
$$\iota_{1s} = \frac{k_1}{r'_b}$$
$$r'_b = k_B - k_2 \cdot \frac{1}{2}$$
$$r'_a = k_A - k_1 \cdot \iota_{1s}.$$

forward reduction:

$$\iota'_1 = \frac{k_1}{v_a} \qquad v_a = k_A$$
$$\iota'_2 = \frac{k_2}{v_b}. \qquad v_b = k_B - k_1 \cdot \iota'_1$$

$$\left.\begin{array}{c}(1086)\\ \text{and}\\ (1087)\end{array}\right.$$

Principal influence coefficients by recursion:

forward:

$$n'_{aa} = \frac{1}{r'_a}$$
$$n'_{bb} = \frac{1}{r'_b} + n'_{aa} \cdot \iota^2_{1s}$$
$$n'_{cc} = \frac{1}{2k_2} + n'_{bb} \cdot \frac{1}{4}.$$

backward:

$$n'_{cc} = \frac{1}{k_2(2 - \iota'_2)}$$
$$n'_{bb} = \frac{1}{v_b} + n'_{cc} \cdot \iota'^2_2$$
$$n'_{aa} = \frac{1}{v_a} + n'_{bb} \cdot \iota'^2_1.$$

$$\left.\begin{array}{c}(1088)\\ \text{and}\\ (1089)\end{array}\right.$$

Principal influence coefficients direct

$$n'_{aa} = 1/r'_a \qquad n'_{bb} = 1/(r'_b + v_b - k_B) \qquad n'_{cc} = 1/k_2(2 - \iota'_2). \quad (1090)$$

Computation scheme for the symmetrical influence coefficient matrix*

$$\left.\begin{array}{c}
\begin{array}{c}\cdot\,\iota_{1s}\\ \\ \cdot\,\dfrac{1}{2}\end{array}
\left|\begin{array}{ccc}
(n'_{aa}) & n'_{ab} & n'_{ac}\\
\downarrow & \uparrow & \uparrow\\
n'_{ba} & (n'_{bb}) & n'_{bc}\\
\downarrow & \downarrow & \uparrow\\
n'_{ca} & n'_{cb} & (n'_{cc})
\end{array}\right|
\begin{array}{c}\cdot\,\iota'_1\\ \\ \cdot\,\iota'_2\end{array}
\end{array}\right\} \quad (1091)$$

---

* Between the elements of matrix (1091) for the left-hand half of the frame and the elements of matrix (846) for the whole frame there exist, according to the "load transposition method", the following relationships: $n'_{aa} = n_{aa} + n_{ae}$, $n'_{ab} = n_{ab} + n_{ad}$ etc., up to $n'_{cc} = 2n_{c\bullet}$.

(The values n' of the column c of (1091) here become twice as large as the values n of the column c of (846)).

Loading Condition 1a

Horizontal member carrying vertical loading of arbitrary magnitude, but symmetrical about center of frame

$$(\mathfrak{R}_4 = \mathfrak{L}_1, \quad \mathfrak{L}_4 = \mathfrak{R}_1; \quad \mathfrak{R}_3 = \mathfrak{L}_2, \quad \mathfrak{L}_3 = \mathfrak{R}_2)$$

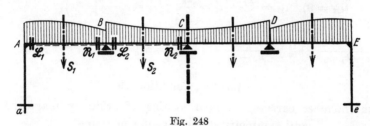

Fig. 248

Beam loading diagram with load terms $\mathfrak{L}$ and $\mathfrak{R}$ for left-hand half of frame

Beam moments ($L_\nu = \mathfrak{L}_\nu k_\nu$ and $R_\nu = \mathfrak{R}_\nu k_\nu$):

	L_1	$(R_1 + L_2)$	R_2
$M_{A1} = M_{E4} =$	$-n'_{aa}$	$+n'_{ab}$	$-n'_{ac}$
$M_B = M_D =$	$+n'_{ba}$	$-n'_{bb}$	$+n'_{bc}$
$M_C =$	$-n'_{ca}$	$+n'_{cb}$	$-n'_{cc}$.

(1092)

Column moments:

$$M_{Aa} = -M_{Ee} = M_{A1} \qquad M_a = -M_e = -M_{Aa} \cdot \varepsilon_a. \qquad (1093)$$

Special case: symmetrical bay loading ($R_\nu = L_\nu$):

$$\begin{aligned} M_{A1} = M_{E4} &= -L_1 \cdot s'_{a1} + L_2 \cdot s'_{a2} \\ M_B = M_D &= -L_1 \cdot s'_{b1} - L_2 \cdot s'_{b2} \\ M_C &= +L_1 \cdot s'_{c1} - L_2 \cdot s'_{c2}; \end{aligned} \qquad (1094)$$

where*:

$$\begin{aligned} s'_{a1} &= +n'_{aa} - n'_{ab} = (s_{a1} - s_{a4}) & s'_{a2} &= \phantom{-n'_{cb} +} n'_{ac} = (s_{a2} - s_{a3}) \\ s'_{b1} &= -n'_{ba} + n'_{bb} = (s_{b1} + s_{b4}) & s'_{b2} &= \phantom{-n'_{cb} +} n'_{bc} = (s_{b2} - s_{b3}) \\ s'_{c1} &= -n'_{ca} + n'_{cb} = (s_{c1} + s_{c4}) & s'_{c2} &= -n'_{cb} + n'_{cc} = (s_{c2} + s_{c3}). \end{aligned} \qquad (1095)$$

---

* The members enclosed in parentheses have been formed with the values s (851).

b) Coefficients for anti-symmetrical load arrangements

Flexibility coefficients, column flexibility and diagonal coefficients (without k_c'): same as formulas (1084) and (1085).

Moment carry-over factors and influence coefficients:[*]

$$(\iota_{2t} = 0) \quad \iota_{1t} = \frac{k_1}{k_B} \quad \iota_1' = \frac{k_1}{k_A} \quad \left(\iota_2' = \frac{k_2}{k_B - k_1 \cdot \iota_1'}\right); \tag{1096}$$

$$n_{aa}'' = \frac{1}{k_A - k_1 \cdot \iota_{1t}} \quad n_{bb}'' = \frac{1}{k_B - k_1 \cdot \iota_1'} \;;\quad n_{ab}'' = n_{ba}'' = n_{aa}'' \cdot \iota_{1t} = n_{bb}'' \cdot \iota_1'. \tag{1097}$$

Loading Condition 1b
Horizontal member carrying vertical loading of arbitrary magnitude, but anti-symmetrical about center of frame

$$(\mathfrak{R}_4 = -\mathfrak{L}_1, \quad \mathfrak{L}_4 = -\mathfrak{R}_1; \quad \mathfrak{R}_3 = -\mathfrak{L}_2)$$

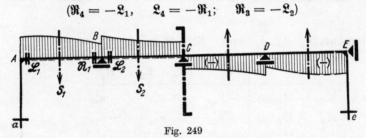

Fig. 249

Beam loading diagram with load terms $\mathfrak{L}$ and $\mathfrak{R}$ for left-hand half of frame

Beam moments $(L_1 = \mathfrak{L}_1 k_1, \quad R_1 = \mathfrak{R}_1 k_1; \quad L_2 = \mathfrak{L}_2 k_2)$:

$$\left.\begin{aligned} M_{A1} &= -M_{E4} = -L_1 \cdot n_{aa}'' + (R_1 + L_2) \cdot n_{ab}'' \\ M_B &= -M_D = +L_1 \cdot n_{ba}'' - (R_1 + L_2) \cdot n_{bb}''; \quad M_C = 0. \end{aligned}\right\} \tag{1098}$$

Column moments:

$$M_{Aa} = M_{Ee} = M_{A1} \qquad M_a = M_e = -M_{Aa} \cdot \varepsilon_a. \tag{1099}$$

Special case: symmetrical bay loading $(R_1 = L_1)$:

$$\left.\begin{aligned} M_{A1} &= -M_{E4} = -L_1 \cdot s_{a1}'' + L_2 \cdot n_{ab}'' \\ M_B &= -M_D = -L_1 \cdot s_{b1}'' - L_2 \cdot n_{bb}''; \quad M_C = 0; \end{aligned}\right\} \tag{1100}$$

where[**]:

$$s_{a1}'' = n_{aa}''(1 - \iota_{1t}) = (s_{a1} + s_{a4}) \qquad s_{b1}'' = n_{bb}''(1 - \iota_1') = (s_{b1} - s_{b4}). \tag{1101}$$

---

[*] Between the values n'' (1097) for the left-hand half of the frame and the values n of matrix (846) for the whole frame there exist, according to the "load transposition method", the following relationships:
$$n_{aa}'' = n_{aa} - n_{ae}, \quad n_{ab}'' = n_{ab} - n_{ad}, \quad n_{bb}'' = n_{bb} - n_{bd}.$$
(The values n'' of the central support c vanish).

[**] The members enclosed in parentheses have been formed with the values s (851).

Loading Conditions 2a and 3a (2b and 3b)*

The two columns carrying arbitrary horizontal loading directed towards center of frame (acting from left to right) and symmetrical (anti-symmetrical) about center of frame, and symmetrical (anti-symmetrical) pair of rotational moments at corners of frame.

Joint load term:

$$(L_a = \mathfrak{L}_a k_a \quad \text{and} \quad R_a = \mathfrak{R}_a k_a); \qquad \mathfrak{B}_A = D_A \cdot b_a + L_a \cdot \varepsilon_a - R_a. \quad (1102)$$

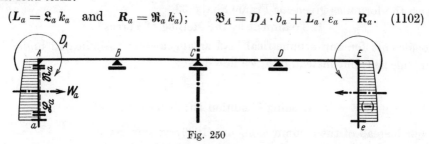

Fig. 250
Loading Conditions 2a and 3a—symmetry

$$(\mathfrak{L}_e = -\mathfrak{L}_a, \quad \mathfrak{R}_e = -\mathfrak{R}_a; \quad D_E = -D_A)$$

Column and beam moments:

$$M_{Aa} = -M_{Ee} = -D_A + M_{A1}$$
$$M_a = -M_e = -(\mathfrak{L}_a + M_{Aa})\varepsilon_a;$$

$$\left.\begin{aligned} M_{A1} = M_{E4} &= +\mathfrak{B}_A \cdot n'_{aa} \\ M_B = M_D &= -\mathfrak{B}_A \cdot n'_{ba} \\ M_C = -M_B/2 &= +\mathfrak{B}_A \cdot n'_{ca} \end{aligned}\right\} \quad \begin{array}{c}(1103)\\ \text{and}\\ (1104)\end{array}$$

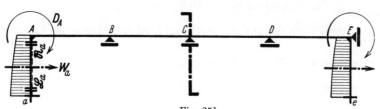

Fig. 251
Loading Conditions 2b and 3b—anti-symmetry

$$(\mathfrak{L}_e = \mathfrak{L}_a, \quad \mathfrak{R}_e = \mathfrak{R}_a; \quad D_E = D_A)$$

Column and beam moments:

$$M_{Aa} = M_{Ee} = -D_A + M_{A1}$$
$$M_a = M_e = -(\mathfrak{L}_a + M_{Aa})\varepsilon_a;$$

$$\left.\begin{aligned} M_{A1} = -M_{E4} &= +\mathfrak{B}_A \cdot n''_{aa} \\ M_B = -M_D &= -\mathfrak{B}_A \cdot n''_{ba} \\ M_C &= 0. \end{aligned}\right\} \quad \begin{array}{c}(1105)\\ \text{and}\\ (1106)\end{array}$$

---

* This arrangement has been adopted in order to save space. For the coefficients to the symmetrical Loading Conditions 2a and 3a see page 238.

Frame Shape 33v

(Frame Shape 33 with laterally unrestrained horizontal member)*

Note: The note relating to Frame Shape 31v, page 217, is applicable here. Only the anti-symmetrical loading conditions for Frame Shape 33v will be considered here, while the symmetrical loading conditions can be analysed with the formulas given for Frame Shape 33.

I. Treatment by the Method of Forces

Coefficients for anti-symmetrical load arrangements: as indicated under b) on page 240, but instead of b_a we now have:

$$m_a = k_a \left(\frac{1}{\varepsilon_a} + 4 \right). \tag{1107}$$

Loading Condition 1b: as on page 240

but instead of the column moments (1099) we now have:

$$(M_{Aa} = M_a) = (M_E = M_e) = M_{A1} = -M_{E4}. \tag{1108}$$

Loading Conditions 2b and 3b (cf. page 219)

Joint load terms: same as formulas (977).

Beam and column moments:

$$\left. \begin{aligned} X_a = M_a = +M_e &= -\mathfrak{B}_a \cdot n''_{aa} + \mathfrak{B}_1 \cdot n''_{ab} \\ X_b = M_B = -M_D &= +\mathfrak{B}_a \cdot n''_{ba} - \mathfrak{B}_1 \cdot n''_{bb}; \end{aligned} \right\} \tag{1109}$$

$$M_C = 0 \qquad M_{Aa} = M_{Ee} = X_a + \mathfrak{S}_{la} \qquad M_{A1} = -M_{E4} = M_{Aa} + D_A. \tag{1110}$$

II. Treatment by the Deformation Method

Coefficients for anti-symmetrical load arrangements

Stiffness coefficients of all the members, and column stiffness:

$$K_a = \frac{J_a}{J_k} \cdot \frac{l_k}{h} \qquad K_1 = \frac{J_1}{J_k} \cdot \frac{l_k}{l_1} \qquad K_2 = \frac{J_2}{J_k} \cdot \frac{l_k}{l_2}; \qquad F_a = \frac{K_a}{1/\varepsilon_a + 4}. \tag{1111}$$

Joint coefficients:

$$K_A^v = 3F_a + 2K_1 \qquad K_B'' = 2K_1 + 3K_2/2. \tag{1112}$$

Rotation carry-over factors and influence coefficients:

$$\left. \begin{aligned} j_1 = \frac{K_1}{K_B''} \qquad j_1' = \frac{K_1}{K_A^v}; \qquad u_{aa} = \frac{1}{K_A^v - K_1 \cdot j_1} \qquad u_{bb} = \frac{1}{K_B'' - K_1 \cdot j_1'}; \\ u_{ab} = u_{ba} = u_{aa} \cdot j_1 = u_{bb} \cdot j_1'. \end{aligned} \right\} \tag{1113}$$

---

*See Fig. 246 and 247 on page 237; in the case of Frame Shape 33v the restraining bearing at E and therefore also the lateral restraining force H are absent.

Loading Condition 4b

Anti-symmetrical overall loading condition
(Superposition of Loading Conditions 1b, 2b and 3b)

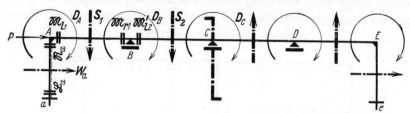

Fig. 252

Loading diagram with fixed-end moments $\mathfrak{M}_l$ and $\mathfrak{M}_r$ of the beam spans, load terms $\mathfrak{L}$ and $\mathfrak{R}$ of the column as well as external rotational moments D, for left-hand half of frame, furthermore with horizontal concentrated load P at level of horizontal member.

Restraining moments for column a and beam span 2:

$$\mathfrak{F}_{ra} = \frac{(Ph/2 + \mathfrak{S}_{la})(1/\varepsilon_a + 1) - (\mathfrak{L}_a + \mathfrak{R}_a)}{1/\varepsilon_a + 4} \ ; \qquad \mathfrak{M}'_{l2} = -\frac{\mathfrak{L}_2}{2} + \frac{D_o}{4}. \qquad (1114)$$

Joint load terms:

$$\mathfrak{R}^v_A = D_a + \mathfrak{F}_{ra} - \mathfrak{M}_{l1} \qquad \mathfrak{R}''_B = D_B + \mathfrak{M}_{r1} - \mathfrak{M}'_{l2}. \qquad (1115)$$

auxiliary moments:

$$Y_A = +\mathfrak{R}^v_A \cdot u_{aa} - \mathfrak{R}''_B \cdot u_{ab} \qquad Y_B = -\mathfrak{R}^v_A \cdot u_{ba} + \mathfrak{R}''_B \cdot u_{bb}. \qquad (1116)$$

Beam moments:

$$\begin{aligned} M_{A1} = -M_{E4} &= \mathfrak{M}_{l1} + (2Y_A + Y_B)K_1 \\ M_{B2} = -M_{D3} &= \mathfrak{M}'_{l2} + Y_B \cdot 3K_2/2; \\ M_{B1} &= -M_{D4} = \mathfrak{M}_{r1} - (Y_A + 2Y_B)K_1 \\ M_{C2} &= -M_{C3} = -D_C/2. \end{aligned} \qquad (1117)$$

Column moments:

$$M_{Aa} = M_{Ee} = \mathfrak{F}_{ra} - Y_A \cdot 3F_a \qquad M_a = M_e = -\frac{Ph}{2} - \mathfrak{S}_{la} + M_{Aa}. \qquad (1118)$$

Check relationships:

$$D_A + M_{Aa} = M_{A1} \qquad D_B + M_{B1} = M_{B2}. \qquad (1119)$$

Column forces:

$$(H_a = Q_{la}) = (H_e = Q_{le}) = W_a + P/2 \qquad (Q_{ra} = Q_{re}) = P/2. \qquad (1120)$$

Frame Shape 34
(Symmetrical form of Frame Shape 30)

Symmetrical single-story three-bay frame with laterally restrained horizontal member, two elastically restrained end columns and two pin-jointed internal supports (or rocker columns)

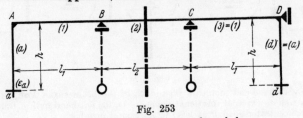

Fig. 253
Frame shape, dimensions and symbols

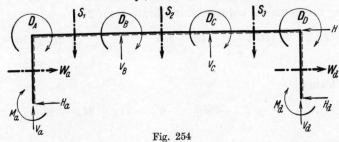

Fig. 254

Definition of positive direction for all external loads on joints and members, for all support reactions, and for the lateral restraining force

> All the dimensions and coefficients for the right-hand half of the frame are the same as those for the left-hand half. In the case of the columns the broken line is always placed on the right-hand side of each column, despite the symmetry of the frame.

Note: For Frame Shape 34 the coefficients and the *Loading Conditions* 1-4 *for Frame Shape* 30 may be employed. (See pages 199-205.) On account of symmetry, the forward reduction becomes the same as the backward reduction, and the backward recursion becomes the same as the forward recursion, both in Method I and Method II; in particular, (877) becomes equal to (876), (879) to (878), (894) to (893), and (896) to (895). Furthermore, the *square matrices of influence coefficients* (881) and (898) are bisymmetrical, i.e., symmetrical about the principal diagonal and the secondary diagonal. If the "load transposition method" is used, however, it is also possible to start from the following symmetrical and anti-symmetrical loading conditions.

[Cf. the treatment presented under Frame Shape 32, which has two bays more than the present Frame Shape.]

I. Treatment by the Method of Forces

a) Coefficients for symmetrical load arrangements

$$k_a = \frac{J_k}{J_a} \cdot \frac{h}{l_k} \qquad k_1 = \frac{J_k}{J_1} \cdot \frac{l_1}{l_k} \qquad k_2 = \frac{J_k}{J_2} \cdot \frac{l_2}{l_k}; \qquad b_a = k_a(2 - \varepsilon_a). \qquad (1121)$$

$$k_A = b_a + 2k_1 \qquad k'_B = 2k_1 + 3k_2. \qquad (1122)$$

$$\left. \begin{array}{c} \iota_{1s} = \dfrac{k_1}{k'_B} \qquad \iota'_1 = \dfrac{k_1}{k_A}; \qquad n'_{aa} = \dfrac{1}{k_A - k_1 \cdot \iota_{1s}} \qquad n'_{bb} = \dfrac{1}{k'_B - k_1 \cdot \iota'_1}; \\ n'_{ab} = n'_{ba} = n'_{aa} \cdot \iota_{1s} = n'_{bb} \cdot \iota'_1.* \end{array} \right\} \quad (1123)$$

$$s'_{a1} = n'_{aa}(1 - \iota_{1s}) = (s_{a1} + s_{a3}) \qquad s'_{b1} = n'_{bb}(1 - \iota'_1) = (s_{b1} - s_{b3}).** \qquad (1124)$$

Loading Condition 1a

Horizontal member carrying vertical loading of arbitrary magnitude symmetrical about center of frame ($\Re_3 = \mathfrak{L}_1$, $\mathfrak{L}_3 = \Re_1$; $\Re_2 = \mathfrak{L}_2$)

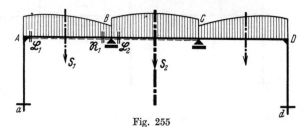

Fig. 255

Beam loading diagram with load terms $\mathfrak{L}$ and $\Re$ for left-hand half of frame

Beam and column moments: ($L_1 = \mathfrak{L}_1 k_1$, $R_1 = \Re_1 k_1$; $L_2 = \mathfrak{L}_2 k_2$):

$$\left. \begin{array}{l} M_{A1} = M_{D3} = -L_1 \cdot n'_{aa} + (R_1 + L_2) \cdot n'_{ab} \\ M_B = M_C = +L_1 \cdot n'_{ba} - (R_1 + L_2) \cdot n'_{bb}; \end{array} \right\} \quad (1125)$$

$$M_{Aa} = -M_{Dd} = M_{A1} \qquad M_a = -M_d = -M_{Aa} \cdot \varepsilon_a. \qquad (1126)$$

Special case: symmetrical bay loading ($R_1 = L_1$):

$$\left. \begin{array}{l} M_{A1} = M_{D3} = -L_1 \cdot s'_{a1} + L_2 \cdot n'_{ab} \\ M_B = M_C = -L_1 \cdot s'_{b1} - L_2 \cdot n'_{bb}. \end{array} \right\} \quad (1127)$$

---

* Between the values n'' (1123) for the left-hand half of the frame and the values n of matrix (881) for the whole frame there exist, according to the "load transposition method", the following relationships:
$$n'_{aa} = n_{aa} - n_{ad}, \quad n'_{ab} = n_{ab} - n_{ac}, \quad n'_{bb} = n_{bb} - n_{bc}.$$

** The members enclosed in parentheses have been formed with the values s (885).

b) Coefficients for anti-symmetrical load arrangements

$$k_a = \frac{J_k}{J_a} \cdot \frac{h}{l_k} \qquad k_1 = \frac{J_k}{J_1} \cdot \frac{l_1}{l_k} \qquad k_2 = \frac{J_k}{J_2} \cdot \frac{l_2}{l_k}; \qquad b_a = k_a(2 - \varepsilon_a). \qquad (1128)$$

$$k_A = b_a + 2k_1 \qquad k_B'' = 2k_1 + k_2. \qquad (1129)$$

$$\left. \begin{array}{c} \iota_{1t} = \dfrac{k_1}{k_B''} \qquad \iota_1' = \dfrac{k_1}{k_A}; \qquad n_{aa}'' = \dfrac{1}{k_A - k_1 \cdot \iota_{1t}} \qquad n_{bb}'' = \dfrac{1}{k_B'' - k_1 \cdot \iota_1'}; \\ n_{ab}'' = n_{ba}'' = n_{aa}'' \cdot \iota_{1t} = n_{bb}'' \cdot \iota_1'.\,* \end{array} \right\} \qquad (1130)$$

$$s_{a1}'' = n_{aa}''(1 - \iota_{1t}) = (s_{a1} - s_{a3}) \qquad s_{b1}'' = n_{bb}''(1 - \iota_1') = (s_{b1} + s_{b3}).\,** \qquad (1131)$$

Loading Condition 1b

Horizontal member carrying vertical loading of arbitrary magnitude anti-symmetrical about center of frame

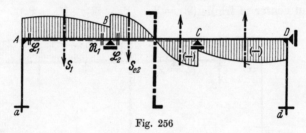

Fig. 256

Beam loading diagram with load terms $\mathfrak{L}$ and $\mathfrak{R}$ for left-hand half of frame

Beam and column moments ($L_1 = \mathfrak{L}_1 k_1, \quad R_1 = \mathfrak{R}_1 k_1; \quad L_2 = \mathfrak{L}_2 k_2$):

$$\left. \begin{array}{l} M_{A1} = -M_{D3} = -L_1 \cdot n_{aa}'' + (R_1 + L_2) \cdot n_{ab}'' \\ M_B = -M_C = +L_1 \cdot n_{ba}'' - (R_1 + L_2) \cdot n_{bb}''; \end{array} \right\} \qquad (1132)$$

$$M_{Aa} = M_{Dd} = M_{A1} \qquad M_a = M_d = -M_{Aa} \cdot \varepsilon_a. \qquad (1133)$$

Special case: symmetrical bay loading ($R_1 = L_1$):

$$\left. \begin{array}{l} M_{A1} = -M_{D3} = -L_1 \cdot s_{a1}'' \\ M_B = -M_C = -L_1 \cdot s_{b1}''. \end{array} \right\} \qquad (1134)$$

---

* Between the values n'' (1130) for the left-hand half of the frame and the values n of matrix (881) for the whole frame there exist, according to the "load transposition method", the following relationships:

$$n_{aa}'' = n_{aa} + n_{ad}, \quad n_{ab}'' = n_{ab} + n_{ac}, \quad n_{bb}'' = n_{bb} + n_{bc}.$$

** The members enclosed in parentheses have been formed with the values s (885).

Loading Conditions 2a and 3a (2b and 3b)*

The two columns carrying arbitrary horizontal loading directed towards center of frame (acting from left to right) and symmetrical (anti-symmetrical) about center of frame, and symmetrical (anti-symmetrical) pair of rotational moments at corners of frame.

Joint load term:

$$(L_a = \mathfrak{L}_a k_a \quad \text{and} \quad R_a = \mathfrak{R}_a k_a); \qquad \mathfrak{B}_A = D_A \cdot b_a + L_a \cdot \varepsilon_a - R_a. \qquad (1135)$$

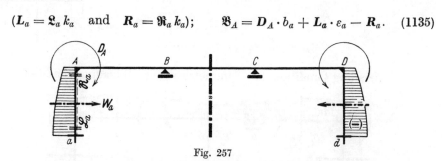

Fig. 257
Loading Conditions 2a and 3a—symmetry

$$(\mathfrak{L}_d = -\mathfrak{L}_a, \quad \mathfrak{R}_d = -\mathfrak{R}_a; \quad D_D = -D_A)$$

Column and beam moments:

$$\left. \begin{array}{ll} M_{A1} = M_{D3} = +\mathfrak{B}_A \cdot n'_{aa} & M_{Aa} = -M_{Dd} = -D_A + M_{A1} \\ M_B = M_C = -\mathfrak{B}_A \cdot n'_{ba}; & M_a = -M_d = -(\mathfrak{L}_a + M_{Aa})\varepsilon_a. \end{array} \right\} \begin{array}{c} (1136) \\ \text{and} \\ (1137) \end{array}$$

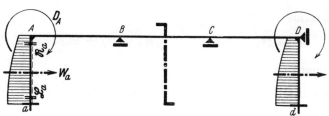

Fig. 258
Loading Conditions 2b and 3b—anti-symmetry

$$(\mathfrak{L}_d = \mathfrak{L}_a, \quad \mathfrak{R}_d = \mathfrak{R}_a; \quad D_D = D_A)$$

Column and beam moments:

$$\left. \begin{array}{ll} M_{A1} = -M_{D3} = +\mathfrak{B}_A \cdot n''_{aa} & M_{Aa} = M_{Dd} = -D_A + M_{A1} \\ M_B = -M_C = -\mathfrak{B}_A \cdot n''_{ba}; & M_a = M_d = -(\mathfrak{L}_a + M_{Aa})\varepsilon_a. \end{array} \right\} \begin{array}{c} (1138) \\ \text{and} \\ (1139) \end{array}$$

---

* This arrangement has been adopted in order to save space. For the coefficients to the symmetrical Loading Conditions 2a and 3a see page 245.

II. Treatment by the Deformation Method

a) Coefficients for symmetrical load arrangements

$$K_a = \frac{J_a}{J_k} \cdot \frac{l_k}{h} \qquad K_1 = \frac{J_1}{J_k} \cdot \frac{l_k}{l_1} \qquad K_2 = \frac{J_2}{J_k} \cdot \frac{l_k}{l_2}; \qquad S_a = \frac{K_a}{2 - \varepsilon_a}. \qquad (1140)$$

$$K_A = 3 S_a + 2 K_1 \qquad K'_B = 2 K_1 + K_2. \qquad (1141)$$

$$\left. \begin{array}{l} j_{1s} = \dfrac{K_1}{K'_B} \qquad j'_1 = \dfrac{K_1}{K_A}; \qquad u'_{aa} = \dfrac{1}{K_A - K_1 \cdot j_{1s}} \qquad u'_{bb} = \dfrac{1}{K'_B - K_1 \cdot j'_1}; \\ u'_{ab} = u'_{ba} = u'_{aa} \cdot j_{1s} = u'_{bb} \cdot j'_1. \end{array} \right\} \quad (1142)$$

Loading Condition 4a—Symmetrical overall loading condition

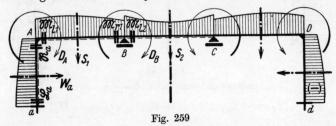

Fig. 259

Loading diagram, with fixed-end moments $\mathfrak{M}_l$ and $\mathfrak{M}_r$ of the beam spans, load terms $\mathfrak{L}$ and $\mathfrak{R}$ for the column, and external rotational moments D, for left-hand half of frame

Column fixed-end moment:

$$\mathfrak{C}_{ra} = - \frac{\mathfrak{R}_a - \mathfrak{L}_a \cdot \varepsilon_a}{2 - \varepsilon_a}. \qquad (1143)$$

Joint load terms:

$$\mathfrak{R}_A = D_A + \mathfrak{C}_{ra} - \mathfrak{M}_{l1} \qquad \mathfrak{R}_B = D_B + \mathfrak{M}_{r1} - \mathfrak{M}_{l2}. \qquad (1144)$$

auxiliary moments:

$$Y'_A = + \mathfrak{R}_A \cdot u'_{aa} - \mathfrak{R}_B \cdot u'_{ab} \qquad Y'_B = - \mathfrak{R}_A \cdot u'_{ba} + \mathfrak{R}_B \cdot u'_{bb}. \qquad (1145)$$

Beam moments:

$$\left. \begin{array}{l} M_{A1} = M_{D3} = \mathfrak{M}_{l1} + (2 Y'_A + Y'_B) K_1 \\ M_{B1} = M_{C3} = \mathfrak{M}_{r1} - (Y'_A + 2 Y'_B) K_1 \qquad M_{B2} = M_{C2} = \mathfrak{M}_{l2} + Y'_B \cdot K_2. \end{array} \right\} \quad (1146)$$

Column moments:

$$M_{Aa} = - M_{Dd} = \mathfrak{C}_{ra} - Y'_A \cdot 3 S_a \qquad M_a = - M_d = - (\mathfrak{L}_a + M_{Aa}) \varepsilon_a. \qquad (1147)$$

Check relationships:

$$D_A + M_{Aa} = M_{A1} \qquad D_B + M_{B1} = M_{B2}. \qquad (1148)$$

b) Coefficients for anti-symmetrical load arrangements

$$K_a = \frac{J_a}{J_k} \cdot \frac{l_k}{h} \qquad K_1 = \frac{J_1}{J_k} \cdot \frac{l_k}{l_1} \qquad K_2 = \frac{J_2}{J_k} \cdot \frac{l_k}{l_2}; \qquad S_a = \frac{K_a}{2 - \varepsilon_a}. \tag{1149}$$

$$K_A = 3 S_a + 2 K_1 \qquad K''_B = 2 K_1 + 3 K_2. \tag{1150}$$

$$\left.\begin{array}{l} j_{1t} = \dfrac{K_1}{K''_B} \quad j'_1 = \dfrac{K_1}{K_A}; \qquad u''_{aa} = \dfrac{1}{K_A - K_1 \cdot j_{1t}} \qquad u''_{bb} = \dfrac{1}{K''_B - K_1 \cdot j'_1}; \\[4pt] \qquad\qquad u''_{ab} = u''_{ba} = u''_{aa} \cdot j_{1t} = u''_{bb} \cdot j'_1. \end{array}\right\} \tag{1151}$$

Loading Condition 4b—Anti-symmetrical overall loading condition

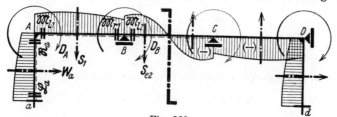

Fig. 260

Loading diagram, with fixed-end moments $\mathfrak{M}_l$ and $\mathfrak{M}_r$ of the beam spans, load terms $\mathfrak{L}$ and $\mathfrak{R}$ for the column, and external rotational moments D, for left-hand half of frame

Column fixed-end moment:

$$\mathfrak{C}_{ra} = -\frac{\mathfrak{R}_a - \mathfrak{L}_a \cdot \varepsilon_a}{2 - \varepsilon_a}. \tag{1152}$$

Joint load terms:

$$\mathfrak{R}_A = D_A + \mathfrak{C}_{ra} - \mathfrak{M}_{l1} \qquad \mathfrak{R}_B = D_B + \mathfrak{M}_{r1} - \mathfrak{M}_{l2}. \tag{1153}$$

auxiliary moments:

$$Y''_A = + \mathfrak{R}_A \cdot u''_{aa} - \mathfrak{R}_B \cdot u''_{ab} \qquad Y''_B = - \mathfrak{R}_A \cdot u''_{ba} + \mathfrak{R}_B \cdot u''_{bb}. \tag{1154}$$

Beam moments:

$$\begin{aligned} M_{A1} &= - M_{D3} = \mathfrak{M}_{l1} + (2 Y''_A + Y''_B) K_1 \\ M_{B1} &= - M_{C3} = \mathfrak{M}_{r1} - (Y''_A + 2 Y''_B) K_1 \\ M_{B2} &= - M_{C2} = \mathfrak{M}_{l2} + Y''_B \cdot 3 K_2. \end{aligned} \tag{1155}$$

Column moments:

$$M_{Aa} = M_{Dd} = \mathfrak{C}_{ra} - Y''_A \cdot 3 S_a \qquad M_a = M_d = -(\mathfrak{L}_a + M_{Aa}) \varepsilon_a. \tag{1156}$$

Check relationships:

$$D_A + M_{Aa} = M_{A1} \qquad D_B + M_{B1} = M_{B2}. \tag{1157}$$

Frame Shape 34v

(Frame Shape 34 with laterally unrestrained horizontal member)*

Note: The note relating to Frame Shape 32v, page 232, is applicable here. Only the anti-symmetrical loading conditions for Frame Shape 34v will be considered here, while the symmetrical loading conditions can be analyzed with the formulas given for Frame Shape 34.

I. Treatment by the Method of Forces

Coefficients for anti-symmetrical load arrangements: as indicated under b) on page 246, but instead of b_a we now have:

$$m_a = k_a \left(\frac{1}{\varepsilon_a} + 4 \right). \tag{1158}$$

Loading Condition 1b: as on page 246

but instead of the column moments (1133) we now have:

$$(M_{Aa} = M_a) = (M_{Dd} = M_d) = M_{A1} = -M_{D3}. \tag{1159}$$

Loading Conditions 2b and 3b (cf. page 234)

Joint load terms: same as formulas (1065).

Beam and column moments:

$$\left. \begin{array}{l} X_a = M_A = +M_d = -\mathfrak{B}_a \cdot n''_{aa} + \mathfrak{B}_1 \cdot n''_{ab} \\ X_b = M_B = -M_C = +\mathfrak{B}_a \cdot n''_{ba} - \mathfrak{B}_1 \cdot n''_{bb}; \end{array} \right\} \tag{1160}$$

$$M_{Aa} = M_{Dd} = X_a + \mathfrak{S}_{la} \qquad M_{A1} = -M_{D3} = M_{Aa} + \boldsymbol{D}_A. \tag{1161}$$

II. Treatment by the Deformation Method
Coefficients for anti-symmetrical load arrangements

Stiffness coefficients of all the members, and column stiffness:

$$K_a = \frac{J_a}{J_k} \cdot \frac{l_k}{h} \qquad K_1 = \frac{J_1}{J_k} \cdot \frac{l_k}{l_1} \qquad K_2 = \frac{J_2}{J_k} \cdot \frac{l_k}{l_2}; \qquad F_a = \frac{K_a}{1/\varepsilon_a + 4}. \tag{1162}$$

Joint coefficients: $\quad K_A^v = 3F_a + 2K_1 \qquad K_B'' = 2K_1 + 3K_2. \tag{1163}$

Rotation carry-over factors and influence coefficients:

$$\left. \begin{array}{l} j_1 = \dfrac{K_1}{K_B''} \qquad j_1' = \dfrac{K_1}{K_A^v}; \qquad u_{aa} = \dfrac{1}{K_A^v - K_1 \cdot j_1} \qquad u_{bb} = \dfrac{1}{K_B'' - K_1 \cdot j_1'} \\ u_{ab} = u_{ba} = u_{aa} \cdot j_1 = u_{bb} \cdot j_1'. \end{array} \right\} \tag{1164}$$

---

* See Fig. 253 and 254 on page 244, in the case of Frame Shape 34v the restraining bearing at D and therefore also the lateral restraining force H are absent.

Loading Condition 4b—Anti-symmetrical overall loading condition

(Superposition of Loading Conditions 1b, 2b, 3b and 5)

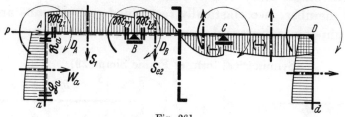

Fig. 261

Loading diagram with fixed-end moments $\mathfrak{M}_l$ and $\mathfrak{M}_r$ of the beam spans, load terms $\mathfrak{L}$ and $\mathfrak{R}$ of the column as well as external rotational moments D, for left-hand half of frame, furthermore with horizontal concentrated load P at level of horizontal member.

Column restraining moment:

$$\mathfrak{F}_{ra} = \frac{(Ph/2 + \mathfrak{S}_{la})(1/\varepsilon_a + 1) - (\mathfrak{L}_a + \mathfrak{R}_a)}{1/\varepsilon_a + 4}. \tag{1165}$$

Joint load terms:

$$\mathfrak{R}_A^v = D_A + \mathfrak{F}_{ra} - \mathfrak{M}_{l1} \qquad \mathfrak{R}_B = D_B + \mathfrak{M}_{r1} - \mathfrak{M}_{l2}. \tag{1166}$$

auxiliary moments:

$$Y_A = +\mathfrak{R}_A^v \cdot u_{aa} - \mathfrak{R}_B \cdot u_{ab} \qquad Y_B = -\mathfrak{R}_A^v \cdot u_{ba} + \mathfrak{R}_B \cdot u_{bb}. \tag{1167}$$

Beam moments:

$$\begin{aligned} M_{A1} &= -M_{D3} = \mathfrak{M}_{l1} + (2Y_A + Y_B)K_1 \\ M_{B1} &= -M_{C3} = \mathfrak{M}_{r1} - (Y_A + 2Y_B)K_1 \\ M_{B2} &= -M_{C2} = \mathfrak{M}_{l2} + Y_B \cdot 3K_2. \end{aligned} \tag{1168}$$

Column moments:

$$M_{Aa} = M_{Dd} = \mathfrak{F}_{ra} - Y_A \cdot 3F_a \qquad M_a = M_d = -\frac{Ph}{2} - \mathfrak{S}_{la} + M_{Aa}. \tag{1169}$$

Check relationships:

$$D_A + M_{Aa} = M_{A1} \qquad D_B + M_{B1} = M_{B2}. \tag{1170}$$

Column forces:

$$(H_a = Q_{la}) = (H_d = Q_{ld}) = W_a + P/2 \qquad (Q_{ra} = Q_{rd}) = P/2. \tag{1171}$$

Frame Shape 35

Symmetrical single-story two-bay frame with laterally restrained horizontal member, two elastically restrained end columns and one pin-jointed internal support (or rocker column)

(Symmetrical form of Frame Shape 29)

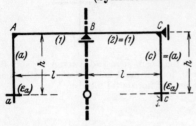

Fig. 262
Frame shape, dimensions and symbols

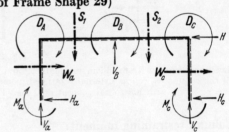

Fig. 263
Definition of positive direction for all external loads on joints and members and for all the support reactions

All the dimensions for the right-hand half of the frame are the same as those for the left-hand half. In the case of the columns the broken line is always placed on the right-hand side of each column, despite the symmetry of the frame.

Note: The analysis of Frame Shape 35 can essentially be carried out with the aid of the formulas given for Frame Shape 29. If the "load transposition method" is employed, however, it is also possible to start from the following symmetrical and anti-symmetrical loading conditions.

I. Treatment by the Method of Forces

(For $J_k = J_1$ and $l_k = l$, hence $k_1 = 1$)

a) Symmetrical load arrangements

Coefficients: exactly as formulas (365) (of Frame Shape 10)

Loading Condition 1a: exactly as formulas (366)—(369)

Loading Conditions 2a and 3a: exactly as formulas (370)—(372)

b) Anti-symmetrical load arrangements

Coefficients

$$(k_1 = 1) \qquad k_a = \frac{J_1}{J_a} \cdot \frac{h}{l} \qquad b_a = k_a(2 - \varepsilon_a) \qquad N = b_a + 2. \qquad (1172)$$

Loading Condition 1b

Both beam spans carrying vertical loading of arbitrary magnitude anti-symmetrical about center of frame ($\mathfrak{R}_2 = -\mathfrak{L}_1,\ \ \mathfrak{L}_2 = -\mathfrak{R}_1$)

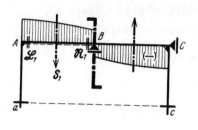

Fig. 264

Beam loading diagram with load terms $\mathfrak{L}$ and $\mathfrak{R}$ for left-hand half of frame

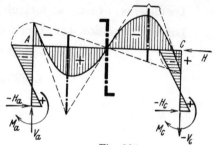

Fig. 265

Anti-symmetrical diagram of bending moments and reactions at supports

Beam and column moments ($L_1 = \mathfrak{L}_1$ because $k_1 = 1$):

$$\begin{aligned}
M_{A1} &= -M_{C2} = -L_1/N & M_B &= 0; \\
M_{Aa} &= M_{Cc} = +M_{A1} & M_a &= M_c = -M_{Aa}\cdot\varepsilon_a.
\end{aligned} \qquad (1173)$$

Note: All the moments remain independent of $\mathfrak{R}_1$.

Reactions at supports

$$V_a = -V_c = (\mathfrak{S}_{r1} - M_{A1})/l \qquad \begin{array}{l} V_b = 0 \qquad H = -2H_a; \\ H_a = H_c = M_{Aa}(1+\varepsilon_a)/h. \end{array} \qquad (1174)$$

Loading Conditions 2b and 3b

Both columns carrying arbitrary loading acting from left to right anti-symmetrical about center of frame, and external rotational moments arranged anti-symmetrically

$(\mathfrak{L}_c = \mathfrak{L}_a,\ \ \mathfrak{R}_c = \mathfrak{R}_a)$

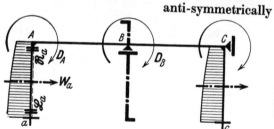

Fig. 266

Loading diagram with load terms $\mathfrak{L}$ and $\mathfrak{R}$ for the column, and external rotational moments D, for left-hand half of frame

Beam and column moments ($L_a = \mathfrak{L}_a k_a,\ R_a = \mathfrak{R}_a k_a$):

$$\begin{aligned}
M_{A1} &= -M_{C2} = (D_A + L_a\cdot\varepsilon_a - R_a + D_B/2)/N & M_{B1} &= -M_{B2} = -D_B/2 \\
M_{Aa} &= M_{Cc} = M_{A1} - D_A & M_a &= M_c = -(\mathfrak{L}_a + M_{Aa})\varepsilon_a.
\end{aligned} \qquad (1175)$$

Reactions at supports

$$\begin{array}{ll} V_a = -V_c = -(M_{A1} + D_D/2)/l & V_b = 0; \\ H_a = H_c = (\mathfrak{S}_{ra} + M_{Aa} - M_a)/h & H = 2(\mathbf{W}_a - H_u). \end{array} \qquad (1176)$$

II. Treatment by the Deformation Method

(For $J_k = J_1$ and $l_k = l$, hence $K_1 = 1$)

a) Symmetrical load arrangements

Coefficients: exactly as formulas (381) (of Frame Shape 10)

Loading Condition 4a: exactly as formulas (382)—(386)

b) Anti-symmetrical load arrangements

Coefficients

$$K_a = \frac{J_a}{J_1} \cdot \frac{l}{h} = \frac{1}{k_a} \quad S_a = \frac{K_a}{2 - \varepsilon_a} = \frac{1}{b_a}; \quad \mu_1 = \frac{1{,}5}{3 S_a + 1{,}5} \quad \mu_a = 1 - \mu_1. \quad (1177)$$

Loading Condition 4b—Anti-symmetrical overall loading condition

(Superposition of Loading Conditions 1b, 2b and 3b)

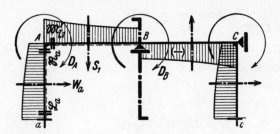

Fig. 267
Loading diagram with restraining moment $\mathfrak{M}'_l$ for the horizontal member, load terms $\mathfrak{L}$ and $\mathfrak{R}$ for the column, and external rotational moments D, for left-hand half of frame

Column and beam restraining moments:

$$\mathfrak{C}_{ra} = -\frac{\mathfrak{R}_a - \mathfrak{L}_a \cdot \varepsilon_a}{2 - \varepsilon_a} \qquad \mathfrak{M}'_{l1} = -\frac{\mathfrak{L}_1}{2} + \frac{D_B}{4}. \quad (1178)$$

Joint load term:

$$\mathfrak{R}_A = D_A + \mathfrak{C}_{ra} - \mathfrak{M}'_{l1}. \quad (1179)$$

Beam and column moments:

$$\left.\begin{array}{ll} M_{A1} = -M_{C2} = \mathfrak{M}'_{l1} + \mathfrak{R}_A \cdot \mu_1 & M_{B1} = -M_{B2} = -D_B/2; \\ M_{Aa} = +M_{Cc} = \mathfrak{C}_{ra} - \mathfrak{R}_A \cdot \mu_a & M_a = M_c = -(\mathfrak{L}_a + M_{\Delta a})\varepsilon_a. \end{array}\right\} \quad (1180)$$

Check relationship:

$$D_A + M_{Aa} = M_{A1}. \quad (1181)$$

Support reactions:

$$\left.\begin{array}{ll} V_a = -V_c = \frac{\mathfrak{C}_{r1}}{l} - \frac{M_{A1} + D_B/2}{l} & V_b = 0; \\ H_a = +H_c = \frac{\mathfrak{C}_{ra}}{h} + \frac{M_{Aa} - M_a}{h} & H = 2(W_a - H_a). \end{array}\right\} \quad (1182)$$

Frame Shape 35v
(**Frame Shape 35 with laterally unrestrained horizontal member**)

Note: The note relating to Frame Shape 31v, page 217, is applicable here. Only the anti-symmetrical loading conditions for Frame Shape 35v will be considered here, while the symmetrical loading conditions can be analysed with the formulas given for Frame Shape 35.

I. Treatment by the Method of Forces
Coefficients for anti-symmetrical load arrangements

(For $J_k = J_1$ and $l_k = l$, hence $k_1 = 1$)

$$(k_1 = 1) \qquad k_a = \frac{J_1}{J_a} \cdot \frac{h}{l} \qquad m_a = k_a\left(\frac{1}{\varepsilon_a} + 4\right) \qquad N^v = m_a + 2. \qquad (1183)$$

Loading Condition 1b

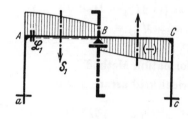
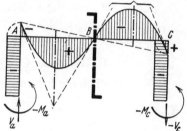

Fig. 268
Beam loading diagram with load term $\mathfrak{L}$

Fig. 269
Diagram of bending moments and reactions at supports

Beam and column moments ($L_1 = \mathfrak{L}_1$ since $k_1 = 1$):

$$\left.\begin{array}{l} X = (M_{Aa} = M_a) = (M_{Cc} = M_c) = -L_1/N^v \qquad M_B = 0 \\ M_{A1} = -M_{C2} = X. \qquad [V_a = -V_c = (\mathfrak{S}_{r1} - X)/l \qquad H_a = H_c = 0]. \end{array}\right\} \quad (1184)$$

Loading Conditions 2b and 5

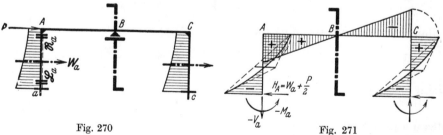

Fig. 270
Loading diagram with P, $\mathfrak{L}$ and $\mathfrak{R}$

Fig. 271
Diagram of bending moments and reactions at supports

Beam and column moments ($L_a = \mathfrak{L}_a k_a$, $R_a = \mathfrak{R}_a k_a$):

$$\left.\begin{array}{l} X = M_a = M_c = -[(\mathfrak{S}_{la} + Ph/2)(3k_a + 2) + (L_a + R_a)]/N^v \\ (M_{Aa} = M_{Cc}) = (M_{A1} = -M_{C2}) = \mathfrak{S}_{la} + Ph/2 + X \qquad M_B = 0. \end{array}\right\} \quad (1185)$$

Loading Condition 3b
External Rotational Moments in anti-symmetrical Arrangement

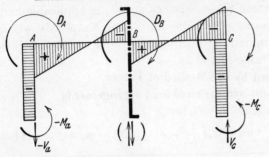

Fig. 272
Loading diagram, bending moments, and reactions at supports

Beam and column moments:

$$X = (M_{Aa} = M_a) = (M_{Cc} = M_c) = (-2D_A + D_B/2)/N^v \\ M_{A1} = -M_{C2} = X + D_A \qquad M_{B1} = -M_{B2} = -D_B/2. \quad \} \quad (1186)$$

II. Treatment by the Deformation Method
Coefficients for anti-symmetrical load arrangements

$$K_a = \frac{J_a}{J_1} \cdot \frac{l}{h} \qquad F_a = \frac{K_a}{1/\varepsilon_a + 4} \qquad \eta_a = \frac{F_a}{F_a + 1/2} \qquad \eta_1 = 1 - \eta_a. \quad (1187)$$

Loading Condition 4b—Anti-symmetrical overall loading condition

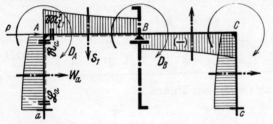

Fig. 273
Loading diagram with fixed moment $\mathfrak{M}'_{l1}$ for the horizontal member, load terms $\mathfrak{L}$ and $\mathfrak{R}$ for the column, and external rotational moments D, for left-hand half of frame; furthermore with horizontal point load P acting at level of horizontal member

Column and beam restraining moments:

$$\mathfrak{F}_{ra} = \frac{(Ph/2 + \mathfrak{S}_{la})(1/\varepsilon_a + 1) - (\mathfrak{L}_a + \mathfrak{R}_a)}{1/\varepsilon_a + 4} \qquad \mathfrak{M}'_{l1} = -\frac{\mathfrak{L}_1}{2} + \frac{D_B}{4}. \quad (1188)$$

Joint load term:

$$\mathfrak{K}^v_A = D_A + \mathfrak{F}_{ra} - \mathfrak{M}'_{l1}. \quad (1189)$$

Beam and column moments:

$$M_{A1} = -M_{C2} = \mathfrak{M}'_{l1} + \mathfrak{K}^v_A \cdot \eta_1 \qquad M_{Aa} = M_{Cc} = \mathfrak{F}_{ra} - \mathfrak{K}^v_A \cdot \eta_a \\ M_{B1} = -M_{B2} = -D_B/2 \qquad M_a = M_c = -Ph/2 - \mathfrak{S}_{la} + M_{Aa}. \quad \} \quad (1190)$$

Check relationship:

$$D_A + M_{Aa} = M_{A1}. \quad (1191)$$

Frame Shape 36

Two-story four-bay frame with laterally restrained horizontal member and five laterally restrained pairs of columns

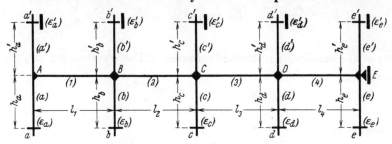

Fig. 274
Frame shape, dimensions and symbols

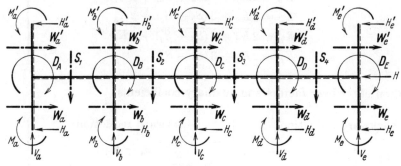

Fig. 275
Definition of positive direction for all external loads on joints and members, for all column reactions, and for the lateral restraining force

Important

All the formula sequences and formula matrices given for Frame Shape 36 can readily be extended to suit frames of similar type having any even number of bays.*

Note: The *collective subscripts* have the following meaning:

$\nu = 1, 2, 3, 4$ denoting the bays (or beam spans)
$n = a, b, c, d, e$ and $n' = a', b', c', d', e'$ denoting the columns
$N = A, B, C, D, E$ denoting the joints

The problem will be treated by the Deformation Method (Method II) only. For the treatment of a similar frame, though with two bays less, according to the Method of Forces (Method I) see Frame Shape 38, page 274-277.

---

* The dotted lines in the sets of formulas define the scope of these formulas as applicable to the two-bay frame. See also Frame Shape 38.

Coefficients:

Stiffness coefficients for all individual members:

$$K_v = \frac{J_v}{J_k} \cdot \frac{l_k}{l_v} \qquad K_n = \frac{J_n}{J_k} \cdot \frac{l_k}{h_n} \qquad K'_n = \frac{J'_n}{J_k} \cdot \frac{l_k}{h'_n}. \tag{1192}$$

Stiffness of the elastically restrained columns (The ϵ values are assumed to be known):

$$S_n = \frac{K_n}{2 - \varepsilon_n} \qquad S'_n = \frac{K'_n}{2 - \varepsilon'_n}. \tag{1193}$$

Joint coefficients:

$$\left.\begin{aligned}
K_A &= 3(S_a + S'_a) + 2K_1 \\
K_B &= 2K_1 + 3(S_b + S'_b) + 2K_2 \\
K_C &= 2K_2 + 3(S_c + S'_c) + 2K_3 \\
K_D &= 2K_3 + 3(S_d + S'_d) + 2K_4 \\
K_E &= 2K_4 + 3(S_e + S'_e).
\end{aligned}\right\} \tag{1194}$$

Rotation carry-over factors and auxiliary coefficients:

backward reduction:	forward reduction:
$j_4 = \dfrac{K_4}{r_e}$ $r_e = K_E$	$j'_1 = \dfrac{K_1}{v_a}$ $v_a = K_A$
$r_d = K_D - K_4 \cdot j_4$	$v_b = K_B - K_1 \cdot j'_1$
$j_3 = \dfrac{K_3}{r_d}$ $r_c = K_C - K_3 \cdot j_3$	$j'_2 = \dfrac{K_2}{v_b}$ $v_c = K_C - K_2 \cdot j'_2$
$j_2 = \dfrac{K_2}{r_c}$ $r_b = K_B - K_2 \cdot j_2$	$j'_3 = \dfrac{K_3}{v_c}$ $v_d = K_D - K_3 \cdot j'_3$
$j_1 = \dfrac{K_1}{r_b}$ $r_a = K_A - K_1 \cdot j_1.$	$j'_4 = \dfrac{K_4}{v_d}$ $v_e = K_E - K_4 \cdot j'_4.$

(1195) and (1196)

Principal influence coefficients by recursion:

forward:	backward:
$u_{aa} = 1/r_a$	$u_{ee} = 1/v_e$
$u_{bb} = 1/r_b + u_{aa} \cdot j_1^2$	$u_{dd} = 1/v_d + u_{ee} \cdot j'^2_4$
$u_{cc} = 1/r_c + u_{bb} \cdot j_2^2$	$u_{cc} = 1/v_c + u_{dd} \cdot j'^2_3$
$u_{dd} = 1/r_d + u_{cc} \cdot j_3^2$	$u_{bb} = 1/v_b + u_{cc} \cdot j'^2_2$
$u_{ee} = 1/r_e + u_{dd} \cdot j_4^2.$	$u_{aa} = 1/v_a + u_{bb} \cdot j'^2_1.$

(1197) and (1198)

Principal influence coefficients direct

$$u_{aa} = \frac{1}{r_a} \qquad u_{bb} = \frac{1}{r_b + v_b - K_B} \qquad u_{cc} = \frac{1}{r_c + v_c - K_C} \\ u_{dd} = \frac{1}{r_d + v_d - K_D} \qquad u_{ee} = \frac{1}{v_e}. \qquad (1199)$$

Computation scheme for the symmetrical influence coefficient matrix

$$\begin{array}{c|ccccc|c}
\cdot j_1 & (u_{aa}) & u_{ab} & u_{ac} & u_{ad} & u_{ae} & \cdot j_1' \\
 & \downarrow & \uparrow & \uparrow & \uparrow & \uparrow & \\
\cdot j_2 & u_{ba} & (u_{bb}) & u_{bc} & u_{bd} & u_{be} & \cdot j_2' \\
 & \downarrow & \downarrow & \uparrow & \uparrow & \uparrow & \\
\cdot j_3 & u_{ca} & u_{cb} & (u_{cc}) & u_{cd} & u_{ce} & \cdot j_3' \\
 & \downarrow & \downarrow & \downarrow & \uparrow & \uparrow & \\
\cdot j_4 & u_{da} & u_{db} & u_{dc} & (u_{dd}) & u_{de} & \cdot j_4' \\
 & \downarrow & \downarrow & \downarrow & \downarrow & \uparrow & \\
 & u_{ea} & u_{eb} & u_{ec} & u_{ed} & (u_{ee}) &
\end{array} \qquad (1200)$$

Loading Condition 4

Overall loading condition (superposition of Loading Conditions, 1, 2 and 3)

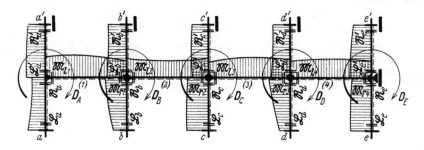

Fig. 276
Loading diagram with fixed end moments $\mathfrak{M}_l$ and $\mathfrak{M}_r$ for the beam spans, load terms $\mathfrak{L}$ and $\mathfrak{R}$ for the columns, and external rotational moments D^*

Column restraining moments at the beam joints:

$$\mathfrak{E}_{rn} = -\frac{\mathfrak{R}_n - \mathfrak{L}_n \cdot \varepsilon_n}{2 - \varepsilon_n} \qquad \mathfrak{E}_{ln}' = -\frac{\mathfrak{L}_n' - \mathfrak{R}_n' \cdot \varepsilon_n'}{2 - \varepsilon_n'}. \qquad (1201)$$

---

* For convenience of representation the resultants S and W have been omitted; cf. Fig. 275.

Joint load terms:

$$\left.\begin{aligned}
\mathfrak{K}_A &= \phantom{\mathfrak{M}_{r1} -{}} -\mathfrak{M}_{l1} + \mathfrak{C}_{ra} - \mathfrak{C}'_{la} + D_A \\
\mathfrak{K}_B &= \mathfrak{M}_{r1} - \mathfrak{M}_{l2} + \mathfrak{C}_{rb} - \mathfrak{C}'_{lb} + D_B \\
\mathfrak{K}_C &= \mathfrak{M}_{r2} - \mathfrak{M}_{l3} + \mathfrak{C}_{rc} - \mathfrak{C}'_{lc} + D_C \\
\mathfrak{K}_D &= \mathfrak{M}_{r3} - \mathfrak{M}_{l4} + \mathfrak{C}_{rd} - \mathfrak{C}'_{ld} + D_D \\
\mathfrak{K}_E &= \mathfrak{M}_{r4} \phantom{-\mathfrak{M}_{l4}} + \mathfrak{C}_{re} - \mathfrak{C}'_{le} + D_E .
\end{aligned}\right\} \quad (1202)$$

auxiliary moments:

	$\mathfrak{K}_A$	$\mathfrak{K}_B$	$\mathfrak{K}_C$	$\mathfrak{K}_D$	$\mathfrak{K}_E$
$Y_A =$	$+u_{aa}$	$-u_{ab}$	$+u_{ac}$	$-u_{ad}$	$+u_{ae}$
$Y_B =$	$-u_{ba}$	$+u_{bb}$	$-u_{bc}$	$+u_{bd}$	$-u_{be}$
$Y_C =$	$+u_{ca}$	$-u_{cb}$	$+u_{cc}$	$-u_{cd}$	$+u_{ce}$
$Y_D =$	$-u_{da}$	$+u_{db}$	$-u_{dc}$	$+u_{dd}$	$-u_{de}$
$Y_E =$	$+u_{ea}$	$-u_{eb}$	$+u_{ec}$	$-u_{ed}$	$+u_{ee}$.

(1203)

Beam moments:

$$\left.\begin{aligned}
M_{A1} &= \mathfrak{M}_{l1} + (2Y_A + Y_B)K_1 & M_{B1} &= \mathfrak{M}_{r1} - (Y_A + 2Y_B)K_1 \\
M_{B2} &= \mathfrak{M}_{l2} + (2Y_B + Y_C)K_2 & M_{C2} &= \mathfrak{M}_{r2} - (Y_B + 2Y_C)K_2 \\
M_{C3} &= \mathfrak{M}_{l3} + (2Y_C + Y_D)K_3 & M_{D3} &= \mathfrak{M}_{r3} - (Y_C + 2Y_D)K_3 \\
M_{D4} &= \mathfrak{M}_{l4} + (2Y_D + Y_E)K_4; & M_{E4} &= \mathfrak{M}_{r4} - (Y_D + 2Y_E)K_4 .
\end{aligned}\right\} \quad (1204)$$

Column moments:

$$\left.\begin{array}{ll}
\text{Lower story: } (n), & \text{Upper story: } (n'), \\
\text{at top } (1205): & \text{at base } (1207): \\
M_{Nn} = \mathfrak{C}_{rn} - Y_N \cdot 3 S_n; & M'_{Nn} = \mathfrak{C}'_{ln} + Y_N \cdot 3 S'_n; \\
\text{at base } (1206): & \text{at top } (1208): \\
M_n = -(\mathfrak{L}_n + M_{Nn})\varepsilon_n. & M'_n = -(\mathfrak{R}'_n + M'_{Nn})\varepsilon'_n.
\end{array}\right\} \begin{array}{c}(1205)\\ \text{to} \\ (1208)\end{array}$$

Check relationships see next page.

Check relationships:

$$\begin{aligned} D_A + M_{Aa} - M'_{Aa} \phantom{+ M_{B1}} - M_{A1} &= 0 \\ D_B + M_{Bb} - M'_{Bb} + M_{B1} - M_{B2} &= 0 \\ D_C + M_{Cc} - M'_{Cc} + M_{C2} - M_{C3} &= 0 \\ D_D + M_{Dd} - M'_{Dd} + M_{D3} - M_{D4} &= 0 \\ D_E + M_{Ee} - M'_{Ee} + M_{E4} \phantom{- M_{E4}} &= 0. \end{aligned} \qquad (1209)$$

Horizontal thrusts, column shears and lateral restraining force

For Loading Conditions 1 (arbitrary loading on horizontal member) and 3 (external moments at joints)

Lower story (n):

$$H_n = +Q_n = M_{Nn}(1+\varepsilon_n)/h_n. \qquad (1210)$$

Upper story (n'):

$$H'_n = -Q'_n = M'_{Nn}(1+\varepsilon'_n)/h'_n. \qquad (1211)$$

Lateral restraining force:

$$H = -(H_a + H_b + H_c \vert + H_d + H_e \vert) - (H'_a + H'_b + H'_c \vert + H'_d + H'_e \vert). \qquad (1212)$$

For Loading Conditions 2 (arbitrary loading on columns) and 4 (overall loading condition)

Lower story (n):

where
$$\begin{aligned} H_n = Q_{ln} = \mathfrak{S}_{rn}/h_n + T_n \qquad Q_{rn} &= -\mathfrak{S}_{ln}/h_n + T_n \\ T_n &= (M_{Nn} - M_n)/h_n. \end{aligned} \qquad (1213)$$

Upper story (n'):

where
$$\begin{aligned} Q'_{ln} = \mathfrak{S}'_{rn}/h'_n + T'_n \qquad H'_n = -Q'_{rn} &= \mathfrak{S}'_{ln}/h'_n - T'_n \\ T'_n &= (M'_n - M'_{Nn})/h'_n. \end{aligned} \qquad (1214)$$

Lateral restraining force:

$$H = -(Q_{ra} + Q_{rb} + Q_{rc} \vert + Q_{rd} + Q_{re} \vert) + (Q'_{la} + Q'_{lb} + Q'_{lc} \vert + Q'_{ld} + Q'_{le} \vert). \qquad (1215)$$

Collective subscripts:

$$N = A, B, C, \vert D, E; \qquad n = a, b, c, \vert d, e; \qquad n' = a', b', c', \vert d', e'.$$

Frame Shape 37

Two-story three-bay frame with laterally restrained horizontal member and four laterally restrained pairs of columns

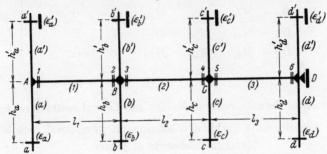

Fig. 277
Frame shape, dimensions and symbols

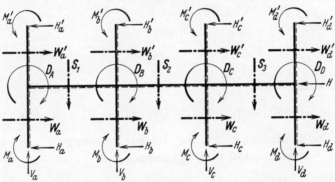

Fig. 278

Definition of positive direction for all external loads on joints and members, for all column reactions, and for the lateral restraining force

Important

All the formula sequences and formula matrices given for Frame Shape 37 can readily be extended to suit frames of similar shape having any odd number of bays.*

Note: The *collective subscripts* have the following meaning:

ν = 1, 2, 3 denoting the bays (or beam spans)
n = a, b, c, d and $n' = a', b', c', d'$ denoting the columns
N = A, B, C, D denoting the joints

The formulas for Frame Shape 37 are in all cases given for Method I as well as Method II.

---

* The dotted lines in the sets of formulas define the scope of these formulas for the single-bay frame. See also Frame Shape 39.

I. Treatment by the Method of Forces

Coefficients

Flexibility coefficients for all individual members:

$$k_v = \frac{J_k}{J_v} \cdot \frac{l_v}{l_k} \qquad k_n = \frac{J_k}{J_n} \cdot \frac{h_n}{l_k} \qquad k_n' = \frac{J_k}{J_n'} \cdot \frac{h_n'}{l_k}. \qquad (1216)$$

Flexibilities of the elastically restrained columns (The ϵ values are assumed to be known):

$$b_n = k_n(2 - \varepsilon_n) \qquad b_n' = k_n'(2 - \varepsilon_n'). \qquad (1217)$$

Flexibilities of the column lines:

$$s_N = \frac{1}{1/b_n + 1/b_n'} = \frac{b_n \cdot b_n'}{b_n + b_n'}. \qquad (1218)$$

Column distribution factors:

$$\eta_n = \frac{s_N}{b_n} \qquad \eta_n' = \frac{s_N}{b_n'} \qquad (\eta_n + \eta_n' = 1). \qquad (1219)$$

Diagonal coefficients:

$$\left. \begin{array}{lll} O_1 = s_A + 2k_1 & O_3 = s_B + 2k_2 & O_5 = s_C + 2k_3 \\ O_2 = 2k_1 + s_B & O_4 = 2k_2 + s_C & O_6 = 2k_3 + s_D. \end{array} \right\} \qquad (1220)$$

Moment carry-over factors and auxiliary coefficients:

backward reduction: forward reduction:

$$\iota_3 = \frac{k_3}{r_6} \quad r_6 = O_6 \qquad\qquad \iota_1' = \frac{k_1}{v_1} \quad v_1 = O_1$$

$$\sigma_C = \frac{s_C}{r_5} \quad r_5 = O_5 - k_3 \cdot \iota_3 \qquad\qquad \sigma_B' = \frac{s_B}{v_2} \quad v_2 = O_2 - k_1 \cdot \iota_1'$$

$$\iota_2 = \frac{k_2}{r_4} \quad r_4 = O_4 - s_C \cdot \sigma_C \qquad\qquad \iota_2' = \frac{k_2}{v_3} \quad v_3 = O_3 - s_B \cdot \sigma_B'$$

$$\sigma_B = \frac{s_B}{r_3} \quad r_3 = O_3 - k_2 \cdot \iota_2 \qquad\qquad \sigma_C' = \frac{s_C}{v_4} \quad v_4 = O_4 - k_2 \cdot \iota_2'$$

$$\iota_1 = \frac{k_1}{r_2} \quad r_2 = O_2 - s_B \cdot \sigma_B \qquad\qquad \iota_3' = \frac{k_3}{v_5} \quad v_5 = O_5 - s_C \cdot \sigma_C'$$

$$\qquad\qquad r_1 = O_1 - k_1 \cdot \iota_1. \qquad\qquad\qquad v_6 = O_6 - k_3 \cdot \iota_3'.$$

(1221) and (1222)

Principal influence coefficients by recursion:

forward:

$$n_{11} = \frac{1}{r_1}$$

$$n_{22} = \frac{1}{r_2} + n_{11} \cdot \iota_1^2$$

$$n_{33} = \frac{1}{r_3} + n_{22} \cdot \sigma_B^2$$

$$n_{44} = \frac{1}{r_4} + n_{33} \cdot \iota_2^2$$

$$n_{55} = \frac{1}{r_5} + n_{44} \cdot \sigma_C^2$$

$$n_{66} = \frac{1}{r_6} + n_{55} \cdot \iota_3^2 .$$

backward:

$$n_{66} = \frac{1}{v_6}$$

$$n_{55} = \frac{1}{v_5} + n_{66} \cdot \iota_3'^2$$

$$n_{44} = \frac{1}{v_4} + n_{55} \cdot \sigma_C'^2$$

$$n_{33} = \frac{1}{v_3} + n_{44} \cdot \iota_2'^2$$

$$n_{22} = \frac{1}{v_2} + n_{33} \cdot \sigma_B'^2$$

$$n_{11} = \frac{1}{v_1} + n_{22} \cdot \iota_1'^2 .$$

(1223) and (1224)

Principal influence coefficients direct

$$n_{11} = \frac{1}{r_1} \qquad n_{22} = \frac{1}{r_2 + v_2 - O_2} \qquad n_{33} = \frac{1}{r_3 + v_3 - O_3}$$

$$n_{44} = \frac{1}{r_4 + v_4 - O_4} \qquad n_{55} = \frac{1}{r_5 + v_5 - O_5} \qquad n_{66} = \frac{1}{v_6} .$$

(1225)

Computation scheme for the symmetrical influence coefficient matrix

$\cdot \iota_1$	$(n_{11})\downarrow$	$n_{12}\uparrow$	$n_{13}\uparrow$	$n_{14}\uparrow$	$n_{15}\uparrow$	$n_{16}\uparrow$	$\cdot \iota_1'$
$\cdot \sigma_B$	$n_{21}\downarrow$	$(n_{22})\downarrow$	$n_{23}\uparrow$	$n_{24}\uparrow$	$n_{25}\uparrow$	$n_{26}\uparrow$	$\cdot \sigma_B'$
$\cdot \iota_2$	$n_{31}\downarrow$	$n_{32}\downarrow$	(n_{33})	$n_{34}\uparrow$	$n_{35}\uparrow$	$n_{36}\uparrow$	$\cdot \iota_2'$
$\cdot \sigma_C$	$n_{41}\downarrow$	$n_{42}\downarrow$	$n_{43}\downarrow$	(n_{44})	$n_{45}\uparrow$	$n_{46}\uparrow$	$\cdot \sigma_C'$
$\cdot \iota_3$	$n_{51}\downarrow$	$n_{52}\downarrow$	$n_{53}\downarrow$	$n_{54}\downarrow$	$(n_{55})\downarrow$	$n_{56}\uparrow$	$\cdot \iota_3'$
	n_{61}	n_{62}	n_{63}	n_{64}	n_{65}	(n_{66})	

(1226)

Loading Condition 1

All the beam spans carrying arbitrary vertical loading

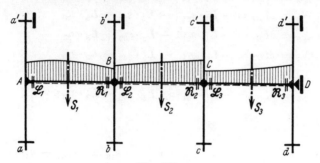

Fig. 279
Beam loading diagram with load terms $\mathfrak{L}$ and $\mathfrak{R}$

Beam moments $(L_\nu = \mathfrak{L}_\nu k_\nu$ and $R_\nu = \mathfrak{R}_\nu k_\nu)$:

	L_1	R_1	L_2	R_2	L_3	R_3
$X_1 = M_{A1} =$	$-n_{11}$	$+n_{12}$	$+n_{13}$	$-n_{14}$	$-n_{15}$	$+n_{16}$
$X_2 = M_{B1} =$	$+n_{21}$	$-n_{22}$	$-n_{23}$	$+n_{24}$	$+n_{25}$	$-n_{26}$
$X_3 = M_{B2} =$	$+n_{31}$	$-n_{32}$	$-n_{33}$	$+n_{34}$	$+n_{35}$	$-n_{36}$
$X_4 = M_{C2} =$	$-n_{41}$	$+n_{42}$	$+n_{43}$	$-n_{44}$	$-n_{45}$	$+n_{46}$
$X_5 = M_{C3} =$	$-n_{51}$	$+n_{52}$	$+n_{53}$	$-n_{54}$	$-n_{55}$	$+n_{56}$
$X_6 = M_{D3} =$	$+n_{61}$	$-n_{62}$	$-n_{63}$	$+n_{64}$	$+n_{65}$	$-n_{66}.$

(1227)

Column moments:

Lower story (n),

at top (1228):
$$M_{Aa} = \quad\quad +X_1 \cdot \eta_a$$
$$M_{Bb} = (-X_2 + X_3)\, \eta_b$$
$$M_{Cc} = (-X_4 + X_5)\, \eta_c$$
$$M_{Dd} = -X_6 \cdot \eta_d;$$

at base (1229):
$$M_n = -M_{Nn} \cdot \varepsilon_n.$$

Upper story (n'),

at base (1230):
$$M'_{Aa} = \quad\quad -X_1 \cdot \eta'_a$$
$$M'_{Bb} = (+X_2 - X_3)\, \eta'_b$$
$$M'_{Cc} = (+X_4 - X_5)\, \eta'_c$$
$$M'_{Dd} = \quad\quad +X_6 \cdot \eta'_d;$$

at top (1231):
$$M'_n = -M'_{Nn} \cdot \varepsilon'_n.$$

(1228) to (1231)

Special case: see page 266.

Special case: All the beam spans symmetrically loaded: $(R_\nu = L_\nu)$:

$$\begin{aligned}
X_1 &= M_{A1} = -L_1 \cdot s_{11} + L_2 \cdot s_{12} - L_3 \cdot s_{13} \\
X_2 &= M_{B1} = -L_1 \cdot s_{21} - L_2 \cdot s_{22} + L_3 \cdot s_{23} \\
X_3 &= M_{B2} = -L_1 \cdot s_{31} - L_2 \cdot s_{32} + L_3 \cdot s_{33} \\
X_4 &= M_{C2} = +L_1 \cdot s_{41} - L_2 \cdot s_{42} - L_3 \cdot s_{43} \\
X_5 &= M_{C3} = +L_1 \cdot s_{51} - L_2 \cdot s_{52} - L_3 \cdot s_{53} \\
X_6 &= M_{D3} = -L_1 \cdot s_{61} + L_2 \cdot s_{62} - L_3 \cdot s_{63};
\end{aligned} \qquad (1232)$$

where

$$\begin{array}{lll}
s_{11} = +n_{11} - n_{12} & s_{12} = +n_{13} - n_{14} & s_{13} = +n_{15} - n_{16} \\
& s_{22} = +n_{23} - n_{24} & s_{23} = +n_{25} - n_{26} \\
s_{21} = -n_{21} + n_{22} & s_{32} = +n_{33} - n_{34} & s_{33} = +n_{35} - n_{36} \\
s_{31} = -n_{31} + n_{32} & & \\
s_{41} = -n_{41} + n_{42} & s_{42} = -n_{43} + n_{44} & s_{43} = +n_{45} - n_{46} \\
s_{51} = -n_{51} + n_{52} & s_{52} = -n_{53} + n_{54} & s_{53} = +n_{55} - n_{56} \\
s_{61} = -n_{61} + n_{62} & s_{62} = -n_{63} + n_{64} & s_{63} = -n_{65} + n_{66}.
\end{array} \qquad (1233)$$

Loading Conditions 2 and 3

**All the columns carrying arbitrary horizontal loading (2);
external rotational moments applied at all the beam joints (3)**

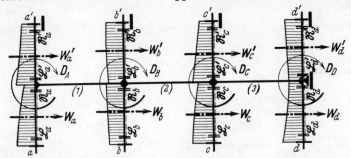

Fig. 280

Column loading diagram with load terms $\mathfrak{L}$ and $\mathfrak{R}$ for the columns and external rotational moments D.

Column load terms $(L_n = \mathfrak{L}_n k_n, \quad R_n = \mathfrak{R}_n k_n; \quad L'_n = \mathfrak{L}'_n k'_n, \quad R'_n = \mathfrak{R}'_n k'_n)$:

$$\mathfrak{B}_n = R_n - L_n \cdot \varepsilon_n \qquad \mathfrak{B}'_n = L'_n - R'_n \cdot \varepsilon'_n. \qquad (1234)$$

Joint load terms:

$$\mathfrak{B}_N = D_N \cdot s_N - \mathfrak{B}_n \cdot \eta_n + \mathfrak{B}'_n \cdot \eta'_n. \qquad (1235)$$

Beam moments*

	$\mathfrak{B}_A$	$\mathfrak{B}_B$	$\mathfrak{B}_C$	$\mathfrak{B}_D$
$X_1 = M_{A1} =$	$+ n_{11}$	$+ w_{1B}$	$- w_{1C}$	$+ n_{16}$
$X_2 = M_{B1} =$	$- n_{21}$	$- w_{2B}$	$+ w_{2C}$	$- n_{26}$
$X_3 = M_{B2} =$	$- n_{31}$	$+ w_{3B}$	$+ w_{3C}$	$- n_{36}$
$X_4 = M_{C2} =$	$+ n_{41}$	$- w_{4B}$	$- w_{4C}$	$+ n_{46}$
$X_5 = M_{C3} =$	$+ n_{51}$	$- w_{5B}$	$+ w_{5C}$	$+ n_{56}$
$X_6 = M_{D3} =$	$- n_{61}$	$+ w_{6B}$	$- w_{6C}$	$- n_{66}$;

(1236)

where

$$\begin{aligned}
w_{1B} &= + n_{12} - n_{13} & w_{1C} &= + n_{14} - n_{15} \\
w_{2B} &= + n_{22} - n_{23} & w_{2C} &= + n_{24} - n_{25} \\
& & w_{3C} &= + n_{34} - n_{35} \\
w_{3B} &= - n_{32} + n_{33} & w_{4C} &= + n_{44} - n_{45} \\
w_{4B} &= - n_{42} + n_{43} & & \\
w_{5B} &= - n_{52} + n_{53} & w_{5C} &= - n_{54} + n_{55} \\
w_{6B} &= - n_{62} + n_{63} & w_{6C} &= - n_{64} + n_{65}.
\end{aligned}$$

(1237)

Support moments for the column lines in themselves (auxiliary values)

$$M_N^0 = - \frac{\mathfrak{B}_n + \mathfrak{B}_n'}{b_n + b_n'}.$$

(1238)

Column moments:

Lower story (n), at top (1239):

$$\begin{aligned}
M_{Aa} &= M_A^0 - (\boldsymbol{D}_A \qquad\quad - X_1)\, \eta_a \\
M_{Bb} &= M_B^0 - (\boldsymbol{D}_B + X_2 - X_3)\, \eta_b \\
M_{Cc} &= M_C^0 - (\boldsymbol{D}_C + X_4 - X_5)\, \eta_c \\
M_{Dd} &= M_D^0 - (\boldsymbol{D}_D + X_6 \qquad\quad)\, \eta_d;
\end{aligned}$$

at base (1240):

$$M_n = -(\mathfrak{L}_n + M_{Nn})\, \varepsilon_n.$$

Upper story (n'), at base (1241):

$$\begin{aligned}
M'_{Aa} &= M_A^0 + (\boldsymbol{D}_A \qquad\quad - X_1)\, \eta'_a \\
M'_{Bb} &= M_B^0 + (\boldsymbol{D}_B + X_2 - X_3)\, \eta'_b \\
M'_{Cc} &= M_C^0 + (\boldsymbol{D}_C + X_4 - X_5)\, \eta'_c \\
M'_{Dd} &= M_D^0 + (\boldsymbol{D}_D + X_6 \qquad\quad)\, \eta'_d;
\end{aligned}$$

at top (1242):

$$M'_n = -(\mathfrak{R}'_n + M'_{Nn})\, \varepsilon'_n.$$

(1239) to (1242)

---

*In the case of the single-bay frame the composite column w under $\mathfrak{B}_B$ must be replaced by the second column n (i.e., n_{12} and n_{22} are substituted for w_{1B} and w_{2B}).

II. Treatment by the Deformation Method
Coefficients

Stiffness coefficients for all individual members:

$$K_v = \frac{J_v}{J_k} \cdot \frac{l_k}{l_v} = \frac{1}{k_v} \qquad K_n = \frac{J_n}{J_k} \cdot \frac{l_k}{h_n} = \frac{1}{k_n} \qquad K'_n = \frac{J'_n}{J_k} \cdot \frac{l_k}{h'_n} = \frac{1}{k'_n}. \qquad (1243)$$

Stiffness of the elastically restrained columns (The values ε are assumed to be known):

$$S_n = \frac{K_n}{2 - \varepsilon_n} = \frac{1}{b_n} \qquad S'_n = \frac{K'_n}{2 - \varepsilon'_n} = \frac{1}{b'_n}. \qquad (1244)$$

Joint coefficients:

$$\left.\begin{aligned} K_A &= \phantom{2K_1 + {}} 3(S_a + S'_a) + 2K_1 \\ K_B &= 2K_1 + 3(S_b + S'_b) + 2K_2 \\ K_C &= 2K_2 + 3(S_c + S'_c) + 2K_3 \\ K_D &= 2K_3 + 3(S_d + S'_d). \end{aligned}\right\} \qquad (1245)$$

Rotation carry-over factors and auxiliary coefficients:

$$\left.\begin{array}{l|l} \text{backward reduction:} & \text{forward reduction:} \\[4pt] \begin{aligned} j_3 &= \frac{K_3}{r_d} & r_d &= K_D \\ j_2 &= \frac{K_2}{r_c} & r_c &= K_C - K_3 \cdot j_3 \\ & & r_b &= K_B - K_2 \cdot j_2 \\ j_1 &= \frac{K_1}{r_b} & r_a &= K_A - K_1 \cdot j_1. \end{aligned} & \begin{aligned} & & v_a &= K_A \\ j'_1 &= \frac{K_1}{v_a} & v_b &= K_B - K_1 \cdot j'_1 \\ j'_2 &= \frac{K_2}{v_b} & v_c &= K_C - K_2 \cdot j'_2 \\ j'_3 &= \frac{K_3}{v_c} & v_d &= K_D - K_3 \cdot j'_3. \end{aligned} \end{array}\right\} \begin{array}{l} (1246) \\ \text{and} \\ (1247) \end{array}$$

Principal influence coefficients by recursion:

$$\left.\begin{array}{l|l} \text{forward:} & \text{backward:} \\[4pt] \begin{aligned} u_{aa} &= 1/r_a \\ u_{bb} &= 1/r_b + u_{aa} \cdot j_1^2 \\ u_{cc} &= 1/r_c + u_{bb} \cdot j_2^2 \\ u_{dd} &= 1/r_d + u_{cc} \cdot j_3^2. \end{aligned} & \begin{aligned} u_{dd} &= 1/v_d \\ u_{cc} &= 1/v_c + u_{dd} \cdot j'^2_3 \\ u_{bb} &= 1/v_b + u_{cc} \cdot j'^2_2 \\ u_{aa} &= 1/v_a + u_{bb} \cdot j'^2_1. \end{aligned} \end{array}\right\} \begin{array}{l} (1248) \\ \text{and} \\ (1249) \end{array}$$

Principal influence coefficients direct:

$$u_{aa} = \frac{1}{r_a} \qquad u_{bb} = \frac{1}{r_b + v_b - K_B} \qquad u_{cc} = \frac{1}{r_c + v_c - K_C} \qquad u_{dd} = \frac{1}{v_d}. \qquad (1250)$$

Computation scheme of the symmetrical influence coefficient matrix:

$$\left. \begin{array}{c|cccc|c} \cdot \hat{j}_1 & (u_{aa}) & u_{ab} & u_{ac} & u_{ad} & \cdot \hat{j}_1' \\ & \downarrow & \uparrow & \uparrow & \uparrow & \\ \hline \cdot \hat{j}_2 & u_{ba} & (u_{bb}) & u_{bc} & u_{bd} & \cdot \hat{j}_2' \\ & \downarrow & \downarrow & \uparrow & \uparrow & \\ \cdot \hat{j}_3 & u_{ca} & u_{cb} & (u_{cc}) & u_{cd} & \cdot \hat{j}_3' \\ & \downarrow & \downarrow & \downarrow & \uparrow & \\ & u_{da} & u_{db} & u_{dc} & (u_{dd}) & \end{array} \right\} \quad (1251)$$

Loading Condition 4

Overall loading condition (superposition of Loading Conditions, 1, 2 and 3)

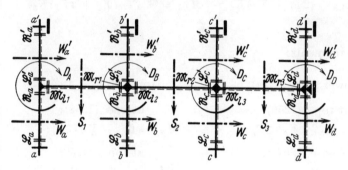

Fig. 281

Loading diagram with fixed moments $\mathfrak{M}_l$ and $\mathfrak{M}_r$ for the beam spans, load terms $\mathfrak{L}$ and $\mathfrak{R}$ for the columns, and external rotational moments D

Column restraining moments at the beam joints:

$$\mathfrak{E}_{rn} = - \frac{\mathfrak{R}_n - \mathfrak{L}_n \cdot \varepsilon_n}{2 - \varepsilon_n} \qquad \mathfrak{E}'_{ln} = - \frac{\mathfrak{L}'_n - \mathfrak{R}'_n \cdot \varepsilon'_n}{2 - \varepsilon'_n}. \qquad (1252)$$

Joint load terms

$$\left. \begin{array}{l} \mathfrak{K}_A = \phantom{\mathfrak{M}_{r1}} - \mathfrak{M}_{l1} + \mathfrak{E}_{ra} - \mathfrak{E}'_{la} + D_A \\ \mathfrak{K}_B = \mathfrak{M}_{r1} - \mathfrak{M}_{l2} + \mathfrak{E}_{rb} - \mathfrak{E}'_{lb} + D_B \\ \mathfrak{K}_C = \mathfrak{M}_{r2} - \mathfrak{M}_{l3} + \mathfrak{E}_{rc} - \mathfrak{E}'_{lc} + D_C \\ \mathfrak{K}_D = \mathfrak{M}_{r3} \phantom{- \mathfrak{M}_{l3}} + \mathfrak{E}_{rd} - \mathfrak{E}'_{ld} + D_D. \end{array} \right\} \qquad (1253)$$

Auxiliary moments

	$\mathfrak{K}_A$	$\mathfrak{K}_B$	$\mathfrak{K}_C$	$\mathfrak{K}_D$
$Y_A =$	$+ u_{aa}$	$- u_{ab}$	$+ u_{ac}$	$- u_{ad}$
$Y_B =$	$- u_{ba}$	$+ u_{bb}$	$- u_{bc}$	$+ u_{bd}$
$Y_C =$	$+ u_{ca}$	$- u_{cb}$	$+ u_{cc}$	$- u_{cd}$
$Y_D =$	$- u_{da}$	$+ u_{db}$	$- u_{dc}$	$+ u_{dd}$

(1254)

Beam moments:

$$M_{A1} = \mathfrak{M}_{l1} + (2Y_A + Y_B)K_1 \qquad M_{B1} = \mathfrak{M}_{r1} - (Y_A + 2Y_B)K_1$$
$$M_{B2} = \mathfrak{M}_{l2} + (2Y_B + Y_C)K_2 \qquad M_{C2} = \mathfrak{M}_{r2} - (Y_B + 2Y_C)K_2$$
$$M_{C3} = \mathfrak{M}_{l3} + (2Y_C + Y_D)K_3; \qquad M_{D3} = \mathfrak{M}_{r3} - (Y_C + 2Y_D)K_3.$$

(1255)

Column moments:

Lower story: (n),

at top (1256):
$$M_{Nn} = \mathfrak{C}_{rn} - Y_N \cdot 3 S_n;$$
at base (1257):
$$M_n = -(\mathfrak{L}_n + M_{Nn})\varepsilon_n.$$

Upper story: (n'),

at base (1258):
$$M'_{Nn} = \mathfrak{C}'_{ln} + Y_N \cdot 3 S'_n;$$
at top (1259):
$$M'_n = -(\mathfrak{R}'_n + M'_{Nn})\varepsilon'_n.$$

(1256) to (1259)

Check relationships:

$$D_A + M_{Aa} - M'_{Aa} \qquad\qquad - M_{A1} = 0$$
$$D_B + M_{Bb} - M'_{Bb} + M_{B1} - M_{B2} = 0$$
$$D_C + M_{Cc} - M'_{Cc} + M_{C2} - M_{C3} = 0$$
$$D_D + M_{Dd} - M'_{Dd} + M_{D3} \qquad = 0.$$

(1260)

Horizontal thrusts, column shears and lateral restraining force

Generally similar to formulas (1210) – (1215) for the collective subscripts $N = A, B, C, D$; $n = a, b, c, d$; and $n' = a', b', c', d'$. In the formulas (1212) and (1215) for H the terms H_e and H'_e as also Q_{re} and Q'_{le} must be omitted.

Frame Shape 37v
(Frame Shape 37 with laterally unrestrained horizontal member)*

First step: Calculation of all the requisite loading conditions as well as lateral restraining force H according to Frame Shape 37.

Second step: Determination of the effect of a horizontal concentrated load W, acting at the level of the horizontal member, with the aid of Loading Condition 5 for Frame Shape 37v, as indicated below.

Third step: Determination of the final moments and forces for Frame Shape 37v by superposition of the results of the first step and those of the second step for $W = H$.

Loading Condition 5

Horizontal point load W at level of horizontal member

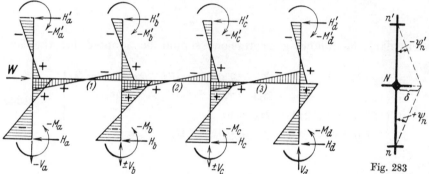

Fig. 282

Loading diagram, bending moment diagram, and column reactions.**) (The column bases $a - d$ and the column tops $a' - d'$ are assumed to be prevented from undergoing horizontal displacement)

Fig. 283

Diagram representing any particular column line n-N-n'; the elastic lateral displacement δ of the horizontal member and the angles of rotation ψ_n and ψ'_n of the columns are indicated

Important

The computation procedure given here for the "influence of a horizontal concentrated load W acting at the level of the laterally unrestrained horizontal member (or for the influence of such a concentrated load in the event of the entire upper story being laterally unrestrained: see Supplementary Note on page 273)" is fundamentally applicable also to all the other two-story frames in this book. It should only be noted that the subscripts n and n' in each case relate to all the rigidly connected columns and that the subscript N relates to all the corresponding beam joints.

---

*See Fig. 277 and 278, in which the restraining bearing at D and the lateral restraining force H should be omitted in the present case. See also Supplementary Note on page 273.

** The bending moment diagram corresponds to the case where the lower story is stiffer than the upper story. It should be borne in mind that for different relative stiffness conditions the bending moment diagram for the horizontal member of the frame may present an entirely different shape.

Column ratios with arbitrary comparison height h_k*):

$$\gamma_n = \frac{h_k}{h_n} \quad \text{and} \quad \gamma'_n = \frac{h_k}{h'_n}. \tag{1261}$$

Specific bending moments m (instead of M):

Either

according to *Method I* (Method of Forces), from the formulas (1236) – (1242), i.e., the moments:

$$X_i = m_{N\nu}; \quad m_N^0; \quad m_{Nn}, \quad m_n; \quad m'_{Nn}, \quad m'_n. \tag{1262}$$

In particular, the following expressions should be adopted for the *load terms:*

$$\left. \begin{array}{ll} \text{in (1236):} & \mathfrak{B}_N = \gamma_n(1+\varepsilon_n)\eta_n - \gamma'_n(1+\varepsilon'_n)\eta'_n; \\ \text{in (1238):} & -(\mathfrak{B}_n + \mathfrak{B}'_n) = \gamma_n(1+\varepsilon_n) + \gamma'_n(1+\varepsilon'_n); \\ \text{in (1239) and (1241):} & D_N = 0; \\ \text{in (1240) and (1242):} & \mathfrak{L}_n = \gamma_n/k_n \quad \mathfrak{R}'_n = \gamma'_n/k'_n. \end{array} \right\} \tag{1263}$$

Or

according to *Method II* (Deformation Method), from the formulas (1254) – (1260), i.e., the moments:

$$y_N \text{ (instead of } Y_N\text{);} \quad m_{N\nu}; \quad m_{Nn}, \quad m_n; \quad m'_{Nn}, \quad m'_n. \tag{1264}$$

In particular, the following expressions should be adopted for the *load terms:*

$$\left. \begin{array}{ll} \text{in (1254):} & \mathfrak{K}_N = \gamma_n(1+\varepsilon_n)S_n - \gamma'_n(1+\varepsilon'_n)S'_n; \\ \text{in (1255):} & \mathfrak{M}_{l\nu} = \mathfrak{M}_{r\nu} = 0; \\ \text{in (1256):} & \mathfrak{E}_{rn} = \gamma_n(1+\varepsilon_n)S_n; \\ \text{in (1258):} & \mathfrak{E}'_{ln} = \gamma'_n(1+\varepsilon'_n)S'_n; \\ \text{in (1257) and (1259):} & \mathfrak{L}_n = \gamma_n K_n \quad \mathfrak{R}'_n = \gamma'_n K'_n. \end{array} \right\} \tag{1265}$$

---

* In the special case where all the columns per story are of equal height, i.e., $(h_a = h_b = h_c = h_d) = h$ and $(h'_a = h'_b = h'_c = h'_d) = h'$, then we may write $\gamma_n = 1$ and $\gamma'_n = \gamma' = h/h'$.

Specific h_k-fold column shears:

$$p_n = (m_{Nn} - m_n)\gamma_n \quad \text{and} \quad p'_n = (m'_n - m'_{Nn})\gamma'_n. \tag{1266}$$

Corresponding specific h_k-fold applied load

$$p = (p_a + p_b + p_c + p_d) - (p'_a + p'_b + p'_c + p'_d). \tag{1267}$$

Conversion moment for a given load W according to Fig. 282:

$$U = \frac{Wh_k}{p}. \tag{1268}$$

Final moments and forces for the given concentrated load W:

Beam moments:

$$\begin{aligned} M_{A1} &= U \cdot m_{A1} & M_{B2} &= U \cdot m_{B2} & M_{C3} &= U \cdot m_{C3} \\ M_{B1} &= U \cdot m_{B1} & M_{C2} &= U \cdot m_{C2} & M_{D3} &= U \cdot m_{D3}. \end{aligned} \tag{1269}$$

Column moments:

Lower story: \qquad Upper story:

$$\begin{aligned} M_{Nn} &= U \cdot m_{Nn} & M'_{Nn} &= U \cdot m'_{Nn} \\ M_n &= U \cdot m_n; & M'_n &= U \cdot m'_n. \end{aligned} \tag{1270}$$

Column shears and horizontal thrusts:

$$\begin{aligned} \text{Upper story:} \qquad Q_n &= +H_n = \frac{U \cdot p_n}{h_k} = W \cdot \frac{p_n}{p}; \\ \text{Lower story:} \qquad Q'_n &= -H'_n = \frac{U \cdot p'_n}{h_k} = W \cdot \frac{p'_n}{p}. \end{aligned} \tag{1271}$$

Collective subscript: $n = a, b, c, d$ and $N = A, B, C, D$.

Supplementary note: If not only the horizontal member but also the entire upper story is laterally unrestrained, i.e., if no angular rotations ψ'_n (as in Fig. 283) occur, then substitute $\gamma'_n = 0$ in all the foregoing formulas.

— 274 —
Frame Shape 38

Two-story two-bay frame with laterally restrained horizontal member and three elastically restrained pairs of columns

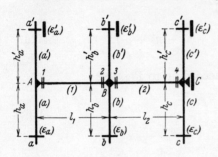

Fig. 284
Frame shape, dimensions and symbols

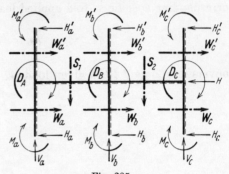

Fig. 285
Definition of positive direction for all external loads on joints and members, for all column reactions, and for the lateral restraining force

---

Important

All the formula sequences and formula matrices given for Frame Shape 38 can readily be extended to suit frames of similar type having any even number of bays.

Note: The *collective subscripts* have the following meaning:

$\nu = 1, 2;$ denoting the bays (or beam spans)
$n = a, b, c$ and $n' = a', b', c';$ denoting the columns
$N = A, B, C;$ denoting the joints

The problem will be treated by the Method of Forces (Method I) only. The formulas for treatment by the Deformation Method (Method II) are already contained in the formulas given for Frame Shape 36, in so far as they relate to the above collective subscripts. All the formulas in which no collective subscripts occur are valid up to the dotted lines; conversely, in the case of backward reduction (1195) and backward recursion (1198) they are valid from those dotted lines onward.

Coefficients

Flexibility coefficients, flexibilities and distribution coefficients: exactly as formulas (1216) – (1219)

Diagonal coefficients

$$\begin{aligned} O_1 &= s_A + 2k_1 & O_3 &= s_B + 2k_2 \\ O_2 &= 2k_1 + s_B & O_4 &= 2k_2 + s_C. \end{aligned} \tag{1272}$$

Moment carry-over factors and auxiliary coefficients:

$$\begin{array}{ll} \text{backward reduction:} & \text{forward reduction:} \\ \end{array}$$

$$\left.\begin{array}{ll}
\iota_2 = \dfrac{k_2}{r_4} \quad \begin{array}{l} r_4 = O_4 \\ r_3 = O_3 - k_2 \cdot \iota_2 \end{array} & \iota_1' = \dfrac{k_1}{v_1} \quad \begin{array}{l} v_1 = O_1 \\ v_2 = O_2 - k_1 \cdot \iota_1' \end{array} \\
\sigma_B = \dfrac{s_B}{r_3} \quad \begin{array}{l} \\ r_2 = O_2 - s_B \cdot \sigma_B \end{array} & \sigma_B' = \dfrac{s_B}{v_2} \quad \begin{array}{l} \\ v_3 = O_3 - s_B \cdot \sigma_B' \end{array} \\
\iota_1 = \dfrac{k_1}{r_2} \quad \begin{array}{l} \\ r_1 = O_1 - k_1 \cdot \iota_1 . \end{array} & \iota_2' = \dfrac{k_2}{v_3} \quad \begin{array}{l} \\ v_4 = O_4 - k_2 \cdot \iota_2'. \end{array}
\end{array}\right\} \begin{array}{c} (1273) \\ \text{and} \\ (1274) \end{array}$$

Principal influence coefficients by recursion:

$$\left.\begin{array}{ll}
\text{forward:} & \text{backward:} \\
n_{11} = 1/r_1 & n_{44} = 1/v_4 \\
n_{22} = 1/r_2 + n_{11} \cdot \iota_1^2 & n_{33} = 1/v_3 + n_{44} \cdot \iota_2'^2 \\
n_{33} = 1/r_3 + n_{22} \cdot \sigma_B^2 & n_{22} = 1/v_2 + n_{33} \cdot \sigma_B'^2 \\
n_{44} = 1/r_4 + n_{33} \cdot \iota_2^2 . & n_{11} = 1/v_1 + n_{22} \cdot \iota_1'^2 .
\end{array}\right\} \begin{array}{c} (1275) \\ \text{and} \\ (1276) \end{array}$$

Principal influence coefficients direct

$$n_{11} = \dfrac{1}{r_1} \qquad n_{22} = \dfrac{1}{r_2 + v_2 - O_2} \qquad n_{33} = \dfrac{1}{r_3 + v_3 - O_3} \qquad n_{44} = \dfrac{1}{v_4}. \qquad (1277)$$

Computation scheme for the symmetrical influence coefficient matrix

$$\left.\begin{array}{c|cccc|c}
 & (n_{11}) & n_{12} & n_{13} & n_{14} & \\
\cdot \iota_1 & \downarrow & \uparrow & \uparrow & \uparrow & \cdot \iota_1' \\
 & n_{21} & (n_{22}) & n_{23} & n_{24} & \\
\cdot \sigma_B & \downarrow & \downarrow & \uparrow & \uparrow & \cdot \sigma_B' \\
 & n_{31} & n_{32} & (n_{33}) & n_{34} & \\
\cdot \iota_2 & \downarrow & \downarrow & \downarrow & \uparrow & \cdot \iota_2' \\
 & n_{41} & n_{42} & n_{43} & (n_{44}) & \\
\end{array}\right\} (1278)$$

Composite influence coefficients:

$$\left.\begin{array}{lll}
s_{11} = +n_{11} - n_{12} & w_{1B} = +n_{12} - n_{13} & s_{12} = +n_{13} - n_{14} \\
s_{21} = -n_{21} + n_{22} & w_{2B} = +n_{22} - n_{23} & s_{22} = +n_{23} - n_{24} \\
s_{31} = -n_{31} + n_{32} & w_{3B} = -n_{32} + n_{33} & s_{32} = +n_{33} - n_{34} \\
s_{41} = -n_{41} + n_{42} & w_{4B} = -n_{42} + n_{43} & s_{42} = -n_{43} + n_{44}.
\end{array}\right\} (1279)$$

Loading Condition 1

Both beam spans carrying arbitrary vertical loading

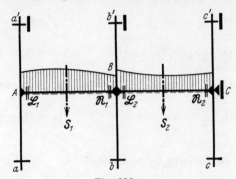

Fig. 286
Beam loading diagram with load terms $\mathfrak{L}$ and $\mathfrak{R}$

Beam moments ($L_\nu = \mathfrak{L}_\nu k_\nu$ and $R_\nu = \mathfrak{R}_\nu k_\nu$):

	L_1	R_1	L_2	R_2
$X_1 = M_{A1} =$	$-n_{11}$	$+n_{12}$	$+n_{13}$	$-n_{14}$
$X_2 = M_{B1} =$	$+n_{21}$	$-n_{22}$	$-n_{23}$	$+n_{24}$
$X_3 = M_{B2} =$	$+n_{31}$	$-n_{32}$	$-n_{33}$	$+n_{34}$
$X_4 = M_{C2} =$	$-n_{41}$	$+n_{42}$	$+n_{43}$	$-n_{44}.$

(1280)

Column moments:

Lower story (n),

at top (1281):
$$M_{Aa} = \qquad +X_1 \cdot \eta_a$$
$$M_{Bb} = (-X_2 + X_3)\,\eta_b$$
$$M_{Cc} = -X_4 \cdot \eta_c;$$

at base (1282):
$$M_n = -M_{Nn} \cdot \varepsilon_n.$$

Upper story (n'),

at base (1283):
$$M'_{Aa} = \qquad -X_1 \cdot \eta'_a$$
$$M'_{Bb} = (+X_2 - X_3)\,\eta'_b$$
$$M'_{Cc} = +X_4 \cdot \eta'_c;$$

at top (1284):
$$M'_n = -M'_{Nn} \cdot \varepsilon'_n.$$

(1281) to (1284)

Special case: Both beam spans symmetrically loaded ($R_\nu = L_\nu$):

$$X_1 = M_{A1} = -L_1 \cdot s_{11} + L_2 \cdot s_{12}$$
$$X_2 = M_{B1} = -L_1 \cdot s_{21} - L_2 \cdot s_{22}$$
$$X_3 = M_{B2} = -L_1 \cdot s_{31} - L_2 \cdot s_{32}$$
$$X_4 = M_{C2} = +L_1 \cdot s_{41} - L_2 \cdot s_{42}.$$

(1285)

Loading Conditions 2 and 3

All the columns carrying arbitrary horizontal loading (2); external rotational moments applied at all the beam joints (3)

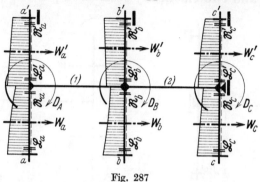

Fig. 287

Column loading diagram with load terms $\mathfrak{L}$ and $\mathfrak{R}$ for the columns and external rotational moments D

Column and joint load terms, and auxiliary values M_n^0: exactly as formulas (1234), (1235) and (1238).

Beam moments:

$$\left.\begin{aligned}
X_1 &= M_{A1} = +\mathfrak{B}_A \cdot n_{11} + \mathfrak{B}_B \cdot w_{1B} - \mathfrak{B}_C \cdot n_{14} \\
X_2 &= M_{B1} = -\mathfrak{B}_A \cdot n_{21} - \mathfrak{B}_B \cdot w_{2B} + \mathfrak{B}_C \cdot n_{24} \\
X_3 &= M_{B2} = -\mathfrak{B}_A \cdot n_{31} + \mathfrak{B}_B \cdot w_{3B} + \mathfrak{B}_C \cdot n_{34} \\
X_4 &= M_{C2} = +\mathfrak{B}_A \cdot n_{41} - \mathfrak{B}_B \cdot w_{4B} - \mathfrak{B}_C \cdot n_{44}
\end{aligned}\right\} \quad (1286)$$

Column moments:

Lower story (n),	Upper story (n').
at top (1287):	at base (1289):
$M_{Aa} = M_A^0 - (\mathbf{D}_A \qquad - X_1)\,\eta_a$	$M'_{Aa} = M_A^0 + (\mathbf{D}_A \qquad - X_1)\,\eta'_a$
$M_{Bb} = M_B^0 - (\mathbf{D}_B + X_2 - X_3)\,\eta_b$	$M'_{Bb} = M_B^0 + (\mathbf{D}_B + X_2 - X_3)\,\eta'_b$
$M_{Cc} = M_C^0 - (\mathbf{D}_C + X_4 \qquad)\,\eta_c;$	$M'_{Cc} = M_C^0 + (\mathbf{D}_C + X_4 \qquad)\,\eta'_c;$
at base (1288):	at top (1290):
$M_n = -(\mathfrak{L}_n + M_{Nn})\,\varepsilon_n.$	$M'_n = -(\mathfrak{R}'_n + M'_{Nn})\,\varepsilon'_n.$

(1287) to (1290)

Horizontal thrusts, column shears and lateral restraining force:

Generally similar to formulas (1210) – (1215) for the collective subscripts $N = A, B, C$; $n = a, b, c$ and $n' = a',\, b',\, c'$. In formulas (1212) and (1215) for H the terms with subscripts d and e must be omitted.

Frame Shape 39

Two-story single-bay frame with laterally restrained horizontal member and two elastically restrained pairs of columns

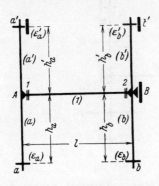

Fig. 288
Frame shape, dimensions and symbols

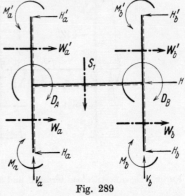

Fig. 289
Definition of positive direction for all external loads on joints and members, for all column reactions, and for the lateral restraining force

Important

All the formulas required for Frame Shape 39 are already comprised in the sets of formulas that have been given for Frame Shape 37, namely, for the collective subscripts: $\nu = 1$; $N = A, B$; $n = a, b$ and $n' = a', b'$. All the sets of formulas without collective subscripts are valid up to the dotted lines; conversely, for backward reductions and backward recursions they are valid from the dotted lines onward.— On account of the considerable practical importance of the single-bay frame, however, the relevant formulas will now be given again as a self-contained whole.

I. Treatment by the Method of Forces

Coefficients, where $J_k = J_1$ and $l_k = l$ hence $k_1 = 1$:

$$k_a = \frac{J_1}{J_a} \cdot \frac{h_a}{l} \qquad k_b = \frac{J_1}{J_b} \cdot \frac{h_b}{l} \qquad k_a' = \frac{J_1}{J_a'} \cdot \frac{h_a'}{l} \qquad k_b' = \frac{J_1}{J_b'} \cdot \frac{h_b'}{l}. \qquad (1291)$$

$$\left.\begin{array}{l} b_a = k_a(2 - \varepsilon_a) \\ b_a' = k_a'(2 - \varepsilon_a'); \end{array}\right. \quad s_A = \frac{1}{1/b_a + 1/b_a'} = \frac{b_a \cdot b_a'}{b_a + b_a'}; \qquad (1292)$$

$$\left.\begin{array}{l} b_b = k_b(2 - \varepsilon_b) \\ b_b' = k_b'(2 - \varepsilon_b'); \end{array}\right. \quad s_B = \frac{1}{1/b_b + 1/b_b'} = \frac{b_b \cdot b_b'}{b_b + b_b'}. \qquad (1293)$$

and

Coefficients (cont.):

$$\eta_a = \frac{s_A}{b_a} \quad \eta_a' = \frac{s_A}{b_a'}; \quad \eta_b = \frac{s_B}{b_b} \quad \eta_b' = \frac{s_B}{b_b'}; \quad \begin{matrix}(\eta_a + \eta_a' = 1)\\(\eta_b + \eta_b' = 1).\end{matrix} \quad (1294)$$

$$\iota_1 = \frac{1}{2+s_B} \quad b_1 = 2 - \iota_1; \quad \iota_1' = \frac{1}{s_A + 2} \quad b_1' = 2 - \iota_1'. \quad (1295)$$

$$n_{11} = \frac{1}{s_A + b_1} \quad n_{22} = \frac{1}{b_1' + s_B}; \quad n_{12} = n_{21} = n_{11} \cdot \iota_1 = n_{22} \cdot \iota_1'. \quad (1296)$$

$$s_{11} = n_{11} - n_{12} = n_{11}(1-\iota_1) \qquad s_{21} = n_{22} - n_{21} = n_{22}(1-\iota_1'). \quad (1297)$$

Loading Condition 1

Beam span carrying arbitrary vertical loading

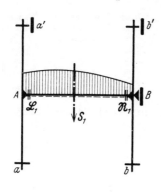 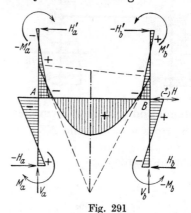

Fig. 290   Fig. 291

Beam loading diagram with load terms $\mathfrak{L}$ and $\mathfrak{R}$ Diagram of moments and support reactions

Beam moments ($L_1 = \mathfrak{L}_1$ and $R_1 = \mathfrak{R}_1$ since $k_1 = 1$):

$$\left.\begin{matrix}X_1 = M_{A1} = -L_1 \cdot n_{11} + R_1 \cdot n_{12} = -(L_1 - R_1 \cdot \iota_1)/(s_A + b_1)\\X_2 = M_{B1} = +L_1 \cdot n_{21} - R_1 \cdot n_{22} = -(R_1 - L_1 \cdot \iota_1')/(b_1' + s_B).\end{matrix}\right\} \quad (1298)$$

Column moments:

directly at the beam joints:

$$\left.\begin{matrix}M_{Aa}' = -X_1 \cdot \eta_a' & M_{Bb}' = +X_2 \cdot \eta_b'\\M_{Aa} = +X_1 \cdot \eta_a & M_{Bb} = -X_2 \cdot \eta_b;\end{matrix}\right\} \quad (1299)$$

at the elastically restrained opposite ends:

$$\left.\begin{matrix}M_a' = -M_{Aa}' \cdot \varepsilon_a' & M_b' = -M_{Bb}' \cdot \varepsilon_b'\\M_a = -M_{Aa} \cdot \varepsilon_a & M_b = -M_{Bb} \cdot \varepsilon_b.\end{matrix}\right\} \quad (1300)$$

Special case (of Loading Condition 1): **Symmetrical beam load**

$$\left. \begin{aligned} X_1 &= M_{A1} = -L_1 \cdot s_{11} = -L_1(1-\iota_1)/(s_A + b_1) \\ X_2 &= M_{B1} = -L_1 \cdot s_{21} = -L_1(1-\iota_1')/(b_1' + s_B) \end{aligned} \right\} \quad (1301)$$

Loading Conditions 2 and 3

All the individual columns carrying arbitrary horizontal loading (2); external rotational moments at the beam joints (3)

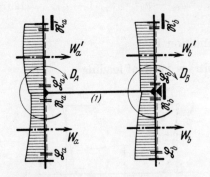

Fig. 292

Column loading diagram with load terms $\mathfrak{L}$ and $\mathfrak{R}$ for the columns and external rotational moments D

Column and joint load terms: exactly as formulas (1234) and (1235) for the subscripts $n = a, b$ and $N = A, B$.

Beam moments

$$\left. \begin{aligned} X_1 &= M_{A1} = +\mathfrak{B}_A \cdot n_{11} + \mathfrak{B}_B \cdot n_{12} = +(\mathfrak{B}_A + \mathfrak{B}_B \cdot \iota_1)/(s_A + b_1) \\ X_2 &= M_{B1} = -\mathfrak{B}_A \cdot n_{21} - \mathfrak{B}_B \cdot n_{22} = -(\mathfrak{B}_A \cdot \iota_1' + \mathfrak{B}_B)/(b_1' + s_B) \end{aligned} \right\} \quad (1302)$$

Support moments for the column lines in themselves (auxiliary values)

$$M_A^0 = -\frac{\mathfrak{B}_a + \mathfrak{B}_a'}{b_a + b_a'} \qquad M_B^0 = -\frac{\mathfrak{B}_b + \mathfrak{B}_b'}{b_b + b_b'}. \qquad (1303)$$

Column moments:

directly at the beam joints:

$$\left. \begin{aligned} M'_{Aa} &= M_A^0 + (D_A - X_1)\eta_a' & M'_{Bb} &= M_B^0 + (D_B + X_2)\eta_b' \\ M_{Aa} &= M_A^0 - (D_A - X_1)\eta_a & M_{Bb} &= M_B^0 - (D_B + X_2)\eta_b; \end{aligned} \right\} \quad (1304)$$

at the elastically restrained opposite ends:

$$\left. \begin{aligned} M_a' &= -(\mathfrak{R}_a' + M'_{Aa})\varepsilon_a' & M_b' &= -(\mathfrak{R}_b' + M'_{Bb})\varepsilon_b' \\ M_a &= -(\mathfrak{L}_a + M_{Aa})\varepsilon_a & M_b &= -(\mathfrak{L}_b + M_{Bb})\varepsilon_b. \end{aligned} \right\} \quad (1305)$$

II. Treatment by the Deformation Method

Coefficients, where $J_k = J_1$ and $l_k = l$, hence $K_1 = 1$:

$$K_a = \frac{J_a}{J_1} \cdot \frac{l}{h_a} \quad K_b = \frac{J_b}{J_1} \cdot \frac{l}{h_b} \quad K_a' = \frac{J_a'}{J_1} \cdot \frac{l}{h_a'} \quad K_b' = \frac{J_b'}{J_1} \cdot \frac{l}{h_b'}. \quad (1306)$$

$$S_a = \frac{K_a}{2-\varepsilon_a} \quad S_b = \frac{K_b}{2-\varepsilon_b} \quad S_a' = \frac{K_a'}{2-\varepsilon_a'} \quad S_b' = \frac{K_b'}{2-\varepsilon_b'}. \quad (1307)$$

$$K_A = 3(S_a + S_a') + 2 \quad K_B = 2 + 3(S_b + S_b'). \quad (1308)$$

$$j_1 = \frac{1}{K_B} \quad j_1'' = \frac{1}{K_A}; \quad u_{aa} = \frac{1}{K_A - j_1} \quad u_{bb} = \frac{1}{K_B - j_1''} \\ u_{ab} = u_{ba} = u_{aa} \cdot j_1 = u_{bb} \cdot j_1''. \quad \Big\} \quad (1309)$$

Loading Condition 4

Overall loading condition (superposition of Loading Conditions, 1, 2 and 3)

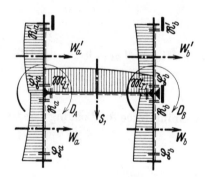

Fig. 293
Loading diagram with fixed-end moments $\mathfrak{M}_l$ and $\mathfrak{M}_r$ for the beam, load terms $\mathfrak{L}$ and $\mathfrak{R}$ for the columns, and external rotational moments D

Column restraining moments at the beam joints:

$$\mathfrak{C}_{la}' = -\frac{\mathfrak{L}_a' - \mathfrak{R}_a' \cdot \varepsilon_a'}{2-\varepsilon_a'} \quad \mathfrak{C}_{lb}' = -\frac{\mathfrak{L}_b' - \mathfrak{R}_b' \cdot \varepsilon_b'}{2-\varepsilon_b'} \\ \mathfrak{C}_{ra} = -\frac{\mathfrak{R}_a - \mathfrak{L}_a \cdot \varepsilon_a}{2-\varepsilon_a} \quad \mathfrak{C}_{rb} = -\frac{\mathfrak{R}_b - \mathfrak{L}_b \cdot \varepsilon_b}{2-\varepsilon_b}. \quad \Big\} \quad (1310)$$

Joint load terms:

$$\mathfrak{R}_A = -\mathfrak{M}_{l1} + \mathfrak{C}_{ra} - \mathfrak{C}_{la}' + D_A \\ \mathfrak{R}_B = +\mathfrak{M}_{r1} + \mathfrak{C}_{rb} - \mathfrak{C}_{lb}' + D_B. \quad \Big\} \quad (1311)$$

auxiliary moments:

$$\left.\begin{aligned}Y_A &= +\mathfrak{K}_A \cdot u_{aa} - \mathfrak{K}_B \cdot u_{ab} = (\mathfrak{K}_A - \mathfrak{K}_B \cdot j_1)/(K_A - j_1)\\Y_B &= -\mathfrak{K}_A \cdot u_{ba} + \mathfrak{K}_B \cdot u_{bb} = (\mathfrak{K}_B - \mathfrak{K}_A \cdot j_1')/(K_B - j_1').\end{aligned}\right\} \quad (1312)$$

Beam moments:

$$M_{A1} = \mathfrak{M}_{l1} + (2Y_A + Y_B) \qquad M_{B1} = \mathfrak{M}_{r1} - (Y_A + 2Y_B). \quad (1313)$$

Column moments:

directly at the beam joints

$$\left.\begin{aligned}M'_{Aa} &= \mathfrak{C}'_{la} + Y_A \cdot 3S'_a & M'_{Bb} &= \mathfrak{C}'_{lb} + Y_B \cdot 3S'_b\\M_{Aa} &= \mathfrak{C}_{ra} - Y_A \cdot 3S_a & M_{Bb} &= \mathfrak{C}_{rb} - Y_B \cdot 3S_b;\end{aligned}\right\} \quad (1314)$$

at the elastically restrained opposite ends

$$\left.\begin{aligned}M'_a &= -(\mathfrak{R}'_a + M'_{Aa})\varepsilon'_a & M'_b &= -(\mathfrak{R}'_b + M'_{Bb})\varepsilon'_b\\M_a &= -(\mathfrak{L}_a + M_{Aa})\varepsilon_a & M_b &= -(\mathfrak{L}_b + M_{Bb})\varepsilon_b.\end{aligned}\right\} \quad (1315)$$

Check relationships:

$$D_A + M_{Aa} - M'_{Aa} - M_{A1} = 0 \qquad D_B + M_{Bb} - M'_{Bb} + M_{B1} = 0. \quad (1316)$$

Horizontal thrusts, column shears and lateral restraining force for Loading Condition

Upper story:

$$\left.\begin{aligned}T'_a &= (M'_a - M'_{Aa})/h'_a; & T'_b &= (M'_b - M'_{Bb})/h'_b;\\H'_a &= -Q'_{ra} = \mathfrak{S}'_{la}/h'_a - T'_a & H'_b &= -Q'_{rb} = \mathfrak{S}'_{lb}/h'_b - T'_b\\Q'_{la} &= \mathfrak{S}'_{ra}/h'_a + T'_a; & Q'_{lb} &= \mathfrak{S}'_{rb}/h'_b + T'_b.\end{aligned}\right\} \quad (1317)$$

Lower story:

$$\left.\begin{aligned}T_a &= (M_{Aa} - M_a)/h_a; & T_b &= (M_{Bb} - M_b)/h_b;\\Q_{ra} &= -\mathfrak{S}_{la}/h_a + T_a & Q_{rb} &= -\mathfrak{S}_{lb}/h_b + T_b\\H_a &= Q_{la} = \mathfrak{S}_{ra}/h_a + T_a & H_b &= Q_{lb} = \mathfrak{S}_{rb}/h_b + T_b.\end{aligned}\right\} \quad (1318)$$

Lateral restraining force:

$$H = -Q_{ra} - Q_{rb} + Q'_{la} + Q'_{lb}. \quad (1319)$$

Frame Shape 40

Symmetrical two-story six-bay frame with laterally restrained horizontal member and seven elastically restrained pairs of columns

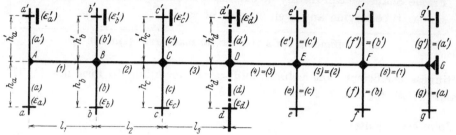

Fig. 294 Frame shape, dimensions and symbols

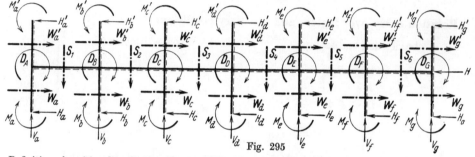

Fig. 295

Definition of positive direction for all external loads on joints and members, for all column reactions, and for the lateral restraining force

All the dimensions and coefficients for the right-hand half of the frame are the same as those for the left-hand half. In the case of the columns the broken line is always placed on the right-hand side of each column, despite the symmetry of the frame.

Note: For Frame Shape 40 the *coefficients and the Loading Condition* 4 for *Frame Shape* 36, enlarged by the addition of two bays, may be employed. On account of symmetry, the reduction sequences (1195)/(1196) and the recursion sequences (1197)/(1198) respectively become equal. Furthermore, the square matrix of influence coefficients (1200)—which in the case of Frame Shape 40 would have 7×7 elements—becomes bisymmetrical, i.e., symmetrical about the principal diagonal and the secondary diagonal. If the "load transposition method" is employed, it is also possible to start from the following symmetrical and anti-symmetrical loading conditions, which in the present case are given for the Deformation Method (Method II) only.*

---

* For the treatment of a similar frame by the Method of Forces (Method I) see Frame Shape 42.

> **Important**
>
> All the instructions, formula sequences and formula matrices given for Frame Shape 40 can readily be extended to suit symmetrical frames of similar type having any even number of bays.

a) Coefficients for symmetrical load arrangements

Stiffness coefficients and column stiffnesses: according to formulas (1192) and (1193), for $\nu = 1, 2, 3$ and $n = a, b, c$.

Joint coefficients:
$$K_A = 3(S_a + S'_a) + 2K_1$$
$$K_B = 2K_1 + 3(S_b + S'_b) + 2K_2 \qquad K_C = 2K_2 + 3(S_c + S'_c) + 2K_3. \tag{1320}$$

Rotation carry-over factors and auxiliary coefficients:

backward reduction: | forward reduction:

$$j_{2s} = \frac{K_2}{r'_c} \qquad \begin{aligned} r'_c &= K_C \\ r'_b &= K_B - K_2 \cdot j_{2s} \end{aligned} \qquad \begin{aligned} j''_1 &= \frac{K_1}{v_a} \\ j''_2 &= \frac{K_2}{v_b} \end{aligned} \qquad \begin{aligned} v_a &= K_A \\ v_b &= K_B - K_1 \cdot j''_1 \\ v_c &= K_C - K_2 \cdot j''_2. \end{aligned} \qquad \begin{matrix} (1321) \\ \text{and} \\ (1322) \end{matrix}$$

$$j_{1s} = \frac{K_1}{r'_b} \qquad r'_a = K_A - K_1 \cdot j_{1s}.$$

Principal influence coefficients by recursion forward and backward:

$$\begin{aligned} u'_{aa} &= 1/r'_a & u'_{cc} &= 1/v_c \\ u'_{bb} &= 1/r'_b + u'_{aa} \cdot j^2_{1s} & u'_{bb} &= 1/v_b + u'_{cc} \cdot j''^2_2 \\ u'_{cc} &= 1/r'_c + u'_{bb} \cdot j^2_{2s}. & u'_{aa} &= 1/v_a + u'_{bb} \cdot j''^2_1. \end{aligned} \qquad \begin{matrix} (1323) \\ \text{and} \\ (1324) \end{matrix}$$

Principal influence coefficients direct

$$u'_{aa} = 1/r'_a \qquad u'_{bb} = 1/(r'_b + v_b - K_B) \qquad u'_{cc} = 1/v_c. \tag{1325}$$

Computation scheme for the symmetrical influence coefficient matrix*

$$\begin{array}{c|ccc|c} & (u'_{aa}) & u'_{ab} & u'_{ac} & \\ \cdot j_{1s} & \downarrow & \uparrow & \uparrow & \cdot j''_1 \\ & u'_{ba} & (u'_{bb}) & u'_{bc} & \\ \cdot j_{2s} & \downarrow & \downarrow & \uparrow & \cdot j''_2 \\ & u'_{ca} & u'_{cb} & (u'_{cc}) & \end{array} \tag{1326}$$

---

* Between the elements of the 3 × 3 matrix (1326) or the 4 × 4 matrix (1333) for the left-hand half of the frame, on the one hand, and the matrix (1200), enlarged to 7 × 7 elements, for the whole frame, on the other hand, there exist, according to the "load transposition method", the following relationships:

$u'_{aa} = u_{aa} - u_{ag}$ etc., up to $u'_{cc} = u_{cc} - u_{ce}$ or $u''_{aa} = u_{aa} + u_{ag}$ etc., up to $u''_{cc} = u_{cc} + u_{ce}$, $u''_{dd} = 2u_{dd}$.

(The values u' with the subscript d for the central column line vanish, while the values u'' of the column d of (1333) become twice as large as the values u of the column d of the 7 × 7 matrix).

b) Coefficients for anti-symmetrical load arrangements

Stiffness coefficients and column stiffnesses: same as for symmetrical load arrangements (see previous page), but additionally for $n = d$.

Joint coefficients:

$$K_A = 3(S_a + S'_a) + 2K_1 \qquad K_C = 2K_2 + 3(S_c + S'_c) + 2K_3$$
$$K_B = 2K_1 + 3(S_b + S'_b) + 2K_2 \qquad K''_D = 2K_3 + 3(S_d + S'_d)/2. \qquad (1327)$$

Rotation carry-over factors and auxiliary coefficients:

backward reduction:

$$j_{3t} = \frac{K_3}{r''_d} \qquad r''_d = K''_D$$
$$j_{2t} = \frac{K_2}{r''_c} \qquad r''_c = K_C - K_3 \cdot j_{3t}$$
$$j_{1t} = \frac{K_1}{r''_b} \qquad r''_b = K_B - K_2 \cdot j_{2t}$$
$$\qquad r''_a = K_A - K_1 \cdot j_{1t}.$$

forward reduction:

$$v_a = K_A$$
$$j'_1 = \frac{K_1}{v_a} \qquad v_b = K_B - K_1 \cdot j'_1$$
$$j'_2 = \frac{K_2}{v_b} \qquad v_c = K_C - K_2 \cdot j'_2$$
$$j'_3 = \frac{K_3}{v_c} \qquad v''_d = K''_D - K_3 \cdot j'_3.$$

$$(1328)$$
and
$$(1329)$$

Principal influence coefficients by recursion:

forward:

$$u''_{aa} = 1/r''_a$$
$$u''_{bb} = 1/r''_b + u''_{aa} \cdot j_{1t}^2$$
$$u''_{cc} = 1/r''_c + u''_{bb} \cdot j_{2t}^2$$
$$u''_{dd} = 1/r''_d + u''_{cc} \cdot j_{3t}^2.$$

backward:

$$u''_{dd} = 1/v''_d$$
$$u''_{cc} = 1/v_c + u''_{dd} \cdot j'^2_3$$
$$u''_{bb} = 1/v_b + u''_{cc} \cdot j'^2_2$$
$$u''_{aa} = 1/v_a + u''_{bb} \cdot j'^2_1.$$

$$(1330)$$
and
$$(1331)$$

Principal influence coefficients direct

$$u''_{aa} = \frac{1}{r''_a} \qquad u''_{bb} = \frac{1}{r''_b + v_b - K_B} \qquad u''_{cc} = \frac{1}{r''_c + v_c - K_C} \qquad u''_{dd} = \frac{1}{v''_d}. \qquad (1332)$$

Computation scheme for the symmetrical influence coefficient matrix*

$$\begin{array}{c|cccc|c}
 & (u''_{aa}) & u''_{ab} & u''_{ac} & u''_{ad} & \\
\cdot j_{1t} & \downarrow \uparrow & \uparrow & \uparrow & \uparrow & \cdot j'_1 \\
 & u''_{ba} & (u''_{bb}) & u''_{bc} & u''_{bd} & \\
\cdot j_{2t} & \downarrow & \downarrow \uparrow & \uparrow & \uparrow & \cdot j'_2 \\
 & u''_{ca} & u''_{cb} & (u''_{cc}) & u''_{cd} & \\
\cdot j_{3t} & \downarrow & \downarrow & \downarrow \uparrow & \uparrow & \cdot j'_3 \\
 & u''_{da} & u''_{db} & u''_{dc} & (u''_{dd}) & \\
\end{array} \qquad (1333)$$

---

* See footnote on previous page

Loading Condition 4a

Symmetrical overall loading condition

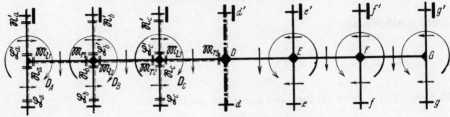

Fig. 296

Loading diagram with fixed end moments $\mathfrak{M}_l$ and $\mathfrak{M}_r$ for the beam spans, load terms $\mathfrak{L}$ and $\mathfrak{R}$ for the columns, and external rotational moments D, for left-hand half of frame. (Symmetrical loading of central column line d-D-d' not shown)

Column restraining moments at the beam joints:

Formulas for $\mathfrak{C}_{rn}$ and $\mathfrak{C}'_{ln}$ as (1201), for $n = a, b, c$.

Joint load terms:

$$\left.\begin{aligned}
\mathfrak{R}_A &= \phantom{\mathfrak{M}_{r1} - {}} -\mathfrak{M}_{l1} + \mathfrak{C}_{ra} - \mathfrak{C}'_{la} + \boldsymbol{D}_A \\
\mathfrak{R}_B &= \mathfrak{M}_{r1} - \mathfrak{M}_{l2} + \mathfrak{C}_{rb} - \mathfrak{C}'_{lb} + \boldsymbol{D}_B \\
\mathfrak{R}_C &= \mathfrak{M}_{r2} - \mathfrak{M}_{l3} + \mathfrak{C}_{rc} - \mathfrak{C}'_{lc} + \boldsymbol{D}_C.
\end{aligned}\right\} \quad (1334)$$

auxiliary moments:

$$\left.\begin{aligned}
Y'_A &= +\mathfrak{R}_A \cdot u'_{aa} - \mathfrak{R}_B \cdot u'_{ab} + \mathfrak{R}_C \cdot u'_{ac} \\
Y'_B &= -\mathfrak{R}_A \cdot u'_{ba} + \mathfrak{R}_B \cdot u'_{bb} - \mathfrak{R}_C \cdot u'_{bc} \\
Y'_C &= +\mathfrak{R}_A \cdot u'_{ca} - \mathfrak{R}_B \cdot u'_{cb} + \mathfrak{R}_C \cdot u'_{cc}.
\end{aligned}\right\} \quad (1335)$$

Beam moments:

$$\left.\begin{aligned}
M_{A1} &= M_{G6} = \mathfrak{M}_{l1} + (2Y'_A + Y'_B)K_1 \\
M_{B2} &= M_{F5} = \mathfrak{M}_{l2} + (2Y'_B + Y'_C)K_2 \\
M_{C3} &= M_{E4} = \mathfrak{M}_{l3} + 2Y'_C \cdot K_3; \\
M_{B1} &= M_{F6} = \mathfrak{M}_{r1} - (Y'_A + 2Y'_B)K_1 \\
M_{C2} &= M_{E5} = \mathfrak{M}_{r2} - (Y'_B + 2Y'_C)K_2 \\
M_{D3} &= M_{D4} = \mathfrak{M}_{r3} - Y'_C \cdot K_3.
\end{aligned}\right\} \quad (1336)$$

Column moments:

Lower story (n):

$$M_{Aa} = -M_{Gg} = \mathfrak{C}_{ra} - Y'_A \cdot 3S_a$$
$$M_{Bb} = -M_{Ff} = \mathfrak{C}_{rb} - Y'_B \cdot 3S_b$$
$$M_{Cc} = -M_{Ee} = \mathfrak{C}_{rc} - Y'_C \cdot 3S_c;$$
$$M_a = -M_g = -(\mathfrak{L}_a + M_{Aa})\varepsilon_a$$
$$M_b = -M_f = -(\mathfrak{L}_b + M_{Bb})\varepsilon_b$$
$$M_c = -M_e = -(\mathfrak{L}_c + M_{Cc})\varepsilon_c;$$
$$M_{Dd} = M_d = 0.$$

Upper story (n'):

$$M'_{Aa} = -M'_{Gg} = \mathfrak{C}'_{la} + Y'_A \cdot 3S'_a$$
$$M'_{Bb} = -M'_{Ff} = \mathfrak{C}'_{lb} + Y'_B \cdot 3S'_b$$
$$M'_{Cc} = -M'_{Ee} = \mathfrak{C}'_{lc} + Y'_C \cdot 3S'_c;$$
$$M'_a = -M'_g = -(\mathfrak{R}'_a + M'_{Aa})\varepsilon'_a$$
$$M'_b = -M'_f = -(\mathfrak{R}'_b + M'_{Bb})\varepsilon'_b$$
$$M'_c = -M'_e = -(\mathfrak{R}'_c + M'_{Cc})\varepsilon'_c;$$
$$M'_{Dd} = M'_d = 0.$$

(1337) and (1338)

Check relationships: as (1344), but without joint D.

Loading Condition 4b

Anti-symmetrical overall loading condition

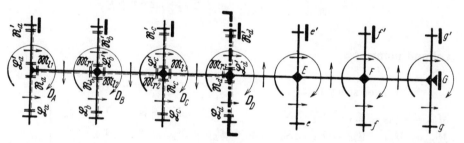

Fig. 297

Loading diagram with fixed-end moment $\mathfrak{M}_l$ and $\mathfrak{M}_r$ for the beam spans, load terms $\mathfrak{L}$ and $\mathfrak{R}$ for the columns, and external rotational moments D, for left-hand half of frame, including central column line d–D–d'

Column restraining moments at the beam joints:

Formulas for $\mathfrak{C}_{rn}$ and $\mathfrak{C}'_{ln}$ as (1201), for $n = a, b, c$.

Joint load terms

$$\begin{aligned}
\mathfrak{K}_A &= \qquad\quad -\mathfrak{M}_{l1} + \mathfrak{C}_{ra} - \mathfrak{C}'_{la} + D_A \\
\mathfrak{K}_B &= \mathfrak{M}_{r1} - \mathfrak{M}_{l2} + \mathfrak{C}_{rb} - \mathfrak{C}'_{lb} + D_B \\
\mathfrak{K}_C &= \mathfrak{M}_{r2} - \mathfrak{M}_{l3} + \mathfrak{C}_{rc} - \mathfrak{C}'_{lc} + D_C \\
\mathfrak{K}''_D &= \mathfrak{M}_{r3} \qquad\quad + (\mathfrak{C}_{rd} - \mathfrak{C}'_{ld} + D_D)/2.
\end{aligned} \qquad (1339)$$

auxiliary moments:

	$\mathfrak{K}_A$	$\mathfrak{K}_B$	$\mathfrak{K}_C$	$\mathfrak{K}''_D$
$Y''_A =$	$+ u''_{aa}$	$- u''_{ab}$	$+ u''_{ac}$	$- u''_{ad}$
$Y''_B =$	$- u''_{ba}$	$+ u''_{bb}$	$- u''_{bc}$	$+ u''_{bd}$
$Y''_C =$	$+ u''_{ca}$	$- u''_{cb}$	$+ u''_{cc}$	$- u''_{cd}$
$Y''_D =$	$- u''_{da}$	$+ u''_{db}$	$- u''_{dc}$	$+ u''_{dd}$.

$$\hspace{10cm} (1340)$$

Beam moments:

$$\begin{aligned}
M_{A1} = -M_{G6} &= \mathfrak{M}_{l1} + (2Y''_A + Y''_B)K_1 \\
M_{B2} = -M_{F5} &= \mathfrak{M}_{l2} + (2Y''_B + Y''_C)K_2 \\
M_{C3} = -M_{E4} &= \mathfrak{M}_{l3} + (2Y''_C + Y''_D)K_3; \\
M_{B1} = -M_{F6} &= \mathfrak{M}_{r1} - (Y''_A + 2Y''_B)K_1 \\
M_{C2} = -M_{E5} &= \mathfrak{M}_{r2} - (Y''_B + 2Y''_C)K_2 \\
M_{D3} = -M_{D4} &= \mathfrak{M}_{r3} - (Y''_C + 2Y''_D)K_3.
\end{aligned} \quad (1341)$$

Column moments:

Lower story (n):

$$\begin{aligned}
M_{Aa} = M_{Gg} &= \mathfrak{E}_{ra} - Y''_A \cdot 3S_a \\
M_{Bb} = M_{Ff} &= \mathfrak{E}_{rb} - Y''_B \cdot 3S_b \\
M_{Cc} = M_{Ee} &= \mathfrak{E}_{rc} - Y''_C \cdot 3S_c \\
M_{Dd} &= \mathfrak{E}_{rd} - Y''_D \cdot 3S_d; \\
M_a = M_g &= -(\mathfrak{L}_a + M_{Aa})\varepsilon_a \\
M_b = M_f &= -(\mathfrak{L}_b + M_{Bb})\varepsilon_b \\
M_c = M_e &= -(\mathfrak{L}_c + M_{Cc})\varepsilon_c \\
M_d &= -(\mathfrak{L}_d + M_{Dd})\varepsilon_d.
\end{aligned}$$

Upper story (n'):

$$\begin{aligned}
M'_{Aa} = M'_{Gg} &= \mathfrak{E}'_{la} + Y''_A \cdot 3S'_a \\
M'_{Bb} = M'_{Ff} &= \mathfrak{E}'_{lb} + Y''_B \cdot 3S'_b \\
M'_{Cc} = M'_{Ee} &= \mathfrak{E}'_{lc} + Y''_C \cdot 3S'_c \\
M'_{Dd} &= \mathfrak{E}'_{ld} + Y''_D \cdot 3S'_d; \\
M'_a = M'_g &= -(\mathfrak{R}'_a + M'_{Aa})\varepsilon'_a \\
M'_b = M'_f &= -(\mathfrak{R}'_b + M'_{Bb})\varepsilon'_b \\
M'_c = M'_e &= -(\mathfrak{R}'_c + M'_{Cc})\varepsilon'_c \\
M'_d &= -(\mathfrak{R}'_d + M'_{Dd})\varepsilon'_d.
\end{aligned}$$

$$(1342) \text{ and } (1343)$$

Check relationships:

$$\begin{aligned}
\boldsymbol{D}_A + M_{Aa} - M'_{Aa} - M_{A1} &= 0 & \boldsymbol{D}_B + M_{Bb} - M'_{Bb} + M_{B1} - M_{B2} &= 0 \\
\boldsymbol{D}_C + M_{Cc} - M'_{Cc} + M_{C2} - M_{C3} &= 0 & \boldsymbol{D}_D + M_{Dd} - M'_{Dd} + 2M_{D3} &= 0.
\end{aligned} \quad (1344)$$

Horizontal thrusts, column shears and lateral restraining force for all loading conditions

Note: All values without the superscript ′ refer to the lower story; all values provided with the superscript ′ refer to the upper story (cf. the notation employed on p. 261).

The formulas given below are applicable, having due regard to the subscripts n and N, to all the symmetrical two-story frames in this book.

For the asymmetrical Loading Conditions 1 and 3:

$$H_n = +Q_n = M_{Nn}(1+\varepsilon_n)/h_n \quad H'_n = -Q'_n = M'_{Nn}(1+\varepsilon'_n)/h'_n; \quad (1345)$$
$$\text{for} \quad n = a,b,c,d,e,f,g \quad \text{and} \quad N = A,B,C,D,E,F,G.$$

$$H = -(H_a + H_b + \cdots + H_g) - (H'_a + H'_b + \cdots + H'_g). \quad (1346)$$

For the asymmetrical Loading Conditions 2 and 4:

$$T_n = \frac{M_{Nn} - M_n}{h_n}; \quad H_n = Q_{ln} = \frac{\mathfrak{S}_{rn}}{h_n} + T_n \quad Q_{rn} = -\frac{\mathfrak{S}_{ln}}{h_n} + T_n;$$
$$T'_n = \frac{M'_n - M'_{Nn}}{h'_n}; \quad Q'_{ln} = \frac{\mathfrak{S}'_{rn}}{h'_n} + T'_n \quad H'_n = -Q'_{rn} = \frac{\mathfrak{S}'_{ln}}{h'_n} - T'_n; \quad (1347)$$
$$\text{for} \quad n = a,b,c,d,e,f,g \quad \text{and} \quad N = A,B,C,D,E,F,G.$$

$$H = -(Q_{ra} + Q_{rb} + \cdots + Q_{rg}) + (Q'_{la} + Q'_{lb} + \cdots + Q'_{lg}). \quad (1348)$$

For the symmetrical Loading Conditions 1a and 3a:

H_n and H'_n as equation (1345), for $n = a,b,c$ and $N = A,B,C$. (1349)

Lateral restraining force $H = 0$. (1350)

For the symmetrical Loading Conditions 2a and 4a:

T_n to H'_n as equation (1347), for $n = a,b,c$ and $N = A,B,C$. (1351)

Lateral restraining force $H = 0$. (1352)

For the anti-symmetrical Loading Conditions 1b and 3b:

H_n and H'_n as equation (1345)
$$\text{for} \quad n = a,b,c,d \quad \text{and} \quad N = A,B,C,D. \quad (1353)$$

$$H = -2(H_a + H_b + H_c) - H_d - 2(H'_a + H'_b + H'_c) - H'_d. \quad (1354)$$

For the anti-symmetrical Loading Conditions 2b and 4b:

$T_n, H_n, Q_{rn}, T'_n, Q'_{ln}, H'_n$ as equation (1347)
$$\text{for} \quad n = a,b,c,d \quad \text{and} \quad N = A,B,C,D. \quad (1355)$$

$$H = -2(Q_{ra} + Q_{rb} + Q_{rc}) - Q_{rd} + 2(Q'_{la} + Q'_{lb} + Q'_{lc}) + Q'_{ld}. \quad (1356)$$

Frame Shape 41

Symmetrical two-story five-bay frame with laterally restrained horizontal member and six elastically restrained pairs of columns

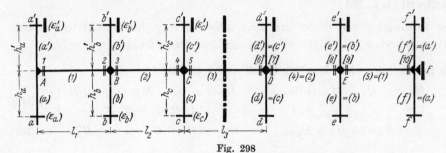

Fig. 298
Frame shape, dimensions and symbols

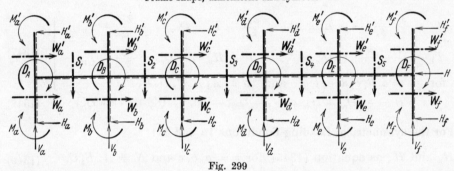

Fig. 299
Definition of positive direction for all external loads on joints and members, for all column reactions, and for the lateral restraining force

> All the dimensions and coefficients for the right-hand half of the frame are the same as those for the left-hand half. In the case of the columns the broken line is always placed on the right-hand side of each column, despite the symmetry of the frame.

Note: For Frame Shape 41 the *coefficients and the Loading Conditions* 1 – 4 for *Frame Shape* 37, enlarged by the addition of two bays, may be employed. On account of symmetry, the forward reduction becomes the same as the backward reduction, and the backward recursion becomes the same as the forward recursion, both in Method I and Method II. Furthermore, the *square matrix of influence coefficients* (with 10 × 10 elements in Method I and 6 × 6 elements in Method II) are bisymmetrical, i.e., symmetrical about the principal diagonal and the secondary diagonal. If the *"load transposition method"* is employed, however, it is also possible to start from the following symmetrical and anti-symmetrical loading conditions.

Important

All the instructions, formula sequences and formula matrices given for Frame Shape 41 can readily be extended to suit symmetrical frames of similar type having any odd number of bays.*

I. Treatment by the Method of Forces
a) Symmetrical load arrangements
Coefficients

Flexibility coefficients, flexibilities of the individual columns and of the column lines, and column distribution coefficients: according to formulas (1216) – (1219)

$$\text{for} \quad v = 1, 2, \vdots 3; \quad n = a, b, \vdots c; \quad N = A, B, \vdots C.$$

Diagonal coefficients **

$$O_1 = s_A + 2k_1 \qquad O_3 = s_B + 2k_2 \quad \vdots \quad O_5' = s_C + 3k_3.$$
$$O_2 = 2k_1 + s_B \quad \vdots \quad O_4 = 2k_2 + s_C \qquad \qquad \qquad \qquad \quad \} \quad (1357)$$

Moment carry-over factors and auxiliary coefficients **

backward reduction: | forward reduction:

$$\sigma_{Cs} = \frac{s_C}{r_5'} \qquad r_5' = O_5'$$
$$\iota_{2s} = \frac{k_2}{r_4'} \qquad r_4' = O_4 - s_C \cdot \sigma_{Cs}$$
$$\sigma_{Bs} = \frac{s_B}{r_3'} \qquad r_3' = O_3 - k_2 \cdot \iota_{2s}$$
$$\iota_{1s} = \frac{k_1}{r_2'} \qquad r_2' = O_2 - s_B \cdot \sigma_{Bs}$$
$$\qquad \qquad r_1' = O_1 - k_1 \cdot \iota_{1s}.$$

$$\iota_1' = \frac{k_1}{v_1} \qquad v_1 = O_1$$
$$\sigma_B' = \frac{s_B}{v_2} \qquad v_2 = O_2 - k_1 \cdot \iota_1'$$
$$\iota_2' = \frac{k_2}{v_3} \qquad v_3 = O_3 - s_B \cdot \sigma_B'$$
$$\sigma_C' = \frac{s_C}{v_4} \qquad v_4 = O_4 - k_2 \cdot \iota_2'$$
$$\qquad \qquad v_5' = O_5' - s_C \cdot \sigma_C'.$$

(1358) and (1359)

Principal influence coefficients by recursion:

forward: | backward:

$$n_{11}' = \frac{1}{r_1'} \qquad\qquad\qquad n_{55}' = \frac{1}{v_5'}$$
$$n_{22}' = \frac{1}{r_2'} + n_{11}' \cdot \iota_{1s}^2 \qquad n_{44}' = \frac{1}{v_4} + n_{55}' \cdot \sigma_C'^2$$
$$n_{33}' = \frac{1}{r_3'} + n_{22}' \cdot \sigma_{Bs}^2 \qquad n_{33}' = \frac{1}{v_3} + n_{44}' \cdot \iota_2'^2$$
$$n_{44}' = \frac{1}{r_4'} + n_{33}' \cdot \iota_{2s}^2 \qquad n_{22}' = \frac{1}{v_2} + n_{33}' \cdot \sigma_B'^2$$
$$n_{55}' = \frac{1}{r_5'} + n_{44}' \cdot \sigma_{Cs}^2. \qquad n_{11}' = \frac{1}{v_1} + n_{22}' \cdot \iota_1'^2.$$

(1360) and (1361)

---

* The dotted lines in the sets of formulas define the scope of these formulas as applicable to the symmetrical three-bay frame. See also Frame Shape 43.
** For Frame Shape 43 the coefficient O_3 is replaced by $O_3' = s_B + 3k_2$.

— 292 —

Principal influence coefficients direct

$$n'_{11} = \frac{1}{r'_1}; \qquad n'_{ii} = \frac{1}{r'_i + v_i - O_i} \quad \text{for} \quad i = 2, 3, 4; \qquad n'_{55} = \frac{1}{v'_5}. \quad (1362)$$

Computation scheme for the symmetrical influence coefficient matrix*

$$\begin{array}{c}
\cdot \iota_{1s} \\
\cdot \sigma_{Bs} \\
\cdot \iota_{2s} \\
\cdot \sigma_{Cs}
\end{array}
\left|
\begin{array}{ccccc}
(n'_{11}) & n'_{12} & n'_{13} & n'_{14} & n'_{15} \\
\downarrow & \uparrow & \uparrow & \uparrow & \uparrow \\
n'_{21} & (n'_{22}) & n'_{23} & n'_{24} & n'_{25} \\
\downarrow & \downarrow & \uparrow & \uparrow & \uparrow \\
n'_{31} & n'_{32} & (n'_{33}) & n'_{34} & n'_{35} \\
\downarrow & \downarrow & \downarrow & \uparrow & \uparrow \\
n'_{41} & n'_{42} & n'_{43} & (n'_{44}) & n'_{45} \\
\downarrow & \downarrow & \downarrow & \downarrow & \uparrow \\
n'_{51} & n'_{52} & n'_{53} & n'_{54} & (n'_{55})
\end{array}
\right|
\begin{array}{c}
\cdot \iota'_1 \\
\cdot \sigma'_B \\
\cdot \iota'_2 \\
\cdot \sigma'_C
\end{array}
\Bigg\} \quad (1363)$$

Composite influence coefficients**

$$\begin{aligned}
s'_{11} &= + n'_{11} - n'_{12} = (s_{11} + s_{15}) & s'_{12} &= + n'_{13} - n'_{14} = (s_{12} + s_{14}) \\
s'_{21} &= - n'_{21} + n'_{22} = (s_{21} - s_{25}) & s'_{22} &= + n'_{23} - n'_{24} = (s_{22} + s_{24}) \\
s'_{31} &= - n'_{31} + n'_{32} = (s_{31} - s_{35}) & s'_{32} &= + n'_{33} - n'_{34} = (s_{32} + s_{34}) \\
s'_{41} &= - n'_{41} + n'_{42} = (s_{41} - s_{45}) & s'_{42} &= - n'_{43} + n'_{44} = (s_{42} - s_{44}) \\
s'_{51} &= - n'_{51} + n'_{52} = (s_{51} - s_{55}) & s'_{52} &= - n'_{53} + n'_{54} = (s_{52} - s_{54});
\end{aligned} \Bigg\} \quad (1364)$$

$$\begin{aligned}
w'_{1B} &= + n'_{12} - n'_{13} = (w_{1B} + w_{1E}) & w'_{1C} &= + n'_{14} - n'_{15} = (w_{1C} + w_{1D}) \\
w'_{2B} &= + n'_{22} - n'_{23} = (w_{2B} + w_{2E}) & w'_{2C} &= + n'_{24} - n'_{25} = (w_{2C} + w_{2D}) \\
w'_{3B} &= - n'_{32} + n'_{33} = (w_{3B} - w_{3E}) & w'_{3C} &= + n'_{34} - n'_{35} = (w_{3C} + w_{3D}) \\
w'_{4B} &= - n'_{42} + n'_{43} = (w_{4B} - w_{4E}) & w'_{4C} &= + n'_{44} - n'_{45} = (w_{4C} + w_{4D}) \\
w'_{5B} &= - n'_{52} + n'_{53} = (w_{5B} - w_{5E}) & w'_{5C} &= - n'_{54} + n'_{55} = (w_{5C} - w_{5D}).
\end{aligned} \Bigg\} \quad (1365)$$

---

* Between the elements of the 5 × 5 matrix (1363) for the left-hand half of the frame and the matrix (1226), enlarged to 10 × 10 elements, there exist, according to the "load transposition method", the following relationships:

$$n'_{11} = n_{11} - n_{1,10}, \quad n'_{12} = n_{12} - n_{19} \text{ etc., up to } n'_{55} = n_{55} - n_{56}.$$

** The expressions in parentheses have been formed with the values s (1233), and the values w (1237), appropriately extended to the case of the five-bay frame.—For Frame Shape 43 the columns s_{i5} and w_{iE} are to be replaced by the columns s_{i3} and w_{iC}.

Loading Condition 1a*

Horizontal member carrying vertical loading of arbitrary magnitude symmetrical about center of frame

$(\mathfrak{R}_5 = \mathfrak{L}_1, \quad \mathfrak{L}_5 = \mathfrak{R}_1; \quad \mathfrak{R}_4 = \mathfrak{L}_2, \quad \mathfrak{L}_4 = \mathfrak{R}_2; \quad \mathfrak{R}_3 = \mathfrak{L}_3)$

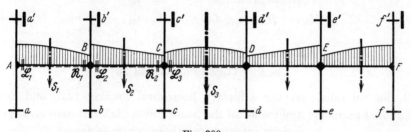

Fig. 300
Beam loading diagram with load terms $\mathfrak{L}$ and $\mathfrak{R}$ for left-hand half of frame

Beam moments $(L_\nu = \mathfrak{L}_\nu k_\nu \text{ and } R_\nu = \mathfrak{R}_\nu k_\nu \text{ for } \nu = 1, 2, 3)$:

	L_1	R_1	L_2	R_2	L_3
$X_1' = M_{A1} = M_{F5} =$	$-n_{11}'$	$+n_{12}'$	$+n_{13}'$	$-n_{14}'$	$-n_{15}'$
$X_2' = M_{B1} = M_{E5} =$	$+n_{21}'$	$-n_{22}'$	$-n_{23}'$	$+n_{24}'$	$+n_{25}'$
$X_3' = M_{B2} = M_{E4} =$	$+n_{31}'$	$-n_{32}'$	$-n_{33}'$	$+n_{34}'$	$+n_{35}'$
$X_4' = M_{C2} = M_{D4} =$	$-n_{41}'$	$+n_{42}'$	$+n_{43}'$	$-n_{44}'$	$-n_{45}'$
$X_5' = M_{C3} = M_{D3} =$	$-n_{51}'$	$+n_{52}'$	$+n_{53}'$	$-n_{54}'$	$-n_{55}'$.

(1366)

Column moments:

Lower story (n),
at top (1367):
$M_{Aa} = -M_{Ff} = \qquad +X_1' \cdot \eta_a$
$M_{Bb} = -M_{Ee} = (-X_2' + X_3') \eta_b$
$M_{Cc} = -M_{Dd} = (-X_4' + X_5') \eta_c;$

at base (1368):
$M_a = -M_f = -M_{Aa} \cdot \varepsilon_a$
$M_b = -M_e = -M_{Bb} \cdot \varepsilon_b$
$M_c = -M_d = -M_{Cc} \cdot \varepsilon_c.$

Upper story (n'),
at base (1369):
$M_{Aa}' = -M_{Ff}' = \qquad -X_1' \cdot \eta_a'$
$M_{Bb}' = -M_{Ee}' = (+X_2' - X_3') \eta_b'$
$M_{Cc}' = -M_{Dd}' = (+X_4' - X_5') \eta_c';$

at top (1370):
$M_a' = -M_f' = -M_{Aa}' \cdot \varepsilon_a'$
$M_b' = -M_e' = -M_{Bb}' \cdot \varepsilon_b'$
$M_c' = -M_d' = -M_{Cc}' \cdot \varepsilon_c'.$

(1367) to (1370)

* For Frame Shape 43 the M subscripts $E, F, e, f, 4, 5$ are to be replaced by the subscripts $C, D, c, d, 2, 3$ in the same sequence.

Special case: symmetrical beam span loading ($R_1 = L_1$ and $R_2 = L_2$)*

$$\left.\begin{aligned}
X'_1 &= M_{A1} = M_{F5} = -L_1 \cdot s'_{11} + L_2 \cdot s'_{12} - L_3 \cdot n'_{15} \\
X'_2 &= M_{B1} = M_{E5} = -L_1 \cdot s'_{21} - L_2 \cdot s'_{22} + L_3 \cdot n'_{25} \\
X'_3 &= M_{B2} = M_{E4} = -L_1 \cdot s'_{31} - L_2 \cdot s'_{32} + L_3 \cdot n'_{35} \\
X'_4 &= M_{C2} = M_{D4} = +L_1 \cdot s'_{41} - L_2 \cdot s'_{42} - L_3 \cdot n'_{45} \\
X'_5 &= M_{C3} = M_{D3} = +L_1 \cdot s'_{51} - L_2 \cdot s'_{52} - L_3 \cdot n'_{55}.
\end{aligned}\right\} \quad (1371)$$

Loading Conditions 2a and 3a**

All the columns carrying arbitrary horizontal loading (2a), and external rotational moments applied at all the beam joints (3a), but load arrangement as a whole is symmetrical about center of frame

$$(\mathfrak{L}_f = -\mathfrak{L}_a \quad \text{to} \quad \mathfrak{R}_d = -\mathfrak{R}_c \quad \text{and} \quad \mathfrak{L}'_f = -\mathfrak{L}'_a \quad \text{to} \quad \mathfrak{R}'_d = -\mathfrak{R}'_c;$$
$$D_F = -D_A, \quad D_E = -D_B, \quad D_D = -D_C)$$

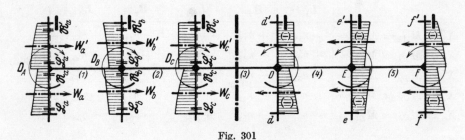

Fig. 301

Column loading diagram with load terms $\mathfrak{L}$ and $\mathfrak{R}$ for the columns, and external rotational moments D, for left-hand half of frame

Column load terms:

$$(L_n = \mathfrak{L}_n k_n, \quad R_n = \mathfrak{R}_n k_n; \quad L'_n = \mathfrak{L}'_n k'_n, \quad R'_n = \mathfrak{R}'_n k'_n):$$
$$\mathfrak{B}_n = R_n - L_n \cdot \varepsilon_n \qquad \mathfrak{B}'_n = L'_n - R'_n \cdot \varepsilon'_n \quad \text{for} \quad n = a, b, |c. \quad (1372)$$

Joint load terms:

$$\left.\begin{aligned}
\mathfrak{B}_N &= D_N \cdot s_N - \mathfrak{B}_n \cdot \eta_n + \mathfrak{B}'_n \cdot \eta'_n \\
&\text{for} \quad N = A, B, |C \quad \text{and} \quad n = a, b, |c.
\end{aligned}\right\} \quad (1373)$$

---

* For Frame Shape 43 the column s'_{i2} is to be replaced by the column n'_{i3}
** With regard to Frame Shape 43 see footnote on previous page.

Beam moments:

	$\mathfrak{B}_A$	$\mathfrak{B}_B$	$\mathfrak{B}_C$
$X'_1 = M_{A1} = M_{F5} =$	$+n'_{11}$	$+w'_{1B}$	$-w'_{1C}$
$X'_2 = M_{B1} = M_{E5} =$	$-n'_{21}$	$-w'_{2B}$	$+w'_{2C}$
$X'_3 = M_{B2} = M_{E4} =$	$-n'_{31}$	$+w'_{3B}$	$+w'_{3C}$
$X'_4 = M_{C2} = M_{D4} =$	$+n'_{41}$	$-w'_{4B}$	$-w'_{4C}$
$X'_5 = M_{C3} = M_{D3} =$	$+n'_{51}$	$-w'_{5B}$	$+w'_{5C}.$

$$\quad (1374)$$

Support moments for the column lines in themselves (auxiliary values):

$$M_N^0 = -\frac{\mathfrak{B}_n + \mathfrak{B}'_n}{b_n + b'_n}; \quad \text{for} \quad N = A, B, C \quad \text{and} \quad n = a, b, c. \quad (1375)$$

Column moments in upper story (n')

at base:
$$\begin{aligned} M'_{Aa} &= -M'_{Ff} = M_A^0 + (\mathbf{D}_A \qquad\quad - X'_1)\eta'_a \\ M'_{Bb} &= -M'_{Ee} = M_B^0 + (\mathbf{D}_B + X'_2 - X'_3)\eta'_b \\ M'_{Cc} &= -M'_{Dd} = M_C^0 + (\mathbf{D}_C + X'_4 - X'_5)\eta'_c; \end{aligned} \quad (1376)$$

at top:
$$\begin{aligned} M'_a &= -M'_f = -(\mathfrak{R}'_a + M'_{Aa})\varepsilon'_a \\ M'_b &= -M'_e = -(\mathfrak{R}'_b + M'_{Bb})\varepsilon'_b \\ M'_c &= -M'_d = -(\mathfrak{R}'_c + M'_{Cc})\varepsilon'_c. \end{aligned} \quad (1377)$$

Column moments in lower story (n)

at top:
$$\begin{aligned} M_{Aa} &= -M_{Ff} = M_A^0 - (\mathbf{D}_A \qquad\quad - X'_1)\eta_a \\ M_{Bb} &= -M_{Ee} = M_B^0 - (\mathbf{D}_B + X'_2 - X'_3)\eta_b \\ M_{Cc} &= -M_{Dd} = M_C^0 - (\mathbf{D}_C + X'_4 - X'_5)\eta_c; \end{aligned} \quad (1378)$$

at base:
$$\begin{aligned} M_a &= -M_f = -(\mathfrak{L}_a + M_{Aa})\varepsilon_a \\ M_b &= -M_e = -(\mathfrak{L}_b + M_{Bb})\varepsilon_b \\ M_c &= -M_d = -(\mathfrak{L}_c + M_{Cc})\varepsilon_c. \end{aligned} \quad (1379)$$

b) Anti-symmetrical load arrangements
Coefficients

Flexibility coefficients, flexibilities of the individual columns and of the column lines, and column distribution coefficients: according to formulas (1216) – (1219)

$$\text{for } v = 1, 2, 3; \quad n = a, b, c; \quad N = A, B, C.$$

Diagonal coefficients*

$$\left.\begin{array}{ll} O_1 = s_A + 2k_1 & O_3 = s_B + 2k_2 \quad O_5'' = s_C + k_3. \\ O_2 = 2k_1 + s_B & O_4 = 2k_2 + s_C \end{array}\right\} \quad (1380)$$

Moment carry-over factors and auxiliary coefficients*

$$\left.\begin{array}{ll} \text{backward reduction:} & \text{forward reduction:} \\[4pt] \begin{array}{l} \sigma_{Ct} = \dfrac{s_C}{r_5''} \\[6pt] \iota_{2t} = \dfrac{k_2}{r_4''} \\[6pt] \sigma_{Bt} = \dfrac{s_B}{r_3''} \\[6pt] \iota_{1t} = \dfrac{k_1}{r_2''} \end{array} \quad \begin{array}{l} r_5'' = O_5'' \\[4pt] r_4'' = O_4 - s_C \cdot \sigma_{Ct} \\[4pt] r_3'' = O_3 - k_2 \cdot \iota_{2t} \\[4pt] r_2'' = O_2 - s_B \cdot \sigma_{Bt} \\[4pt] r_1'' = O_1 - k_1 \cdot \iota_{1t}. \end{array} & \begin{array}{l} \iota_1' = \dfrac{k_1}{v_1} \\[6pt] \sigma_B' = \dfrac{s_B}{v_2} \\[6pt] \iota_2' = \dfrac{k_2}{v_3} \\[6pt] \sigma_C' = \dfrac{s_C}{v_4} \end{array} \quad \begin{array}{l} v_1 = O_1 \\[4pt] v_2 = O_2 - k_1 \cdot \iota_1' \\[4pt] v_3 = O_3 - s_B \cdot \sigma_B' \\[4pt] v_4 = O_4 - k_2 \cdot \iota_2' \\[4pt] v_5'' = O_5'' - s_C \cdot \sigma_C'. \end{array} \end{array}\right\} \begin{array}{c} (1381) \\ \text{and} \\ (1382) \end{array}$$

Principal influence coefficients by recursion:

$$\left.\begin{array}{ll} \text{forward:} & \text{backward:} \\[6pt] n_{11}'' = \dfrac{1}{r_1''} & n_{55}'' = \dfrac{1}{v_5''} \\[8pt] n_{22}'' = \dfrac{1}{r_2''} + n_{11}'' \cdot \iota_{1t}^2 & n_{44}'' = \dfrac{1}{v_4} + n_{55}'' \cdot \sigma_C'^2 \\[8pt] n_{33}'' = \dfrac{1}{r_3''} + n_{22}'' \cdot \sigma_{Bt}^2 & n_{33}'' = \dfrac{1}{v_3} + n_{44}'' \cdot \iota_2'^2 \\[8pt] n_{44}'' = \dfrac{1}{r_4''} + n_{33}'' \cdot \iota_{2t}^2 & n_{22}'' = \dfrac{1}{v_2} + n_{33}'' \cdot \sigma_B'^2 \\[8pt] n_{55}'' = \dfrac{1}{r_5''} + n_{44}'' \cdot \sigma_{Ct}^2. & n_{11}'' = \dfrac{1}{v_1} + n_{22}'' \cdot \iota_1'^2. \end{array}\right\} \begin{array}{c} (1383) \\ \text{and} \\ (1384) \end{array}$$

---

* For Frame Shape 43, O_3 is to be replaced by coefficients $O_3'' = s_B + k_2$.

Principal influence coefficients direct

$$n''_{11} = \frac{1}{r''_1}; \qquad n''_{ii} = \frac{1}{r''_i + v_i - O_i} \quad \text{for} \quad i = 2, 3, 4; \qquad n''_{55} = \frac{1}{v''_5}. \tag{1385}$$

Computation scheme for the symmetrical influence coefficient matrix*

$$\left. \begin{array}{l} \cdot \iota_{1t} \\ \cdot \sigma_{Bt} \\ \cdot \iota_{2t} \\ \cdot \sigma_{Ct} \end{array} \right| \begin{array}{ccccc} (n''_{11}) & n''_{12} & n''_{13} & n''_{14} & n''_{15} \\ \downarrow & \uparrow & \uparrow & \uparrow & \uparrow \\ n''_{21} & (n''_{22}) & n''_{23} & n''_{24} & n''_{25} \\ \downarrow & \downarrow & \uparrow & \uparrow & \uparrow \\ n''_{31} & n''_{32} & (n''_{33}) & n''_{34} & n''_{35} \\ \downarrow & \downarrow & \downarrow & \uparrow & \uparrow \\ n''_{41} & n''_{42} & n''_{43} & (n''_{44}) & n''_{45} \\ \downarrow & \downarrow & \downarrow & \downarrow & \uparrow \\ n''_{51} & n''_{52} & n''_{53} & n''_{54} & (n''_{55}) \end{array} \left| \begin{array}{l} \cdot \iota'_1 \\ \cdot \sigma'_B \\ \cdot \iota'_2 \\ \cdot \sigma'_C \end{array} \right\} \tag{1386}$$

Composite influence coefficients**

$$\begin{aligned} s''_{11} &= +n''_{11} - n''_{12} = (s_{11} - s_{15}) \\ s''_{21} &= -n''_{21} + n''_{22} = (s_{21} + s_{25}) \\ s''_{31} &= -n''_{31} + n''_{32} = (s_{31} + s_{35}) \\ s''_{41} &= -n''_{41} + n''_{42} = (s_{41} + s_{45}) \\ s''_{51} &= -n''_{51} + n''_{52} = (s_{51} + s_{55}) \end{aligned} \quad \begin{aligned} s''_{12} &= +n''_{13} - n''_{14} = (s_{12} - s_{14}) \\ s''_{22} &= +n''_{23} - n''_{24} = (s_{22} - s_{24}) \\ s''_{32} &= +n''_{33} - n''_{34} = (s_{32} - s_{34}) \\ s''_{42} &= -n''_{43} + n''_{44} = (s_{42} + s_{44}) \\ s''_{52} &= -n''_{53} + n''_{54} = (s_{52} + s_{54}); \end{aligned} \right\} \tag{1387}$$

$$\begin{aligned} w''_{1B} &= +n''_{12} - n''_{13} = (w_{1B} - w_{1E}) \\ w''_{2B} &= +n''_{22} - n''_{23} = (w_{2B} - w_{2E}) \\ w''_{3B} &= -n''_{32} + n''_{33} = (w_{3B} + w_{3E}) \\ w''_{4B} &= -n''_{42} + n''_{43} = (w_{4B} + w_{4E}) \\ w''_{5B} &= -n''_{52} + n''_{53} = (w_{5B} + w_{5E}) \end{aligned} \quad \begin{aligned} w''_{1C} &= +n''_{14} - n''_{15} = (w_{1C} - w_{1D}) \\ w''_{2C} &= +n''_{24} - n''_{25} = (w_{2C} - w_{2D}) \\ w''_{3C} &= +n''_{34} - n''_{35} = (w_{3C} - w_{3D}) \\ w''_{4C} &= +n''_{44} - n''_{45} = (w_{4C} - w_{4D}) \\ w''_{5C} &= -n''_{54} + n''_{55} = (w_{5C} + w_{5D}). \end{aligned} \right\} \tag{1388}$$

* Between the elements of the 5 × 5 matrix (1386) for the left-hand half of the frame and the matrix (1226), enlarged to 10 × 10 elements, there exist, according to the "load transposition method", the following relationships:

$$n''_{11} = n_{11} + n_{1,10}, \quad n''_{12} = n_{12} + n_{19} \text{ etc., up to } n''_{55} = n_{55} + n_{56}.$$

** The expressions in parentheses have been formed with the values s (1233), and the values w (1237), appropriately extended to the case of the five-bay frame.—For Frame Shape 43 the columns s_{i5} and w_{iE} are to be replaced by the columns s_{i3} and w_{iC}.

Loading Condition 1b*

Horizontal member carrying vertical loading of arbitrary magnitude anti-symmetrical about center of frame

$(\mathfrak{R}_5 = -\mathfrak{L}_1, \quad \mathfrak{L}_5 = -\mathfrak{R}_1; \quad \mathfrak{R}_4 = -\mathfrak{L}_2, \quad \mathfrak{L}_4 = -\mathfrak{R}_2; \quad \mathfrak{R}_3 = -\mathfrak{L}_3)$

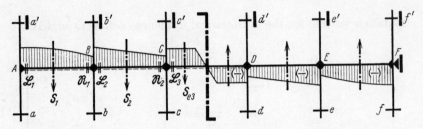

Fig. 302

Beam loading diagram with load terms $\mathfrak{L}$ and $\mathfrak{R}$ for left-hand half of frame

Beam moments $(L_\nu = \mathfrak{L}_\nu k_\nu$ and $R_\nu = \mathfrak{R}_\nu k_\nu$ for $\nu = 1, 2, 3)$:

	L_1	R_1	L_2	R_2	L_3
$X_1'' = M_{A1} = -M_{F5} =$	$-n_{11}''$	$+n_{12}''$	$+n_{13}''$	$-n_{14}''$	$-n_{15}''$
$X_2'' = M_{B1} = -M_{E5} =$	$+n_{21}''$	$-n_{22}''$	$-n_{23}''$	$+n_{24}''$	$+n_{25}''$
$X_3'' = M_{B2} = -M_{E4} =$	$+n_{31}''$	$-n_{32}''$	$-n_{33}''$	$+n_{34}''$	$+n_{35}''$
$X_4'' = M_{C2} = -M_{D4} =$	$-n_{41}''$	$+n_{42}''$	$+n_{43}''$	$-n_{44}''$	$-n_{45}''$
$X_5'' = M_{C3} = -M_{D3} =$	$-n_{51}''$	$+n_{52}''$	$+n_{53}''$	$-n_{54}''$	$-n_{55}''$

(1389)

Column moments:

Lower story (n),

at top (1390):
$M_{Aa} = M_{Ff} = + X_1'' \cdot \eta_a$
$M_{Bb} = M_{Ee} = (-X_2'' + X_3'') \eta_b$
$M_{Cc} = M_{Dd} = (-X_4'' + X_5'') \eta_c;$

at base (1391):
$M_a = M_f = -M_{Aa} \cdot \varepsilon_a$
$M_b = M_e = -M_{Bb} \cdot \varepsilon_b$
$M_c = M_d = -M_{Cc} \cdot \varepsilon_c.$

Upper story (n'),

at base (1392):
$M'_{Aa} = M'_{Ff} = - X_1'' \cdot \eta'_a$
$M'_{Bb} = M'_{Ee} = (+X_2'' - X_3'') \eta'_b$
$M'_{Cc} = M'_{Dd} = (+X_4'' - X_5'') \eta'_c;$

at top (1393):
$M'_a = M'_f = -M'_{Aa} \cdot \varepsilon'_a$
$M'_b = M'_e = -M'_{Bb} \cdot \varepsilon'_b$
$M'_c = M'_d = -M'_{Cc} \cdot \varepsilon'_c.$

(1390) to (1393)

---

* For Frame Shape 43 the M subscripts $E, F, e, f, 4, 5$ are to be replaced by the subscripts $C, D, c, d, 2, 3$ in the same sequence.

Special case: symmetrical beam span loading ($R_1 = L_1$ and $R_2 = L_2$)*:

$$\begin{aligned}
X_1'' &= M_{A1} = -M_{F5} = -L_1 \cdot s_{11}'' + L_2 \cdot s_{12}'' \\
X_2'' &= M_{B1} = -M_{E5} = -L_1 \cdot s_{21}'' - L_2 \cdot s_{22}'' \\
X_3'' &= M_{B2} = -M_{E4} = -L_1 \cdot s_{31}'' - L_2 \cdot s_{32}'' \\
X_4'' &= M_{C2} = -M_{D4} = +L_1 \cdot s_{41}'' - L_2 \cdot s_{42}'' \\
X_5'' &= M_{C3} = -M_{D3} = +L_1 \cdot s_{51}'' - L_2 \cdot s_{52}''.
\end{aligned} \quad (1394)$$

Loading Conditions 2b and 3b *

All the columns carrying arbitrary horizontal loading (2b), and external rotational moments applied at all the beam joints (3b), but load arrangement as a whole is anti-symmetrical about center of frame

$$(\mathfrak{L}_f = \mathfrak{L}_a \text{ to } \mathfrak{R}_d = \mathfrak{R}_c \text{ and } \mathfrak{L}_f' = \mathfrak{L}_a' \text{ to } \mathfrak{R}_d' = \mathfrak{R}_c';$$
$$D_F = D_A, \quad D_E = D_B, \quad D_D = D_C)$$

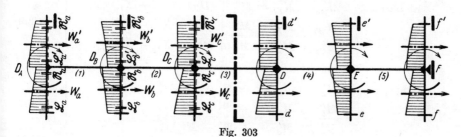

Fig. 303

Column loading diagram with load terms $\mathfrak{L}$ and $\mathfrak{R}$ for the columns, and external rotational moments D, for left-hand half of frame

Column load terms:

$$(L_n = \mathfrak{L}_n k_n \quad R_n = \mathfrak{R}_n k_n \quad L_n' = \mathfrak{L}_n' k_n' \quad R_n' = \mathfrak{R}_n' k_n'):$$
$$\mathfrak{B}_n = R_n - L_n \cdot \varepsilon_n \quad \mathfrak{B}_n' = L_n' - R_n' \cdot \varepsilon_n' \quad \text{for} \quad n = a, b, c. \quad (1395)$$

Joint load terms:

$$\mathfrak{B}_N = D_N \cdot s_N - \mathfrak{B}_n \cdot \eta_n + \mathfrak{B}_n' \cdot \eta_n'$$
$$\text{for} \quad N = A, B, C \quad \text{and} \quad n = a, b, c. \quad (1396)$$

---

* For Frame Shape 43 see footnote on page 298.

Beam moments:

	$\mathfrak{B}_A$	$\mathfrak{B}_B$	$\mathfrak{B}_C$
$X_1'' = M_{A1} = -M_{F5} =$	$+n_{11}''$	$+w_{1B}''$	$-w_{1C}''$
$X_2'' = M_{B1} = -M_{E5} =$	$-n_{21}''$	$-w_{2B}''$	$+w_{2C}''$
$X_3'' = M_{B2} = -M_{E4} =$	$-n_{31}''$	$+w_{3B}''$	$+w_{3C}''$
$X_4'' = M_{C2} = -M_{D4} =$	$+n_{41}''$	$-w_{4B}''$	$-w_{4C}''$
$X_5'' = M_{C3} = -M_{D3} =$	$+n_{51}''$	$-w_{5B}''$	$+w_{5C}''$.

(1397)

Support moments for the column lines in themselves (auxiliary values):

$$M_N^0 = -\frac{\mathfrak{B}_n + \mathfrak{B}_n'}{b_n + b_n'}; \quad \text{for} \quad N = A, B, C \quad \text{and} \quad n = a, b, c. \tag{1398}$$

Column moments in upper story (n')

at base:
$$\begin{aligned} M_{Aa}' &= M_{Ff}' = M_A^0 + (\boldsymbol{D}_A \qquad\quad - X_1'')\,\eta_a' \\ M_{Bb}' &= M_{Ee}' = M_B^0 + (\boldsymbol{D}_B + X_2'' - X_3'')\,\eta_b' \\ M_{Cc}' &= M_{Dd}' = M_C^0 + (\boldsymbol{D}_C + X_4'' - X_5'')\,\eta_c'; \end{aligned} \tag{1399}$$

at top:
$$\begin{aligned} M_a' &= M_f' = -(\mathfrak{R}_a' + M_{Aa}')\,\varepsilon_a' \\ M_b' &= M_e' = -(\mathfrak{R}_b' + M_{Bb}')\,\varepsilon_b' \\ M_c' &= M_d' = -(\mathfrak{R}_c' + M_{Cc}')\,\varepsilon_c'. \end{aligned} \tag{1400}$$

Column moments in lower story (n)

at top:
$$\begin{aligned} M_{Aa} &= M_{Ff} = M_A^0 - (\boldsymbol{D}_A \qquad\quad - X_1'')\,\eta_a \\ M_{Bb} &= M_{Ee} = M_B^0 - (\boldsymbol{D}_B + X_2'' - X_3'')\,\eta_b \\ M_{Cc} &= M_{Dd} = M_C^0 - (\boldsymbol{D}_C + X_4'' - X_5'')\,\eta_c; \end{aligned} \tag{1401}$$

at base:
$$\begin{aligned} M_a &= M_f = -(\mathfrak{L}_a + M_{Aa})\,\varepsilon_a \\ M_b &= M_e = -(\mathfrak{L}_b + M_{Bb})\,\varepsilon_b \\ M_c &= M_d = -(\mathfrak{L}_c + M_{Cc})\,\varepsilon_c. \end{aligned} \tag{1402}$$

II. Treatment by the Deformation Method

a) Coefficients for symmetrical load arrangements

Stiffness coefficients and column stiffnesses: according to formulas (1243) and (1244), for $v = 1, 2, 3$ and $n = a, b, c$.

Joint coefficients*

$$K_B = 2K_1 + 3(S_b + S_b') + 2K_2 \quad \bigg| \quad \begin{aligned} K_A &= 3(S_a + S_a') + 2K_1 \\ K_C' &= 2K_2 + 3(S_c + S_c') + K_3 \end{aligned} \quad \bigg\} \quad (1403)$$

Rotation carry-over factors and auxiliary coefficients*

backward reduction: | forward reduction:

$$\begin{aligned} j_{2s} &= \frac{K_2}{r_c'} \\ j_{1s} &= \frac{K_1}{r_b'} \end{aligned} \quad \begin{aligned} r_c' &= K_C' \\ r_b' &= K_B - K_2 \cdot j_{2s} \\ r_a' &= K_A - K_1 \cdot j_{1s} \end{aligned} \quad \bigg| \quad \begin{aligned} j_1' &= \frac{K_1}{v_a} \\ j_2' &= \frac{K_2}{v_b} \end{aligned} \quad \begin{aligned} v_a &= K_A \\ v_b &= K_B - K_1 \cdot j_1' \\ v_c' &= K_C' - K_2 \cdot j_2' \end{aligned} \quad \bigg\} \quad \begin{aligned} (1404) \\ \text{and} \\ (1405) \end{aligned}$$

Principal influence coefficients by recursion:

forward: | backward:

$$\begin{aligned} u_{aa}' &= 1/r_a' \\ u_{bb}' &= 1/r_b' + u_{aa}' \cdot j_{1s}^2 \\ u_{cc}' &= 1/r_c' + u_{bb}' \cdot j_{2s}^2 \end{aligned} \quad \bigg| \quad \begin{aligned} u_{cc}' &= 1/v_c' \\ u_{bb}' &= 1/v_b + u_{cc}' \cdot j_2'^2 \\ u_{aa}' &= 1/v_a + u_{bb}' \cdot j_1'^2 \end{aligned} \quad \bigg\} \quad \begin{aligned} (1406) \\ \text{and} \\ (1407) \end{aligned}$$

Principal influence coefficients direct

$$u_{aa}' = \frac{1}{r_a'} \qquad u_{bb}' = \frac{1}{r_b' + v_b - K_B} \qquad u_{cc}' = \frac{1}{v_c'} . \qquad (1408)$$

Computation scheme for the symmetrical influence coefficient matrix**

$$\begin{array}{c|ccc|c} \cdot j_{1s} & (u_{aa}') & u_{ab}' & u_{ac}' & \cdot j_1' \\ & \downarrow & \uparrow & \uparrow & \\ \hline & u_{ba}' & (u_{bb}') & u_{bc}' & \\ \cdot j_{2s} & \downarrow & \downarrow & \uparrow & \cdot j_2' \\ & u_{ca}' & u_{cb}' & (u_{cc}') & \end{array} \qquad (1409)$$

---

* For Frame Shape 43 the coefficient K_B is to be replaced by $K_B' = 2K_1 + 3(S_b + S_b') + K_2$.
** Between the elements of the 3 × 3 matrix (1409) for the left-hand half of the frame and those of the matrix (1251) enlarged to 6 × 6 elements, for the whole frame, there exist, according to the "load transposition method", the following relationships:

$$u_{aa}' = u_{aa} + u_{af} \text{ etc., up to } u_{cc}' = u_{cc} + u_{cd}.$$

b) Coefficients for anti-symmetrical load arrangements

Stiffness coefficients and column stiffnesses: according to formulas (1243) and (1244), for $v = 1, 2, 3$ and $n = a, b, c$.

Joint coefficients*

$$K_A = 3(S_a + S'_a) + 2K_1$$
$$K_B = 2K_1 + 3(S_b + S'_b) + 2K_2 \quad K''_C = 2K_2 + 3(S_c + S'_c) + 3K_3. \quad \} \quad (1410)$$

Rotation carry-over factors and auxiliary coefficients*

backward reduction: | forward reduction:

$$j_{2t} = \frac{K_2}{r''_c} \qquad r''_c = K''_C$$
$$\qquad\qquad r''_b = K_B - K_2 \cdot j_{2t}$$
$$j_{1t} = \frac{K_1}{r''_b} \qquad r''_a = K_A - K_1 \cdot j_{1t}.$$

$$j'_1 = \frac{K_1}{v_a} \qquad v_a = K_A$$
$$\qquad\qquad v_b = K_B - K_1 \cdot j'_1$$
$$j'_2 = \frac{K_2}{v_b} \qquad v''_c = K''_C - K_2 \cdot j'_2.$$

(1411) and (1412)

Principal influence coefficients by recursion:

forward: | backward:

$$u''_{aa} = 1/r''_a \qquad\qquad u''_{cc} = 1/v''_c$$
$$u''_{bb} = 1/r''_b + u''_{aa} \cdot j^2_{1t} \qquad u''_{bb} = 1/v_b + u''_{cc} \cdot j'^2_2$$
$$u''_{cc} = 1/r''_c + u''_{bb} \cdot j^2_{2t}. \qquad u''_{aa} = 1/v_a + u''_{bb} \cdot j'^2_1.$$

(1413) and (1414)

Principal influence coefficients direct

$$u''_{aa} = \frac{1}{r''_a} \qquad u''_{bb} = \frac{1}{r''_b + v_b - K_B} \qquad u''_{cc} = \frac{1}{v''_c}. \qquad (1415)$$

Computation scheme for the symmetrical influence coefficient matrix**

$$\begin{array}{c|ccc|c} \cdot j_{1t} & (u''_{aa}) & u''_{ab} & u''_{ac} & \cdot j'_1 \\ & \downarrow & \uparrow & \uparrow & \\ & u''_{ba} & (u''_{bb}) & u''_{bc} & \\ \cdot j_{2t} & \downarrow & \downarrow & \uparrow & \cdot j'_2 \\ & u''_{ca} & u''_{cb} & (u''_{cc}) & \end{array} \qquad (1416)$$

---

* For Frame Shape 43 the coefficient K_B is to be replaced by $K'_B = 2K_1 + 3(S_b + S'_b) + 3K_2$

** Between the elements of the 3 × 3 matrix (1416) for the left-hand half of the frame and those of the matrix (1251) enlarged to 6 × 6 elements, for the whole frame, there exist, according to the "load transposition method", the following relationships:

$$u''_{aa} = u_{aa} - u_{af} \text{ etc., up to } u''_{cc} = u_{cc} - u_{cd}.$$

Loading Condition 4a *

Symmetrical overall loading condition

(Superposition of Loading Conditions 1a, 2a and 3a)

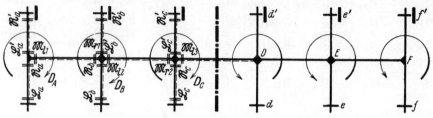

Fig. 304

Loading diagram with fixed end moments $\mathfrak{M}_l$ and $\mathfrak{M}_r$ for the beam spans, load terms $\mathfrak{L}$ and $\mathfrak{R}$ for the elastically restrained columns, and external rotational moments D, for the left-hand half of frame

Column restraining moments at the beam joints:

$$\mathfrak{C}_{rn} = -\frac{\mathfrak{R}_n - \mathfrak{L}_n \cdot \varepsilon_n}{2 - \varepsilon_n}, \qquad \mathfrak{C}'_{ln} = -\frac{\mathfrak{L}'_n - \mathfrak{R}'_n \cdot \varepsilon'_n}{2 - \varepsilon'_n}, \qquad \text{for} \quad n = a, b, c. \qquad (1417)$$

Joint load terms:

$$\left.\begin{aligned}\mathfrak{K}_A &= \phantom{\mathfrak{M}_{r1}} - \mathfrak{M}_{l1} + \mathfrak{C}_{ra} - \mathfrak{C}'_{la} + D_A \\ \mathfrak{K}_B &= \mathfrak{M}_{r1} - \mathfrak{M}_{l2} + \mathfrak{C}_{rb} - \mathfrak{C}'_{lb} + D_B \\ \mathfrak{K}_C &= \mathfrak{M}_{r2} - \mathfrak{M}_{l3} + \mathfrak{C}_{rc} - \mathfrak{C}'_{lc} + D_C .\end{aligned}\right\} \qquad (1418)$$

auxiliary moments:

$$\left.\begin{aligned}Y'_A &= +\mathfrak{K}_A \cdot u'_{aa} - \mathfrak{K}_B \cdot u'_{ab} + \mathfrak{K}_C \cdot u'_{ac} \\ Y'_B &= -\mathfrak{K}_A \cdot u'_{ba} + \mathfrak{K}_B \cdot u'_{bb} - \mathfrak{K}_C \cdot u'_{bc} \\ Y'_C &= +\mathfrak{K}_A \cdot u'_{ca} - \mathfrak{K}_B \cdot u'_{cb} + \mathfrak{K}_C \cdot u'_{cc} .\end{aligned}\right\} \qquad (1419)$$

Beam moments:

$$\left.\begin{aligned}M_{A1} = M_{F5} &= \mathfrak{M}_{l1} + (2Y'_A + Y'_B) K_1 \\ M_{B2} = M_{E4} &= \mathfrak{M}_{l2} + (2Y'_B + Y'_C) K_2 \, **) \\ M_{C3} = M_{D3} &= \mathfrak{M}_{l3} + Y'_C \cdot K_3 ; \\ M_{B1} = M_{E5} &= \mathfrak{M}_{r1} - (Y'_A + 2Y'_B) K_1 \\ M_{C2} = M_{D4} &= \mathfrak{M}_{r2} - (Y'_B + 2Y'_C) K_2 .\end{aligned}\right\} \qquad (1420)$$

---

* For Frame Shape 43 the M subscripts $E, F, e, f, 4, 5$ are to be replaced by the subscripts $C, D, c, d, 2, 3$ in the same sequence.

** For Frame Shape 43 we here have $M_{B2} = M_{C2} = \mathfrak{M}_{l2} + Y'_B \cdot K_2$.

Column moments *

Lower story (n):

$$M_{Aa} = -M_{Ff} = \mathfrak{C}_{ra} - Y'_A \cdot 3S_a$$
$$M_{Bb} = -M_{Ee} = \mathfrak{C}_{rb} - Y'_B \cdot 3S_b$$
$$\boxed{M_{Cc} = -M_{Dd} = \mathfrak{C}_{rc} - Y'_C \cdot 3S_c;}$$
$$M_a = -M_f = -(\mathfrak{L}_a + M_{Aa})\varepsilon_a$$
$$M_b = -M_e = -(\mathfrak{L}_b + M_{Bb})\varepsilon_b$$
$$M_c = -M_d = -(\mathfrak{L}_c + M_{Cc})\varepsilon_c.$$

Upper story (n'):

$$M'_{Aa} = -M'_{Ff} = \mathfrak{C}'_{la} + Y'_A \cdot 3S'_a$$
$$M'_{Bb} = -M'_{Ee} = \mathfrak{C}'_{lb} + Y'_B \cdot 3S'_b$$
$$\boxed{M'_{Cc} = -M'_{Dd} = \mathfrak{C}'_{lc} + Y'_C \cdot 3S'_c;}$$
$$M'_a = -M'_f = -(\mathfrak{R}'_a + M'_{Aa})\varepsilon'_a$$
$$M'_b = -M'_e = -(\mathfrak{R}'_b + M'_{Bb})\varepsilon'_b$$
$$M'_c = -M'_d = -(\mathfrak{R}'_c + M'_{Cc})\varepsilon'_c.$$

(1421) and (1422)

Check relationships:

$$D_A + M_{Aa} - M'_{Aa} - M_{A1} = 0 \quad \begin{array}{l} D_B + M_{Bb} - M'_{Bb} + M_{B1} - M_{B2} = 0 \\ \overline{D_C + M_{Cc} - M'_{Cc} + M_{C2} - M_{C3} = 0.} \end{array} \Bigg\} \quad (1423)$$

Loading Condition 4b*

Anti-symmetrical loading condition

(Superposition of Loading Conditions 1b, 2b and 3b)

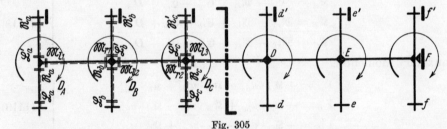

Fig. 305

Loading diagram with fixed-end moments $\mathfrak{M}_l$ and $\mathfrak{M}_r$ for the beam spans, load terms $\mathfrak{L}$ and $\mathfrak{R}$ for the columns, and external rotational moments D, for left-hand half of frame.

Column restraining moments and joint load terms: exactly as formulas (1417) and (1418).

auxiliary moments:

$$\begin{aligned} Y''_A &= +\mathfrak{K}_A \cdot u''_{aa} - \mathfrak{K}_B \cdot u''_{ab} + \mathfrak{K}_C \cdot u''_{ac} \\ Y''_B &= -\mathfrak{K}_A \cdot u''_{ba} + \mathfrak{K}_B \cdot u''_{bb} - \mathfrak{K}_C \cdot u''_{bc} \\ Y''_C &= +\mathfrak{K}_A \cdot u''_{ca} - \mathfrak{K}_B \cdot u''_{cb} + \mathfrak{K}_C \cdot u''_{cc}. \end{aligned} \Bigg\} \quad (1424)$$

---

* With reference to Frame Shape 43 see footnote * on previous page.

Beam moments:

$$\left.\begin{array}{l} M_{A1} = -M_{F5} = \mathfrak{M}_{l1} + (2Y_A'' + Y_B'')K_1 \\ \underline{M_{B2} = -M_{E4} = \mathfrak{M}_{l2} + (2Y_B'' + Y_C'')K_2} \ ^{*} \\ M_{C3} = -M_{D3} = \mathfrak{M}_{l3} + Y_C'' \cdot 3K_3; \\ \qquad \underline{M_{B1} = -M_{E5} = \mathfrak{M}_{r1} - (Y_A'' + 2Y_B'')K_1} \\ \qquad M_{C2} = -M_{D4} = \mathfrak{M}_{r2} - (Y_B'' + 2Y_C'')K_2. \end{array}\right\} \quad (1425)$$

Column moments:

$$\left.\begin{array}{l|l} \text{Lower story } (n): & \text{Upper story } (n'): \\ M_{Aa} = M_{Ff} = \mathfrak{E}_{ra} - Y_A'' \cdot 3S_a & M_{Aa}' = M_{Ff}' = \mathfrak{E}_{la}' + Y_A'' \cdot 3S_a' \\ M_{Bb} = M_{Ee} = \mathfrak{E}_{rb} - Y_B'' \cdot 3S_b & M_{Bb}' = M_{Ee}' = \mathfrak{E}_{lb}' + Y_B'' \cdot 3S_b' \\ \underline{M_{Cc} = M_{Dd} = \mathfrak{E}_{rc} - Y_C'' \cdot 3S_c;} & \underline{M_{Cc}' = M_{Dd}' = \mathfrak{E}_{lc}' + Y_C'' \cdot 3S_c';} \\ M_a = M_f = -(\mathfrak{L}_a + M_{Aa})\varepsilon_a & M_a' = M_f' = -(\mathfrak{R}_a' + M_{Aa}')\varepsilon_a' \\ M_b = M_e = -(\mathfrak{L}_b + M_{Bb})\varepsilon_b & M_b' = M_e' = -(\mathfrak{R}_b' + M_{Bb}')\varepsilon_b' \\ M_c = M_d = -(\mathfrak{L}_c + M_{Cc})\varepsilon_c. & M_c' = M_d' = -(\mathfrak{R}_c' + M_{Cc}')\varepsilon_c'. \end{array}\right\} \begin{array}{c} (1426) \\ \text{and} \\ (1427) \end{array}$$

Check relationships: exactly as formulas (1423).

Horizontal thrusts, column shears and lateral restraining force for all loading conditions

For asymmetrical loading conditions: according to formulas (1345) and (1347) respectively, for $n = a, b, c, d, e, f$ and $N = A, B, C, D, E, F$.

$$\left.\begin{array}{l} H = -(H_a + H_b + \cdots + H_f) - (H_a' + H_b' + \cdots + H_f'); \\ \text{or} \quad H = -(Q_{ra} + Q_{rb} + \cdots + Q_{rf}) + (Q_{la}' + Q_{lb}' + \cdots + Q_{lf}'). \end{array}\right\} \quad (1428)$$

For symmetrical loading conditions: according to formulas (1349)/(1350) and (1351)/(1352) respectively, for $n = a, b, c$ and $N = A, B, C$.

For anti-symmetrical loading conditions: according to formulas (1353) and (1355) respectively, for $n = a, b, c$ and $N = A, B, C$.

$$\left.\begin{array}{l} H = -2(H_a + H_b + H_c) - 2(H_a' + H_b' + H_c'); \\ H = -2(Q_{ra} + Q_{rb} + Q_{rc}) + 2(Q_{la}' + Q_{lb}' + Q_{lc}'). \end{array}\right\} \quad (1429)$$

---

* For Frame Shape 43 we here have $M_{B2} = -M_{C2} = \mathfrak{M}_{l2} + Y_B'' \cdot 3K_2$.

Frame Shape 42

Symmetrical two-story four-bay frame with laterally restrained horizontal member and five elastically restrained pairs of columns

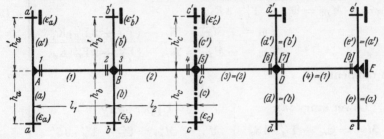

Fig. 306   Frame shape, dimensions and symbols

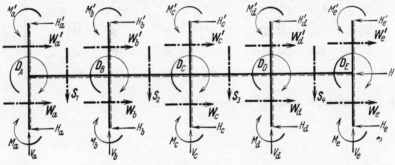

Fig. 307

Definition of positive direction for all external loads on joints and members, for all column reactions, and for the lateral restraining force

> All the dimensions and coefficients for the right-hand half of the frame are the same as those for the left-hand half. In the case of the columns the broken line is always placed on the right-hand side of each column, despite the symmetry of the frame.

Note: Frame Shape 42 can be analysed with the formulas given for *Frame Shape* 36 (using Method II) or, alternatively, it can be analysed with the formulas for *Frame Shape* 38 (using Method I) appropriately enlarged by the addition of two spans. On account of symmetry, the forward reduction becomes the same as the backward reduction, and the backward recursion becomes the same as the forward recursion, both in Method I and Method II. Furthermore, the square matrices of influence coefficients become bisymmetrical, i.e., symmetrical about the principal diagonal and the secondary diagonal. If the "load transposition method" is used, however, it is also possible to start from the following symmetrical and anti-symmetrical loading conditions, which in the present case are given only for the Method of Forces (Method I)*.

---

* For the treatment of a similar frame by the Deformation Method (Method II) see Frame Shape 40.

> **Important**
> All the dimensions and coefficients for the right-hand half of the frame are the same as those for the left-hand half. In the case of the columns the broken line is always placed on the right-hand side of each column, despite the symmetry of the frame.

a) Coefficients for symmetrical load arrangements

Flexibility coefficients, flexibilities of the individual columns and of the column lines, and column distribution coefficients: according to formulas (1216) – (1219)

$$\text{for} \quad v = 1, 2, \quad n = a, b, \quad N = A, B,$$

Diagonal coefficients:

$$O_1 = s_A + 2k_1 \qquad O_2 = 2k_1 + s_B \qquad O_3 = s_B + 2k_2 \qquad (O_4 = 2k_2). \tag{1430}$$

Moment carry-over factors and auxiliary coefficients:

backward reduction:

$$\iota_{2s} = \frac{1}{2}$$
$$\sigma_{Bs} = \frac{s_B}{r'_3} \qquad r'_3 = O_3 - k_2 \cdot \frac{1}{2}$$
$$\iota_{1s} = \frac{k_1}{r'_2} \qquad r'_2 = O_2 - s_B \cdot \sigma_{Bs}$$
$$\qquad r'_1 = O_1 - k_1 \cdot \iota_{1s}.$$

forward reduction:

$$\iota'_1 = \frac{k_1}{v_1} \qquad v_1 = O_1$$
$$\sigma'_B = \frac{s_B}{v_2} \qquad v_2 = O_2 - k_1 \cdot \iota'_1$$
$$\iota'_2 = \frac{k_2}{v_3}. \qquad v_3 = O_3 - s_B \cdot \sigma'_B$$

$$\tag{1431}$$
and
$$\tag{1432}$$

Principal influence coefficients by recursion:

forward:

$$n'_{11} = 1/r'_1$$
$$n'_{22} = 1/r'_2 + n'_{11} \cdot \iota_{1s}^2$$
$$n'_{33} = 1/r'_3 + n'_{22} \cdot \sigma_{Bs}^2$$
$$n'_{44} = 1/2 k_2 + n'_{33} \cdot 1/4.$$

backward:

$$n'_{44} = 1/k_2(2 - \iota'_2)$$
$$n'_{33} = 1/v_3 + n'_{44} \cdot \iota'^2_2$$
$$n'_{22} = 1/v_2 + n'_{33} \cdot \sigma'^2_B$$
$$n'_{11} = 1/v_1 + n'_{22} \cdot \iota'^2_1.$$

$$\tag{1433}$$
and
$$\tag{1434}$$

Computation scheme for the symmetrical influence coefficient matrix*

$$\begin{array}{r|cccc|l}
 & (n'_{11}) & n'_{12} & n'_{13} & n'_{14} & \\
\cdot \iota_{1s} & \downarrow & \uparrow & \uparrow & \uparrow & \cdot \iota'_1 \\
 & n'_{2i} & (n'_{22}) & n'_{23} & n'_{24} & \\
\cdot \sigma_{Bs} & \downarrow & \downarrow & \uparrow & \uparrow & \cdot \sigma'_B \\
 & n'_{31} & n'_{32} & (n'_{33}) & n'_{34} & \\
\cdot \frac{1}{2} & \downarrow & \downarrow & \downarrow & \uparrow & \cdot \iota'_2 \\
 & n'_{41} & n'_{42} & n'_{43} & (n'_{44}) &
\end{array} \tag{1435}$$

---

*Between the elements of the 4 × 4 matrix (1435) for the left-hand half of the frame and the matrix (1278), enlarged to 8 × 8 elements, there exist, according to the "load transposition method", the following relationships:

$$n'_{11} = n_{11} + n_{18}, \quad n'_{12} = n_{12} + n_{17} \text{ etc., up to } n'_{44} = n_{44} + n_{45}.$$

Principal influence coefficients direct

$$n'_{11} = \frac{1}{r'_1}; \qquad n'_{ii} = \frac{1}{r'_i + v_i - O_i} \quad \text{for} \quad i = 2, 3; \quad n'_{44} = \frac{1}{k_2(2 - \iota'_2)}. \tag{1436}$$

Composite influence coefficients*

$$\begin{aligned}
s'_{11} &= +n'_{11} - n'_{12} = (s_{11} - s_{14}) & s'_{12} &= & n'_{14} &= (s_{12} - s_{13}) \\
 & & s'_{22} &= & n'_{24} &= (s_{22} - s_{23}) \\
s'_{21} &= -n'_{21} + n'_{22} = (s_{21} + s_{24}) & s'_{32} &= & n'_{34} &= (s_{32} - s_{33}) \\
s'_{31} &= -n'_{31} + n'_{32} = (s_{31} + s_{34}) \\
s'_{41} &= -n'_{41} + n'_{42} = (s_{41} + s_{44}) & s'_{42} &= -n'_{43} + n'_{44} = (s_{42} + s_{43});
\end{aligned} \tag{1437}$$

$$\begin{aligned}
w'_{1B} &= +n'_{12} - n'_{13} = (w_{1B} - w_{1D}) \\
w'_{2B} &= +n'_{22} - n'_{23} = (w_{2B} - w_{2D}) \\
w'_{3B} &= -n'_{32} + n'_{33} = (w_{3B} + w_{3D}) \\
w'_{4B} &= -n'_{42} + n'_{43} = (w_{4B} + w_{4D}).
\end{aligned} \tag{1438}$$

b) Coefficients for anti-symmetrical load arrangements

Flexibility members: same as under (a), page 307, for $v = 1, 2$; $n = a, b, c$ and $N = A, B, C$.

Diagonal coefficients:

$$O_1 = s_A + 2k_1 \quad O_2 = 2k_1 + s_B \quad O_3 = s_B + 2k_2 \quad O''_4 = 2k_2 + 2s_C. \tag{1439}$$

Moment carry-over factors and auxiliary coefficients:

backward reduction: | forward reduction:

$$\begin{aligned}
\iota_{2t} &= \frac{k_2}{r''_4} & r''_4 &= O''_4 & \iota'_1 &= \frac{k_1}{v_1} & v_1 &= O_1 \\
 & & r''_3 &= O_3 - k_2 \cdot \iota_{2t} & & & v_2 &= O_2 - k_1 \cdot \iota'_1 \\
\sigma_{Bt} &= \frac{s_B}{r''_3} & & & \sigma'_B &= \frac{s_B}{v_2} & & \\
 & & r''_2 &= O_2 - s_B \cdot \sigma_{Bt} & & & v_3 &= O_3 - s_B \cdot \sigma'_B \\
\iota_{1t} &= \frac{k_1}{r''_2} & & & \iota'_2 &= \frac{k_2}{v_3} & & \\
 & & r''_1 &= O_1 - k_1 \cdot \iota_{1t}. & & & v''_4 &= O''_4 - k_2 \cdot \iota'_2.
\end{aligned} \tag{1440} \text{ and } (1441)$$

Principal influence coefficients direct

$$n''_{11} = \frac{1}{r''_1} \quad n''_{22} = \frac{1}{r''_2 + v_2 - O_2} \quad n''_{33} = \frac{1}{r''_3 + v_3 - O_3} \quad n''_{44} = \frac{1}{v''_4}. \tag{1442}$$

---

* The expressions in parentheses have been formed with the values s and w (1279), appropriately extended to the case of the four-bay frame.

Principal influence coefficients by recursion:

forward:
$$n''_{11} = \frac{1}{r''_1}$$
$$n''_{22} = \frac{1}{r''_2} + n''_{11} \cdot \iota^2_{1t}$$
$$n''_{33} = \frac{1}{r''_3} + n''_{22} \cdot \sigma^2_{Bt}$$
$$n''_{44} = \frac{1}{r''_4} + n''_{33} \cdot \iota^2_{2t}.$$

backward:
$$n''_{44} = \frac{1}{v'_4}$$
$$n''_{33} = \frac{1}{v_3} + n''_{44} \cdot \iota'^2_2$$
$$n''_{22} = \frac{1}{v_2} + n''_{33} \cdot \sigma'^2_B$$
$$n''_{11} = \frac{1}{v_1} + n''_{22} \cdot \iota'^2_1.$$

(1443) and (1444)

Computation scheme for the symmetrical influence coefficient matrix*

$$
\begin{array}{c|cccc|c}
 & (n''_{11}) & n''_{12} & n''_{13} & n''_{14} & \\
\cdot\,\iota_{1t} & \downarrow & \uparrow & \uparrow & \uparrow & \cdot\,\iota'_1 \\
 & n''_{21} & (n''_{22}) & n''_{23} & n''_{24} & \\
\cdot\,\sigma_{Bt} & \downarrow & \downarrow & \uparrow & \uparrow & \cdot\,\sigma'_B \\
 & n''_{31} & n''_{32} & (n''_{33}) & n''_{34} & \\
\cdot\,\iota_{2t} & \downarrow & \downarrow & \downarrow & \uparrow & \cdot\,\iota'_2 \\
 & n''_{41} & n''_{42} & n''_{43} & (n''_{44}) & \\
\end{array}
$$
(1445)

Composite influence coefficients**

$$s''_{11} = +n''_{11} - n''_{12} = (s_{11} + s_{14})$$
$$s''_{21} = -n''_{21} + n''_{22} = (s_{21} - s_{24})$$
$$s''_{31} = -n''_{31} + n''_{32} = (s_{31} - s_{34})$$
$$s''_{41} = -n''_{41} + n''_{42} = (s_{41} - s_{44})$$

$$s''_{12} = +n''_{13} - n''_{14} = (s_{12} + s_{13})$$
$$s''_{22} = +n''_{23} - n''_{24} = (s_{22} + s_{23})$$
$$s''_{32} = +n''_{33} - n''_{34} = (s_{32} + s_{33})$$
$$s''_{42} = -n''_{43} + n''_{44} = (s_{42} - s_{43});$$

(1446)

$$w'_{1B} = +n''_{12} - n''_{13} = (w_{1B} + w_{1D})$$
$$w'_{2B} = +n''_{22} - n''_{23} = (w_{2B} + w_{2D})$$
$$w'_{3B} = -n''_{32} + n''_{33} = (w_{3B} - w_{3D})$$
$$w'_{4B} = -n''_{42} + n''_{43} = (w_{4B} - w_{4D}).$$

(1447)

---

* Between the elements of the 4 × 4 matrix (1445) for the left-hand half of the frame and the matrix enlarged to 8 × 8 elements, there exist, according to the "load transposition method", the following relationships:

$$n''_{11} = n_{11} - n_{18}, \quad n''_{12} = n_{12} - n_{17} \text{ etc., up to } n''_{44} = n_{44} - n_{45}.$$

** The expressions in parentheses have been formed with the values s and w (1279), appropriately extended to the case of the four-bay frame.

Loading Condition 1a

Horizontal member carrying vertical loading of arbitrary magnitude symmetrical about center of frame

$$(\mathfrak{R}_4 = \mathfrak{L}_1, \quad \mathfrak{L}_4 = \mathfrak{R}_1; \quad \mathfrak{R}_3 = \mathfrak{L}_2, \quad \mathfrak{L}_3 = \mathfrak{R}_2)$$

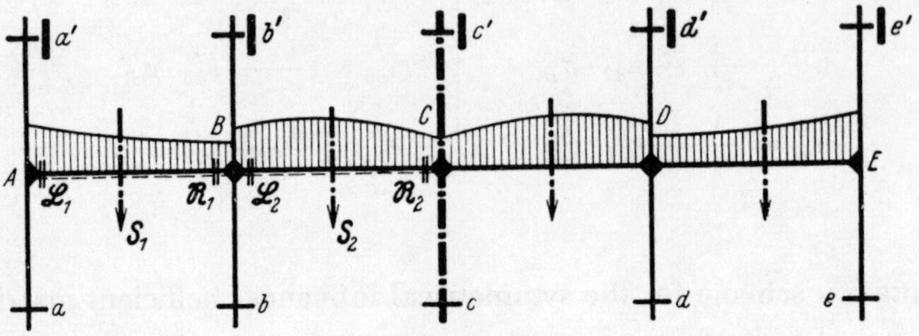

Fig. 308

Beam loading diagram with load terms $\mathfrak{L}$ and $\mathfrak{R}$ for left-hand half of frame

Beam moments ($L_1 = \mathfrak{L}_1 k_1, \quad R_1 = \mathfrak{R}_1 k_1; \quad L_2 = \mathfrak{L}_2 k_2, \quad R_2 = \mathfrak{R}_2 k_2$):

	L_1	R_1	L_2	R_2
$X'_1 = M_{A1} = M_{E4} =$	$-n'_{11}$	$+n'_{12}$	$+n'_{13}$	$-n'_{14}$
$X'_2 = M_{B1} = M_{D4} =$	$+n'_{21}$	$-n'_{22}$	$-n'_{23}$	$+n'_{24}$
$X'_3 = M_{B2} = M_{D3} =$	$+n'_{31}$	$-n'_{32}$	$-n'_{33}$	$+n'_{34}$
$X'_4 = M_{C2} = M_{C3} =$	$-n'_{41}$	$+n'_{42}$	$+n'_{43}$	$-n'_{44}.$

(1448)

Column moments:

Lower story (n),

at top (1449):
$$M_{Aa} = -M_{Ee} = \quad\quad + X'_1 \cdot \eta_a$$
$$M_{Bb} = -M_{Dd} = (-X'_2 + X'_3) \eta_b$$
$$M_{Cc} = 0;$$

at base (1450):
$$M_a = -M_e = -M_{Aa} \cdot \varepsilon_a$$
$$M_b = -M_d = -M_{Bb} \cdot \varepsilon_b$$
$$M_c = 0.$$

Upper story (n'),

at base (1451):
$$M'_{Aa} = -M'_{Ee} = \quad\quad - X'_1 \cdot \eta'_a$$
$$M'_{Bb} = -M'_{Dd} = (+X'_2 - X'_3) \eta'_b$$
$$M'_{Cc} = 0;$$

at top (1452):
$$M'_a = -M'_e = -M'_{Aa} \cdot \varepsilon'_a$$
$$M'_b = -M'_d = -M'_{Bb} \cdot \varepsilon'_b$$
$$M'_c = 0.$$

(1449) to (1452)

Special case: see next page.

Special case: symmetrical beam span loading ($R_1 = L_1$ and $R_2 = L_2$):

$$\left.\begin{aligned}
X'_1 &= M_{A1} = M_{E4} = -L_1 \cdot s'_{11} \dotplus L_2 \cdot s'_{12} \\
X'_2 &= M_{B1} = M_{D4} = -L_1 \cdot s'_{21} - L_2 \cdot s'_{22} \\
X'_3 &= M_{B2} = M_{D3} = -L_1 \cdot s'_{31} - L_2 \cdot s'_{32} \\
X'_4 &= M_{C2} = M_{C3} = +L_1 \cdot s'_{41} - L_2 \cdot s'_{42}
\end{aligned}\right\} \quad (1453)$$

Loading Conditions 2a and 3a

All the columns carrying arbitrary horizontal loading (2a), and external rotational moments applied at all the beam joints (3a), but load arrangement as a whole is symmetrical about center of frame

$$(\mathfrak{L}_e = -\mathfrak{L}_a \text{ to } \mathfrak{R}_d = -\mathfrak{R}_b \text{ and } \mathfrak{L}'_e = -\mathfrak{L}'_a \text{ to } \mathfrak{R}'_d = -\mathfrak{R}'_b;$$
$$D_E = -D_A, \quad D_D = -D_B)$$

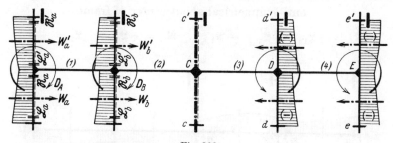

Fig. 309
Column loading diagram with load terms $\mathfrak{L}$ and $\mathfrak{R}$ for the columns, and external rotational moments D, for left-hand half of frame. (Symmetrical loading of central column c-C-c' not shown.)

Column and joint load terms, and support moments for the column lines (auxiliary values): according to formulas (1234), (1235) and (1238), for $n = a, b$ and $N = A, B$.

Beam moments:

$$\left.\begin{aligned}
X'_1 &= M_{A1} = M_{E4} = +\mathfrak{B}_A \cdot n'_{11} + \mathfrak{B}_B \cdot w'_{1B} \\
X'_2 &= M_{B1} = M_{D4} = -\mathfrak{B}_A \cdot n'_{21} - \mathfrak{B}_B \cdot w'_{2B} \\
X'_3 &= M_{B2} = M_{D3} = -\mathfrak{B}_A \cdot n'_{31} + \mathfrak{B}_B \cdot w'_{3B} \\
X'_4 &= M_{C2} = M_{C3} = +\mathfrak{B}_A \cdot n'_{41} - \mathfrak{B}_B \cdot w'_{4B} = -X'_3/2
\end{aligned}\right\} \quad (1454)$$

Column moments in upper story (n')

at base:
$$M'_{Aa} = -M'_{Ee} = M_A^0 + (\boldsymbol{D}_A \qquad - X'_1)\eta'_a$$
$$M'_{Bb} = -M'_{Dd} = M_B^0 + (\boldsymbol{D}_B + X'_2 - X'_3)\eta'_b \qquad M'_{Cc} = 0; \quad (1455)$$

at top:
$$M'_a = -M'_e = -(\mathfrak{R}'_a + M'_{Aa})\varepsilon'_a$$
$$M'_b = -M'_d = -(\mathfrak{R}'_b + M'_{Bb})\varepsilon'_b \qquad M'_c = 0. \quad (1456)$$

Column moments in lower story (n)

at top:
$$M_{Aa} = -M_{Ee} = M_A^0 - (\boldsymbol{D}_A \qquad - X'_1)\eta_a$$
$$M_{Bb} = -M_{Dd} = M_B^0 - (\boldsymbol{D}_B + X'_2 - X'_3)\eta_b \qquad M_{Cc} = 0; \quad (1457)$$

at base:
$$M_a = -M'_e = -(\mathfrak{L}_a + M_{Aa})\varepsilon_a$$
$$M_b = -M'_d = -(\mathfrak{L}_b + M_{Bb})\varepsilon_b \qquad M_c = 0. \quad (1458)$$

Loading Condition 1b

Horizontal member carrying vertical loading of arbitrary magnitude anti-symmetrical about center of frame

$$(\mathfrak{R}_4 = -\mathfrak{L}_1, \quad \mathfrak{L}_4 = -\mathfrak{R}_1; \quad \mathfrak{R}_3 = -\mathfrak{L}_2, \quad \mathfrak{L}_3 = -\mathfrak{R}_2)$$

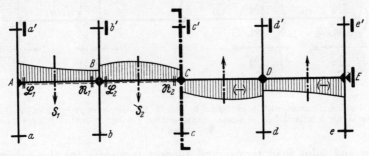

Fig. 310
Beam loading diagram with load terms $\mathfrak{L}$ and $\mathfrak{R}$ for left-hand half of frame

Beam moments $(L_1 = \mathfrak{L}_1 k_1, \quad R_1 = \mathfrak{R}_1 k_1; \quad L_2 = \mathfrak{L}_2 k_2, \quad R_2 = \mathfrak{R}_2 k_2)$:

	L_1	R_1	L_2	R_2
$X''_1 = M_{A1} = -M_{E4} =$	$-n''_{11}$	$+n''_{12}$	$+n''_{13}$	$-n''_{14}$
$X''_2 = M_{B1} = -M_{D4} =$	$+n''_{21}$	$-n''_{22}$	$-n''_{23}$	$+n''_{24}$
$X''_3 = M_{B2} = -M_{D3} =$	$+n''_{31}$	$-n''_{32}$	$-n''_{33}$	$+n''_{34}$
$X''_4 = M_{C2} = -M_{C3} =$	$-n''_{41}$	$+n''_{42}$	$+n''_{43}$	$-n''_{44}$.

(1459)

Column moments:

Lower story (n),

at top (1460):
$$M_{Aa} = M_{Ee} = \qquad + X_1'' \cdot \eta_a$$
$$M_{Bb} = M_{Dd} = (-X_2'' + X_3'')\eta_b$$
$$M_{Cc} = -2X_4'' \cdot \eta_c;$$

at base (1461):
$$M_a = M_e = -M_{Aa} \cdot \varepsilon_a$$
$$M_b = M_d = -M_{Bb} \cdot \varepsilon_b$$
$$M_c = -M_{Cc} \cdot \varepsilon_c.$$

Upper story (n'),

at base (1462):
$$M'_{Aa} = M'_{Ee} = \qquad - X_1'' \cdot \eta'_a$$
$$M'_{Bb} = M'_{Dd} = (+X_2'' - X_3'')\eta'_b$$
$$M'_{Cc} = +2X_4'' \cdot \eta'_c;$$

at top (1463):
$$M'_a = M'_e = -M'_{Aa} \cdot \varepsilon'_a$$
$$M'_b = M'_d = -M'_{Bb} \cdot \varepsilon'_b$$
$$M'_c = -M'_{Cc} \cdot \varepsilon'_c.$$

(1460) to (1463)

Special case: symmetrical beam span loading ($R_1 = L_1$ and $R_2 = L_2$):

$$X_1'' = M_{A1} = -M_{E4} = -L_1 \cdot s_{11}'' + L_2 \cdot s_{12}''$$
$$X_2'' = M_{B1} = -M_{D4} = -L_1 \cdot s_{21}'' - L_2 \cdot s_{22}''$$
$$X_3'' = M_{B2} = -M_{D3} = -L_1 \cdot s_{31}'' - L_2 \cdot s_{32}''$$
$$X_4'' = M_{C2} = -M_{C3} = +L_1 \cdot s_{41}'' - L_2 \cdot s_{42}''.$$

(1464)

Loading Conditions 2b and 3b

All the columns carrying arbitrary horizontal loading (2b), and external rotational moments applied at all the beam joints (3b), but load arrangement as a whole is anti-symmetrical about center of frame

($\mathfrak{L}_e = \mathfrak{L}_a$ to $\mathfrak{R}_d = \mathfrak{R}_b$ and $\mathfrak{L}'_e = \mathfrak{L}'_a$ to $\mathfrak{R}'_d = \mathfrak{R}'_b$; $D_E = D_A$, $D_D = D_B$)

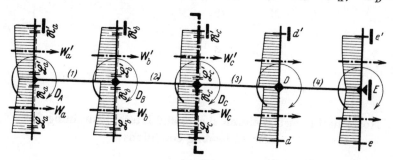

Fig. 311

Column loading diagram with load terms $\mathfrak{L}$ and $\mathfrak{R}$ for the columns, and external rotational moments D, for left-hand half of frame, including central column c-C-c'

Column and joint load terms, and support moments for the column lines (auxiliary values): according to formulas (1234), (1235) and (1238), for $n = a, b, c$ and $N = A, B, C$.

Beam moments ($\mathfrak{B}''_C = \mathfrak{B}_C/2$):

$$\left. \begin{aligned} X''_1 &= M_{A1} = -M_{E4} = +\mathfrak{B}_A \cdot n''_{11} + \mathfrak{B}_B \cdot w''_{1B} - \mathfrak{B}''_C \cdot n''_{14} \\ X''_2 &= M_{B1} = -M_{D4} = -\mathfrak{B}_A \cdot n''_{21} - \mathfrak{B}_B \cdot w''_{2B} + \mathfrak{B}''_C \cdot n''_{24} \\ X''_3 &= M_{B2} = -M_{D3} = -\mathfrak{B}_A \cdot n''_{31} + \mathfrak{B}_B \cdot w''_{3B} + \mathfrak{B}''_C \cdot n''_{34} \\ X''_4 &= M_{C2} = -M_{C3} = +\mathfrak{B}_A \cdot n''_{41} - \mathfrak{B}_B \cdot w''_{4B} - \mathfrak{B}''_C \cdot n''_{44}. \end{aligned} \right\} \quad (1465)$$

Column moments in upper story (n')

at base:
$$\left. \begin{aligned} M'_{Aa} &= M'_{Ee} = M^0_A + (\mathbf{D}_A \quad\quad - X''_1)\eta'_a \\ M'_{Bb} &= M'_{Dd} = M^0_B + (\mathbf{D}_B + X''_2 - X''_3)\eta'_b \\ M'_{Cc} &= M^0_C + (\mathbf{D}_C + 2X''_4)\eta'_c; \end{aligned} \right\} \quad (1466)$$

at top:
$$\left. \begin{aligned} M'_a &= M'_e = -(\mathfrak{R}'_a + M'_{Aa})\varepsilon'_a \\ M'_b &= M'_d = -(\mathfrak{R}'_b + M'_{Bb})\varepsilon'_b \\ M'_c &= -(\mathfrak{R}'_c + M'_{Cc})\varepsilon'_c. \end{aligned} \right\} \quad (1467)$$

Column moments in lower story (n)

at top:
$$\left. \begin{aligned} M_{Aa} &= M_{Ee} = M^0_A - (\mathbf{D}_A \quad\quad - X''_1)\eta_a \\ M_{Bb} &= M_{Dd} = M^0_B - (\mathbf{D}_B + X''_2 - X''_3)\eta_b \\ M_{Cc} &= M^0_C - (\mathbf{D}_C + 2X''_4)\eta_c; \end{aligned} \right\} \quad (1468)$$

at base:
$$\left. \begin{aligned} M_a &= M_e = -(\mathfrak{L}_a + M_{Aa})\varepsilon_a \\ M_b &= M_d = -(\mathfrak{L}_b + M_{Bb})\varepsilon_b \\ M_c &= -(\mathfrak{L}_c + M_{Cc})\varepsilon_c. \end{aligned} \right\} \quad (1469)$$

Horizontal thrusts, column shears and lateral restraining force for all loading conditions

All the relevant expressions are essentially as formulas (1345) – (1356), but with the subscripts $n = a, b, c, d, e$ and $N = A, B, C, D, E$; or with $n = a, b$ and $N = A, B$ for symmetry; or with $n = a, b, c$ and $N = A, B, C$ for anti-symmetry

Frame Shape 43

Symmetrical two-story three-bay frame with laterally restrained horizontal member and four elastically restrained pairs of columns (symmetrical form of Frame Shape 37)

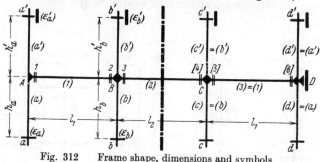

Fig. 312   Frame shape, dimensions and symbols

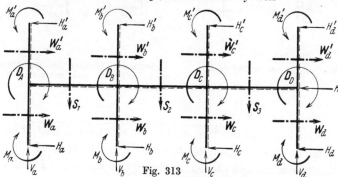

Fig. 313

Definition of positive direction for all external loads on joints and members, for all column reactions, and for the lateral restraining force

See boxed statement for Frame Shape 41, page 290.

Note: Frame Shape 43 can, in principle, be analysed with the aid of the formulas for *Frame Shape* 37. On account of symmetry, the reduction and recursion sequences (1221)/(1222) and (1223)/(1224), and (1246)/(1247) and (1248)/(1249), respectively become equal. Furthermore, the square matrices of influence coefficients (1226) and (1251) become bisymmetrical, i.e., symmetrical about the principal diagonal and secondary diagonal. If the "load transposition method" is employed, it is also possible to start from symmetrical and anti-symmetrical loading conditions. All the requisite formulas have already been given under *Frame Shape 41*, provided that the collective subscripts applied are $\nu = 1, 2$; $N = A, B$; $n = a, b$. All the sets of formulas without collective subscripts are valid, up to the dotted lines; conversely, for backward reductions and backward recursions they are valid from the dotted lines onward. Any departures from this general statement are indicated in footnotes.

Frame Shape 44

Symmetrical two-story two-bay frame with laterally restrained horizontal member and three elastically restrained pairs of columns
(Symmetrical form of Frame Shape 38)

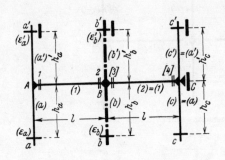

Fig. 314
Frame shape, dimensions and symbols

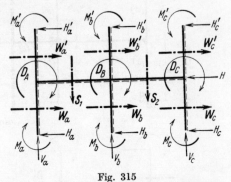

Fig. 315
Definition of positive direction for all external loads on joints and members, for all column reactions, and for the lateral restraining force

---

All the dimensions and coefficients for the right-hand half of the frame are the same as those for the left-hand half. In the case of the columns the broken line is always placed on the right-hand side of each column, despite the symmetry of the frame.

Note: Frame Shape 44 can, in principle, be analysed with the aid of the formulas for *Frame Shape* 38. On account of symmetry, the reduction sequences (1273)/(1274) and the recursion sequences (1275)/(1276) respectively become equal. Furthermore, the square matrix of influence coefficients (1278) becomes bisymmetrical, i.e.; symmetrical about the principal diagonal and secondary diagonal. If the "load transposition method" is employed, it is also possible to start from the following symmetrical and anti-symmetrical loading conditions.*

---

* The reader could, alternatively, have been simply referred to Frame Shape 40 (for treatment by Method II) or to Frame Shape 42 (for treatment by Method I), as the formulas given for those frames are generally valid for symmetrical continuous frames with any number of bays.

I. Treatment by the Method of Forces

(Where $J_k = J_1$ and $l_k = l$, hence $k_1 = 1$)

a) Coefficients for symmetrical load arrangements

$$\left.\begin{aligned}
& k_a = \frac{J_1}{J_a} \cdot \frac{h_a}{l} \quad k'_a = \frac{J_1}{J'_a} \cdot \frac{h'_a}{l}; \quad b_a = k_a(2 - \varepsilon_a) \quad b'_a = k'_a(2 - \varepsilon'_a); \\
& s_A = \frac{1}{1/b_a + 1/b'_a} = \frac{b_a \cdot b'_a}{b_a + b'_a}; \quad \eta_a = \frac{s_A}{b_a} \quad \eta'_a = \frac{s_A}{b'_a} \quad (\eta_a + \eta'_a = 1).
\end{aligned}\right\} \quad (1470)$$

$$\left.\begin{aligned}
& \left(\iota_{1s} = \frac{1}{2}\right) \quad \iota'_1 = \frac{1}{s_A + 2}; \quad n'_{11} = \frac{1}{s_A + 3/2} \quad n'_{22} = \frac{1}{b'_1 = (2 - \iota'_1)} \\
& \qquad\qquad (n'_{12} = n'_{21}) = (n'_{11}/2 = n'_{22} \cdot \iota'_1); \\
& s'_{11} = n'_{11} - n'_{12} = n'_{11}/2 \quad s'_{21} = n'_{22} - n'_{21} = n'_{22}(1 - \iota'_1).
\end{aligned}\right\} \quad (1471)$$

b) Coefficients for anti-symmetrical load arrangements

$$\left.\begin{aligned}
& k_a = \frac{J_1}{J_a} \cdot \frac{h_a}{l} \quad k'_a = \frac{J_1}{J'_a} \cdot \frac{h'_a}{l} \quad k_b = \frac{J_1}{J_b} \cdot \frac{h_b}{l} \quad k'_b = \frac{J_1}{J'_b} \cdot \frac{h'_b}{l}; \\
& b_a = k_a(2 - \varepsilon_a) \quad b'_a = k'_a(2 - \varepsilon'_a) \quad b_b = k_b(2 - \varepsilon_b) \quad b'_b = k'_b(2 - \varepsilon'_b); \\
& s_A = \frac{1}{1/b_a + 1/b'_a} = \frac{b_a \cdot b'_a}{b_a + b'_a} \quad s_B = \frac{1}{1/b_b + 1/b'_b} = \frac{b_b \cdot b'_b}{b_b + b'_b}; \\
& \eta_a = \frac{s_A}{b_a} \quad \eta'_a = 1 - \eta_a \quad \eta_b = \frac{s_B}{b_b} \quad \eta'_b = 1 - \eta_b.
\end{aligned}\right\} \quad (1472)$$

$$\left.\begin{aligned}
& \iota_{1t} = \frac{1}{2 + 2s_B} \quad b_{1t} = 2 - \iota_{1t} \quad \iota'_1 = \frac{1}{s_A + 2} \quad b'_1 = 2 - \iota'_1; \\
& n''_{11} = \frac{1}{s_A + b_{1t}} \quad n''_{22} = \frac{1}{b'_1 + 2s_B} \quad (n''_{12} = n''_{21}) = (n''_{11} \cdot \iota_{1t} = n''_{22} \cdot \iota'_1); \\
& s''_{11} = n''_{11} - n''_{12} = n''_{11}(1 - \iota_{1t}) \quad s''_{21} = n''_{22} - n''_{21} = n''_{22}(1 - \iota'_1).
\end{aligned}\right\} \quad (1473)$$

Note: For completeness it should be mentioned that, by virtue of the "load transposition method", the following relations exist between the values n' and n'', on the one hand, and the values n of the 4×4 matrix (1278), on the other hand:

$$\left.\begin{aligned}n'_{11} \\ n''_{11}\end{aligned}\right\} = n_{11} \pm n_{14} \qquad \left.\begin{aligned}n'_{12} \\ n''_{12}\end{aligned}\right\} = n_{12} \pm n_{13} \qquad \left.\begin{aligned}n'_{22} \\ n''_{22}\end{aligned}\right\} = n_{22} \pm n_{23}.$$

Loading Condition 1a

Both beam spans carrying vertical loading of arbitrary magnitude, but symmetrical about centre of frame

($\Re_2 = \mathfrak{L}_1$ and $\mathfrak{L}_2 = \Re_1$)

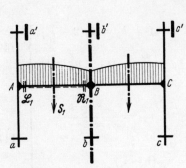

Fig. 316
Beam loading diagram with load terms $\mathfrak{L}$ and $\Re$ for left-hand half of frame

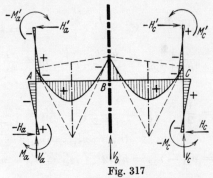

Fig. 317
Diagram of symmetrical bending moments, and support reactions

Beam moments ($L_1 = \mathfrak{L}_1$ and $R_1 = \Re_1$ since $k_1 = 1$):

$$\left. \begin{aligned} X'_1 &= M_{A1} = M_{C2} = -L_1 \cdot n'_{11} + R_1 \cdot n'_{12} = -\frac{L_1 - R_1/2}{s_A + 3/2} \\ X'_2 &= M_{B1} = M_{B2} = +L_1 \cdot n'_{21} - R_1 \cdot n'_{22} = -\frac{R_1 - L_1 \cdot \iota'_1}{b'_1} \end{aligned} \right\} \quad (1474)$$

Column moments:

Lower story:	Upper story:
at top:	at base:
$M_{Aa} = -M_{Cc} = +X'_1 \cdot \eta_a$	$M'_{Aa} = -M'_{Cc} = -X'_1 \cdot \eta'_a$
$M_{Bb} = 0;$	$M'_{Bb} = 0;$
at base:	at top:
$M_a = -M_c = -M_{Aa} \cdot \varepsilon_a$	$M'_a = -M'_c = -M'_{Aa} \cdot \varepsilon'_a$
$M_b = 0.$	$M'_b = 0.$

(1475)

Special case: Span load symmetrical ($R_1 = L_1$):

$$\left. \begin{aligned} X'_1 &= M_{A1} = M_{C2} = -L_1 \cdot s'_{11} = -\frac{L_1/2}{s_A + 3/2} \\ X'_2 &= M_{B1} = M_{B2} = -L_1 \cdot s'_{21} = -\frac{L_1(1 - \iota'_1)}{b'_1} \end{aligned} \right\} \quad (1476)$$

Loading Condition 1b
Both beam spans carrying vertical loading of arbitrary magnitude, but anti-symmetrical about centre of frame

$$(\mathfrak{R}_2 = -\mathfrak{L}_1 \text{ and } \mathfrak{L}_2 = -\mathfrak{R}_1)$$

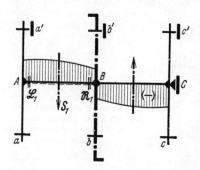

Fig. 318
Beam loading diagram with load terms $\mathfrak{L}$ and $\mathfrak{R}$ for left-hand half of frame

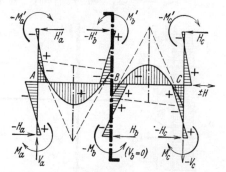

Fig. 319
Anti-symmetrical bending moment diagram, and support reaction

Beam moments ($L_1 = \mathfrak{L}_1$ and $R_1 = \mathfrak{R}_1$ since $k_1 = 1$):

$$\left. \begin{array}{l} X_1'' = M_{A1} = -M_{C2} = -L_1 \cdot n_{11}'' + R_1 \cdot n_{12}'' = -\dfrac{L_1 - R_1 \cdot \iota_{1t}}{s_A + b_{1t}} \\[1em] X_2'' = M_{B1} = -M_{B2} = +L_1 \cdot n_{21}'' - R_1 \cdot n_{22}'' = -\dfrac{R_1 - L_1 \cdot \iota_1'}{b_1' + 2 s_B} \end{array} \right\} \quad (1477)$$

Column moments:

Lower story:	Upper story:	
at top:	at base:	
$M_{Aa} = M_{Cc} = +X_1'' \cdot \eta_a$	$M'_{Aa} = M'_{Cc} = -X_1'' \cdot \eta_a'$	
$M_{Bb} = -2 X_2'' \cdot \eta_b;$	$M'_{Bb} = +2 X_2'' \cdot \eta_b';$	(1478)
at base:	at top:	
$M_a = M_c = -M_{Aa} \cdot \varepsilon_a$	$M'_a = M'_c = -M'_{Aa} \cdot \varepsilon_a'$	
$M_b = -M_{Bb} \cdot \varepsilon_b.$	$M'_b = -M'_{Bb} \cdot \varepsilon_b'.$	

Special case: Span load symmetrical ($R_1 = L_1$):

$$\left. \begin{array}{l} X_1'' = M_{A1} = -M_{C2} = -L_1 \cdot s_{11}'' = -\dfrac{L_1(1 - \iota_{1t})}{s_A + b_{1t}} \\[1em] X_2'' = M_{B1} = -M_{B2} = -L_1 \cdot s_{21}'' = -\dfrac{L_1(1 - \iota_1')}{b_1' + 2 s_B} \end{array} \right\} \quad (1479)$$

Loading Conditions 2a and 3a

The two external pairs of columns carrying horizontal loading of arbitrary magnitude symmetrical about center of frame (2a), and symmetrical pair of rotational moments at external joints (3a)

$$(\mathfrak{L}_c = -\mathfrak{L}_a, \quad \mathfrak{R}_c = -\mathfrak{R}_a; \quad \mathfrak{L}'_c = -\mathfrak{L}'_a, \quad \mathfrak{R}'_c = -\mathfrak{R}'_a; \quad D_C = -D_A)$$

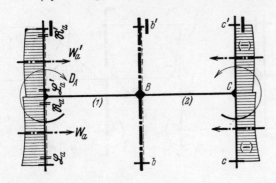

Fig. 320
Column loading diagram with load terms $\mathfrak{L}$ and $\mathfrak{R}$ for the columns, and external rotational moments D, for left-hand half of frame. (Symmetrical loading of central column line not shown.)

Column load terms:

$$(L_a = \mathfrak{L}_a k_a, \quad R_a = \mathfrak{R}_a k_a; \quad L'_a = \mathfrak{L}'_a k'_a, \quad R'_a = \mathfrak{R}'_a k'_a).$$

$$\mathfrak{B}_a = R_a - L_a \cdot \varepsilon_a \qquad \mathfrak{B}'_a = L'_a - R'_a \cdot \varepsilon'_a. \tag{1480}$$

Joint load terms:

$$\mathfrak{B}_A = D_A \cdot s_A - \mathfrak{B}_a \cdot \eta_a + \mathfrak{B}'_a \cdot \eta'_a. \tag{1481}$$

Beam moments:

$$\left.\begin{array}{l} X'_1 = M_{A1} = M_{C2} = +\mathfrak{B}_A \cdot n'_{11} = +\dfrac{\mathfrak{B}_A}{s_A + 3/2} \\[2mm] X'_2 = M_{B1} = M_{B2} = -\mathfrak{B}_A \cdot n'_{21} = -X'_1/2. \end{array}\right\} \tag{1482}$$

Support moment of column line in itself (auxiliary value)

$$M_A^0 = -\frac{\mathfrak{B}_a + \mathfrak{B}'_a}{b_a + b'_a}. \tag{1483}$$

Column moments:

$$\left.\begin{array}{ll} \text{Lower story:} & \text{Upper story:} \\ \text{at top:} & \text{at base:} \\ M_{Aa} = -M_{Cc} = M_A^0 - (D_A - X'_1)\eta_a & M'_{Aa} = -M'_{Cc} = M_A^0 + (D_A - X'_1)\eta'_a \\ M_{Bb} = 0; & M'_{Bb} = 0; \\ \text{at base:} & \text{at top:} \\ M_a = -M_c = -(\mathfrak{L}_a + M_{Aa})\varepsilon_a & M'_a = -M'_c = -(\mathfrak{R}'_a + M'_{Aa})\varepsilon'_a \\ M_b = 0. & M'_b = 0. \end{array}\right\} \tag{1484}$$

Loading Conditions 2b and 3b

All the columns carrying arbitrary horizontal loading (2b), and external rotational moments at all the beam joints (3b), but load arrangement as a whole is anti-symmetrical about center of frame

$$(\mathfrak{L}_c = \mathfrak{L}_a, \quad \mathfrak{R}_c = \mathfrak{R}_a; \quad \mathfrak{L}'_c = \mathfrak{L}'_a, \quad \mathfrak{R}'_c = \mathfrak{R}'_a; \quad D_C = D_A)$$

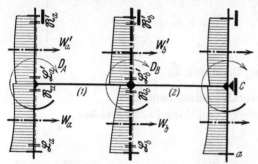

Fig. 321

Column loading diagram with load terms $\mathfrak{L}$ and $\mathfrak{R}$ for the columns, and external rotational moments D, for left-hand half of frame, including central column line b–B–b'

Column load terms $(n = a, b)$:

$$(L_n = \mathfrak{L}_n k_n \quad R_n = \mathfrak{R}_n k_n \quad L'_n = \mathfrak{L}'_n k'_n \quad R'_n = \mathfrak{R}'_n k'_n).$$
$$\mathfrak{B}_n = R_n - L_n \cdot \varepsilon_n \qquad \mathfrak{B}'_n = L'_n - R'_n \cdot \varepsilon'_n. \tag{1485}$$

Joint load terms $(N = A, B; \; n = a, b)$:

$$\mathfrak{B}_N = D_N \cdot s_N - \mathfrak{B}_n \cdot \eta_n + \mathfrak{B}'_n \cdot \eta'_n \qquad \mathfrak{B}''_B = \mathfrak{B}_B/2. \tag{1486}$$

Beam moments:

$$\left.\begin{aligned}
X''_1 = M_{A1} = -M_{C2} = +\mathfrak{B}_A \cdot n''_{11} + \mathfrak{B}''_B \cdot n''_{12} = +\frac{\mathfrak{B}_A + \mathfrak{B}''_B \cdot \iota_{1t}}{s_A + b_{1t}} \\
X''_2 = M_{B1} = -M_{B2} = -\mathfrak{B}_A \cdot n''_{21} - \mathfrak{B}''_B \cdot n''_{22} = -\frac{\mathfrak{B}_A \cdot \iota'_1 + \mathfrak{B}''_B}{b'_1 + 2 s_B}.
\end{aligned}\right\} \tag{1487}$$

Support moments of column lines in themselves (auxiliary values: $n = a, b$) $\Big\}$ $\quad M^0_N = -\dfrac{\mathfrak{B}_n + \mathfrak{B}'_n}{b_n + b'_n}.$ (1488)

Column moments:

$$\left.\begin{array}{ll}
\text{Lower story:} & \text{Upper story:} \\
M_{Aa} = M_{Cc} = M^0_A - (D_A - X''_1)\eta_a & M'_{Aa} = M'_{Cc} = M^0_A + (D_A - X''_1)\eta'_a \\
M_{Bb} = M^0_B - (D_B + 2X''_2)\eta_b; & M'_{Bb} = M^0_B + (D_B + 2X''_2)\eta'_b; \\
M_a = M_c = -(\mathfrak{L}_a + M_{Aa})\varepsilon_a & M'_a = M'_c = -(\mathfrak{R}'_a + M'_{Aa})\varepsilon'_a \\
M_b = -(\mathfrak{L}_b + M_{Bb})\varepsilon_b. & M'_b = -(\mathfrak{R}'_b + M'_{Bb})\varepsilon'_b.
\end{array}\right\} \tag{1489}$$

II. Treatment by the Deformation Method

(Where $J_k = J_1$ and $l_k = l$, hence $K_1 = 1$)

a) Coefficients for symmetrical load arrangements

$$K_a = \frac{J_a}{J_1} \cdot \frac{l}{h_a} \quad K'_a = \frac{J'_a}{J_1} \cdot \frac{l}{h'_a}; \quad S_a = \frac{K_a}{2-\varepsilon_a} \quad S'_a = \frac{K'_a}{2-\varepsilon'_a};$$
$$K_A = 3S_a + 3S'_a + 2. \quad (\mu_a + \mu'_a + \mu_1 = 1);$$
$$\mu_a = 3S_a/K_A \quad \mu'_a = 3S'_a/K_A \quad \mu_1 = 2/K_A. \tag{1490}$$

Loading Condition 4a

Symmetrical overall loading condition
(Superposition of Loading Conditions 1a, 2a and 3a)

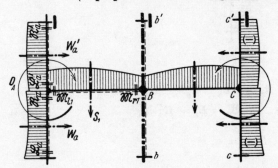

Fig. 322
Loading diagram with fixed-end moments $\mathfrak{M}_l$ and $\mathfrak{M}_r$ for the beam spans, load terms $\mathfrak{L}$ and $\mathfrak{R}$ for the external columns, and external rotational moments D, for left-hand half of frame. (Symmetrical loading of central line not shown.)

Column restraining moment at the beam joint:

$$\mathfrak{E}_{ra} = -\frac{\mathfrak{R}_a - \mathfrak{L}_a \cdot \varepsilon_a}{2-\varepsilon_a} \qquad \mathfrak{E}'_{la} = -\frac{\mathfrak{L}'_a - \mathfrak{R}'_a \cdot \varepsilon'_a}{2-\varepsilon'_a}. \tag{1491}$$

Joint load term:

$$\mathfrak{R}_A = \mathfrak{E}_{ra} - \mathfrak{E}'_{la} - \mathfrak{M}_{l1} + D_A. \tag{1492}$$

Beam moments:

$$M_{A1} = M_{C2} = \mathfrak{M}_{l1} + \mathfrak{R}_A \cdot \mu_1 \qquad M_{B1} = M_{B2} = \mathfrak{M}_{r1} - \mathfrak{R}_A \cdot \mu_1/2. \tag{1493}$$

Column moments:

Lower story:	Upper story:
$M_{Aa} = -M_{Cc} = \mathfrak{E}_{ra} - \mathfrak{R}_A \cdot \mu_a$	$M'_{Aa} = -M'_{Cc} = \mathfrak{E}'_{la} + \mathfrak{R}_A \cdot \mu'_a$
$M_a = -M_c = -(\mathfrak{L}_a + M_{Aa})\varepsilon_a$	$M'_a = -M'_c = -(\mathfrak{R}'_a + M'_{Aa})\varepsilon'_a$
$M_{Bb} = M_b = 0.$	$M'_{Bb} = M'_b = 0.$

$$\tag{1494}$$

Check relationship:

$$D_A + M_{Aa} - M'_{Aa} - M_{A1} = 0. \tag{1495}$$

Loading Condition 4b*
Anti-symmetrical overall loading condition
(Superposition of Loading Conditions 1b, 2b and 3b)

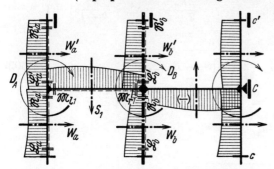

Fig. 323
Loading diagram with fixed-end moments $\mathfrak{M}_l$ and $\mathfrak{M}_r$ for the beam span, load term $\mathfrak{L}$ and $\mathfrak{R}$ for the columns and external rotational moments D, for left-hand half of frame including central column line b-B-b'

Column restraining moment at the beam joint $(n = a, b)$:

$$\mathfrak{C}_{rn} = -\frac{\mathfrak{R}_n - \mathfrak{L}_n \cdot \varepsilon_n}{2 - \varepsilon_n} \qquad \mathfrak{C}'_{ln} = -\frac{\mathfrak{L}'_n - \mathfrak{R}'_n \cdot \varepsilon'_n}{2 - \varepsilon'_n}. \qquad (1496)$$

Joint load terms:

$$\mathfrak{K}_A = \mathfrak{C}_{ra} - \mathfrak{C}'_{la} - \mathfrak{M}_{l1} + D_A \qquad \mathfrak{K}''_B = \mathfrak{M}_{r1} + (\mathfrak{C}_{rb} - \mathfrak{C}'_{lb} + D_B)/2. \quad (1497)$$

auxiliary moments:

$$\left.\begin{aligned} Y''_A &= +\mathfrak{K}_A \cdot u''_{aa} - \mathfrak{K}''_B \cdot u''_{ab} \\ Y''_B &= -\mathfrak{K}_A \cdot u''_{ba} + \mathfrak{K}''_B \cdot u''_{bb}. \end{aligned}\right\} \qquad (1498)$$

Beam moments:

$$\left.\begin{aligned} M_{A1} &= -M_{C2} = \mathfrak{M}_{l1} + (2Y''_A + Y''_B) \\ M_{B1} &= -M_{B2} = \mathfrak{M}_{r1} - (Y''_A + 2Y''_B). \end{aligned}\right\} \qquad (1499)$$

Column moments:

$$\left.\begin{array}{l|l} \text{Lower story:} & \text{Upper story:} \\ M_{Aa} = M_{Cc} = \mathfrak{C}_{ra} - Y''_A \cdot 3S_a & M'_{Aa} = M'_{Cc} = \mathfrak{C}'_{la} + Y''_A \cdot 3S'_a \\ M_{Bb} = \mathfrak{C}_{rb} - Y''_B \cdot 3S_b; & M'_{Bb} = \mathfrak{C}'_{lb} + Y''_B \cdot 3S'_b; \\ M_a = M_c = -(\mathfrak{L}_a + M_{Aa})\varepsilon_a & M'_a = M'_c = -(\mathfrak{R}'_a + M'_{Aa})\varepsilon'_a \\ M_b = -(\mathfrak{L}_b + M_{Bb})\varepsilon_b. & M'_b = -(\mathfrak{R}'_b + M'_{Bb})\varepsilon'_b. \end{array}\right\} \quad (1500)$$

Check relationships:

$$D_A + M_{Aa} - M'_{Aa} - M_{A1} = 0 \qquad D_B + M_{Bb} - M'_{Bb} + 2M_{B1} = 0. \quad (1501)$$

---

* Coefficients for anti-symmetrical load arrangement are given on next page.

b) **Coefficients for anti-symmetrical load arrangements**
(for Loading Conditions 4b, page 323).

$$K_n = \frac{J_n}{J_1} \cdot \frac{l}{h_n} \quad K'_n = \frac{J'_n}{J_1} \cdot \frac{l}{h'_n}; \quad S_n = \frac{K_n}{2 - \varepsilon_n} \quad S'_n = \frac{K'_n}{2 - \varepsilon'_n};$$

$$\text{for} \quad n = a, b.$$

$$K_A = 3(S_a + S'_a) + 2 \quad K''_B = 2 + 3(S_b + S'_b)/2.$$

$$j_{1t} = \frac{1}{K''_B} \quad j'_1 = \frac{1}{K_A}; \quad u''_{aa} = \frac{1}{K_A - j_{1t}} \quad u''_{bb} = \frac{1}{K''_B - j'_1}$$

$$(u''_{ab} = u''_{ba}) = (u''_{aa} \cdot j_{1t} = u''_{bb} \cdot j'_1).$$

(1502)

Horizontal thrusts, column shears and lateral restraining force for all loading conditions

For asymmetrical loading conditions: use formulas (1345) – (1348), for the collective subscripts $n = a, b, c$ and $N = A, B, C$. In particular:

$$H = -(H_a + H_b + H_c) - (H'_a + H'_b + H'_c);$$
$$H = -(Q_{ra} + Q_{rb} + Q_{rc}) + (Q'_{la} + Q'_{lb} + Q'_{lc}).$$

(1503)

For the symmetrical Loading Conditions 1a and 3a, or 2a and 4a: H_n and H'_n given by formulas (1345), T_n to H'_n are given by formulas (1347), for $n = a$ and $N = A$.

(1504)

Furthermore $\quad H = 0$

For the anti-symmetrical Loading Conditions 1b and 3b, or 2b and 4b: H_n and H'_n as equations (1345) and (1347), respectively, for $n = a, b$ and $N = A, B$.

Furthermore
$$H = -2H_a - H_b - 2H'_a - H'_b$$
$$H = -2Q_{ra} - Q_{rb} + 2Q'_{la} + Q'_{lb}.$$

(1505)

Frame Shape 45
Symmetrical two-story single-bay frame with laterally restrained horizontal member and two elastically restrained pairs of columns

(Symmetrical form of Frame Shape 39)

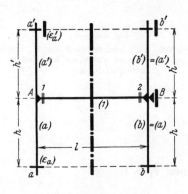

Fig. 324
Frame shape, dimensions and symbols

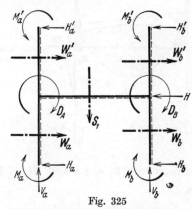

Fig. 325
Definition of positive direction for all external loads on joints and members, for all column reactions, and for the lateral restraining force

---

All the dimensions and coefficients for the right-hand half of the frame are the same as those for the left-hand half. In the case of the columns the broken line is always placed on the right-hand side of each column, despite the symmetry of the frame.

Note: Frame Shape 45 can, in principle, be analysed with the aid of the formulas for *Frame Shape* 39. However, various simplifications can be introduced thanks to the symmetry of the frame. Having regard to the practical importance of Frame Shape 45, all the relevant formulas have therefore been reproduced as a self-contained whole under the present heading.

I. Treatment by the Method of Forces

Coefficients, where $J_k = J_1$ and $l_k = l$, , hence $k_1 = 1$:

$$\left.\begin{aligned}
k_a &= \frac{J_1}{J_a} \cdot \frac{h}{l} \quad k_a' = \frac{J_1}{J_a'} \cdot \frac{h'}{l}; \quad b_a = k_a(2 - \varepsilon_a) \quad b_a' = k_a'(2 - \varepsilon_a'); \\
s_A &= \frac{1}{1/b_a + 1/b_a'} = \frac{b_a \cdot b_a'}{b_a + b_a'}; \quad \eta_a = \frac{s_A}{b_a} \quad \eta_a' = 1 - \eta_a. \\
& \qquad N_1 = s_A + 1 \quad N_2 = s_A + 2 \quad N_3 = s_A + 3.
\end{aligned}\right\} \quad (1506)$$

Loading Condition 1
Beam carrying arbitrary vertical loading

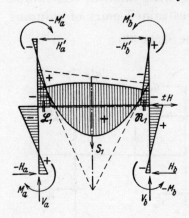

Fig. 326
Loading diagram, with load terms $\mathfrak{L}$ and $\mathfrak{R}$, bending moment diagram, column reactions and lateral restraining force

Beam moments ($L_1 = \mathfrak{L}_1$ and $R_1 = \mathfrak{R}_1$, since $k_1 = 1$):

$$X_1 = M_{A1} = -\frac{L_1 N_2 - R_1}{N_1 N_3} \qquad X_2 = M_{B1} = -\frac{R_1 N_2 - L_1}{N_1 N_3}. \qquad (1507)$$

Column moments and column shears:

$$\left.\begin{array}{lll} M_{Aa} = +X_1 \cdot \eta_a & M_a = -M_{Aa} \cdot \varepsilon_a & H_a = M_{Aa}(1+\varepsilon_a)/h \\ M'_{Aa} = -X_1 \cdot \eta'_a & M'_a = -M'_{Aa} \cdot \varepsilon'_a & H'_a = M'_{Aa}(1+\varepsilon'_a)/h' \\ M_{Bb} = -X_2 \cdot \eta_a & M_b = -M_{Bb} \cdot \varepsilon_a & H_b = M_{Bb}(1+\varepsilon_a)/h \\ M'_{Bb} = +X_2 \cdot \eta'_a; & M'_b = -M'_{Bb} \cdot \varepsilon'_a; & H'_b = M'_{Bb}(1+\varepsilon'_a)/h'. \end{array}\right\} \begin{array}{c}(1508)\\ \text{to}\\ (1510)\end{array}$$

Lateral restraining force:

$$H = -H_a - H'_a - H_b - H'_b = \frac{(L_1 - R_1)}{h N_1}\left(\eta_a(1+\varepsilon_a) - \eta'_a(1+\varepsilon'_a)\frac{h}{h'}\right). \qquad (1511)$$

Support reactions:

$$V_a = \frac{\mathfrak{S}_{r1}}{l} + \frac{(L_1 - R_1)}{l N_1} \qquad V_b = \frac{\mathfrak{S}_{l1}}{l} - \frac{(L_1 - R_1)}{l N_1}; \qquad (V_a + V_b = S_1). \qquad (1512)$$

Special case 1a: Beam span symmetrically loaded ($\mathfrak{R}_1 = \mathfrak{L}_1$):

$$X'_1 = M_{A1} = M_{B1} = -L_1/N_3. \qquad (1513)$$
$$M_{Aa} = -M_{Bb} = +X'_1 \cdot \eta_a \qquad M'_{Aa} = -M'_{Bb} = -X'_1 \cdot \eta'_a; \qquad (1514)$$
$$M_a = -M_b = -M_{Aa} \cdot \varepsilon_a \qquad M'_a = -M'_b = -M'_{Aa} \cdot \varepsilon'_a. \qquad (1515)$$
$$H_b = -H_a \qquad H'_b = -H'_a; \qquad H = 0; \qquad V_a = V_b = S_1/2. \qquad (1516)$$

Loading Condition 1b (Special case 1b of 1)
Beam span carrying arbitrary anti-symmetrical loading

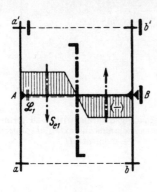

Fig. 327
Beam span loading, with load term $\mathfrak{L}$

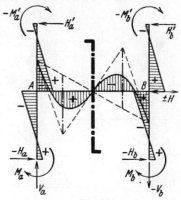

Fig. 328
Anti-symmetrical bending moment diagram, and support reaction

Beam moments: $\qquad X_1'' = M_{A1} = -M_{B1} = -L_1/N_1.$ \hfill (1517)

Column moments:

$$\begin{aligned}
M_{Aa} &= M_{Bb} = +X_1'' \cdot \eta_a & M'_{Aa} &= M'_{Bb} = -X_1'' \cdot \eta'_a \\
M_a &= M_b = -M_{Aa} \cdot \varepsilon_a & M'_a &= M'_b = -M'_{Aa} \cdot \varepsilon'_a.
\end{aligned} \Bigg\} \quad (1518)$$

Column shears and lateral restraining force:

$$H_a = H_b = M_{Aa}(1+\varepsilon_a)/h \qquad H'_a = H'_b = M'_{Aa}(1+\varepsilon'_a)/h'; \quad (1519)$$

$$H = -2H_a - 2H'_a = \frac{2L_1}{h\,N_1}\left(\eta_a(1+\varepsilon_a) - \eta'_a(1+\varepsilon'_a)\frac{h}{h'}\right). \quad (1520)$$

Support reactions:

$$V_a = -V_b = \frac{\mathfrak{S}_{r1}}{l} + \frac{2L_1}{l\,N_1}. \quad (1521)$$

Loading Conditions 2 and 3

All the individual columns carrying arbitrary horizontal loading (2), and external rotational moments at the beam joints (3)

Column load terms:

$$(L_n = \mathfrak{L}_n k_a, \qquad R_n = \mathfrak{R}_n k_a; \qquad L'_n = \mathfrak{L}'_n k'_a, \qquad R'_n = \mathfrak{R}'_n k'_a; \qquad \text{for} \quad n = a, b).$$

$$\begin{aligned}
\mathfrak{B}_a &= R_a - L_a \cdot \varepsilon_a & \mathfrak{B}'_a &= L'_a - R'_a \cdot \varepsilon'_a \\
\mathfrak{B}_b &= R_b - L_b \cdot \varepsilon_a & \mathfrak{B}'_b &= L'_b - R'_b \cdot \varepsilon'_a.
\end{aligned} \Bigg\} \quad (1522)$$

Joint load terms:

$$\left.\begin{aligned}\mathfrak{B}_A &= \boldsymbol{D}_A \cdot s_A - \mathfrak{B}_a \cdot \eta_a + \mathfrak{B}'_a \cdot \eta'_a \\ \mathfrak{B}_B &= \boldsymbol{D}_B \cdot s_A - \mathfrak{B}_b \cdot \eta_a + \mathfrak{B}'_b \cdot \eta'_a .\end{aligned}\right\} \quad (1523)$$

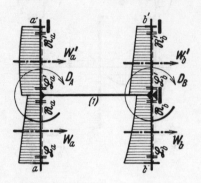

Fig. 329

Column loading diagram with load terms $\mathfrak{L}$ and $\mathfrak{R}$ for the columns and external rotational moments D

Beam moments:

$$X_1 = M_{A1} = + \frac{\mathfrak{B}_A N_2 + \mathfrak{B}_B}{N_1 N_3} \qquad X_2 = M_{B1} = - \frac{\mathfrak{B}_A + \mathfrak{B}_B N_2}{N_1 N_3} . \quad (1524)$$

Support moments for the column lines in themselves (auxiliary values):

$$M^0_A = - \frac{\mathfrak{B}_a + \mathfrak{B}'_a}{b_a + b'_a} \qquad M^0_B = - \frac{\mathfrak{B}_b + \mathfrak{B}'_b}{b_a + b'_a} . \quad (1525)$$

Column moments:

directly at the beam joints

$$\left.\begin{aligned}M'_{Aa} &= M^0_A + (\boldsymbol{D}_A - X_1)\,\eta'_a & M'_{Bb} &= M^0_B + (\boldsymbol{D}_B + X_2)\,\eta'_a \\ M_{Aa} &= M^0_A - (\boldsymbol{D}_A - X_1)\,\eta_a & M_{Bb} &= M^0_B - (\boldsymbol{D}_B + X_2)\,\eta_a ;\end{aligned}\right\} \quad (1526)$$

at the elastically restrained opposite ends:

$$\left.\begin{aligned}M'_a &= - (\mathfrak{R}'_a + M'_{Aa})\,\varepsilon'_a & M'_b &= - (\mathfrak{R}'_b + M'_{Bb})\,\varepsilon'_a \\ M_a &= - (\mathfrak{L}_a + M_{Aa})\,\varepsilon_a & M_b &= - (\mathfrak{L}_b + M_{Bb})\,\varepsilon_a .\end{aligned}\right\} \quad (1527)$$

Horizontal thrusts, column shears and lateral restraining force: according to formulas (1317), (1318) and (1319), where $(h'_a = h'_b) = h'$ and $(h_a = h_b) = h$.

Support reactions:

$$V_b = - V_a = (M_{A1} - M_{B1})/l . \quad (1528)$$

Loading Conditions 2a and 3a, and 2b and 3b
Symmetrical and anti-symmetrical special cases of Loading Conditions 2 and 3

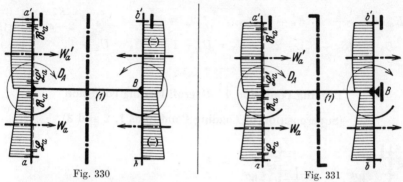

Fig. 330
Loading Conditions 2a and 3a

Fig. 331
Loading Conditions 2b and 3b

Loading diagrams with load terms $\mathfrak{L}$ and $\mathfrak{R}$ of the columns, and external rotational moment D, for left-hand half of frame

Load terms $\mathfrak{B}_a$, $\mathfrak{B}'_a$; $\mathfrak{B}_A$; in each case the first formula of (1522), (1523) and (1525) is applicable.

Beam moments:

$$X'_1 = M_{A1} = M_{B1} = \mathfrak{B}_A/N_3. \quad (1529) \quad \bigg| \quad X''_1 = M_{A1} = -M_{B1} = \mathfrak{B}_A/N_1. \quad (1530)$$

Column moments for 2a and 3a (symmetry):

$$\left. \begin{array}{ll} M'_{Aa} = -M'_{Bb} = M^0_A + (D_A - X'_1)\,\eta'_a & M'_a = -M'_b = -(\mathfrak{R}'_a + M'_{Aa})\,\varepsilon'_a \\ M_{Aa} = -M_{Bb} = M^0_A - (D_A - X'_1)\,\eta_a & M_a = -M_b = -(\mathfrak{L}_a + M_{Aa})\,\varepsilon_a. \end{array} \right\} \quad (1531)$$

Column moments for 2b and 3b (anti-symmetry):

$$\left. \begin{array}{ll} M'_{Aa} = M'_{Bb} = M^0_A + (D_A - X''_1)\,\eta'_a & M'_a = M'_b = -(\mathfrak{R}'_a + M'_{Aa})\,\varepsilon'_a \\ M_{Aa} = M_{Bb} = M^0_A - (D_A - X''_1)\,\eta_a & M_a = M_b = -(\mathfrak{L}_a + M_{Aa})\,\varepsilon_a. \end{array} \right\} \quad (1532)$$

Horizontal thrusts, column shears for left-hand column line: in each case the left-hand sets of formulas of (1317) and (1318) are applicable, where $h_a = h$ and $h'_a = h'$.

Lateral restraining force and support reactions:

$$\left. \begin{array}{l} H = 0; \\ V_a = V_b = 0. \end{array} \right\} \quad (1533) \quad \bigg| \quad \left. \begin{array}{l} H = -2Q_{ra} + 2Q'_{la}; \\ V_b = -V_a = 2X''_1/l. \end{array} \right\} \quad (1534)$$

II. Treatment by the Deformation Method

Coefficients, for $J_k = J_1$ and $l_k = l$, hence $K_1 = 1$:

$$K_a = \frac{J_a}{J_1} \cdot \frac{l}{h} \qquad K'_a = \frac{J'_a}{J_1} \cdot \frac{l}{h'}; \qquad S_a = \frac{K_a}{2-\varepsilon_a} \qquad S'_a = \frac{K'_a}{2-\varepsilon'_a}.$$
$$U_2 = 3(S_a + S'_a) + 2 \qquad U_1 = U_2 - 1 \qquad U_3 = U_2 + 1. \qquad (1535)$$

Loading Condition 4—Overall loading condition
(Superposition of Loading Conditions 1, 2 and 3)

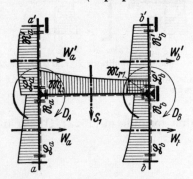

Fig. 332

Loading diagram with fixed-end moments $\mathfrak{M}_l$ and $\mathfrak{M}_r$ for the beam span, load terms $\mathfrak{L}$ and $\mathfrak{R}$ for the elastically restrained columns, and external rotational moments D

Column restraining moments at the beam joints:

$$\mathfrak{C}_{rn} = -\frac{\mathfrak{R}_n - \mathfrak{L}_n \cdot \varepsilon_n}{2 - \varepsilon_n} \qquad \mathfrak{C}'_{ln} = -\frac{\mathfrak{L}'_n - \mathfrak{R}'_n \cdot \varepsilon'_n}{2 - \varepsilon'_n} \qquad \text{for} \quad n = a, b. \qquad (1536)$$

Joint load terms:

$$\mathfrak{R}_A = -\mathfrak{M}_{l1} + \mathfrak{C}_{ra} - \mathfrak{C}'_{la} + D_A \qquad \mathfrak{R}_B = +\mathfrak{M}_{r1} + \mathfrak{C}_{rb} - \mathfrak{C}'_{lb} + D_B. \qquad (1537)$$

Auxiliary moments:

$$Y_A = \frac{+\mathfrak{R}_A U_2 - \mathfrak{R}_B}{U_1 U_3} \qquad Y_B = \frac{-\mathfrak{R}_A + \mathfrak{R}_B U_2}{U_1 U_3}. \qquad (1538)$$

Beam moments: same as formulas (1313).

Column moments: same as formulas (1314) and (1315), for
$$S'_b = S'_a, \ S_b = S_a; \ \varepsilon'_b = \varepsilon'_a, \ \varepsilon_b = \varepsilon_a.$$

Check relationships: same as formulas (1316).

Horizontal thrusts, column shears and lateral restraining force: same as formulas (1317) – (1319), where $(h'_a = h'_b) = h'$ and $(h_a = h_b) = h$.

Loading Conditions 4a and 4b

Symmetrical and anti-symmetrical special case of Loading Condition 4

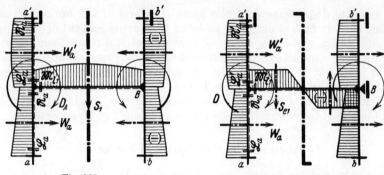

Fig. 333 Fig. 334
Loading Condition 4a Loading Condition 4b

Loading diagrams with fixed-end moment $\mathfrak{M}_l$ of the beam, load terms $\mathfrak{L}$ and $\mathfrak{R}$ of the elastically restrained columns, and external rotational moment D, for left-hand half of frame

Column restraining moment: as formulas (1536) for $n = a$

Joint load term: as first formula (1537).

Moment distribution factors at joint:

$$\mu_{as} = 3S_a/U_1 \qquad \mu'_{as} = 3S'_a/U_1 \quad \bigg| \quad \mu_{at} = 3S_a/U_3 \qquad \mu'_{at} = 3S'_a/U_3 \qquad (1539)$$
$$\mu_{1s} = 1 - \mu_{as} - \mu'_{as}. \quad \bigg| \quad \mu_{1t} = 1 - \mu_{at} - \mu'_{at}. \qquad \text{and} \quad (1540)$$

Beam moments:

$$M_{A1} = M_{B1} = \mathfrak{M}_{l1} + \mathfrak{R}_A \cdot \mu_{1s}. \quad \bigg| \quad M_{A1} = -M_{B1} = \mathfrak{M}_{l1} + \mathfrak{R}_A \cdot \mu_{1t}. \qquad \begin{array}{c}(1541)\\(1542)\end{array}$$

Column moments:

$$\begin{array}{l}
M_{Aa} = -M_{Bb} = \mathfrak{E}_{ra} - \mathfrak{R}_A \cdot \mu_{as} \\
M'_{Aa} = -M'_{Bb} = \mathfrak{E}'_{la} + \mathfrak{R}_A \cdot \mu'_{as}; \\
M_a = -M_b = -(\mathfrak{L}_a + M_{Aa})\,\varepsilon_a \\
M'_a = -M'_b = -(\mathfrak{R}'_a + M'_{Aa})\,\varepsilon'_a.
\end{array} \quad \bigg| \quad \begin{array}{l}
M_{Aa} = M_{Bb} = \mathfrak{E}_{ra} - \mathfrak{R}_A \cdot \mu_{at} \\
M'_{Aa} = M'_{Bb} = \mathfrak{E}'_{la} + \mathfrak{R}_A \cdot \mu'_{at}; \\
M_a = M_b = -(\mathfrak{L}_a + M_{Aa})\,\varepsilon_a \\
M'_a = M'_b = -(\mathfrak{R}'_a + M'_{Aa})\,\varepsilon'_a.
\end{array} \quad \begin{array}{c}(1543)\\ \text{and} \\(1544)\end{array}$$

Horizontal thrusts and shears for the left-hand column line: in each the left-hand sets of formulas of (1317) and (1318) are applicable, where $h_a = h$ and $h'_a = h'$.

Lateral restraining force and support reactions:

$$\left.\begin{array}{l} H = 0; \\ V_a = V_b = S_1/2. \end{array}\right\} (1545) \quad \bigg| \quad \left.\begin{array}{l} H = -2Q_{ra} + 2Q'_{la}; \\ V_a = -V_b = (\mathfrak{S}_{r1} - 2M_{A1})/l. \end{array}\right\} (1546)$$

Frame Shape 45v

(Frame Shape 45 with laterally restrained upper story)

Note: Horizontal displacement of the beam A-B and of the horizontal connecting line a'-b' (which is assumed to be of unvarying length) of the upper story column heads will occur only under asymmetrical or anti-symmetrical loading conditions (for all symmetrical loading conditions of Frame Shape 45 we have $H = 0$ and $H'_a + H'_b = 0$). Since any asymmetrical loading condition can, with the aid of the "load transposition method", be split up into a symmetrical and an anti-symmetrical loading condition, only the formulas for the anti-symmetrical loading conditions are given here.

I. Treatment by the Method of Forces

Coefficients, where $J_k = J_1$ and $l_k = l$, hence $k_1 = 1$:

$$k_a = \frac{J_1}{J_a} \cdot \frac{h}{l} \quad k'_a = \frac{J_1}{J'_a} \cdot \frac{h'}{l}; \quad m_a = k_a\left(\frac{1}{\varepsilon_a} + 4\right) \quad m'_a = k'_a\left(4 + \frac{1}{\varepsilon'_a}\right); \quad (1547)$$
$$N = m_a m'_a + (m_a + m'_a).$$

Loading Conditions 1b and 3b

Horizontal member carrying arbitrary anti-symmetrical vertical loading, and anti-symmetrical pair of rotational moments at the joints

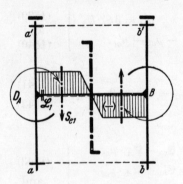

Fig. 335
Loading diagram with load terms $\mathfrak{L}$ and $\mathfrak{R}$ for left-hand half of frame

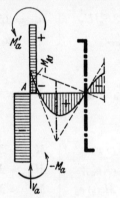

Fig. 336
Loading Condition 1b

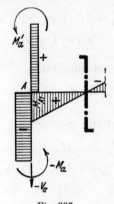

Fig. 337
Loading Condition 3b

Anti-symmetrical bending moment diagram and support reactions for left-hand half of frame.

Column and beam moments ($L_1 = \mathfrak{L}_1$ since $k_1 = 1$):

$$M_a = M_{Aa} = -\frac{(L_1 + D_A) m'_a}{N} \qquad M'_a = M'_{Aa} = +\frac{(L_1 + D_A) m_a}{N}; \quad (1548)$$
$$M_{A1} = -M_{B1} = M_a - M'_a + D_A. \quad [V_a = -V_b = (\mathfrak{S}_{r1} - 2M_{A1})/l]. \quad [1549]$$

Loading Condition 2b

Both lower columns carrying horizontal loading of arbitrary magnitude anti-symmetrical about center of frame

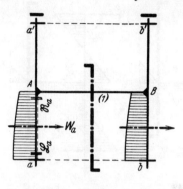

Fig. 338
Loading diagram with load terms $\mathfrak{L}$ and $\mathfrak{R}$ for left-hand half of frame

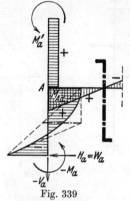

Fig. 339
Loading Condition 2b
Anti-symmetrical bending moment diagram and support reactions for left-hand half of frame.

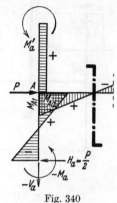

Fig. 340
Special Case 5

Column moments ($L_a = \mathfrak{L}_a k_a$ and $R_a = \mathfrak{R}_a k_a$):

$$\left. \begin{aligned} M'_a = M'_{Aa} = M'_b = M'_{Bb} &= \frac{\mathfrak{S}_{la} \cdot k_a(1/\varepsilon_a + 1) - (L_a + R_a)}{N}; \\ M_{Aa} = M_{Bb} = M'_a(1 + m'_a) \qquad M_a &= M_b = -\mathfrak{S}_{la} + M_{Aa}. \end{aligned} \right\} \quad (1550)$$

Beam moments:

$$M_{A1} = -M_{B1} = M'_a \cdot m'_a. \tag{1551}$$

Support reaction:

$$V_b = -V_a = \frac{2 M_{A1}}{l}. \tag{1552}$$

Special Case 5

Horizontal concentrated load P acting at level of horizontal member
(See Fig. 340)

In the above formulas (1550) write $\mathfrak{S}_{la} = Ph/2$ and $L_a = R_a = 0$:

$$\left. \begin{aligned} M'_a = M'_{Aa} = M'_b = M'_{Bb} &= \frac{Ph\, k_a(1/\varepsilon_a + 1)}{2N}; \\ M_{Aa} = M_{Bb} = M'_a(1 + m'_a) \qquad M_a &= M_b = -Ph/2 + M_{Aa}. \end{aligned} \right\} \quad (1553)$$

Loading Condition 2b′

Both upper columns carrying horizontal loading of arbitrary magnitude anti-symmetrical about center of frame

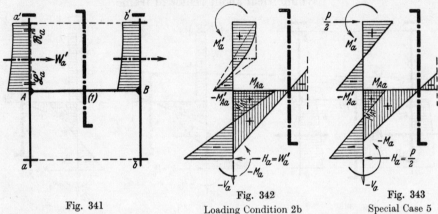

Fig. 341
Loading diagram with load terms $\mathfrak{L}$ and $\mathfrak{R}$ for left-hand half of frame

Fig. 342
Loading Condition 2b

Fig. 343
Special Case 5

Anti-symmetrical bending moment diagram and support reactions for left-hand half of frame.

Column load terms ($L'_a = \mathfrak{L}'_a k'_a$ and $R'_a = \mathfrak{R}'_a k'_a$):

$$\begin{aligned} \mathfrak{B}_a &= W'_a h (3k_a + 1) + \mathfrak{S}'_{la} \\ \mathfrak{B}'_a &= W'_a h + \mathfrak{S}'_{la}(1 + 3k'_a) - (L'_a + R'_a). \end{aligned} \qquad (1554)$$

Column moments:

$$M_a = M_b = -\frac{\mathfrak{B}_a(1 + m'_a) - \mathfrak{B}'_a}{N} \qquad M'_a = M'_b = +\frac{\mathfrak{B}'_a(m_a + 1) - \mathfrak{B}_a}{N}; \qquad (1555)$$

$$M_{Aa} = M_{Bb} = +W'_a h + M_a \qquad M'_{Aa} = M'_{Bb} = -\mathfrak{S}'_{la} + M'_a.$$

Beam moments:

$$M_{A1} = -M_{B1} = M_{Aa} - M'_{Aa} = W'_a h + \mathfrak{S}'_{la} + M_a - M'_a. \qquad (1556)$$

Support reactions:

$$V_b = -V_a = 2 M_{A1}/l. \qquad (1557)$$

Special Case 5′

Horizontal concentrated load $P/2$ acting at each upper column end a′ and b′ respectively.

(See Fig. 343)

In the above formulas (1554) and (1557) write:

$$W'_a h = Ph/2 \qquad \mathfrak{S}'_{la} = Ph'/2 \qquad L_a = R_a = 0. \qquad (1558)$$

II. Treatment by the Deformation Method

Coefficients, where $J_k = J_1$ and $l_k = l$, hence $K_1 = 1$:

$$\left.\begin{array}{l} K_a = \dfrac{J_a}{J_1} \cdot \dfrac{l}{h} \quad K'_a = \dfrac{J'_a}{J_1} \cdot \dfrac{l}{h'}; \quad F_a = \dfrac{K_a}{1/\varepsilon_a + 4} \quad F'_a = \dfrac{K'_a}{4 + 1/\varepsilon'_a}; \\ S_A = F_a + F'_a + 1; \\ (\eta_a + \eta'_a + \eta_1 = 1). \quad\quad\quad \eta_a = \dfrac{F_a}{S_A} \quad \eta'_a = \dfrac{F'_a}{S_A} \quad \eta_1 = \dfrac{1}{S_A}. \end{array}\right\} \quad (1559)$$

Loading Condition 4b

Anti-symmetrical overall loading condition

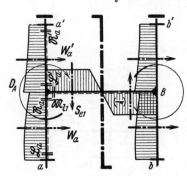

Fig. 344
Loading diagram with load terms $\mathfrak{L}$ and $\mathfrak{R}$ for the columns, fixed-end moment $\mathfrak{M}_l$ for the horizontal member, and external rotational moment D, for left-hand half of frame

Column restraining moments at the beam joint:

$$\left.\begin{array}{l} \mathfrak{F}_{ra} = + \dfrac{(\mathfrak{S}_{la} + W'_a h)(1/\varepsilon_a + 1) - (\mathfrak{L}_a + \mathfrak{R}_a)}{1/\varepsilon_a + 4} \\ \mathfrak{F}'_{la} = - \dfrac{\mathfrak{S}'_a(1 + 1/\varepsilon'_a) + (\mathfrak{L}'_a + \mathfrak{R}'_a)}{4 + 1/\varepsilon'_a}. \end{array}\right\} \quad (1560)$$

Joint load term:

$$\mathfrak{R}_A = D_A + \mathfrak{F}_{ra} - \mathfrak{F}'_{la} - \mathfrak{M}_{l1}. \quad (1561)$$

Column moments:

$$\left.\begin{array}{ll} M_{Aa} = M_{Bb} = \mathfrak{F}_{ra} - \mathfrak{R}_A \cdot \eta_a & M'_{Aa} = M'_{Bb} = \mathfrak{F}'_{la} + \mathfrak{R}_A \cdot \eta'_a \\ M_a = M_b = -\mathfrak{S}_{la} - W'_a h + M_{Aa}; & M'_a = M'_b = +\mathfrak{S}'_{la} + M'_{Aa}. \end{array}\right\} \quad (1562)$$

Beam moments:

$$M_{A1} = -M_{B1} = \mathfrak{M}_{l1} + \mathfrak{R}_A \cdot \eta_1. \quad (1563)$$

Check relationship:

$$D_A + M_{Aa} - M'_{Aa} - M_{A1} = 0. \quad (1564)$$

Frame Shape 46

Two-story five-bay frame with laterally restrained horizontal member, elastically restrained at each end, and four elastically restrained pairs of columns

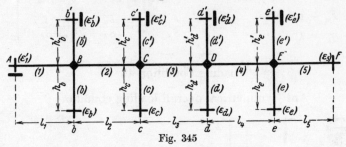

Fig. 345

Frame shape, dimensions and symbols

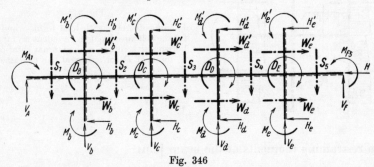

Fig. 346

Definition of positive direction for all external loads on joints and members, for all column reactions, and for the lateral restraining force

---

Important

All the formula sequences and formula matrices given for Frame Shape 46 can readily be extended to suit frames of similar type having any (odd) number of bays.*

Note: The *collective subscripts* have the following meaning:

$\nu = 1, 2, 3, 4, 5$ denoting the bays (or beam spans)
$n = b, c, d, e$ and $n' = b', c', d', e'$ denoting the columns
$N = B, C, D, E$ denoting the joints

The problem will be treated by the Deformation Method (Method II) only. For the treatment of a similar frame according to the Method of Forces (Method I) see Frame Shape 47.

---

* The dotted lines in the sets of formulas define the scope of these formulas as applicable to the three-bay frame.

Coefficients:

Stiffness coefficients for all individual members:

$$K_\nu = \frac{J_\nu}{J_k} \cdot \frac{l_k}{l_\nu} \qquad K_n = \frac{J_n}{J_k} \cdot \frac{l_k}{h_n} \qquad K_n' = \frac{J_n'}{J_k} \cdot \frac{l_k}{h_n'}. \tag{1565}$$

Stiffnesses of the Beam End Spans and the Individual Columns:

$$S_1 = \frac{K_1}{2-\varepsilon_1'} \qquad S_5 = \frac{K_5}{2-\varepsilon_5} * \: ; \qquad S_n = \frac{K_n}{2-\varepsilon_n} \qquad S_n' = \frac{K_n'}{2-\varepsilon_n'}. \tag{1566}$$

Joint coefficients:

$$\left. \begin{aligned} K_B &= 2K_1 + 3(S_b + S_b') + 2K_2 \\ K_C &= 2K_2 + 3(S_c + S_c') + 2K_3 \\ K_D &= 2K_3 + 3(S_d + S_d') + 2K_4 \\ K_E &= 2K_4 + 3(S_e + S_e') + 2K_5. \end{aligned} \right\} \tag{1567}$$

Rotation carry-over factors and auxiliary coefficients:

$$\left. \begin{array}{ll} \text{backward reduction:} & \text{forward reduction:} \\[4pt] \begin{array}{ll} j_4 = \dfrac{K_4}{r_e} & r_e = K_E \\[6pt] j_3 = \dfrac{K_3}{r_d} & r_d = K_D - K_4 \cdot j_4 \\[6pt] j_2 = \dfrac{K_2}{r_c} & r_c = K_C - K_3 \cdot j_3 \\[6pt] & r_b = K_B - K_2 \cdot j_2. \end{array} & \begin{array}{ll} & v_b = K_B \\[6pt] j_2' = \dfrac{K_2}{v_b} & \\[6pt] j_3' = \dfrac{K_3}{v_c} & v_c = K_C - K_2 \cdot j_2' \\[6pt] j_4' = \dfrac{K_4}{v_d} & v_d = K_D - K_3 \cdot j_3' \\[6pt] & v_e = K_E - K_4 \cdot j_4'. \end{array} \end{array} \right\} \begin{array}{l} (1568) \\ \text{and} \\ (1569) \end{array}$$

Principal influence coefficients by recursion:

$$\left. \begin{array}{ll} \text{forward:} & \text{backward:} \\[4pt] u_{bb} = 1/r_b & u_{ee} = 1/v_e \\ u_{cc} = 1/r_c + u_{bb} \cdot j_2^2 & u_{dd} = 1/v_d + u_{ee} \cdot j_4'^2 \\ u_{dd} = 1/r_d + u_{cc} \cdot j_3^2 & u_{cc} = 1/v_c + u_{dd} \cdot j_3'^2 \\ u_{ee} = 1/r_e + u_{dd} \cdot j_4^2. & u_{bb} = 1/v_b + u_{cc} \cdot j_2'^2. \end{array} \right\} \begin{array}{l} (1570) \\ \text{and} \\ (1571) \end{array}$$

Principal influence coefficients direct

$$u_{bb} = \frac{1}{r_b} \qquad u_{cc} = \frac{1}{r_c + v_c - K_C} \qquad u_{dd} = \frac{1}{r_d + v_d - K_D} \qquad u_{ee} = \frac{1}{v_e}. \tag{1572}$$

---

* For the three-bay frame the subscript 5 is replaced by 3.

Computation scheme for the symmetrical influence coefficient matrix:

$$\begin{array}{c|cccc|c}
\cdot j_2 & (u_{bb}) \downarrow & u_{bc} \uparrow & u_{bd} \uparrow & u_{be} \uparrow & \cdot j_2' \\
\hline
\cdot j_3 & u_{cb} \downarrow & (u_{cc}) \downarrow & u_{cd} \uparrow & u_{ce} \uparrow & \cdot j_3' \\
\hline
\cdot j_4 & u_{db} \downarrow & u_{dc} \downarrow & (u_{dd}) \downarrow & u_{de} \uparrow & \cdot j_4' \\
 & u_{eb} & u_{ec} & u_{ed} & (u_{ee}) &
\end{array}\Bigg\} \quad (1573)$$

Loading Condition 4—Overall loading condition
(Superposition of Loading Conditions 1, 2 and 3)

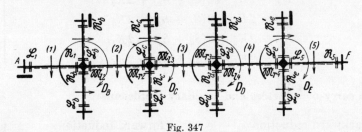

Fig. 347

Loading diagram with fixed-end moments $\mathfrak{M}_l$ and $\mathfrak{M}_r$ for the internal spans, load terms $\mathfrak{L}$ and $\mathfrak{R}$ for the semi-rigidly fixed external spans and the columns, and external rotational moments D

Restraining moments of the end spans of the horizontal member at the joints:

$$\mathfrak{C}_{r1} = -\frac{\mathfrak{R}_1 - \mathfrak{L}_1 \cdot \varepsilon_1'}{2 - \varepsilon_1'} \qquad \mathfrak{C}_{l5} = -\frac{\mathfrak{L}_5 - \mathfrak{R}_5 \cdot \varepsilon_5}{2 - \varepsilon_5} * \,. \qquad (1574)$$

Restraining moments of the columns at the joints:

$$\mathfrak{C}_{rn} = -\frac{\mathfrak{R}_n - \mathfrak{L}_n \cdot \varepsilon_n}{2 - \varepsilon_n} \qquad \mathfrak{C}_{ln}' = -\frac{\mathfrak{L}_n' - \mathfrak{R}_n' \cdot \varepsilon_n'}{2 - \varepsilon_n'} \,. \qquad (1575)$$

Joint load terms:

$$\begin{aligned}
\mathfrak{K}_B &= \mathfrak{C}_{r1} - \mathfrak{M}_{l2} + \mathfrak{C}_{rb} - \mathfrak{C}_{lb}' + D_B \\
\mathfrak{K}_C &= \mathfrak{M}_{r2} - \mathfrak{M}_{l3} + \mathfrak{C}_{rc} - \mathfrak{C}_{lc}' + D_C \\
\mathfrak{K}_D &= \mathfrak{M}_{r3} - \mathfrak{M}_{l4} + \mathfrak{C}_{rd} - \mathfrak{C}_{ld}' + D_D \\
\mathfrak{K}_E &= \mathfrak{M}_{r4} - \mathfrak{C}_{l5} + \mathfrak{C}_{re} - \mathfrak{C}_{le}' + D_E \,.
\end{aligned} \Bigg\} \quad (1576)$$

---

* See footnote on previous page.

auxiliary moments:

	$\mathfrak{K}_B$	$\mathfrak{K}_C$	$\mathfrak{K}_D$	$\mathfrak{K}_E$
$Y_B =$	$+ u_{bb}$	$- u_{bc}$	$+ u_{bd}$	$- u_{be}$
$Y_C =$	$- u_{cb}$	$+ u_{cc}$	$- u_{cd}$	$+ u_{ce}$
$Y_D =$	$+ u_{db}$	$- u_{dc}$	$+ u_{dd}$	$- u_{de}$
$Y_E =$	$- u_{eb}$	$+ u_{ec}$	$- u_{ed}$	$+ u_{ee}$

(1577)

Beam moments:

$$M_{A1} = -(\mathfrak{L}_1 + M_{B1})\,\varepsilon_1' \qquad M_{B1} = \mathfrak{C}_{r1} - Y_B \cdot 3S_1$$
$$M_{B2} = \mathfrak{M}_{l2} + (2Y_B + Y_C)\,K_2 \qquad M_{C2} = \mathfrak{M}_{r2} - (Y_B + 2Y_C)\,K_2$$
$$M_{C3} = \mathfrak{M}_{l3} + (2Y_C + Y_D)\,K_3 \qquad M_{D3} = \mathfrak{M}_{r3} - (Y_C + 2Y_D)\,K_3$$
$$M_{D4} = \mathfrak{M}_{l4} + (2Y_D + Y_E)\,K_4 \qquad M_{E4} = \mathfrak{M}_{r4} - (Y_D + 2Y_E)\,K_4$$
$$M_{E5} = \mathfrak{C}_{l5} + Y_E \cdot 3S_5 \qquad M_{F5} = -(\mathfrak{R}_5 + M_{E5})\,\varepsilon_5 *$$

(1578)

Column moments:

Lower story (n),

at top (1579):
$$M_{Nn} = \mathfrak{C}_{rn} - Y_N \cdot 3S_n;$$

at base (1580):
$$M_n = -(\mathfrak{L}_n + M_{Nn})\,\varepsilon_n.$$

Upper story (n'),

at base (1581):
$$M'_{Nn} = \mathfrak{C}_{ln} + Y_N \cdot 3S'_n;$$

at top (1582):
$$M'_n = -(\mathfrak{R}'_n + M'_{Nn})\,\varepsilon'_n.$$

(1579) to (1582)

Check relationships:

$$D_B + M_{B1} - M_{B2} + M_{Bb} - M'_{Bb} = 0$$
$$D_C + M_{C2} - M_{C3} + M_{Cc} - M'_{Cc} = 0$$
$$D_D + M_{D3} - M_{D4} + M_{Dd} - M'_{Dd} = 0$$
$$D_E + M_{E4} - M_{E5} + M_{Ee} - M'_{Ee} = 0.$$

(1583)

Horizontal thrusts, column shears and lateral restraining force: according to formulas (1210) – (1215), for the collective subscripts as indicated on page 261. In the formulas (1212) and (1215) for H all the terms with the subscript a are to be omitted.

---

*For the three-bay frame the subscripts in this line should be altered as follows: 5 to be replaced by 3, E by C, and F by D.

Frame Shape 47

Two-story four-bay frame with laterally restrained horizontal member, elastically restrained fixed at each end, and three elastically restrained pairs of columns

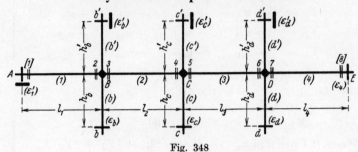

Fig. 348
Frame shape, dimensions and symbols

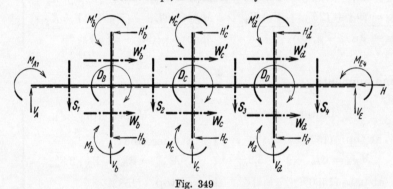

Fig. 349
Definition of positive direction for all external loads on joints and members, for all column reactions, and for the lateral restraining force

---
Important

All the formula sequences and formula matrices given for Frame Shape 47 can readily be extended to suit frames of similar type having any (even) number of bays.*

Note: The *collective subscripts* have the following meaning:
  $\nu = 1, 2, 3, 4$ denoting the bays (or beam spans)
  $n = b, c, d$ and $n' = b', c', d'$ denoting the columns
  $N = B, C, D$ denoting the joints

The problem will be treated by the Method of Forces (Method I) only. For the treatment of a similar frame according to the Deformation Method (Method II) see Frame Shape 46.

---
* The dotted lines in the sets of formulas define the scope of these formulas as applicable to the three-bay frame.

Coefficients

Flexibility coefficients for all individual members:

$$k_v = \frac{J_k}{J_v} \cdot \frac{l_v}{l_k} \qquad k_n = \frac{J_k}{J_n} \cdot \frac{h_n}{l_k} \qquad k'_n = \frac{J_k}{J'_n} \cdot \frac{h'_n}{l_k}. \tag{1584}$$

Flexibilities of the beam end spans and the individual columns:

$$b_1 = k_1(2 - \varepsilon'_1) \qquad \begin{aligned} b'_n &= k'_n(2 - \varepsilon'_n) \\ b_n &= k_n(2 - \varepsilon_n) \end{aligned} \qquad b_4 = k_4(2 - \varepsilon_4) * \; . \; \biggr\} \tag{1585}$$

Flexibilities of the column lines:

$$s_N = \frac{1}{1/b_n + 1/b'_n} = \frac{b_n \cdot b'_n}{b_n + b'_n}. \tag{1586}$$

Column distribution factors:

$$\eta_n = \frac{s_N}{b_n} \qquad \eta'_n = \frac{s_N}{b'_n}; \qquad (\eta_n + \eta'_n = 1). \tag{1587}$$

Diagonal coefficients:

$$\begin{aligned} O_2 &= b_1 + s_B & O_4 &= 2k_2 + s_C & O_6 &= 2k_3 + s_D \\ O_3 &= s_B + 2k_2 & O_5 &= s_C + 2k_3 * & O_7 &= s_D + b_4. \end{aligned} \biggr\} \tag{1588}$$

Moment carry-over factors and auxiliary coefficients:

backward reduction: | forward reduction:

$$\sigma_d = \frac{s_D}{r_7} \qquad r_7 = O_7$$
$$\iota_3 = \frac{k_3}{r_6} \qquad r_6 = O_6 - s_D \cdot \sigma_d$$
$$\sigma_c = \frac{s_C}{r_5} \qquad r_5 = O_5 - k_3 \cdot \iota_3$$
$$\iota_2 = \frac{k_2}{r_4} \qquad r_4 = O_4 - s_C \cdot \sigma_c$$
$$\sigma_b = \frac{s_B}{r_3} \qquad r_3 = O_3 - k_2 \cdot \iota_2$$
$$\qquad r_2 = O_2 - s_B \cdot \sigma_b.$$

$$\sigma'_b = \frac{s_B}{v_2} \qquad v_2 = O_2$$
$$\iota'_2 = \frac{k_2}{v_3} \qquad v_3 = O_3 - s_B \cdot \sigma'_b$$
$$\sigma'_c = \frac{s_C}{v_4} \qquad v_4 = O_4 - k_2 \cdot \iota'_2$$
$$\iota'_3 = \frac{k_3}{v_5} \qquad v_5 = O_5 - s_C \cdot \sigma'_c$$
$$\sigma'_d = \frac{s_D}{v_6} \qquad v_6 = O_6 - k_3 \cdot \iota'_3$$
$$\qquad v_7 = O_7 - s_D \cdot \sigma'_d.$$

(1589) and (1590)

---

* For the three-bay frame the subscript 4 in b_4 must be replaced by 3, and in O_5 the term $2k_3$ must be replaced by b_3.

Principal influence coefficients by recursion:

$$\left.\begin{array}{ll}\text{forward:} & \text{backward:} \\[4pt]
n_{22} = \dfrac{1}{r_2} & n_{77} = \dfrac{1}{v_7} \\[6pt]
n_{33} = \dfrac{1}{r_3} + n_{22} \cdot \sigma_b^2 & n_{66} = \dfrac{1}{v_6} + n_{77} \cdot \sigma_d'^2 \\[6pt]
n_{44} = \dfrac{1}{r_4} + n_{33} \cdot \iota_2^2 & n_{55} = \dfrac{1}{v_5} + n_{66} \cdot \iota_3'^2 \\[6pt]
n_{55} = \dfrac{1}{r_5} + n_{44} \cdot \sigma_c^2 & n_{44} = \dfrac{1}{v_4} + n_{55} \cdot \sigma_c'^2 \\[6pt]
n_{66} = \dfrac{1}{r_6} + n_{55} \cdot \iota_3^2 & n_{33} = \dfrac{1}{v_3} + n_{44} \cdot \iota_2'^2 \\[6pt]
n_{77} = \dfrac{1}{r_7} + n_{66} \cdot \sigma_d^2. & n_{22} = \dfrac{1}{v_2} + n_{33} \cdot \sigma_b'^2.
\end{array}\right\} \quad \begin{array}{c}(1591)\\ \text{and}\\ (1592)\end{array}$$

Principal influence coefficients direct

$$\left.\begin{array}{lll}
n_{22} = \dfrac{1}{r_2} & n_{33} = \dfrac{1}{r_3 + v_3 - O_3} & n_{44} = \dfrac{1}{r_4 + v_4 - O_4} \\[8pt]
n_{55} = \dfrac{1}{r_5 + v_5 - O_5} & n_{66} = \dfrac{1}{r_6 + v_6 - O_6} & n_{77} = \dfrac{1}{v_7}.
\end{array}\right\} \quad (1593)$$

Computation scheme for the symmetrical influence coefficient matrix

$$\left.\begin{array}{c|cccccc|c}
\cdot \sigma_b & (n_{22})\downarrow & n_{23}\uparrow & n_{24}\uparrow & n_{25}\uparrow & n_{26}\uparrow & n_{27}\uparrow & \cdot \sigma_b' \\
\cdot \iota_2 & n_{32}\downarrow & (n_{33})\downarrow & n_{34}\uparrow & n_{35}\uparrow & n_{36}\uparrow & n_{37}\uparrow & \cdot \iota_2' \\
\cdot \sigma_c & n_{42}\downarrow & n_{43}\downarrow & (n_{44})\downarrow & n_{45}\uparrow & n_{46}\uparrow & n_{47}\uparrow & \cdot \sigma_c' \\
\cdot \iota_3 & n_{52}\downarrow & n_{53}\downarrow & n_{54}\downarrow & (n_{55})\downarrow & n_{56}\uparrow & n_{57}\uparrow & \cdot \iota_3' \\
\cdot \sigma_d & n_{62}\downarrow & n_{63}\downarrow & n_{64}\downarrow & n_{65}\downarrow & (n_{66})\downarrow & n_{67}\uparrow & \cdot \sigma_d' \\
 & n_{72} & n_{73} & n_{74} & n_{75} & n_{76} & (n_{77}) &
\end{array}\right\} \quad (1594)$$

Loading Condition 1

All the beam spans carrying arbitrary vertical loading

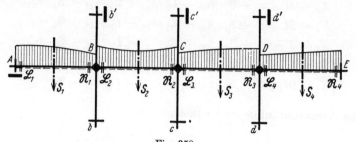

Fig. 350
Beam loading diagram with load terms $\mathfrak{L}$ and $\mathfrak{R}$

Beam load terms $\quad (L_\nu = \mathfrak{L}_\nu k_\nu \quad R_\nu = \mathfrak{R}_\nu k_\nu):$

$$\mathfrak{B}_1 = R_1 - L_1 \cdot \varepsilon_1' \qquad \mathfrak{B}_4 = L_4 - R_4 \cdot \varepsilon_4 * . \tag{1595}$$

Beam moments:

	$\mathfrak{B}_1$	L_2	R_2	L_3*)	R_3	$\mathfrak{B}_4$
$X_2 = M_{B1} =$	$-n_{22}$	$-n_{23}$	$+n_{24}$	$+n_{25}$	$-n_{26}$	$-n_{27}$
$X_3 = M_{B2} =$	$-n_{32}$	$-n_{33}$	$+n_{34}$	$+n_{35}$	$-n_{36}$	$-n_{37}$
$X_4 = M_{C2} =$	$+n_{42}$	$+n_{43}$	$-n_{44}$	$-n_{45}$	$+n_{46}$	$+n_{47}$
$X_5 = M_{C3} =$	$+n_{52}$	$+n_{53}$	$-n_{54}$	$-n_{55}$	$+n_{56}$	$+n_{57}$
$X_6 = M_{D3} =$	$-n_{62}$	$-n_{63}$	$+n_{64}$	$+n_{65}$	$-n_{66}$	$-n_{67}$
$X_7 = M_{D4} =$	$-n_{72}$	$-n_{73}$	$+n_{74}$	$+n_{75}$	$-n_{76}$	$-n_{77}$.

$$M_{A1} = -(\mathfrak{L}_1 + M_{B1})\varepsilon_1' \qquad M_{E4} = -(\mathfrak{R}_4 + M_{D4})\varepsilon_4 * . \tag{1596}$$

Column moments:

Lower story (n),

at top (1597):
$$M_{Bb} = (-X_2 + X_3)\eta_b$$
$$M_{Cc} = (-X_4 + X_5)\eta_c$$
$$M_{Dd} = (-X_6 + X_7)\eta_d;$$

at base (1598):
$$M_n = -M_{Nn} \cdot \varepsilon_n .$$

Upper story (n'),

at base (1599):
$$M'_{Bb} = (+X_2 - X_3)\eta'_b$$
$$M'_{Cc} = (+X_4 - X_5)\eta'_c$$
$$M'_{Dd} = (+X_6 - X_7)\eta'_d;$$

at top (1600):
$$M'_n = -M'_{Nn} \cdot \varepsilon'_n .$$

(1597) to (1600)

---

* For the three-bay frame the subscript 4 in $\mathfrak{B}_4$ must be replaced by 3, in (1596) L_3 must be replaced by $\mathfrak{B}_3$ and for M_{E4} the subscript 4 must be replaced by 3, E by D, and D by C.

Special case: all the beam spans symmetrically loaded $(R_\nu = L_\nu)^*$

$$\left.\begin{aligned}
X_2 &= M_{B1} = -\mathfrak{B}_1 \cdot n_{22} - L_2 \cdot s_{22} + L_3 \cdot s_{23} - \mathfrak{B}_4 \cdot n_{27} \\
X_3 &= M_{B2} = -\mathfrak{B}_1 \cdot n_{32} - L_2 \cdot s_{32} + L_3 \cdot s_{33} - \mathfrak{B}_4 \cdot n_{37} \\
X_4 &= M_{C2} = +\mathfrak{B}_1 \cdot n_{42} - L_2 \cdot s_{42} - L_3 \cdot s_{43} + \mathfrak{B}_4 \cdot n_{47} \\
X_5 &= M_{C3} = +\mathfrak{B}_1 \cdot n_{52} - L_2 \cdot s_{52} - L_3 \cdot s_{53} + \mathfrak{B}_4 \cdot n_{57} \\
X_6 &= M_{D3} = -\mathfrak{B}_1 \cdot n_{62} + L_2 \cdot s_{62} - L_3 \cdot s_{63} - \mathfrak{B}_4 \cdot n_{67} \\
X_7 &= M_{D4} = -\mathfrak{B}_1 \cdot n_{72} + L_2 \cdot s_{72} - L_3 \cdot s_{73} - \mathfrak{B}_4 \cdot n_{77};
\end{aligned}\right\} \quad (1601)$$

where the composite influence coefficients are:

$$\left.\begin{aligned}
s_{22} &= +n_{23} - n_{24} & s_{23} &= +n_{25} - n_{26} \\
s_{32} &= +n_{33} - n_{34} & s_{33} &= +n_{35} - n_{36} \\
s_{42} &= -n_{43} + n_{44} & s_{43} &= +n_{45} - n_{46} \\
s_{52} &= -n_{53} + n_{54} & s_{53} &= +n_{55} - n_{56} \\
s_{62} &= -n_{63} + n_{64} & s_{63} &= -n_{65} + n_{66} \\
s_{72} &= -n_{73} + n_{74} & s_{73} &= -n_{75} + n_{76}.
\end{aligned}\right\} \quad (1602)$$

Loading Conditions 2 and 3

All the columns carrying arbitrary horizontal loading (2); external rotational moments applied at all the beam joints (3)

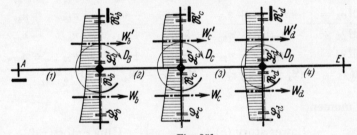

Fig. 351

Column loading diagram with load terms $\mathfrak{L}$ and $\mathfrak{R}$ for the columns and external rotational moments D.

Column load terms $(L_n = \mathfrak{L}_n k_n, \quad R_n = \mathfrak{R}_n k_n; \quad L'_n = \mathfrak{L}'_n k'_n, \quad R'_n = \mathfrak{R}'_n k'_n)$:

$$\mathfrak{B}_n = R_n - L_n \cdot \varepsilon_n \qquad \mathfrak{B}'_n = L'_n - R'_n \cdot \varepsilon'_n. \tag{1603}$$

Joint load terms:

$$\mathfrak{B}_N = D_N \cdot s_N - \mathfrak{B}_n \cdot \eta_n + \mathfrak{B}'_n \cdot \eta'_n. \tag{1604}$$

---
* For the three-bay frame replace L_3 by $\mathfrak{B}_3$ and the column s_{i_3} by the column n_{i5}.

Beam moments:

$$\left.\begin{aligned}
X_2 &= M_{B1} = -\mathfrak{B}_B \cdot w_{2B} + \mathfrak{B}_C \cdot w_{2C} - \mathfrak{B}_D \cdot w_{2D} \\
X_3 &= M_{B2} = +\mathfrak{B}_B \cdot w_{3B} + \mathfrak{B}_C \cdot w_{3C} - \mathfrak{B}_D \cdot w_{3D} \\
X_4 &= M_{C2} = -\mathfrak{B}_B \cdot w_{4B} - \mathfrak{B}_C \cdot w_{4C} + \mathfrak{B}_D \cdot w_{4D} \\
X_5 &= M_{C3} = -\mathfrak{B}_B \cdot w_{5B} + \mathfrak{B}_C \cdot w_{5C} + \mathfrak{B}_D \cdot w_{5D} \\
X_6 &= M_{D3} = +\mathfrak{B}_B \cdot w_{6B} - \mathfrak{B}_C \cdot w_{6C} - \mathfrak{B}_D \cdot w_{6D} \\
X_7 &= M_{D4} = +\mathfrak{B}_B \cdot w_{7B} - \mathfrak{B}_C \cdot w_{7C} + \mathfrak{B}_D \cdot w_{7D};
\end{aligned}\right\} \quad (1605)$$

where the composite influence coefficients are

$$\left.\begin{aligned}
w_{2B} &= +n_{22} - n_{23} & w_{2C} &= +n_{24} - n_{25} & w_{2D} &= +n_{26} - n_{27} \\
w_{3B} &= -n_{32} + n_{33} & w_{3C} &= +n_{34} - n_{35} & w_{3D} &= +n_{36} - n_{37} \\
w_{4B} &= -n_{42} + n_{43} & w_{4C} &= +n_{44} - n_{45} & w_{4D} &= +n_{46} - n_{47} \\
w_{5B} &= -n_{52} + n_{53} & w_{5C} &= -n_{54} + n_{55} & w_{5D} &= +n_{56} - n_{57} \\
w_{6B} &= -n_{62} + n_{63} & w_{6C} &= -n_{64} + n_{65} & w_{6D} &= +n_{66} - n_{67} \\
w_{7B} &= -n_{72} + n_{73} & w_{7C} &= -n_{74} + n_{75} & w_{7D} &= -n_{76} + n_{77}.
\end{aligned}\right\} \quad (1606)$$

Support moment of the column line (auxiliary value)
$$M_N^0 = -\frac{\mathfrak{B}_n + \mathfrak{B}_n'}{b_n + b_n'}. \quad (1607)$$

Column moments:

Lower story (n),
at top (1608):
$$M_{Bb} = M_B^0 - (D_B + X_2 - X_3)\eta_b$$
$$M_{Cc} = M_C^0 - (D_C + X_4 - X_5)\eta_c$$
$$M_{Dd} = M_D^0 - (D_D + X_6 - X_7)\eta_d;$$
at base (1609):
$$M_n = -(\mathfrak{L}_n + M_{Nn})\varepsilon_n.$$

Upper story (n'),
at base (1610):
$$M'_{Bb} = M_B^0 + (D_B + X_2 - X_3)\eta_b'$$
$$M'_{Cc} = M_C^0 + (D_C + X_4 - X_5)\eta_c'$$
$$M'_{Dd} = M_D^0 + (D_D + X_6 - X_7)\eta_d';$$
at top (1611):
$$M'_n = -(\mathfrak{R}_n' + M'_{Nn})\varepsilon_n'.$$

(1608) to (1611)

Horizontal thrusts, column shears and lateral restraining force for all loading conditions: as formulas (1210) – (1215), for the collective subscripts indicated on page 340. In the formulas (1212) and (1215) for H all the terms with the subscripts a and e are to be omitted.

Frame Shape 48

Laterally restrained cruciform frame consisting of an elastically restrained pair of beam spans and an elastically restrained pair of columns

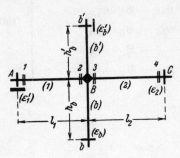

Fig. 352
Frame shape, dimensions and symbols

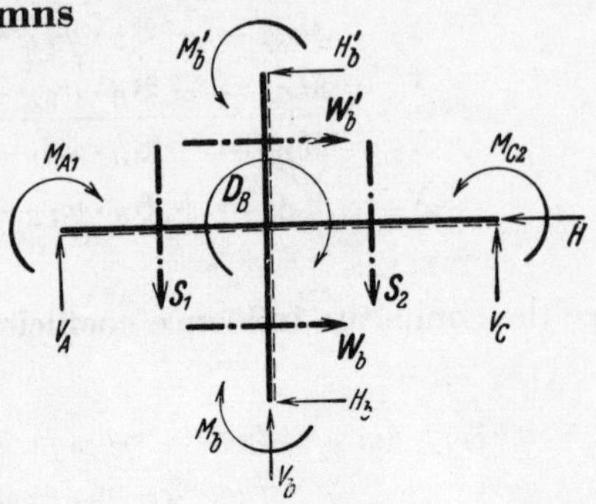

Fig. 353
Definition of positive direction of for all loads on members, for the external rotational moment at the joint, and for all reactions at ends of members

I. Treatment by the Method of Forces

Coefficients

Flexibility coefficients for all individual members:

$$k_1 = \frac{J_k}{J_1} \cdot \frac{l_1}{l_k} \qquad k_2 = \frac{J_k}{J_2} \cdot \frac{l_2}{l_k} \qquad k_b = \frac{J_k}{J_b} \cdot \frac{h}{l_k} \qquad k_b' = \frac{J_k}{J_b'} \cdot \frac{h'}{l_k}. \tag{1612}$$

Flexibilities of the elastically restrained columns*

$$b_b = k_b(2 - \varepsilon_b) \qquad b_b' = k_b'(2 - \varepsilon_b'). \tag{1613}$$

Flexibility of the column line:

$$s_B = \frac{1}{1/b_b + 1/b_b'} = \frac{b_b \cdot b_b'}{b_b + b_b'}. \tag{1614}$$

Column distribution factors:

$$\eta = \frac{s_B}{b_b} \qquad \eta' = \frac{s_B}{b_b'}; \qquad (\eta + \eta' = 1). \tag{1615}$$

Diagonal coefficients*

$$O_1 = \frac{k_1}{\varepsilon_1'} \qquad O_2 = 2k_1 + s_B \qquad O_3 = s_B + 2k_2 \qquad O_4 = \frac{k_2}{\varepsilon_2}. \tag{1616}$$

---

* The values ε are assumed to be known.

Moment carry-over factors and auxiliary coefficients:

backward reduction:

$$\sigma_b = \frac{s_B}{r_3}$$

$$\iota_1 = \frac{k_1}{r_2}$$

$$r_3 = O_3 - k_2 \cdot \varepsilon_2$$
$$r_2 = O_2 - s_B \cdot \sigma_b$$
$$r_1 = O_1 - k_1 \cdot \iota_1.$$

forward reduction:

$$\sigma_b' = \frac{s_B}{v_2}$$

$$\iota_2' = \frac{k_2}{v_3}$$

$$v_2 = O_2 - k_1 \cdot \varepsilon_1'$$
$$v_3 = O_3 - s_B \cdot \sigma_b'$$
$$v_4 = O_4 - k_2 \cdot \iota_2'.$$

(1617) and (1618)

Principal influence coefficients by recursion:

forward:

$$n_{11} = \frac{1}{r_1} = \frac{1}{k_1(1/\varepsilon_1' - \iota_1)}$$
$$n_{22} = 1/r_2 + n_{11} \cdot \iota_1^2$$
$$n_{33} = 1/r_3 + n_{22} \cdot \sigma_b^2$$
$$n_{44} = 1/O_4 + n_{33} \cdot \varepsilon_2^2.$$

backward:

$$n_{44} = \frac{1}{v_4} = \frac{1}{k_2(1/\varepsilon_2 - \iota_2')}$$
$$n_{33} = 1/v_3 + n_{44} \cdot \iota_2'^2$$
$$n_{22} = 1/v_2 + n_{33} \cdot \sigma_b'^2$$
$$n_{11} = 1/O_1 + n_{22} \cdot \varepsilon_1'^2.$$

(1619) and (1620)

Principal influence coefficients direct

$$n_{11} = \frac{1}{r_1} \qquad n_{22}' = \frac{1}{r_2 + v_2 - O_2} \qquad n_{33} = \frac{1}{r_3 + v_3 - O_3} \qquad n_{44} = \frac{1}{v_4}. \qquad (1621)$$

Computation scheme for the symmetrical influence coefficient matrix

$$\begin{array}{c|cccc|c}
\cdot \iota_1 & (n_{11}) \downarrow & n_{12} \uparrow & n_{13} \uparrow & n_{14} \uparrow & \cdot \varepsilon_1' \\
\cdot \sigma_b & n_{21} \downarrow & (n_{22}) \downarrow & n_{23} \uparrow & n_{24} \uparrow & \cdot \sigma_b' \\
\cdot \varepsilon_2 & n_{31} \downarrow & n_{32} \downarrow & (n_{33}) \downarrow & n_{34} \uparrow & \cdot \iota_2' \\
 & n_{41} & n_{42} & n_{43} & (n_{44}) &
\end{array} \qquad (1622)$$

Composite influence coefficients:

$$\begin{aligned}
s_{11} &= +n_{11} - n_{12} & w_{1B} &= +n_{12} - n_{13} & s_{12} &= +n_{13} - n_{14} \\
s_{21} &= -n_{21} + n_{22} & w_{2B} &= +n_{22} - n_{23} & s_{22} &= +n_{23} - n_{24} \\
s_{31} &= -n_{31} + n_{32} & w_{3B} &= -n_{32} + n_{33} & s_{32} &= +n_{33} - n_{34} \\
s_{41} &= -n_{41} + n_{42} & w_{4B} &= -n_{42} + n_{43} & s_{42} &= -n_{43} + n_{44}.
\end{aligned} \qquad (1623)$$

Loading Condition 1

Both beam spans carrying arbitrary vertical loading

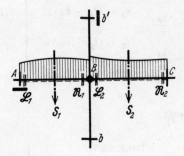

Fig. 354
Beam loading diagram with load terms $\mathfrak{L}$ and $\mathfrak{R}$

Beam moments ($L_1 = \mathfrak{L}_1 k_1$, $R_1 = \mathfrak{R}_1 k_1$; $L_2 = \mathfrak{L}_2 k_2$, $R_2 = \mathfrak{R}_2 k_2$):

$$\left. \begin{array}{l|cccc}
 & L_1 & R_1 & L_2 & R_2 \\ \hline
X_1 = M_{A1} = & -n_{11} & +n_{12} & +n_{13} & -n_{14} \\
X_2 = M_{B1} = & +n_{21} & -n_{22} & -n_{23} & +n_{24} \\
X_3 = M_{B2} = & +n_{31} & -n_{32} & -n_{33} & +n_{34} \\
X_4 = M_{C2} = & -n_{41} & +n_{42} & +n_{43} & -n_{44}.
\end{array} \right\} \quad (1624)$$

Column moments:

Lower column

at top (1625):
$$M_{Bb} = (-X_2 + X_3)\eta;$$

at base (1626):
$$M_b = -M_{Bb} \cdot \varepsilon_b.$$

Upper column

at base (1627):
$$M'_{Bb} = (+X_2 - X_3)\eta';$$

at top (1628):
$$M'_b = -M'_{Bb} \cdot \varepsilon'_b.$$

$$\left. \begin{array}{r} (1625) \\ \text{to} \\ (1628) \end{array} \right.$$

Special case: Both beam spans symmetrically loaded: ($R_1 = L_1$; $R_2 = L_2$):

$$\left. \begin{aligned}
X_1 = M_{A1} &= -L_1 \cdot s_{11} + L_2 \cdot s_{12} \\
X_2 = M_{B1} &= -L_1 \cdot s_{21} - L_2 \cdot s_{22} \\
X_3 = M_{B2} &= -L_1 \cdot s_{31} - L_2 \cdot s_{32} \\
X_4 = M_{C2} &= +L_1 \cdot s_{41} - L_2 \cdot s_{42}.
\end{aligned} \right\} \quad (1629)$$

Loading Conditions 2 and 3

Both columns carrying arbitrary horizontal loading (2); external rotational moment at joint (3)

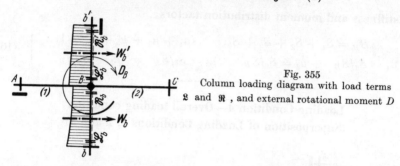

Fig. 355
Column loading diagram with load terms $\mathfrak{L}$ and $\mathfrak{R}$, and external rotational moment D

Column load terms $(L_b = \mathfrak{L}_b k_b, \quad R_b = \mathfrak{R}_b k_b; \quad L'_b = \mathfrak{L}'_b k'_b, \quad R'_b = \mathfrak{R}'_b k'_b)$:

$$\mathfrak{B}_b = R_b - L_b \cdot \varepsilon_b \qquad \mathfrak{B}'_b = L'_b - R'_b \cdot \varepsilon'_b. \tag{1630}$$

Joint load term:

$$\mathfrak{B}_B = D_B \cdot s_B - \mathfrak{B}_b \cdot \eta + \mathfrak{B}'_b \cdot \eta'. \tag{1631}$$

Beam moments:

$$\begin{aligned} X_1 &= M_{A1} = +\mathfrak{B}_B \cdot w_{1B} & X_3 &= M_{B2} = +\mathfrak{B}_B \cdot w_{3B} \\ X_2 &= M_{B1} = -\mathfrak{B}_B \cdot w_{2B} & X_4 &= M_{C2} = -\mathfrak{B}_B \cdot w_{4B}. \end{aligned} \tag{1632}$$

Support moment of the column line (auxiliary value)

$$M_B^0 = -\frac{\mathfrak{B}_b + \mathfrak{B}'_b}{b_b + b'_b}. \tag{1633}$$

Column moments:

Lower story (b),
at top (1634):
$$M_{Bb} = M_B^0 - (D_B + X_2 - X_3)\eta;$$
at base (1635):
$$M_b = -(\mathfrak{L}_b + M_{Bb})\varepsilon_b.$$

Upper story (b'),
at base (1636):
$$M'_{Bb} = M_B^0 + (D_B + X_2 - X_3)\eta';$$
at top (1637):
$$M'_b = -(\mathfrak{R}'_b + M'_{Bb})\varepsilon'_b.$$

(1634) to (1637)

II. Treatment by the Deformation Method
Coefficients

Stiffness coefficients for all individual members:

$$K_1 = \frac{J_1}{J_k} \cdot \frac{l_k}{l_1} \qquad K_2 = \frac{J_2}{J_k} \cdot \frac{l_k}{l_2} \qquad K_b = \frac{J_b}{J_k} \cdot \frac{l_k}{h} \qquad K'_b = \frac{J'_b}{J_k} \cdot \frac{l_k}{h'}. \tag{1638}$$

Stiffnesses of the elastically restrained individual members:

$$S_1 = \frac{K_1}{2-\varepsilon_1'} \qquad S_2 = \frac{K_2}{2-\varepsilon_2} \qquad S_b = \frac{K_b}{2-\varepsilon_b} \qquad S_b' = \frac{K_b'}{2-\varepsilon_b'}. \tag{1639}$$

Joint stiffness and moment distribution factors:

$$\left.\begin{array}{l} S_B = S_1 + S_2 + S_b + S_b'; \qquad (\mu_1 + \mu_2 + \mu_b + \mu_b' = 1). \\ \mu_1 = S_1/S_B \qquad \mu_2 = S_2/S_B \qquad \mu_b = S_b/S_B \qquad \mu_b' = S_b'/S_B. \end{array}\right\} \tag{1640}$$

Loading Condition 4—Overall loading condition
(Superposition of Loading Conditions 1, 2 and 3)

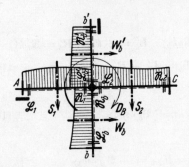

Fig. 356
Loading diagram with load terms $\mathfrak{L}$ and $\mathfrak{R}$ for all the individual members, and external rotational moment D

Restraining moments of the individual members at the joint of the frame:

$$\left.\begin{array}{ll} \mathfrak{E}_{r1} = -(\mathfrak{R}_1 - \mathfrak{L}_1 \cdot \varepsilon_1')/(2-\varepsilon_1') & \mathfrak{E}_{l2} = -(\mathfrak{L}_2 - \mathfrak{R}_2 \cdot \varepsilon_2)/(2-\varepsilon_2); \\ \mathfrak{E}_{rb} = -(\mathfrak{R}_b - \mathfrak{L}_b \cdot \varepsilon_b)/(2-\varepsilon_b) & \mathfrak{E}_{lb}' = -(\mathfrak{L}_b' - \mathfrak{R}_b' \cdot \varepsilon_b')/(2-\varepsilon_b'). \end{array}\right\} \tag{1641}$$

Joint load term:

$$\mathfrak{R}_B = \mathfrak{E}_{r1} - \mathfrak{E}_{l2} + \mathfrak{E}_{rb} - \mathfrak{E}_{lb}' + D_B. \tag{1642}$$

Beam moments:

$$\left.\begin{array}{ll} M_{B1} = \mathfrak{E}_{r1} - \mathfrak{R}_B \cdot \mu_1 & M_{B2} = \mathfrak{E}_{l2} + \mathfrak{R}_B \cdot \mu_2 \\ M_{A1} = -(\mathfrak{L}_1 + M_{B1})\varepsilon_1'; & M_{C2} = -(\mathfrak{R}_2 + M_{B2})\varepsilon_2. \end{array}\right\} \tag{1643}$$

Column moments:

$$\left.\begin{array}{ll} M_{Bb} = \mathfrak{E}_{rb} - \mathfrak{R}_B \cdot \mu_b & M_{Bb}' = \mathfrak{E}_{lb}' + \mathfrak{R}_B \cdot \mu_b' \\ M_b = -(\mathfrak{L}_b + M_{Bb})\varepsilon_b; & M_b' = -(\mathfrak{R}_b' + M_{Bb}')\varepsilon_b'. \end{array}\right\} \tag{1644}$$

Check relationship:

$$D_B + M_{B1} - M_{B2} + M_{Bb} - M_{Bb}' = 0. \tag{1645}$$

Frame Shape 48v
(Frame Shape 48 with laterally unrestrained upper story)

Coefficients according to Deformation Method (Method II):

Stiffness coefficients K_1, K_2, K_b, K_b' : as formulas (1638).

$$S_1 = \frac{K_1}{2-\varepsilon_1'} \qquad S_2 = \frac{K_2}{2-\varepsilon_2} ; \qquad F_b = \frac{K_b}{1/\varepsilon_b + 4} \qquad F_b' = \frac{K_b'}{4 + 1/\varepsilon_b'} ; \qquad (1646)$$

$$\left.\begin{array}{l} S_B^v = S_1 + S_2 + F_b + F_b'; \qquad (\eta_1 + \eta_2 + \eta_b + \eta_b' = 1). \\ \eta_1 = S_1/S_B^v \qquad \eta_2 = S_2/S_B^v \qquad \eta_b = F_b/S_B^v \qquad \eta_b' = F_b'/S_B^v. \end{array}\right\} \quad (1647)$$

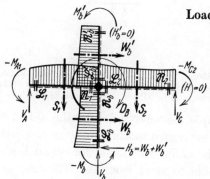

Loading Condition 4—Overall loading condition

Fig. 357
Loading diagram with load terms $\mathfrak{L}$ and $\mathfrak{R}$ for all the individual members, the external rotational moment D, and all support reactions

Restraining moments of the individual members at the joint of the frame:

$$\left.\begin{array}{ll} \mathfrak{C}_{r1} = -\dfrac{\mathfrak{R}_1 - \mathfrak{L}_1 \cdot \varepsilon_1'}{2-\varepsilon_1'} & \mathfrak{F}_{rb} = \dfrac{(\mathfrak{S}_{lb} + W_b' h)(1/\varepsilon_b + 1) - (\mathfrak{L}_b + \mathfrak{R}_b)}{1/\varepsilon_b + 4} \\[2ex] \mathfrak{C}_{l2} = -\dfrac{\mathfrak{L}_2 - \mathfrak{R}_2 \cdot \varepsilon_2}{2-\varepsilon_2} & \mathfrak{F}_{lb}' = -\dfrac{\mathfrak{S}_{lb}'(1+1/\varepsilon_b') + (\mathfrak{L}_b' + \mathfrak{R}_b')}{4+1/\varepsilon_b'}. \end{array}\right\} \quad (1648)$$

Joint load term:

$$\mathfrak{R}_B^v = \mathfrak{C}_{r1} - \mathfrak{C}_{l2} + \mathfrak{F}_{rb} - \mathfrak{F}_{lb}' + D_B. \qquad (1649)$$

Beam moments:

$$\left.\begin{array}{ll} M_{B1} = \mathfrak{C}_{r1} - \mathfrak{R}_B^v \cdot \eta_1 & M_{B2} = \mathfrak{C}_{l2} + \mathfrak{R}_B^v \cdot \eta_2 \\ M_{A1} = -(\mathfrak{L}_1 + M_{B1})\varepsilon_1'; & M_{C2} = -(\mathfrak{R}_2 + M_{B2})\varepsilon_2. \end{array}\right\} \quad (1650)$$

Column moments:

$$\left.\begin{array}{ll} M_{Bb} = \mathfrak{F}_{rb} - \mathfrak{R}_B^v \cdot \eta_b & M_{Bb}' = \mathfrak{F}_{lb}' + \mathfrak{R}_B^v \cdot \eta_b' \\ M_b = -\mathfrak{S}_{lb} - W_b' h + M_{Bb}; & M_b' = \mathfrak{S}_{lb}' + M_{Bb}'. \end{array}\right\} \quad (1651)$$

Check relationship:

$$D_B + M_{B1} - M + M_{Bb} - M_{Bb}' = 0. \qquad (1652)$$

Frame Shape 49
Symmetrical laterally restrained cruciform frame consisting of an elastically restrained pair of beam spans and an elastically restrained pair of columns
(Symmetrical form of Frame Shape 48)

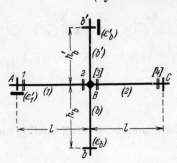

Fig. 358
Frame shape, dimensions and symbols

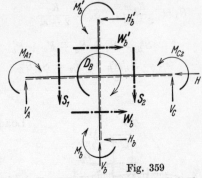

Fig. 359
Definition of positive direction of for all loads on members, for the external rotational moment at the joint, and for all reactions at ends of members

Note: Frame Shape 49 can, in principle, be analysed with the aid of the formulas for Frame Shape 48, having due regard to the symmetry properties. If the "load transposition method" is employed for asymmetrical loading conditions, it is, however, also possible to start from the following symmetrical and anti-symmetrical loading conditions.

a) Symmetrical Loading Condition 1a

Both beam spans carrying vertical loading of arbitrary magnitude, but equal and symmetrical about center of frame ($\Re_2 = \mathfrak{L}_1$, $\mathfrak{L}_2 = \Re_1$)

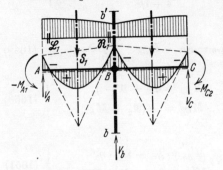

Fig. 360
Loading diagram with load terms $\mathfrak{L}$ and $\Re$ for left-hand half of frame; also bending moment diagram and reactions at ends of members

Beam moments (all column moments are zero):

$$M_{B1} = M_{B2} = -\frac{\Re_1 - \mathfrak{L}_1 \cdot \varepsilon_1'}{2 - \varepsilon_1'} \qquad M_{A1} = M_{C2} = -(\mathfrak{L}_1 + M_{B1})\varepsilon_1'. \quad (1653)$$

Special case ($\Re_1 = \mathfrak{L}_1$): $\qquad M_{B1} = M_{B2} = -\mathfrak{L}_1(1 - \varepsilon_1')/(2 - \varepsilon_1').$ \hfill (1654)

b) Anti-symmetrical loads

Coefficients according to Method of Forces (Method I),

where $J_k = J_1$ and $l_k = l$, , hence $k_1 = 1$:

$$\left.\begin{array}{l} k_b = \dfrac{J_1}{J_b} \cdot \dfrac{h}{l} \quad k'_b = \dfrac{J_1}{J'_b} \cdot \dfrac{h'}{l}; \qquad b_b = k_b(2 - \varepsilon_b) \quad b'_b = k'_b(2 - \varepsilon'_b); \\[4pt] s_B = \dfrac{1}{1/b_b + 1/b'_b} = \dfrac{b_b \cdot b'_b}{b_b + b'_b}; \qquad \eta = \dfrac{s_B}{b_b} \qquad \eta' = \dfrac{s_B}{b'_b} = 1 - \eta; \\[4pt] \iota_{1t} = \dfrac{1}{2 + 2s_B}; \qquad n''_{11} = \dfrac{1}{1/\varepsilon'_1 - \iota_{1t}} \qquad n''_{22} = \dfrac{1}{1/\iota_{1t} - \varepsilon'_1} \\[4pt] \qquad\qquad n''_{12} = n''_{21} = n''_{11} \cdot \iota_{1t} = n''_{22} \cdot \varepsilon'_1. \end{array}\right\} \quad (1655)$$

Loading Condition 1b

Both beam spans carrying vertical loading of arbitrary magnitude anti-symmetrical about center of frame ($\mathfrak{R}_2 = -\mathfrak{L}_1$, $\mathfrak{L}_2 = -\mathfrak{R}_1$)

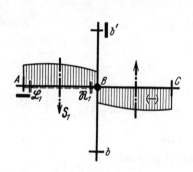

Fig. 361
Loading diagram with load terms $\mathfrak{L}$ and $\mathfrak{R}$ for left-hand beam span

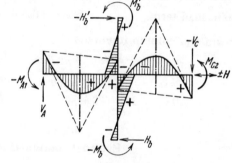

Fig. 362
Bending moment diagram and reactions at ends of members

Beam and column moments:

$$\left.\begin{array}{l} X''_1 = M_{A1} = -M_{C2} = -\mathfrak{L}_1 \cdot n''_{11} + \mathfrak{R}_1 \cdot n''_{12} \\ X''_2 = M_{B1} = -M_{B2} = +\mathfrak{L}_1 \cdot n''_{21} - \mathfrak{R}_1 \cdot n''_{22}; \end{array}\right\} \quad (1656)$$

$$\left.\begin{array}{ll} M_{Bb} = -2X''_2 \cdot \eta & M'_{Bb} = +2X''_2 \cdot \eta' \\ M_b = -M_{Bb} \cdot \varepsilon_b & M'_b = -M'_{Bb} \cdot \varepsilon'_b. \end{array}\right\} \quad (1657)$$

Special case: Symmetrical beam span loads ($\mathfrak{R}_1 = \mathfrak{L}_1$):

$$X''_1 = -\mathfrak{L}_1 \cdot n''_{11}(1 - \iota_{1t}) \qquad X''_2 = -\mathfrak{R}_1 \cdot n''_{22}(1 - \varepsilon'_1). \qquad (1658)$$

Loading Condition 2b

Both columns carrying arbitrary horizontal loading

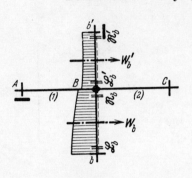

Fig. 363

Loading diagram for columns, with load terms $\mathfrak{L}$ and $\mathfrak{R}$

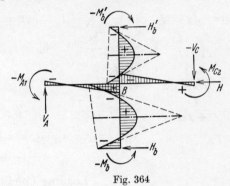

Fig. 364

Anti-symmetrical bending moment diagram and reactions at ends of members

Column load terms $\mathfrak{B}_b$ and $\mathfrak{B}'_b$ and **auxiliary moment** M_B^0: according to formulas (1630) and (1633)

Joint load term: $\qquad \mathfrak{B}''_B = (\mathfrak{B}_b \cdot \eta - \mathfrak{B}'_b \cdot \eta')/2.$ \hfill (1659)

Beam and column moments:

$$M_{B1} = -M_{B2} = \mathfrak{B}''_B \cdot n''_{22} \qquad M_{A1} = -M_{C2} = -M_{B1} \cdot \varepsilon'_1; \qquad (1660)$$

$$\left. \begin{array}{ll} M_{Bb} = M_B^0 - 2 M_{B1} \cdot \eta & M_b = -(\mathfrak{L}_b + M_{Bb})\, \varepsilon_b \\ M'_{Bb} = M_B^0 + 2 M_{B1} \cdot \eta' & M'_b = -(\mathfrak{R}'_b + M'_{Bb})\, \varepsilon'_b. \end{array} \right\} \quad (1661)$$

Loading Condition 3b
External rotational moment at joint

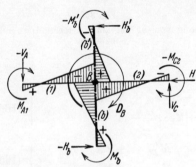

Fig. 365

Loading diagram with anti-symmetrical bending moment diagram and reactions at ends of members

Beam and column moments:

$$M_{B1} = -M_{B2} = -D_B \cdot s_B \cdot n''_{22}/2 \qquad M_{A1} = -M_{C2} = -M_{B1} \cdot \varepsilon'_1; \qquad (1662)$$

$$\left. \begin{array}{ll} M_{Bb} = -(D_B + 2 M_{B1})\, \eta & M_b = -M_{Bb} \cdot \varepsilon_b \\ M'_{Bb} = +(D_B + 2 M_{B1})\, \eta' & M'_b = -M'_{Bb} \cdot \varepsilon'_b. \end{array} \right\} \quad (1663)$$

Coefficients according to Deformation Method (Method II),

where $J_k = J_1$ and $l_k = l$, hence $K_1 = 1$:

$$K_b = \frac{J_b}{J_1} \cdot \frac{l}{h} \qquad K_b' = \frac{J_b'}{J_1} \cdot \frac{l}{h'}; \qquad S_b = \frac{K_b}{2 - \varepsilon_b} \qquad S_b' = \frac{K_b'}{2 - \varepsilon_b'}$$
$$S_1 = 1/(2 - \varepsilon_1'); \qquad\qquad S_B = 2S_1 + S_b + S_b';$$
$$\mu_1 = S_1/S_B \qquad \mu_b = S_b/S_B \qquad \mu_b' = S_b'/S_B; \qquad (2\mu_1 + \mu_b + \mu_b' = 1).$$
(1664)

Loading Condition 4b

Anti-symmetrical loading condition

$$(\mathfrak{R}_2 = -\mathfrak{L}_1, \quad \mathfrak{L}_2 = -\mathfrak{R}_1; \quad \mathfrak{C}_{l2} = -\mathfrak{C}_{r1})$$

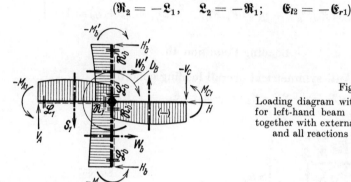

Fig. 366

Loading diagram with load terms $\mathfrak{L}$ and $\mathfrak{R}$ for left-hand beam span and column line, together with external rotational moment D and all reactions at ends of members

Restraining moments of the individual members at the joint of the frame:

$$\mathfrak{C}_{r1} = -\frac{\mathfrak{R}_1 - \mathfrak{L}_1 \cdot \varepsilon_1'}{2 - \varepsilon_1'}; \qquad \mathfrak{C}_{rb} = -\frac{\mathfrak{R}_b - \mathfrak{L}_b \cdot \varepsilon_b}{2 - \varepsilon_b} \qquad \mathfrak{C}_{lb}' = -\frac{\mathfrak{L}_b' - \mathfrak{R}_b' \cdot \varepsilon_b'}{2 - \varepsilon_b'}. \qquad (1665)$$

Joint load term:

$$\mathfrak{R}_B = 2\mathfrak{C}_{r1} + \mathfrak{C}_{rb} - \mathfrak{C}_{lb}' + D_B. \qquad (1666)$$

Beam moments:

$$M_{B1} = -M_{B2} = \mathfrak{C}_{r1} - \mathfrak{R}_B \cdot \mu_1 \qquad M_{A1} = -M_{C2} = -(\mathfrak{L}_1 + M_{B1}) \varepsilon_1'. \qquad (1667)$$

Column moments:

$$M_{Bb} = \mathfrak{C}_{rb} - \mathfrak{R}_B \cdot \mu_b \qquad M_{Bb}' = \mathfrak{C}_{lb}' + \mathfrak{R}_B \cdot \mu_b'$$
$$M_b = -(\mathfrak{L}_b + M_{Bb}) \varepsilon_b; \qquad M_b' = -(\mathfrak{R}_b' + M_{Bb}') \varepsilon_b'. \qquad (1668)$$

Check relationship:

$$D_B + 2M_{B1} + M_{Bb} - M_{Bb}' = 0. \qquad (1669)$$

Frame Shape 49v

(Frame Shape 49 with laterally unrestrained upper story)

a) Symmetrical loads

Loading Condition 1a: as for Frame Shape 49.

b) Anti-symmetrical loads

Coefficients according to Deformation Method (Method II), where $J_k = J_1$ and $l_k = l$, hence $K_1 = 1$:

$$\left.\begin{aligned}
K_b &= \frac{J_b}{J_1} \cdot \frac{l}{h} & K'_b &= \frac{J'_b}{J_1} \cdot \frac{l}{h'}; & F_b &= \frac{K_b}{1/\varepsilon_b + 4} & F'_b &= \frac{K'_b}{4 + 1/\varepsilon'_b}; \\
S_1 &= 1/(2-\varepsilon'_1); & & & S^v_B &= 2S_1 + F_b + F'_b; \\
\eta_1 &= S_1/S^v_B & \eta_b &= F_b/S^v_B & \eta'_b &= F'_b/S^v_B; & (2\eta_1 + \eta_b + \eta'_b = 1).
\end{aligned}\right\} \quad (1670)$$

Loading Condition 4b

Anti-symmetrical overall loading condition

$$(\mathfrak{R}_2 = -\mathfrak{L}_1, \quad \mathfrak{L}_2 = -\mathfrak{R}_1; \quad \mathfrak{E}_{l2} = -\mathfrak{E}_{r1})$$

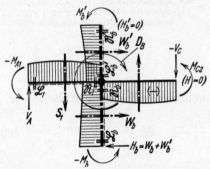

Fig. 367
Loading diagram with load terms $\mathfrak{L}$ and $\mathfrak{R}$ for left-hand beam span and column line, together with external rotational moment D and all reactions at ends of members

Restraining moments of the individual members at the joint:
$\mathfrak{E}_{r1}$, $\mathfrak{F}_{rb}$ and $\mathfrak{F}'_{lb}$ according to formulas (1648)

Joint load term:
$$\mathfrak{R}^v_B = 2\mathfrak{E}_{r1} + \mathfrak{F}_{rb} - \mathfrak{F}'_{lb} + D_B. \tag{1671}$$

Beam and column moments:
$$M_{B1} = -M_{B2} = \mathfrak{E}_{r1} - \mathfrak{R}^v_B \cdot \eta_1 \qquad M_{A1} = -M_{C2} = -(\mathfrak{L}_1 + M_{B1})\varepsilon'_1. \tag{1672}$$

$$\left.\begin{aligned}
M_{Bb} &= \mathfrak{F}_{rb} - \mathfrak{R}^v_B \cdot \eta_b & M'_{Bb} &= \mathfrak{F}'_{lb} + \mathfrak{R}^v_B \cdot \eta'_b \\
M_b &= -\mathfrak{S}_{lb} - W'_b h + M_{Bb}; & M'_b &= \mathfrak{S}_{lb} + M'_{Bb}.
\end{aligned}\right\} \quad (1673)$$

Check relationship:
$$D_B + 2M_{B1} + M_{Bb} - M'_{Bb} = 0. \tag{1674}$$

Frame Shape 50

Two-story frame comprising an elastically and laterally restrained horizontal member and a pair of columns

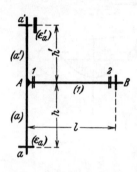

Fig. 368
Frame shape, dimensions and symbols

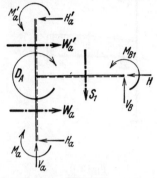

Fig. 369
Definition of positive direction of for all loads on members, for the external rotational moment at the joint, and for all reactions at ends of members

Note: The formulas relate to the Method of Forces (Method I*).

Coefficients, where $J_k = J_1$ and $l_k = l$, hence $k_1 = 1$.

Flexibility coefficients and flexibilities of the individual columns:

$$k_a = \frac{J_1}{J_a} \cdot \frac{h}{l} \qquad k'_a = \frac{J_1}{J'_a} \cdot \frac{h'}{l}; \qquad b_a = k_a(2 - \varepsilon_a) \qquad b'_a = k'_a(2 - \varepsilon'_a). \qquad (1675)$$

Flexibility of the column line and column distribution factors:

$$s_A = \frac{1}{1/b_a + 1/b'_a} = \frac{b_a \cdot b'_a}{b_a + b'_a}; \qquad \eta_a = \frac{s_A}{b_a} \qquad \eta'_a = \frac{s_A}{b'_a} = 1 - \eta_a. \qquad (1676)$$

Moment carry-over factor, beam flexibility and influence coefficients:

$$\left.\begin{array}{l} \iota'_1 = \dfrac{1}{s_A + 2}; \qquad b_1 = 2 - \varepsilon_1; \qquad n_{11} = \dfrac{1}{s_A + b_1} \qquad n_{22} = \dfrac{1}{1/\varepsilon_1 - \iota'_1} \\[2mm] n_{12} = n_{21} = n_{11} \cdot \varepsilon_1 = n_{22} \cdot \iota'_1; \qquad (\sigma_a = s_A \cdot n_{11}); \\[2mm] s_{11} = n_{11} - n_{12} = n_{11}(1 - \varepsilon_1) \qquad s_{21} = n_{22} - n_{21} = n_{22}(1 - \iota'_1). \end{array}\right\} \quad (1677)$$

---

* The appropriate formulas for the Deformation Method (Method II) can be obtained from formulas (1638)–(1645) for Frame Shape 48 by substituting $J_1 = 0$ and $\mathfrak{L}_1 = \mathfrak{R}_1 = 0$. And then $K_1 = S_1 = \mu_1 = 0$ and $\mathfrak{G}_{r1} = 0$. At the same time it will be necessary to replace subscript 2 by 1, b by a, b' by a', B by A and C by B.

Loading Condition 1

Beam span carrying arbitrary vertical loading

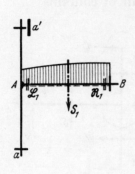

Fig. 370

Loading diagram for columns, with load terms $\mathfrak{L}$ and $\mathfrak{R}$

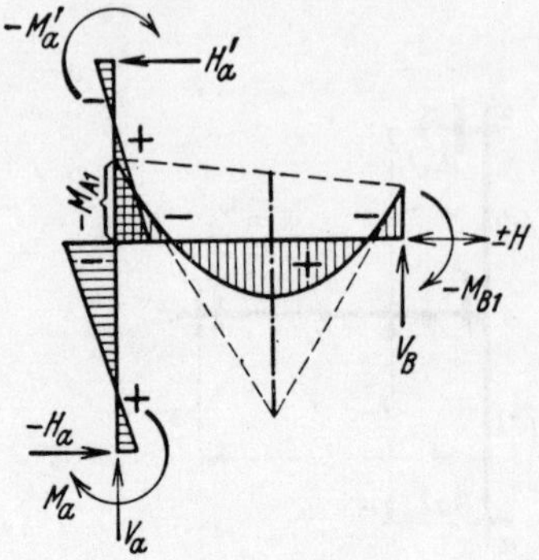

Fig. 371

Bending moment diagram and reactions at ends of members

Beam moments:*

$$\left.\begin{aligned} X_1 = M_{A1} &= -\mathfrak{L}_1 \cdot n_{11} + \mathfrak{R}_1 \cdot n_{12} = -\frac{\mathfrak{L}_1 - \mathfrak{R}_1 \cdot \varepsilon_1}{s_A + b_1} \\ X_2 = M_{B1} &= +\mathfrak{L}_1 \cdot n_{21} - \mathfrak{R}_1 \cdot n_{22} = -\frac{\mathfrak{R}_1 - \mathfrak{L}_1 \cdot \iota_1'}{1/\varepsilon_1 - \iota_1'} . \end{aligned}\right\} \tag{1678}$$

Moments in upper column:

$$M'_{Aa} = -X_1 \cdot \eta'_a \qquad M'_a = -M'_{Aa} \cdot \varepsilon'_a . \tag{1679}$$

Moments in lower column:

$$M_{Aa} = +X_1 \cdot \eta_a \qquad M_a = -M_{Aa} \cdot \varepsilon_a . \tag{1680}$$

Special case: Beam span symmetrically loaded ($\mathfrak{R}_1 = \mathfrak{L}_1$)*

$$\left.\begin{aligned} X_1 = M_{A1} &= -\mathfrak{L}_1 \cdot s_{11} = -\mathfrak{L}_1 \cdot \frac{1 - \varepsilon_1}{s_A + b_1} \\ X_2 = M_{B1} &= -\mathfrak{L}_1 \cdot s_{21} = -\mathfrak{L}_1 \cdot \frac{1 - \iota_1'}{1/\varepsilon_1 - \iota_1'} . \end{aligned}\right\} \tag{1681}$$

---

* Since $k_1 = 1$, we may also write $\mathfrak{L}_1 = \boldsymbol{L}_1$ and $\mathfrak{R}_1 = \boldsymbol{R}_1$.

Loading Conditions 2 and 3

Both individual columns carrying arbitrary horizontal loading (2); external rotational moment at joint (3)

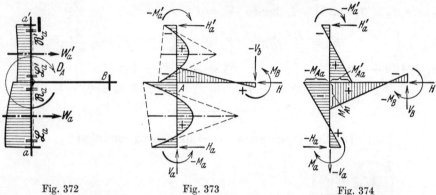

Fig. 372
Loading diagram with load terms $\mathfrak{L}$ and $\mathfrak{R}$ for the columns, and external rotational moment D

Fig. 373
Loading Condition 2

Bending moment diagram and reactions at ends of members

Fig. 374
Loading Condition 3

Column load terms:

$$(L_a = \mathfrak{L}_a \cdot k_a, \quad R_a = \mathfrak{R}_a \cdot k_a; \quad L'_a = \mathfrak{L}'_a \cdot k'_a, \quad R'_a = \mathfrak{R}'_a \cdot k'_a):$$
$$\mathfrak{B}_a = R_a - L_a \cdot \varepsilon_a \qquad \mathfrak{B}'_a = L'_a - R'_a \cdot \varepsilon'_a. \tag{1682}$$

Joint load term:

$$\mathfrak{B}_A = D_A \cdot s_A - \mathfrak{B}_a \cdot \eta_a + \mathfrak{B}'_a \cdot \eta'_a. \tag{1683}$$

Beam moments:

$$X_1 = M_{A1} = +\mathfrak{B}_A \cdot n_{11} \qquad X_2 = M_{B1} = -\mathfrak{B}_A \cdot n_{21} = -X_1 \cdot \varepsilon_1. \tag{1684}$$

Support moment of the column line in itself (auxiliary value):

$$M^0_A = -\frac{\mathfrak{B}_a + \mathfrak{B}'_a}{b_a + b'_a}. \tag{1685}$$

Moments in upper column:

$$M'_{Aa} = M^0_A + (D_A - X_1)\,\eta'_a \qquad M'_a = -(\mathfrak{R}'_a + M'_{Aa})\,\varepsilon'_a. \tag{1686}$$

Moments in lower column:

$$M_{Aa} = M^0_A - (D_A - X_1)\,\eta_a \qquad M_a = -(\mathfrak{L}_a + M_{Aa})\,\varepsilon_a. \tag{1687}$$

Frame Shape 50v

(Frame Shape 50 with laterally unrestrained upper story)

Coefficients according to Deformation Method (Method II), J_1 and $l_k = l$, where $J_k = J_1$ and $l_k = l$, hence $K_1 = 1$:

$$\left.\begin{aligned}
K_a &= \frac{J_a}{J_1} \cdot \frac{l}{h} \qquad K'_a = \frac{J'_a}{J_1} \cdot \frac{l}{h'}; \qquad F_a = \frac{K_a}{1/\varepsilon_a + 4} \qquad F'_a = \frac{K'_a}{4 + 1/\varepsilon'_a}; \\
S_1 &= 1/(2 - \varepsilon_1); \qquad\qquad\qquad\qquad S^v_A = F_a + F'_a + S_1; \\
\eta_a &= F_a/S^v_A \qquad \eta'_a = F'_a/S^v_A \qquad \eta_1 = S_1/S^v_A; \qquad (\eta_a + \eta'_a + \dot\eta_1 = 1).
\end{aligned}\right\} \quad (1688)$$

Loading Condition 4—Overall loading condition

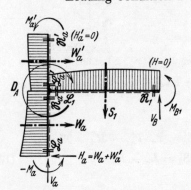

Fig. 375

Loading diagram with load terms $\mathfrak{L}$ and $\mathfrak{R}$ for all the individual members, external rotational moment D, and all reactions at ends of members

Restraining moments of the individual members, and joint load term:

$$\left.\begin{aligned}
\mathfrak{F}_{ra} &= \frac{(\mathfrak{S}_{la} + W'_a h)(1/\varepsilon_a + 1) - (\mathfrak{L}_a + \mathfrak{R}_a)}{1/\varepsilon_a + 4} \qquad \mathfrak{E}_{l1} = -\frac{\mathfrak{L}_1 - \mathfrak{R}_1 \cdot \varepsilon_1}{2 - \varepsilon_1} \\
\mathfrak{F}'_{la} &= -\frac{\mathfrak{S}'_{la}(1 + 1/\varepsilon'_a) + (\mathfrak{L}'_a + \mathfrak{R}'_a)}{4 + 1/\varepsilon'_a}. \qquad \mathfrak{R}^v_A = D_A + \mathfrak{F}_{ra} - \mathfrak{F}'_{la} - \mathfrak{E}_{l1}.
\end{aligned}\right\} \begin{matrix}(1689)\\ \text{and}\\ (1690)\end{matrix}$$

Column moments:

$$\begin{aligned}
M_{Aa} &= \mathfrak{F}_{ra} - \mathfrak{R}^v_A \cdot \eta_a & M'_{Aa} &= \mathfrak{F}'_{la} + \mathfrak{R}^v_A \cdot \eta'_a \\
M_a &= -\mathfrak{S}_{la} - W'_a h + M_{Aa}; & M'_a &= \mathfrak{S}'_{la} + M'_{Aa}.
\end{aligned} \qquad (1691)$$

Beam moments:

$$M_{A1} = \mathfrak{E}_{l1} + \mathfrak{R}^v_A \cdot \eta_1 \qquad\qquad M_{B1} = -(\mathfrak{R}_1 + M_{A1})\varepsilon_1. \qquad (1692)$$

Check relationship:

$$D_A + M_{Aa} - M'_{Aa} - M_{A1} = 0. \qquad (1693)$$

Appendix
Explanatory Notes, Derivations, and Numerical Examples
A. General Considerations

The structural analysis of any multi-membered framed structure may, in principle, be based upon either of two fundamentally different "exact" methods, namely, the Method of Forces (here called "Method I") or the Deformation Method (here called "Method II").[1] The complexity of the calculations relating to a framework will generally depend upon its degree of "statical indeterminacy" (or "redundancy") in a case where the Method of Forces is adopted, and upon its degree of "geometrical indeterminacy" in a case where the Deformation Method is adopted.[2]

The first step, when dealing with a framework which is statically indeterminate (or redundant) to the nth degree, is to determine the values of the n redundants (or statically indeterminate quantities)—forces or moments —from n independent elastic equations. These elastic equations express geometrical compatibility conditions, deduced from a series of geometrically independent self-equilibrating states of stress, which represent conditions of equilibrium that exist without the action of the given internal or external loading of the structure.

Once the n redundants have been determined, the next step is to determine the values of all the other forces and moments in the same manner as when dealing with a statically determinate structure. In some cases the final displacements of the structure will have to be ascertained.

In the present volume, the end moments of the beam spans have consistently been adopted as the statically indeterminate quantities X.* This being so, the so-called "statically determinate primary system" is obtained by introducing an appropriate number of pin-joints ("hinges") into the structure.

The concept "geometrical indeterminacy" is not—as might readily be supposed—exactly analogous to the concept "statical indeterminacy", inasmuch as the solution of a "geometrically indeterminate" structural system cannot be achieved solely by means of geometrical compatibility conditions

---

[1] So-called iteration methods (e.g., Hardy Cross's moment distribution method) or semi-rigorous method (e.g., the method of transverse sections) will not be considered in the present treatment.

[2] An excellent discussion of the concepts "statical and geometrical indeterminacy" is given in: *Die Methoden der Rahmenstatik, Aufbau, Zusammenfassung und Kritik*, by Dr.-Ing. habil. Otto Luetkens, publ. Springer-Verlag, Berlin/Göttingen/Heidelberg, 1949.

* By "beam span" is meant the portion of "the beam"—i.e., the continuous horizontal member —spanning between two successive columns. (Translator's note)

(cf. Luetkens' book referred to in footnote 2). However, it is always necessary that the determining equations for the geometrical unknowns should fulfil the conditions of statical equilibrium obtained from independent sets of self-equilibrating states of strain. These self-equilibrating states of strain represent deformation conditions which exist irrespective of the deformations due to internal or external loading of the structure.

For all the "laterally restrained" rigid frames (i.e., frames in which "side-sway"—lateral displacement of joints—is prevented) considered in this book, the geometrically indeterminate quantities Y adopted are the angles of rotation φ of the joints, expressed dimensionally as moments. The "geometrically determinate primary system" is obtained by assuming that the joints of the frame are immovably fixed, though they are actually capable of elastic rotation. The end moments of the members are then determined from the fixed-end moments and the unknowns Y, followed by the determination of all the accompanying forces.

When using either Method I or Method II, it is possible to establish, for all laterally restrained multi-bay frames dealt with in this book, trinomial elastic equations which are easily solved by means of reduction or recursion formulas.

Actually, in the case of "laterally unrestrained" types of framed structures (i.e., structures in which "side-sway" can occur), which are indicated by the subscript "v", the degree of static indeterminacy is one degree lower than for the corresponding laterally restrained structure, whereas the degree of geometrical indeterminacy is one degree higher than for the latter. In the case of unrestrained frames, however, the elastic equations directly established by either of the Methods I and II are no longer trinomial. Nevertheless, in order not to forgo the advantages of trinomial equations when dealing with laterally unrestrained framed structures, it is always assumed in the first stage of the calculations that the continuous beam, or the top of each two-story column, is horizontally fixed so as to prevent side-sway, and the "restraining force" H (unbalanced horizontal shear) is calculated. A horizontal concentrated load $W = H$ is then imposed on the system, and the displacement effect of this load is taken into account in establishing an additional elastic equation (displacement equation). By superposition of the results of the two stages of analysis—the horizontally fixed and the horizontally movable condition—the required final result for the actual (i.e., laterally unrestrained) frame is obtained. This is always a practicable procedure, which furthermore appears to be appreciably simpler than other methods when dealing with statical or geometrical indeterminacy of the third or higher degree.

Finally, it should also be mentioned that the second stage of the calculations is characterized by the emergence of the angles of rotation ψ of the columns as a special group of geometrical unknowns. In consequence of the attachment of the elastically movable column ends to the continuous beam, only one such angle ψ can be geometrically independent in the case of a one-story multi-bay frame. For a two-story frame there are, of course, two such angles. These angles of rotation of the members concerned can, however, be ignored during the first stage of the analysis, on the assumption that sidesway is prevented, as explained above.

B. Straight Members and Joints of Frames

1. Frame member, with constant moment of inertia, subjected to arbitrary load

a. The general slope deflection equation and special cases

(α) Fig. 376a shows an arbitrarily loaded frame member, rotated to the "normal position" (as envisaged in the book *Belastungsglieder*, Section A, Par. 3, Sub-Section a, page 2)* for convenience of representation. The two adjacent frame joints are indicated schematically. In Fig. 376b the member LR has been separated from the rest of the framed structure by the application of two complete cuts and has been provided with the six requisite "stress resultants" (the equilibrating internal forces and moments), namely, N_l, Q_l, M_L and N_r, Q_r, M_R: All the stress resultants, except Q_r, are assumed positive. The resultant P of the loading acting directly on the member will in general be oblique in relation to the axis of the member, but can always be resolved into mutually perpendicular components S and W.

Fig. 376c shows the basic shape of the moment diagram, divided into the parts due to M_L, M_R and M_x^0. The shear diagram is not shown. The fundamental statical relationships existing between loads, shear forces and bending moments will be assumed to be known.[1]

For determining the six stress resultants, only the three conditions of equilibrium in the plane of the structure are available, namely, $\Sigma V = 0$, $\Sigma H = 0$ and $\Sigma M = 0$. Hence the frame member under consideration is statically indeterminate to the third degree, that is to say, three additional geometrical elastic equations are required for solving the problem.

In Fig. 376d the general deformation of the member is drawn to an exaggerated scale. The member has been displaced from its original rectilinear position LR to the new curved position $L_1 R_2$. This change can be described in general terms of linear and angular displacements:

---

[1] A concise treatment of these statical relationships is given in the auxiliary book *Belastungsglieder*, 8th ed. (see footnote in Section 1 of the "Introduction" of the present volume), Section B, page 9 – 22, publ. Wilhelm Ernst & Sohn.—A very exhaustive discussion of the subject is given in the book entitled *Durchlaufträger*, 7th ed., Vol. I (see footnote in Section 4 of "Introduction" of the present volume), Section B, page 8 – 51, publ. Wilhelm Ernst & Sohn, Berlin, 1949.

* "Beam Formulas", Appendix I, Section B. 3.

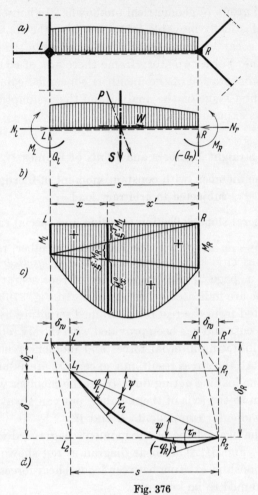

Fig. 376
Arbitrary frame member considered as an individual entity

δ_w displacement of the member in the direction of its longitudinal axis;

δ_L, δ_R displacement of the ends of the member (i.e., the joints of the frame) at right angles to the axis of the member;

$\delta = \delta_R - \delta_L$ relative displacement of the ends of the member at right angles to the axis of the member;

ψ angle of rotation of the member, i.e., the angle between the chord of the deflection curve and the original position of the member;

τ_l, τ_r tangent angle, i.e., the angle between the end tangent of the deflection curve and its chord;

φ_L, φ_R angle of rotation of joint, i.e., the angle between the end tangent of the deflection curve and the original position of the member.*

All translational displacements and angles of rotation are reckoned as positive, in accordance with the sign conventions—with the exception of φ_R.

Having regard to Fig. 376d, two geometrical relationships can immediately be established:

for L_1: $\quad \varphi_L = \psi + \tau_l,$ $\qquad$ for R_2: $\quad \tau_r = \psi + (-\varphi_R),$ $\qquad$ (1694)

or, on rearranging:

$$\tau_l - \varphi_L + \psi = 0 \quad \text{and} \quad \tau_r + \varphi_R - \psi = 0. \qquad (1695)$$

The tangent angles τ_l and τ_r are dependent only on the total moment diagram, divided by EJ, of the member under consideration.[1] From equations (24) and (25) in *Belastungsglieder* (page 24/25**) we obtain:

$$\tau_l = \frac{s}{6EJ} \cdot \mathfrak{L} \quad \text{and} \quad \tau_r = \frac{s}{6EJ} \cdot \mathfrak{R}, \qquad (1696)$$

where, with reference to Fig. 376c,   and   denote the load terms obtained from the total moment diagram. The portions due to the triangular areas with the end ordinates M_L and M_R can be directly written with the aid of Load Conditions 92 and 93 in *Belastungsglieder* (page 188). We thus finally obtain:

$$\tau_l = \frac{s}{6EJ}(2M_L + M_R + \mathfrak{L}), \qquad \tau_r = \frac{s}{6EJ}(M_L + 2M_R + \mathfrak{R}). \qquad (1697)$$

In these expressions $\mathfrak{L}$ and $\mathfrak{R}$ refer only to the segment-shaped portion of the moment diagram (with ordinates M_x^o) due to the loading acting directly on the member. See also equation (159), page 63, in *Durchlaufträger*, Vol. I.

Substituting the expressions (1697) into equation (1695) and multiplying both sides of the equations by

$$\mathfrak{T} = \frac{6EJ}{s} \qquad (1698)$$

we finally obtain the equations for the ends of the member:

left: $\qquad 2M_L + M_R + \mathfrak{L} - \mathfrak{T}\varphi_L + \mathfrak{T}\psi = 0,$

right: $\qquad M_L + 2M_R + \mathfrak{R} + \mathfrak{T}\varphi_R - \mathfrak{T}\psi = 0.$ $\qquad (1699)$

The geometrical equations (1695), or the equations in their elastogeometrical form (1699) in which the expressions for the tangent angles have

---

[1] A concise treatment of these elastic relationships is given in *Belastungsglieder*, Section C, page 22 – 29; a very exhaustive discussion of the subject is given in *Durchlaufträger*, Vol. I, Section B, page 51 – 64.

* 1. For convenience the term "rotation" may be employed to denote "angle of rotation".
  2. In so far as the frames considered in this book are concerned, rotation of the type ψ occurs only in the case of columns of laterally unrestrained structures (i.e., those in which side-sway occurs). (Translator's notes)

** *Beam Formulas*, page 123.

been substituted, constitute two of the three additional relations required. One more relation, with reference to the elastic displacements of the end points of the member, would really be required. This relation may be ignored, however, because it has been assumed (in the section "Design Postulates") that the effect of axial forces on the deformation of the framework, and therefore on the statically indeterminate quantities, can be neglected. With respect to Fig. 376b, the effect of this assumption is that the two axial forces N_l and N_r can always be determined from considerations of simple statics, once the moments (and therefore also all the shear forces) for the whole framework have been calculated.[1] The degree of statical indeterminacy of the arbitrary frame member, as envisaged in Fig. 376, is thus reduced from three to two.

It is not possible to establish elasticity relations for the angular quantities φ_L, φ_R and ψ similar to the relations (1697) for the angular quantities τ_l and τ_r. The interrelation of the five angles, as in equations (1695), is based on purely geometrical considerations. Besides, the angles of rotation of the joints and of the members are dependent on the geometrical shape of the framework as a whole, of which the member under consideration is only a part. Hence, if the three rotations φ_L, φ_R and ψ are considered as geometrical unknowns, then the arbitrary member envisaged in Fig. 376 will be geometrically indeterminate to the third degree. By making use of the three statical equations of plane equilibrium, the forces acting upon the frame member in question will be uniquely determined. However, in view of the considerations given in Section A, the rotation of the member ψ need not be determined in the first stage of the calculations (joints assumed to be fixed, i.e., incapable of rotation). The effect of ψ or —what amounts to the same thing—the effect of a relative displacement of the ends of the member at right angles to its axis can be introduced at any subsequent stage through the direct load terms $\mathfrak{L}$ and $\mathfrak{R}$; we obtain from equations (1699):

$$\mathfrak{L} = +\mathfrak{T}\cdot\psi \quad \text{and} \quad \mathfrak{R} = -\mathfrak{T}\cdot\psi. \tag{1700}$$

From Fig. 376d we furthermore obtain:

$$\psi = \frac{\delta_R - \delta_L}{s} = \frac{\delta}{s}. \tag{1701}$$

(See also the formulas in the book *Belastungsglieder*, Loading Condition 135, page 226, and the derivations given in *Durchlaufträger*, Vol. II, page 355 et seq.).

If the angle of rotation of the member is eliminated, the degree of geometrical indeterminacy of the member is reduced from three to two. In the following treatment of the problem the angle ψ is indeed included in the

---

[1] For further information see *Durchlaufträger*, 7th ed., Vol. II, Section K, page 334 et seq., and also Section N, page 393 et seq., publ. Wilhelm Ernst & Sohn, Berlin, 1956.

equations; but, if desired, it can be discarded or it can—as a known quantity—be conceived as being included in $\mathfrak{L}$ and $\mathfrak{R}$.

(β) The elasto-geometrical equations (1699), when solved for M_L and M_R, yield the Slope Deflection Equations:

$$M_L = -\frac{2}{3}\mathfrak{L} + \frac{1}{3}\mathfrak{R} + \mathfrak{T}\left(\frac{2}{3}\varphi_L + \frac{1}{3}\varphi_R - \psi\right)$$
$$M_R = +\frac{1}{3}\mathfrak{L} - \frac{2}{3}\mathfrak{R} - \mathfrak{T}\left(\frac{1}{3}\varphi_L + \frac{2}{3}\varphi_R - \psi\right). \tag{1702}$$

Substituting:

$$\mathfrak{M}_l = -\frac{2}{3}\mathfrak{L} + \frac{1}{3}\mathfrak{R} \quad \text{and} \quad \mathfrak{M}_r = +\frac{1}{3}\mathfrak{L} - \frac{2}{3}\mathfrak{R}, \tag{1703}$$

the Slope Deflection Equations are obtained in the standard form:

$$M_L = \mathfrak{M}_l + \frac{2EJ}{s}(2\varphi_L + \varphi_R - 3\psi)$$
$$M_R = \mathfrak{M}_r - \frac{2EJ}{s}(\varphi_L + 2\varphi_R - 3\psi). \tag{1704}$$

For $\varphi_L = \varphi_R = 0$ and $\psi = 0$, i.e., for the special case where no rotation and no relative displacement of the joints at the ends of the member occur, the end moments as obtained from equations (1704) become $M_L = \mathfrak{M}_l$ and $M_R = \mathfrak{M}_r$. This means that the composite load terms $\mathfrak{M}_l$ and $\mathfrak{M}_r$ as expressed in equations (1703), represent the fixed end moments due to any given direct load on the single-span beam with full fixity at both ends and incapable of lateral displacement. $\mathfrak{M}_l$ and $\mathfrak{M}_r$ may also be called the restraining moments of the laterally restrained frame member subjected to an arbitrary direct load.

In the case of laterally restrained frames, and in the first stage of analysis of laterally unrestrained frames (assuming side-sway to be prevented), external loading does not produce any angular rotation of the members as such (i.e., the type of rotation expressed by the angle ψ and involving relative displacement of the two ends of a frame member). For this important case the Slope Deflection Equations are rewritten as follows:

$$M_L = \mathfrak{M}_l + K(2Y_L + Y_R) \qquad M_R = \mathfrak{M}_r - K(Y_L + 2Y_R). \tag{1705}$$

where:

$$Y_L = \frac{2EJ_k}{l_k} \cdot \varphi_L = \frac{\mathfrak{T}_k}{3} \cdot \varphi_L \qquad Y_R = \frac{2EJ_k}{l_k} \cdot \varphi_R = \frac{\mathfrak{T}_k}{3} \cdot \varphi_R \tag{1706}$$

and furthermore:

$$K = \frac{J}{J_k} \cdot \frac{l_k}{s} \quad \text{and} \quad \mathfrak{T}_k = \frac{6EJ_k}{l_k}. \tag{1707}$$
$$\tag{1708}$$

J_K denotes an arbitrary reference moment of inertia and l_k denotes an arbitrary reference length. $\mathfrak{T}_k$ is the so-called distortion factor of the system (it has the dimension of a bending moment) and K is the stiffness coefficient of the member under consideration (see below). Finally, the Y terms are the rotations of the joints (φ) expressed dimensionally as moments.

(γ) For a frame member having an external support, the two extreme cases are full end fixity, on the one hand, and full rotational freedom as obtained by the provision of a pin-joint (or hinge), on the other. The extreme cases represented in Fig. 377 and 378 call for no special comment. The condition $\varphi_R = 0$ or $\varphi_L = 0$ can at once be satisfied in equations (1702), (1704) or (1705).

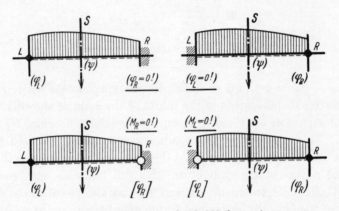

Fig. 377, 378 (top) and 379, 380 (bottom)
Special cases of the arbitrary frame member for full fixity or for hinged bearing at one end

For the extreme cases represented in Fig. 379 and 380 the conditions $M_R = 0$ or $M_L = 0$ must obviously be satisfied. The corresponding rotation at the external support—respectively φ_R and φ_L for the two cases under consideration—is no longer independent of the two other angles involved. Substituting $M_R = 0$ in the first of the two equations (1699), or alternatively substituting $M_L = 0$ in the second equation, we immediately obtain:

$$M_L = \mathfrak{M}'_l + \mathfrak{T} \frac{\varphi_L - \psi}{2}, \quad \text{where} \quad \mathfrak{M}'_l = -\frac{\mathfrak{L}}{2},$$

or alternatively

$$M_R = \mathfrak{M}'_r - \mathfrak{T} \frac{\varphi_R - \psi}{2}, \quad \text{where} \quad \mathfrak{M}'_r = -\frac{\mathfrak{R}}{2}. \tag{1709}$$

In these expressions $\mathfrak{M}'_l$ denotes the restraining moment for the member rigidly fixed at the left-hand end and pin-jointed at the right-hand end; similarly, $\mathfrak{M}'_r$ denotes the restraining moment for the member rigidly fixed at the right-hand end and pin-jointed at the left-hand end.

b. Elastically restrained end conditions, the spring constant and its practical substitute— the moment carry-over factor

(α) Between the two extreme cases for the end conditions of a frame member, namely, full end-fixity and full freedom of rotation as at a frictionless pin-point, there is an intermediate range of end conditions presenting elastic rotational restraint. These end conditions are characterized by the fact that the elastic rotation of the end of the member under consideration is proportional to the corresponding end moment of the member—irrespective of how the member itself is loaded, what the conditions at the other end are, and whether or not the structure is laterally restrained. The following simple relation is valid for every elastically restrained end of a member:

$$\varphi = f \cdot M. \tag{1710}$$

In this expression φ is the rotation of the point of elastic restraint, or the abutment of the member, conceived as the rotation of a joint; M is the end moment of the member at that point of elastic restraint; f denotes the "spring constant".

Analytically, φ is a non-dimensional value, and M is expressed dimensionally as a moment (e.g., ft-lb). Hence the spring constant is expressed dimensionally as the reciprocal of a moment (1/ft-lb), i.e., f can be considered as the rotation produced by a unit moment.

For full end-fixity we have $\varphi = 0$, and therefore $f = 0$. In the case of a pin-jointed end, φ generally has a finite value, and furthermore we have the condition $M = 0$. In the purely mathematical sense, equation (1710) will then assume the form $\varphi = \infty \cdot 0$. The limiting value $f \to \infty$ signifies that a pin-jointed end connection offers no resistance to rotation. It will be evident from this brief consideration of the problem that, on account of its infinitely large range of variation, namely:

$$0 \leq f \leq \infty, \tag{1711}$$

the spring constant f is not very suitable for practical purposes.

(β) In view of what has been said above, let us now consider the frame member, shown in Fig. 376, as being an end member with an elastically restrained connection on the right. The following relationship is valid for the connection to the right-hand portion of the framework (whatever its form):

$$\varphi_R = f_R \cdot M_R, \tag{1712}$$

in which the spring constant f_R can be taken to be a known quantity. The member under consideration is again represented in Fig. 381a. Furthermore, in Fig. 381b the new arrangement at the end R of the member is indicated, introducing the elastically rotating "abutment" to which the relationship (1712) refers.

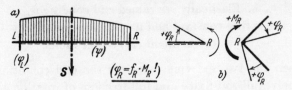

Fig. 381
Member with elastic restraint at right-hand end

Substituting φ_R as expressed by (1712) into equation (1699), we obtain:

$$\begin{aligned} 2M_L + M_R &= -\mathfrak{L} - \mathfrak{T}(\psi - \varphi_L) \\ M_L + (2 + \mathfrak{T} f_R) M_R &= -\mathfrak{R} + \mathfrak{T}\psi. \end{aligned} \quad (1713)$$

For $\mathfrak{R} = 0$ and $\psi = 0$, i.e., for the member subjected to direct external loading and in the absence of side-sway, the second equation (1713) reduces to:

$$M_L + (2 + \mathfrak{T} f_R) M_R = 0,$$

whence we obtain:

$$M_R = -\frac{1}{2 + \mathfrak{T} f_R} M_L = -\iota \cdot M_L. \quad (1714)$$

The new ratio introduced here, namely:

$$\iota = \frac{1}{2 + \mathfrak{T} \cdot f_R} \quad (1715)$$

can serve as a suitable substitute for the spring constant. Whereas f can vary over an infinite range of values, according to (1711), the range of variation for ι is as follows:

$$\boxed{1/2 \geq \iota \geq 0.} \quad (1716)$$

It should be clearly understood that the limiting value $\iota = \tfrac{1}{2}$ (corresponding to $f = 0$) applies to the case of full end-fixity and that the limiting value $\iota = 0$ (corresponding to $f = \infty$) applies to the pin-jointed end.

The ratio ι is known as the moment carry-over factor (also called the moment decay factor) for the left-to-right direction. Incidentally, f_R can, if desired, be expressed in terms of ι:

$$f_R = \frac{1 - 2\iota}{\mathfrak{T} \cdot \iota} = \frac{s(1 - 2\iota)}{6EJ \cdot \iota}. \quad (1717)$$

Since $(2 + \mathfrak{T} \cdot f_R) = 1/\iota$, equations (1713) can be rewritten as follows:

$$\begin{aligned} 2M_L + M_R &= -\mathfrak{L} - \mathfrak{T}\psi + \mathfrak{T}\varphi_L \\ M_L + M_R \cdot 1/\iota &= -\mathfrak{R} + \mathfrak{T}\psi. \end{aligned} \quad (1718)$$

Solving these equations for M_L and M_R, we obtain:

$$M_L = \frac{-\mathfrak{L} + \mathfrak{R}\,\iota}{2 - \iota} + \mathfrak{T}\left(\frac{\varphi_L}{2 - \iota} - \frac{1 + \iota}{2 - \iota} \cdot \psi\right), \tag{1719}$$

$$M_R = \frac{-2\mathfrak{R} + \mathfrak{L}}{2 - \iota} \cdot \iota - \mathfrak{T}\left(\frac{\varphi_L \cdot \iota}{2 - \iota} - \frac{3\iota}{2 - \iota} \cdot \psi\right). \tag{1720}$$

For the first term of the right-hand side of equation (1719) we may set:

$$\boxed{\mathfrak{E}_l = -\frac{\mathfrak{L} - \mathfrak{R} \cdot \iota}{2 - \iota},} \tag{1721}$$

while the first term of the right-hand side of equation (1920) can be modified as follows:

$$\frac{-2\mathfrak{R} + \mathfrak{L}}{2 - \iota} \cdot \iota + \frac{\mathfrak{R}\iota - \mathfrak{R}\iota}{2 - \iota} \cdot \iota = -(\mathfrak{E}_l + \mathfrak{R})\,\iota. \tag{1722}$$

The term containing ψ in equation (1720) can similarly be modified, so that equations (1719) and (1720) can thus be simplified to:

$$\boxed{M_L = \mathfrak{E}_l + \frac{\mathfrak{T}\varphi_L}{2 - \iota} - \mathfrak{T}\psi \cdot \frac{1 + \iota}{2 - \iota},} \tag{1723}$$

$$M_R = -(M_L + \mathfrak{R} - \mathfrak{T}\psi) \cdot \iota. \tag{1724}$$

For $\varphi_L = 0$ and $\psi = 0$, i.e., for the laterally restrained member LR with full end-fixity at L and elastically restrained connection at R, the end moments become $M_L = \mathfrak{E}_l\,\iota$ and $M_R = -(M_L + \mathfrak{R}).\iota$. The newly introduced load term $\mathfrak{E}$ may be called the "restraining moment". For $\iota = \tfrac{1}{2}$, i.e., for full fixity at both ends of the member, $\mathfrak{E}_l$ from equation (1721) becomes $\mathfrak{M}_l$ of the first equation (1703); similarly the expression $-(\mathfrak{E}_l + \mathfrak{R})\,\iota$ of equation (1722) becomes $\mathfrak{M}_r$ of the second equation (1703). For $\iota = 0$, i.e., for full end-fixity at L and a pin-joint at R, we have $\mathfrak{E}_l = -\mathfrak{L}/2 = \mathfrak{M}'_l$, while M_R vanishes.

Omitting the rotation ψ of the member and rewriting the slope deflection equations (1723) and (1724) so as to obtain an arrangement similar to (1705), we arrive at the following expressions:

$$\boxed{M_L = \mathfrak{E}_l + 3S \cdot Y_L \quad \text{where} \quad S = \frac{K}{2 - \iota},} \tag{1725}$$

$$M_R = -(M_L + \mathfrak{R}) \cdot \iota. \tag{1726}$$

The newly introduced value S is the (left-hand) stiffness of the member LR (see below).

(γ) Fig. 382 represents a situation which is the reverse of that envisaged in Fig. 381. The derivations are similar and need not be dealt with in detail; suffice it to give the results.

The symbols employed have the following significance:

$\varphi_L = -f_L . M_L$ the "elastic" relationship for the end of the member at L

$M_L = -\iota' . M_R$ moment carry-over for laterally restrained member.
$\iota' = 1/(2 + f_L)$ moment carry-over factor in the right-to-left direction.

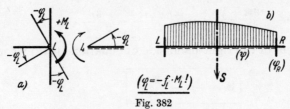

Fig. 382

Member with elastic restraint at left-hand end

The equations corresponding to (1718) are:

$$M_L \cdot 1/\iota' + M_R = -\mathfrak{L} - \mathfrak{T}\psi \atop M_L + 2M_R = -\mathfrak{R} + \mathfrak{T}\psi - \mathfrak{T}\psi_R. \Biggr\} \quad (1727)$$

Whence we obtain the end moments:

$$M_R = \mathfrak{E}_r - \frac{\mathfrak{T}\varphi_R}{2 - \iota'} + \mathfrak{T}\psi \frac{1 + \iota'}{2 - \iota'}, \quad (1728)$$

$$M_L = -(M_R + \mathfrak{L} + \mathfrak{T}\psi) \cdot \iota', \quad (1729)$$

with the restraining moment:

$$\mathfrak{E}_r = -\frac{\mathfrak{R} - \mathfrak{L} \cdot \iota'}{2 - \iota'}. \quad (1730)$$

For $\iota' = \tfrac{1}{2}$, i.e., for full fixity at both ends of the member, $\mathfrak{E}_r$ becomes $\mathfrak{M}_r$ and the expression $-(\mathfrak{E}_r + \mathfrak{L}) \iota'$ becomes $\mathfrak{M}_l$. For $\iota' = 0$, i.e., for a pin-joint at L, $\mathfrak{E}_r$ becomes $\mathfrak{M}_r'$ while M_L vanishes.

Omitting ψ and rewriting so as to obtain an arrangement similar to (1705):

$$M_R = \mathfrak{E}_r - 3S' \cdot Y_R \text{ where } S' = \frac{K}{2 - \iota'}, \quad (1731)$$

$$M_L = -(M_R + \mathfrak{L}) \cdot \iota' \quad (1732)$$

(δ) How the elastically restrained end condition arises has not yet been defined in the foregoing consideration of the subject. The existence of a "resilient abutment" was merely mentioned. The spring constant f of such an "abutment", expressed by the more convenient carry-over factor for the member connected to the abutment, may, in any given practical case, be estimated, measured by experimental means or known in some other way; alternately, it may be determined by rigorous calculation of the behaviour of the "abutment". In order to be able to make use of the "degree of fixity" (or "degree of restraint") thus obtained, the symbols ϵ and ϵ' may be substituted for ι and ι' in the formulas, i.e.:

$$\varepsilon \triangleq \iota \text{ and } \varepsilon' \triangleq \iota'. \quad (1733)$$

This means to say that the carry-over factors ϵ and ϵ' are assumed to be known quantities when using the formulas concerned.

Note: The expressions (1733) are strictly applicable to the end spans of the beams of multi-bay frames without end columns. On the other hand, for formal reasons, a departure from this rule has been introduced in the case of the lower and upper columns of multi-bay frames; in this case the following convention has been adopted: $\epsilon \triangleq \iota'$ and $\epsilon' \triangleq \iota^1$.

c. Frame member with elastic end restraints and side-sway

Note: In this book members of this kind occur only in those framed structures (denoted by the subscript "v") which are directly analysed as "laterally unrestrained" frames, i.e., single-story and symmetrical two-story types of frame. Hence we shall here consider only the two relevant cases of columns with side-sway.

(α) Fig. 383a shows one column of a multi-bay rigid frame, subjected to an arbitrary horizontal loading. The lower end L of the member under consideration is immovably gripped, whereas the upper end R is free to undergo horizontal displacement (side-sway). This means that rotation ψ will occur (as defined in Section B.1.a.α) in consequence of the deflection of the member.

Irrespective of the degree of fixity of the ends L and R of the column,

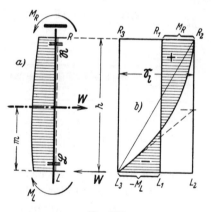

Fig. 383
Column of frame capable of side-sway

the corresponding moment diagram must essentially have the form represented in Fig. 383b, i.e., the following statical relationships will always hold true:

$$Q_l = W \quad Q_r = 0; \quad M_R - M_L = (\mathfrak{S}_l = W \cdot m). \tag{1734}$$

---

[1] This new method of notation has unfortunately not yet been introduced into the formulas for the elastically restrained beam types 4, 7, 12, 15, 20, 23, 28, 31 and 35 in *Durchlaufträger*, Vol. I.

Only the position of the vertical "closer" L_1R_1 will depend on the conditions of elastic restraint. In the extreme case where there is full rotational freedom at R (in addition to freedom of horizontal displacement), i.e., in the case of a cantilever with elastically restrained connection at L, the closing line will have shifted its position from L_1R_1 to L_2R_2 and the moment diagram will have become entirely negative. In the other extreme case where there is a pin-joint at L and elastically restrained connection (in conjunction with freedom of horizontal displacement) at R, the closing line L_1R_1 will have shifted to L_3R_3 and the moment diagram will have become entirely positive.

(β) For an upper column the relationship (1712) must be fulfilled, and hence the pair of equations (1718) is applicable, in which—as explained in the note at the end of Section b — ι should be replaced by ϵ'. Adding the two equations (1718), we obtain the following elastic relationship, which is no longer dependent upon the rotation of the member:

$$3M_L + (1 + 1/\epsilon')M_R = -(\mathfrak{L} + \mathfrak{R}) + \mathfrak{T}\varphi_L. \qquad (1735)$$

Substituting

$$\boxed{M_R = +\mathfrak{S}_l + M_L} \qquad (1736)$$

as obtained from the statical relationship for the moment (1734), into equation (1735), we finally obtain:

$$M_L = -\frac{\mathfrak{S}_l(1 + 1/\epsilon') + (\mathfrak{L} + \mathfrak{R}) - \mathfrak{T}\varphi_L}{4 + 1/\epsilon'}.$$

And putting:

$$\boxed{\mathfrak{F}'_l = -\frac{\mathfrak{S}_l(1 + 1/\epsilon') + (\mathfrak{L} + \mathfrak{R})}{4 + 1/\epsilon'}}, \qquad (1737)$$

we find:

$$\boxed{M_L = \mathfrak{F}'_l + \frac{\mathfrak{T}\cdot\varphi_L}{4 + 1/\epsilon'}.} \qquad (1738)$$

For $\varphi_L = 0$, i.e., for full end-fixity at L, we have $M_L = \mathfrak{F}'_l$, while equation (1736) remains unchanged. Hence it follows that $\mathfrak{F}'_l$ is the restraining moment at the base of the upper column (i.e., at the beam joint) when, at the same time, the elastically restrained top of the column is free to undergo horizontal displacement.

For $\epsilon' = \frac{1}{2}$, i.e., for the column LR fully fixed at both ends but free to undergo horizontal displacement at R, we have:

$$\mathfrak{F}'_l = -\frac{\mathfrak{S}_l}{2} - \frac{(\mathfrak{L} + \mathfrak{R})}{6}. \qquad (1739)$$

For $\epsilon' = 0$, i.e., for the cantilever with full end-fixity at L, $\mathfrak{F}'_l$ becomes $\mathfrak{S}_l$.[1]

(γ) For a lower column the relationship $\varphi_L = -f_L \cdot M_L$ must be fulfilled, and hence the pair of equations (1727) is applicable, except that ϵ must be substituted for ι'. By adding the two equations we again obtain a relationship that is independent of ψ:

$$(1/\varepsilon + 1) M_L + 3 M_R = -(\mathfrak{L} + \mathfrak{R}) - \mathfrak{T}\varphi_R. \tag{1740}$$

Substituting:

$$\boxed{M_L = -\mathfrak{S}_l + M_R,} \tag{1741}$$

we obtain:

$$\boxed{M_R = \mathfrak{F}_r - \frac{\mathfrak{T}\cdot\varphi_R}{1/\varepsilon + 4},} \tag{1742}$$

where:

$$\boxed{\mathfrak{F}_r = +\frac{\mathfrak{S}_l(1/\varepsilon + 1) - (\mathfrak{L} + \mathfrak{R})}{1/\varepsilon + 4}.} \tag{1743}$$

For $\varphi_R = 0$, i.e., for full end-fixity at R, we have—according to equation (1742)—$M_R = \mathfrak{F}_r$, while equation (1741) remains unchanged. Hence $\mathfrak{F}_r$ is the restraining moment at the top of the lower column, when, at the same time, this column is free to undergo horizontal displacement and the base of the column is elastically restrained.

For $\epsilon = \frac{1}{2}$, i.e., for the column LR fully fixed at both ends but free to undergo horizontal displacement at R, we have:

$$\mathfrak{F}_r = +\frac{\mathfrak{S}_l}{2} - \frac{(\mathfrak{L} + \mathfrak{R})}{6}. \tag{1744}$$

For $\epsilon = 0$, i.e., for the column with a pin-joint (hinged bearing) at L and free to undergo horizontal displacement at R, $\mathfrak{F}_r$ becomes $+\mathfrak{S}_l$.[1]

d. Flexibility and rigidity

(α) The statical-geometrical significance of equation (1714), which is valid for the case of the member not subjected to direct external loading and in the absence of side-sway, is represented schematically in Fig. 384a. It shows that moment carry-over takes place from left to right, this being governed by the value ι. Point L could appropriately be called the "active support", and R the "passive abutment". The active end moment M_L represents the "loading influence", which is exercised by the portion of the structure situated to the left of L. The passive end moment M_R represents the elastic rotational resistance (due to elastic restraint) offered by the portion of the structure situated to the right of R. In view of this interpretation, rotational freedom can be assumed to exist at the support L (i.e., it can be conceived as a pin-joint).

---

[1] As a matter of interest it may be noted that there are in each case really four fixed-end moments $\mathfrak{F}$, each with the subscripts l and r, both with and without the superscript $'$.

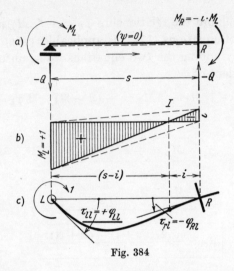

Fig. 384

For the unit value $M_L = +1$ of the applied moment, the carry-over moment becomes $M_R = -\iota$; this special case is represented in Fig. 384b. In this diagram, I denotes the right-hand "fixed point" and i is its distance from R. (See *Durchlaufträger*, Vol. I, page 74 and Fig. 49).

In consequence of the loading condition constituted by $M_L = +1$, the deflection curve indicated in Fig. 384c is produced, with the tangent angles τ_{ll} and τ_{rl}. For the sake of completeness, the relationship of each of these tangent angles with the corresponding rotation of the joint (φ) is likewise indicated in the diagram (in this case the absolute values of the tangent angle and angle of rotation are identical). Using Mohr's Third Theorem (see *Durchlaufträger*, Vol. I, page 58) the tangent angles are found to be:

$$\tau_{ll} = \frac{s(2-\iota)}{6EJ} = b^w \quad \text{and} \quad \tau_{rl} = \frac{s(1-2\iota)}{6EJ}. \quad (1745)(1746)$$

The tangent angle τ_{ll} due to the applied moment $M_L = +1$, i.e., the rotation of the pin-jointed end L of the member due to a unit moment, is designated as the "true flexibility" of the left-hand end of the frame member LR. In order not to be committed to the concept "tangent angle", however, the letter "b" has been adopted to denote "flexibility", while the superscript "w" is introduced to indicate "true".

For practical reasons it is convenient to work with a "distorted" or "magnified" value of the flexibility, consisting of $\mathfrak{T}_k$ times the true flexibility, i.e.:

$$b = \mathfrak{T}_k \cdot b^w = k(2-\iota), \quad (1747)$$

where $\mathfrak{T}_k$ is the distortion factor of the system, and k is the "flexibility coefficient":

$$\mathfrak{T}_k = \frac{6EJ_k}{l_k} \quad \text{and} \quad k = \frac{J_k}{J} \cdot \frac{s}{l_k}. \tag{1748}$$
$$\tag{1749}$$

The mirror-inverted situation is represented in Fig. 385. I' is the left-hand fixed point and i' is its distance from L. Applying $M_R = +1$, we obtain:

$$\tau_{lr} = \frac{s(1-2\iota')}{6EJ} \quad \text{and} \quad \boxed{\tau_{rr} = \frac{s(2-\iota')}{6EJ} = b'^w}. \tag{1750}$$
$$\tag{1751}$$

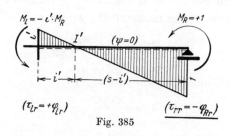

Fig. 385

Here the tangent angle τ_{rr} denotes the "true flexibility" of the right-hand end R of the frame member LR. And furthermore we similarly have:

$$b' = \mathfrak{T}_k \cdot b'^w = k(2-\iota'). \tag{1752}$$

Hence every frame member has two flexibility values which—on reducing them to non-dimensional reference values—have the following form (for the left-hand end L and the right-hand end R respectively):

$$\boxed{b = k(2-\iota) \quad \text{and} \quad b' = k(2-\iota').} \tag{1753}$$

(β) Dividing (1746) by (1745), we obtain (with reference to Fig. 384c):

$$\frac{\tau_{rl}}{\tau_{ll}} = -\frac{\varphi_{Rl}}{\varphi_{Ll}} = \boxed{j = \frac{1-2\iota}{2-\iota}} \tag{1754}$$

and furthermore:

$$\boxed{\varphi_{Rl} = -j \cdot \varphi_{Ll}.} \tag{1755}$$

Dividing (1750) by (1751), we similarly obtain:

$$\frac{\tau_{lr}}{\tau_{rr}} = -\frac{\varphi_{Lr}}{\varphi_{Rr}} = \boxed{j' = \frac{1-2\iota'}{2-\iota'}} \tag{1756}$$

and furthermore:

$$\boxed{\varphi_{Lr} = -j' \cdot \varphi_{Rr}.} \tag{1757}$$

The proportionality factors j and j' which appear in equations (1755) and (1757) are termed "rotation carry-over factors". In calculations based on the Deformation Method (Method II) these factors j play a role similar to that of the factors ι in calculations based on the Method of Forces (Method I).

The factors j and ι (and similarly j' and ι') are mutually interchangeable in equations (1754) and (1756):

$$\iota = \frac{1-2j}{2-j} \quad \text{and} \quad \iota' = \frac{1-2j'}{2-j'}. \tag{1758}$$

(γ) The tangent angle, or the rotation φ_{Ll} of the joint, as expressed by equation (1745) and indicated in Fig. 384c, pertains to $M_L = +1$. Hence for an arbitrary value of the moment M_L we have:

$$\varphi_L = M_L \cdot \tau_{ll} = \frac{M_L s(2-\iota)}{6EJ}. \tag{1759}$$

To satisfy the condition $\varphi_L = 1$, i.e., for:

$$1 = \frac{M_L s(2-\iota)}{6EJ},$$

M_L acquires the fictitious particular value:

$$M_L^{(\varphi=1)} = \frac{6EJ}{s(2-\iota)} = \frac{1}{\tau_{ll}} = \frac{1}{b^w} = S^w. \tag{1760}$$

The fictitious end moment S^w at the end L (assumed to be pin-jointed) which would produce the unit rotation at that end of the member[1] is designated as the "true rotational stiffness" of the left-hand end L of the frame member LR.

Having regard to equations (1760) and (1745), the true rotational stiffness is the reciprocal of the true flexibility.

Again, for practical reasons it is convenient to work with a "distorted" value of the stiffness, consisting of $1/\mathfrak{T}_k$ times the true stiffness, i.e.:

$$S = \frac{1}{\mathfrak{T}_k} \cdot S^w = \frac{1}{k(2-\iota)} = \frac{K}{2-\iota} = \frac{1}{b}, \tag{1761}$$

where $\mathfrak{T}_k$ is the distortion factor of the system (as already envisaged above), $K = 1/k$ is the "stiffness coefficient".

For the mirror-inverted situation (see Fig. 385) we similarly have:

$$-\varphi_R = M_R \cdot \tau_{rr} = \frac{M_R s(2-\iota')}{6EJ} \tag{1762}$$

---

[1] The moment $M(Q=1) = S^w$ is a fictitious (i.e., an imaginary) value just as, for example, the stress-strain proportionality factor designated as the modulus of elasticity E; in reality, the material would be structurally destroyed long before these quantities had even remotely developed the characteristic values attributed to them.

And furthermore, for the condition $\varphi_R = -1$:

$$M_R^{(\varphi=-1)} = \frac{6EJ}{s(2-\iota')} = \frac{1}{\tau_{rr}} = \frac{1}{b'^w} = S'^w. \qquad (1763)$$

The fictitious end moment S'^w is the "true rotational stiffness" of the right-hand end R of the frame member LR.

Hence every frame member has two rotational stiffness values (which, for convenience, can simply be referred to as "stiffnesses"). On reducing them to non-dimensional reference values, they have the following form (for the left-hand end L and the right-hand end R respectively):

$$S = \frac{K}{2-\iota} \quad \text{and} \quad S' = \frac{K}{2-\iota'}. \qquad (1764)$$

(δ) The concepts and relationships dealt with in the foregoing sections (α) to (γ) may be summarized as follows:

True flexibility $b^w \triangleq$ angle of rotation for unit moment
True stiffness $S^w \triangleq$ rotational moment for unit angle

(Distorted) flexibility $b \triangleq$ angle of rotation for moment $\mathfrak{T}_k$
(Distorted) stiffness $S \triangleq$ rotational moment for angle $1/\mathfrak{T}_k$

The values "b" and "S" are provided with a prime (thus: b', S') when they refer to the right-hand end of a member.

The values ϵ as defined in (1733) are used in the case of end members with elastic restraint and having values ι that can be considered to be known quantities (as envisaged in Section B.1.b.δ).

For any particular member the flexibilities and stiffnesses are always reciprocals of each other:

$$\begin{array}{ll} b^w \cdot S^w = 1 & \text{and} \quad b'^w \cdot S'^w = 1 \\ b \cdot S = 1 & \text{and} \quad b' \cdot S' = 1 \\ & k \cdot K = 1 \end{array} \qquad (1765)$$

and furthermore:

In the theoretical case of the freely suspended member LR, as represented in Fig. 368, for the "unit-moment condition $M = 1$", the true relative rotation of the two ends of the member can be obtained by applying Mohr's First Theorem (see *Durchlaufträger*, Vol. I, page 52):

$$\omega_1 = 2\tau_1 = \frac{s}{EJ}, \quad (1/\text{ft–lb}) \qquad (1766)$$

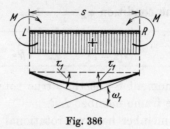

Fig. 386

The "distorted" value of the relative rotation, obtained by multiplying the true relative rotation by a factor EJ_k/l_k, becomes:

$$\frac{EJ_k}{l_k} \cdot \omega_1 = \frac{EJ_k}{l_k} \cdot \frac{s}{EJ} = \boxed{k = \frac{J_k}{J} \cdot \frac{s}{l_k}} \tag{1767}$$

and is called the "flexibility coefficient" (k). These coefficients k provide a means of comparing the flexural capacities of different members, without having to take into account the type of support or the end-fixity conditions.

The reciprocal of the flexibility coefficient, namely,

$$\boxed{K = \frac{1}{k} = \frac{J}{J_k} \cdot \frac{l_k}{s},} \tag{1768}$$

is called the "stiffness coefficient". With reference to Fig. 386, this coefficient corresponds to a pair of end moments of such magnitude as to produce a relative rotation of the two ends of the member through an angle equal to the unit angle of rotation multiplied by l_k/EJ_k.

Note: In this connection the reader is referred to *Durchlaufträger*, Vol. I, Section C.1.g, page 85/86. The discussion of flexibility and stiffness given there are essentially also applicable to the "b" and "S" values.

Replacing the actual rotation φ_L of the joint in equation (1759)—or φ_R in equation (1762)—by the general symbol φ^w, and ignoring the algebraic sign, the rotation can be expressed in terms of the end moment M at the same end of the member, in conjunction with the flexibility b^w or the stiffness S^w, as follows:

true rotation $\qquad\qquad \varphi^w = M \cdot b^w = M/S^w, \qquad\qquad (1769)$

distorted rotation

$$\boxed{\mathfrak{T}_k \cdot \varphi^w = \varphi = M \cdot b = M/S.} \tag{1770}$$

Finally, for the sake of completeness, the following relationships should be noted:

$$\left. \begin{array}{l} b = k(2-\iota) = \dfrac{3k}{2-j} = \dfrac{3}{K(2-j)} = \dfrac{2-\iota}{K} = \dfrac{1}{S} \\[2mm] S = \dfrac{K}{2-\iota} = \dfrac{K(2-j)}{3} = \dfrac{2-j}{3k} = \dfrac{1}{k(2-\iota)} = \dfrac{1}{b} \end{array} \right\} \tag{1771}$$

(ε) The extreme or limiting values for the flexibility b and the stiffness S can be obtained by substituting the relevant particular values of ι and ι' in the general equations (1753) and (1764). For full rotational freedom of the opposite end of the member, we have $\iota = 0$, or $\iota' = 0$, and therefore:

$$b = b' = 2k \quad \text{and} \quad S = S' = \frac{1}{2} K. \tag{1772}$$

For full fixity of the opposite end of the member, we have $\iota = \frac{1}{2}$, or $\iota' = \frac{1}{2}$, and therefore:

$$b = b' = \frac{3}{2} k \quad \text{and} \quad S = S' = \frac{2}{3} K. \tag{1773}$$

If the member LR under consideration is a symmetrical member of a symmetrical framework, the carry-over processes in opposite directions, for symmetrical and also for anti-symmetrical loading of the structure as a whole, can be superimposed into a single process.[1] While a detailed discussion would be outside the scope of the present volume, suffice it to note that "symmetrical carry-over" would correspond to the particular value $\iota = \iota' = -1$ and that "anti-symmetrical carry-over" would correspond to $\iota = \iota' = +1$, so that we have:

$$\text{Symmetry:} \quad b = b' = 3k \quad \text{and} \quad S = S' = \frac{1}{3} K; \tag{1774}$$

$$\text{Anti-symmetry:} \quad b = b' = k \quad \text{and} \quad S = S' = K. \tag{1775}$$

The flexibility and stiffness values relating to all cases that are likely to be encountered in practice are summarized in Table I. (With regard to Table II, see Section e).

e. Flexibility and rigidity in the case of a member with one end free to undergo lateral displacement

(α) The notions of "flexibility" and "stiffness", as conceived in Section d, are applicable only to frame members whose two ends are laterally restrained in the sense that they cannot undergo displacement in relation to each other (no side-sway) ($\psi = 0$). If this limiting condition is set aside, i.e., if the support hitherto conceived as pin-jointed is now assumed also to possess freedom of lateral displacement (that is, displacement in a direction at right angles to the axis of the member under consideration), then, for the same applied moments, the member will rotate through a certain definite angle ψ, as defined in Section B.1.a.α.

Fig. 387a now represents a cantilevered beam elastically restrained at its right-hand end and subjected to a moment M_L acting at its free left-hand end (cf. Fig. 384a). The moment M_L remains constant over the whole length

---

[1] See also the discussion in *Durchlaufträger*, Vol. I, page 116 and 117, Vol. II, page 385 – 393.

Table 1. Flexibility and rigidity of straight members

end condition	deflected shape	moment diagram	flexibility b or b'	stiffness S or S'
elastic restraint, right			$(2-\iota)\cdot k$	$\dfrac{K}{2-\iota}$
full end-fixity, right			$\dfrac{3}{2}\cdot k$	$\dfrac{2}{3}\cdot K$
full rotational freedom, right			$2\cdot k$	$\dfrac{1}{2}\cdot K$
elastic restraint, left			$(2-\iota')\cdot k$	$\dfrac{K}{2-\iota'}$
full end-fixity, left			$\dfrac{3}{2}\cdot k$	$\dfrac{2}{3}\cdot K$
full rotational freedom, left			$2\cdot k$	$\dfrac{1}{2}\cdot K$
symmetry			$3\cdot k$	$\dfrac{1}{3}\cdot K$
anti-symmetry			$1\cdot k$	$1\cdot K$

Table II. Free flexibility and free rigidity of straight members

end condition	deflected shape	moment diagram	free flexibility m or m'	free stiffness F or F'
elastic restraint, right			$(1/\iota + 4)\cdot k$	$\dfrac{K}{1/\iota + 4}$
full end-fixity, right			$6\cdot k$	$\dfrac{1}{6}\cdot K$
elastic restraint, left			$(4 + 1/\iota')\cdot k$	$\dfrac{K}{4 + 1/\iota'}$
full end-fixity, left			$6\cdot k$	$\dfrac{1}{6}\cdot K$

of the member to the right-hand end, where $M_R = M_L$. The moment diagram is a rectangle. There is now obviously no fixed point I as in the case of Fig. 384. No shear force occurs in any part of the member.

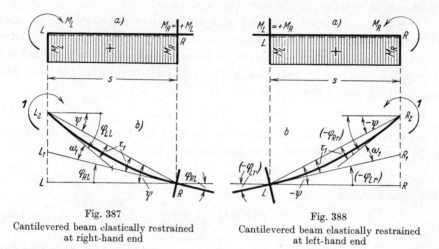

Fig. 387
Cantilevered beam elastically restrained at right-hand end

Fig. 388
Cantilevered beam elastically restrained at left-hand end

For the unit value $M_L = +1$ we obtain the deflection curve represented in Fig. 387b. If there was full fixity at R, the only angular rotations that would arise—in consequence of the bending of the member itself—would be the tangent angles τ_1 and the total angle of rotation ω_1, which angles must comply with equation (1766). The support at R has been assumed to be elastically rotatable, however, and will itself therefore undergo a rotation through an angle [see equation (1712)]:

$$\varphi_{Rl} = 1 \cdot f_R \qquad (1776)$$

Then, having regard to Fig. 387b, the total rotation of the free end L of the member will be:

$$\varphi_{Ll} = \omega_1 + f_R \qquad (1777)$$

In this equation ω_1 can be replaced by the expression (1766) multiplied by $6\iota/6\iota$, and the spring constant f_R can be replaced by the expression (1717):

$$\boxed{\varphi_{Ll} = \frac{s}{EJ} \cdot \frac{6\iota}{6\iota} + \frac{s(1-2\iota)}{6EJ \cdot \iota} = \frac{s(4\iota+1)}{6EJ \cdot \iota} = \frac{4+1/\iota}{\mathfrak{T}} = m^w.} \qquad (1778)$$

In analogy with the definition of $\tau_{ll} = b^w$, as expressed by equation (1745), we shall now designate the rotation φ_{Ll} due to the action of the moment $M_L = +1$ (i.e., the rotation of the free end of the member in consequence of the unit moment applied to it) as the "true free flexibility" of the left-hand end of the frame member LR.

And similarly the "distorted" value of the free flexibility, i.e., multiplied by a factor $\mathfrak{T}_k$ will be:

$$\mathfrak{T}_k \cdot m^w = m = k(4 + 1/\iota). \qquad (1779)$$

For $\iota = \frac{1}{2}$, i.e., with full fixity at the right-hand end of the member, we have $m = 6k$. For $\iota = 0$, i.e., for a pin-joint at the right-hand end, we have $m \to \infty$, which means that in this extreme case the support R offers no resistance to rotation of the member as a whole.

The mirror-inverted situation is represented in Fig. 388. Under these circumstances a positive bending moment M_R will obviously produce negative rotations φ, and these angles have accordingly been provided with a negative sign in Fig. 388. The geometrical relationship for the total rotation of the free end R of the member is:

$$(-\varphi_{Rr}) = \omega_1 + (-\varphi_{Lr}). \tag{1780}$$

The expression (1766) again holds true for ω_1, while the rotation of the semi-rigid support L is (according to Section B.b.γ):

$$(-\varphi_{Lr}) = 1 \cdot f_L. \tag{1781}$$

We finally obtain:

$$\boxed{-\varphi_{Rr} = \frac{s(4\iota' + 1)}{6EJ \cdot \iota'} = \frac{4 + 1/\iota'}{\mathfrak{T}} = m'^w.} \tag{1782}$$

In analogy with the foregoing, m'^w is designated as the "true free flexibility" of the right-hand end of the frame member LR. For practical purposes we shall again adopt the "distorted" value, multiplied by $\mathfrak{T}_k$:

$$\mathfrak{T}_k \cdot m'^w = m' = k(4 + 1/\iota'). \tag{1783}$$

Hence every frame member has two "free" flexibility values which, on multiplying them by a factor $\mathfrak{T}_k$, respectively have the following form for the left-hand L and for the right-hand end R of the member:

$$\boxed{m = (1/\iota + 4)k \quad \text{and} \quad m' = (4 + 1/\iota')k.} \tag{1784}$$

(β) Having regard to Section B.d.γ, it is evident that the reciprocal values of the free flexibilities can be regarded as the "free stiffnesses". Hence, in virtue of equations (1778) and (1782) [see also equations (1760) and (1763)] we have:

$$\boxed{\begin{aligned} M_L^{(\varphi=1)} &= \frac{6EJ \cdot \iota}{s(4\iota+1)} = \frac{\mathfrak{T}}{1/\iota + 4} = \frac{1}{m^w} = F^w, \\ M_R^{(\varphi=-1)} &= \frac{6EJ \cdot \iota'}{s(1+4\iota')} = \frac{\mathfrak{T}}{4 + 1/\iota'} = \frac{1}{m'^w} = F'^w. \end{aligned}} \tag{1785}$$

Hence every laterally unrestrained frame member (i.e., a member whose two ends can undergo displacement in relation to each other) has two "free" stiffness values which, on multiplying them by a factor $1/\mathfrak{T}_k$ respectively have the following form for the left-hand end L and the right-hand end R of the member:

$$\boxed{F = \frac{K}{1/\iota + 4} \quad \text{and} \quad F' = \frac{K}{4 + 1/\iota'}.} \tag{1786}$$

The free flexibilities m, m' and the free stiffnesses F, F', together with the special cases for full end-fixity, are summarized in Table II above.

2. Joints in rigid frames

a. The concept of "joint stiffness"

Fig. 389a represents a joint constituting the junction of three members forming part of a multi-member framework. The ends A, B and C (or, in general, U) of the members will be assumed to be laterally restrained but elastically rotatable joints. Hence, for geometrical reasons, the elastically rotatable joint O must likewise be laterally restrained. The values 1, 2, 3 (or, in general, ν) in brackets are member identification numbers and are used as indices for all the quantities relating to any particular member of the structure (e.g., s_ν is the length of the member, J_ν is the moment of inertia of the member, ι_ν is the carry-over factor, etc.).

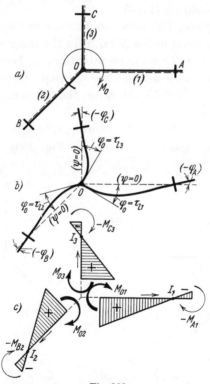

Fig. 389
Joint with three intersecting members
(a) External rotational moment acting upon the joint. (b) Deformation diagram. (c) Bending moment diagram

If an external rotational moment M_O is applied directly to the joint O, then this joint will obviously rotate through an angle φ_O (in the direction of rotation of the moment), and all three members meeting at the joint will undergo flexural deformation as indicated in Fig. 389b. The significant thing about this deformation is that the (as yet unknown) rotation φ_O of the joint will be of equal magnitude for all the three members concerned.

Any individual member OU at the joint O will obviously display a behaviour similar to that of the arbitrary frame member LR represented in Fig. 384. In Fig. 389c the joint O has been "cut open". The rotational or joint moment M_O acting at the joint is resolved into the component end moments $M_{O\nu}$ of the individual members, with the same direction of rotation. Hence we have:

$$M_O = \sum M_{O\nu} = M_{O1} + M_{O2} + M_{O3}. \tag{1787}$$

Finally, each moment $M_{O\nu}$ is transmitted ("carried over") to the other end U of the member, namely:

$$M_{U\nu} = -\iota_\nu \cdot M_{O\nu}, \tag{1788}$$

in conformity with equation (1714).

Note: To be complete, Fig. 389c would have to include the shear forces Q and the "direct" or axial forces N both for the inner end O_ν and the outer end U_ν of each member. These have intentionally been omitted, however, for the sake of clarity, as these forces are not required in the investigation of laterally restrained systems. All the same, the following should be noted for the sake of completeness.

In the present case, Fig. 389c representing the "cut-open" joint O, would have to include the forces shown in Fig. 390a, where all the shear

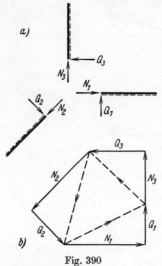

Fig. 390
Shear forces and axial forces at the cut-open joint O

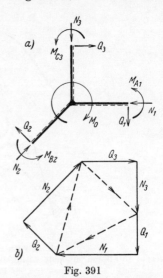

Fig. 391
Stress resultants at the ends of the members farthest from the joint under consideration

forces and axial forces have been given the direction of positive forces in accordance with the sign conventions adopted in the Introduction (although in the case of a positive $M_{Ov} \triangleq M_L$ the shear force Q in the member would really be negative). By virtue of the conditions of equilibrium $\Sigma V = 0$ and $\Sigma H = 0$, the geometrical sum of all the forces Q_v and N_v must be zero, which is reflected in the closed polygon of forces as represented in Fig. 390b.

Furthermore, the system of three intersecting frame members in Fig. 389a would have to be completed by the addition of three stress resultants (which are the reactions induced by M_O) at the end of each of the members, as indicated in Fig. 391a. These stress resultants have here again in all cases been represented as positive quantities. Since there are no loads acting directly upon the members in the case under consideration, Q_v and N_v are of the same magnitude as in Fig. 390a, though acting in opposite directions. For this reason the polygon of forces in Fig. 391b is identical with that in Fig. 390b, except that the directions of all the forces are reversed.

From equation (1770) we obtain for each of the members meeting at O the following general expression:

$$M_{Ov} = \varphi_0 \cdot S_v. \qquad (1789)$$

where v represents the relevant index (subscript) for each particular member. Substituting the individual values into equation (1787), we obtain:

$$M_O = \varphi_0 \cdot S_1 + \varphi_0 \cdot S_2 + \varphi_0 \cdot S_3 = \varphi_0 \cdot \sum S_v, \qquad (1790)$$

and, on dividing by φ_0:

$$\boxed{S_O = \sum S_v = S_1 + S_2 + S_3 = \frac{M_O}{\varphi_0}.} \qquad (1791)$$

The quantity S_O, which has been obtained with the aid of equation (1770), is termed the "rotational stiffness of the joint O" or, more conveniently, the "joint stiffness". The joint stiffness is thus defined as the sum of the stiffnesses of all the members meeting at, and rigidly connected to, the joint concerned. Thus in general:

$$(M_O \text{ for } \varphi_0 = 1/\mathfrak{T}_k) = S_O = \sum S_v. \qquad (1792)$$

It will readily be seen, with reference to equation (1760), that furthermore:

$$M_O^{(\varphi=1)} = S_O^w = \sum S_v^w. \qquad (1793)$$

Equation (1793) signifies that the "true joint stiffness" is defined as an imaginary joint moment (moment acting on the joint) which produces a rotation of the joint analytically equal to the unit angle, and that this is equal to the sum of the true stiffnesses of all the members meeting at, and rigidly connected to, the joint concerned.

Writing in general:
$$M_{Ov} = \mu_v \cdot M_O \quad \text{or} \quad \mu_v = \frac{M_{Ov}}{M_O}, \tag{1794}$$

and substituting the first expression (1794) into equation (1787), we obtain, on dividing all the terms of the latter by M_O:
$$1 = \sum \mu_v = \mu_1 + \mu_2 + \mu_3. \tag{1795}$$

The ratios μ_v are called "moment distribution factors". If, instead of the moments, we substitute the stiffnesses multiplied by φ_O—as expressed by equations (1789) or (1791)—into the second expression (1794), we obtain:
$$\mu_1 = \frac{S_1}{S_0}, \quad \mu_2 = \frac{S_2}{S_0}, \quad \mu_3 = \frac{S_3}{S_0}, \quad \text{or in general} \quad \mu_v = \frac{S_v}{S_0}. \tag{1796}$$

From the foregoing considerations it is also immediately evident that:
$$M_{O1} : M_{O2} : M_{O3} : M_O = S_1 : S_2 : S_3 : S_0, \tag{1797}$$

i.e., an external joint moment is distributed over the members at the joint in proportion to the stiffnesses of those members.

The system of members represented in Fig. 389a is in itself geometrically indeterminate to the 4th degree, because the four joints are all of the kind presenting elastic rotational restraint. However, as the three moment carry-over factors $\iota_1, \iota_2, \iota_3$ and therefore—by virtue of the equation (1754)—the three rotation carry-over factors j_1, j_2, j_3 can be assumed to be known, the rotation φ_O of the joint remains as the only true geometrical unknown. From equation (1791) this rotation is found to be:
$$\varphi_O = \frac{M_O}{S_0}, \tag{1798}$$

while the rotations of the opposite ends of the members, according to equation (1755), would be:
$$\varphi_A = -j_1 \cdot \varphi_O \quad \varphi_B = -j_2 \cdot \varphi_O \quad \varphi_C = -j_3 \cdot \varphi_O, \tag{1799}$$

or in general:
$$\varphi_U = -j_v \cdot \varphi_O. \tag{1800}$$

Alternatively, it would have been possible to proceed as follows:

The angle φ_O through which the joint O rotates in consequence of M_O can be expressed as three (or in general v) different ratios, in accordance with equation (1770), namely:
$$\varphi_O = \frac{M_{O1}}{S_1} = \frac{M_{O2}}{S_2} = \frac{M_{O3}}{S_3}. \tag{1801}$$

From equation (1801) it directly follows that:
$$\varphi_O = \frac{M_{O1} + M_{O2} + M_{O3}}{S_1 + S_2 + S_3} = \frac{\sum M_{Ov}}{\sum S_v} = \frac{M_O}{S_0}. \tag{1802}$$

From (1801) and (1802) we obtain the multiple equation:

$$\frac{M_{O1}}{S_1} = \frac{M_{O2}}{S_2} = \frac{M_{O3}}{S_3} = \frac{M_O}{S_O} \tag{1803}$$

whence it follows that:

$$M_{O1} = M_O \cdot \frac{S_1}{S_O} = M_O \cdot \mu_1, \quad \text{etc., see equation (1796)}$$

From equation (1803) it is also possible directly to obtain the continued proportion (1797), the meaning of which has already been explained.

b. The concept of "joint flexibility"

In analogy with the concepts of "flexibility" and "stiffness" of an individual member of a framework, the reciprocal b_O of the joint stiffness S_O may be designated as the "joint flexibility", i.e.:

$$b_O = \frac{1}{S_O} = \frac{1}{S_1 + S_2 + S_3} = \frac{1}{\dfrac{1}{b_1} + \dfrac{1}{b_2} + \dfrac{1}{b_3}} = \frac{1}{\sum \dfrac{1}{b_\nu}}. \tag{1804}$$

Whereas the total stiffness of a joint constituting the junction of a number of members is simply equal to the sum of the individual stiffnesses of all the members meeting at that joint, this simple relationship does not hold true for the flexibilities. Instead, according to equation (1804), the total flexibility of a multi-member joint is equal to the reciprocal sum of the reciprocal individual flexibilities of all the members meeting at the joint.

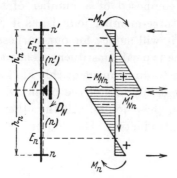

Fig. 392
External rotational moment acting at joint N of a two-story column line

Furthermore, the distribution factors would have the cumbersome form:

$$\mu_\nu = \frac{1/b_\nu}{\sum 1/b_\nu}. \tag{1805}$$

The concept of "joint flexibility" is used in this book only in connection with the formulas for two-story multi-bay framed structures in accordance with Method I in so far as the upper-story and lower-story column are united structurally into one continuous "column line". The situation is represented in Fig. 392. By similarly employing the relationships (1804) and (1805) we obtain the "column line flexibility" s_N and the "column distribution factors" η:

$$s_N = \frac{1}{S_n + S'_n} = \frac{1}{\frac{1}{b_n} + \frac{1}{b'_n}} = \frac{b_n \cdot b'_n}{b_n + b'_n}; \tag{1806}$$

$$\eta_n = \frac{S_n}{S_n + S'_n} = \frac{1/b_n}{1/b_n + 1/b'_n} = \frac{s_N}{b_n}; \quad \eta'_n = \frac{S'_n}{S_n + S'_n} = \frac{s_N}{b'_n}. \tag{1807}$$

In these expressions b_n and b'_n are the flexibilities of the individual columns, in accordance with the formulas (1753).

A rotational moment D_N acting upon the column line only, is distributed over the two columns—with due regard to the algebraic signs—as follows:

$$M_{Nn} = -D_N \cdot \eta_n \qquad M'_{Nn} = +D_N \cdot \eta'_n. \tag{1808}$$

The moment at the foot or base of the lower and the moment at the head or top of the upper column will be, respectively:

$$M_n = -M_{Nn} \cdot \varepsilon_n \qquad M'_n = -M'_{Nn} \cdot \varepsilon'_n. \tag{1809}$$

The points of contraflexure E_n and E'_n in Fig. 392a are the outer fixed points of the column line.

c. The laterally restrained four-member joint

(α) The fifty frame shapes with lateral restraint—i.e., with the continuous horizontal member, or beam, so restrained as to prevent side-sway of the frame—for which formulas are given in the main part of this book can also be conceived as being composed of a number of three-member or four-member joints. As the three-member joint is, as it were, comprised within the four-member joint, it will suffice for our purpose to consider only the latter type of joint as the typical constituent element of a framework.

The symbols relating to an element of this kind are shown in Fig. 393. M, N, O are three successive joints of the beam, while μ and ν denote the two beam spans between these joints. All the other symbols have already been explained in the Introduction. The "outer" carry-over factors ε_n and

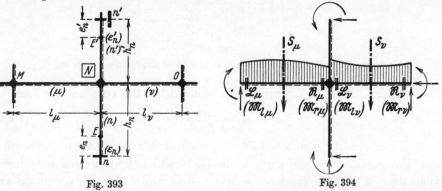

Fig. 393
Four-member joint, showing notation employed

Fig. 394
Arbitrary vertical loading on the two beam spans

ε'_n of the columns (indicated in brackets) are assumed to be known quantities. Further we have, in general:

$$\varepsilon = \frac{e}{h-e} \quad \text{or} \quad e = h \cdot \frac{\varepsilon}{1+\varepsilon}. \tag{1810}$$

The general loading of the members meeting at the joint consists of the arbitrary vertical loading on the beam spans as shown in Fig. 394, the arbitrary horizontal loading on the upper and lower columns as shown in Fig. 395, and an external rotational moment D_N acting upon the joint N as

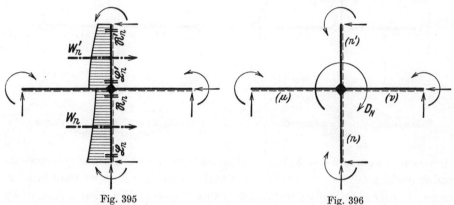

Fig. 395
Arbitrary horizontal loading on the two columns

Fig. 396
Rotation moment at joint N

shown in Fig. 396. In Fig. 394, in addition to the load terms $\mathfrak{L}$ and $\mathfrak{R}$, the fixed end moments $\mathfrak{M}_l$ and $\mathfrak{M}_r$ (for full end-fixity conditions) are indicated, these moments being used in calculations according to Method II. The stress resultants at all the ends of the members opposite to the joint N under consideration are moreover indicated in the three load diagrams. At the ends of the columns these stress resultants are, in fact, the reactions at the supports.

(β) In treating the problem *according to Method I*, each member is assigned a flexibility coefficient k, each column is assigned a flexibility b, and the column line (i.e., the combined upper and lower column regarded as a single continuous whole) is assigned the flexibility s_N:

$$k_\mu = \frac{J_k}{J_\mu} \cdot \frac{l_\mu}{l_k} \quad k_\nu = \frac{J_k}{J_\nu} \cdot \frac{l_\nu}{l_k}; \quad k_n = \frac{J_k}{J_n} \cdot \frac{h_n}{l_k} \quad k'_n = \frac{J_k}{J'_n} \cdot \frac{h'_n}{l_k}; \tag{1811}$$

$$\tag{1812}$$
$$b_n = k_n(2 - \varepsilon_n) \quad b'_n = k'_n(2 - \varepsilon'_n); \quad s_N = \frac{b_n \cdot b'_n}{b_n + b'_n}. \tag{1813}$$

The analysis of a two-story multi-bay framework which is statically indeterminate to the rth degree is based on the consideration of an s times statically indeterminate main system (i.e., statically indeterminate to the sth degree), where s denotes the number of column lines in the structure. As indicated in Fig. 397, pin-joints are introduced at all the rigid connections between the beam span ends and the column line—$(r\text{-}s)$ pin-joints in all—while each column line is retained as a continuous unit. As the elastic restraint conditions at the top and base of the upper and lower columns are considered to be known, each column line thus isolated is statically indeterminate to the first degree. (The four pin-joints in Fig. 397 have been given arbitrary

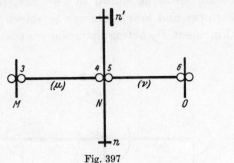

Fig. 397
Main system statically indeterminate to s^{th} degree

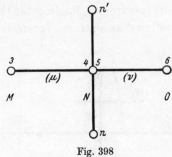

Fig. 398
Statically determinate basic system as auxiliary system

consecutive numbers $i = 3, 4, 5, 6$; this has been done to simplify the presentation of the analysis). In conjunction with the main system we shall furthermore consider a statically determinate primary system (Fig. 398), representing an auxiliary system.

The problem of establishing the actual elastic equations with the aid of the self-equilibrating states of stress $X_i = 1$ for the s times statically indeterminate main system will be discussed in Section C. We shall here merely consider the effects of the known values X_i. For this purpose we can make use of the self-equilibrating states of stress $X_i = 1$ relating to the $(r\text{-}1)$ times statically indeterminate system: the four conditions pertaining to the members meeting at the joint under consideration are represented in Fig. 399–402. The moment influence arising from each of these conditions extends in each case throughout the entire multi-bay framework. On the other hand, the same conditions applied to the associated auxiliary system extend their influence only over two individual members.

In the case represented in Fig. 399 moment carry-over occurs only from left to right; in the case represented in Fig. 400 it occurs only from right to left. In each case the four end moments of the beam spans—adopting simplified notation—will be as follows:

for Fig. 399:
$$\begin{aligned} M_3 &= +X_3 &&= +1 \\ M_4 &= -M_3 \cdot \iota_\mu &&= -\iota_\mu \\ M_5 &= +M_4 \cdot \sigma_N &&= -\iota_\mu \cdot \sigma_N \\ M_6 &= -M_5 \cdot \iota_\nu &&= +\iota_\mu \cdot \sigma_N \cdot \iota_\nu ; \end{aligned}$$

for Fig. 400:
$$\begin{aligned} M_3 &= -M_4 \cdot \iota'_\mu &&= +\iota'_\nu \cdot \sigma'_N \cdot \iota'_\mu \\ M_4 &= +M_5 \cdot \sigma'_N &&= -\iota'_\nu \cdot \sigma'_N \\ M_5 &= -M_6 \cdot \iota'_\nu &&= -\iota'_\nu \\ M_6 &= +X_6 &&= +1 . \end{aligned}$$

(1814) and (1815)

In the cases represented in Fig. 401 and 402 moment carry-over or moment decay occurs in both directions, as follows:

for Fig. 401:

$\uparrow M_3 = -M_4 \cdot \iota'_\mu\ \ = -\iota'_\mu$
$\dashv M_4 = +X_4\ \ \ \ \ \ \ \ \ = +1$
$\ \ M_5 = +M_4 \cdot \sigma_N = +\sigma_N$
$\downarrow M_6 = -M_5 \cdot \iota_\nu\ \ = -\sigma_N \cdot \iota_\nu;$

for Fig. 402:

$\uparrow M_3 = -M_4 \cdot \iota'_\mu\ \ = -\sigma'_N \cdot \iota'_\mu$
$\ \ M_4 = +M_5 \cdot \sigma'_N = +\sigma'_N$
$\dashv M_5 = +X_5\ \ \ \ \ \ \ \ \ = +1$
$\downarrow M_6 = -M_5 \cdot \iota_\nu\ \ = -\iota_\nu.$

(1816) and (1817)

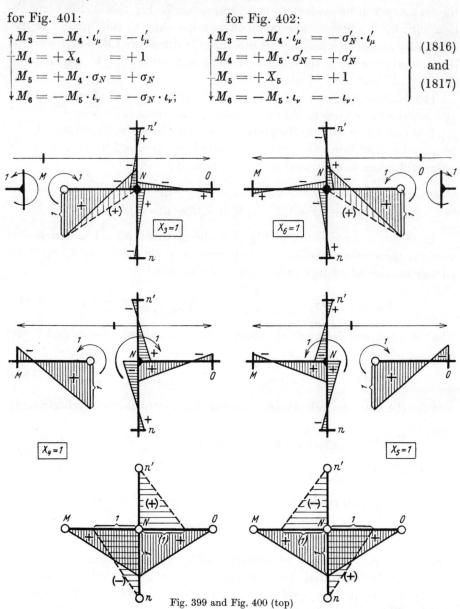

Fig. 399 and Fig. 400 (top)

Self-equilibrating states of stress $X_3 = 1$ and $X_6 = 1$ for the $(r-1)$ times indeterminate system
(The similar states for the statically determinate auxiliary system are shown by broken lines.)

Fig. 401 and Fig. 402 (centre)

Self-equilibrating states of stress $X_4 = 1$ and $X_5 = 1$ for the $(r-1)$ times indeterminate system

Fig. 403 and Fig. 404 (bottom)

Self-equilibrating states of stress $X_4 = 1$ and $X_5 = 1$ for the statically determinate auxiliary system

In addition to the carry-over factors ι and ι', the equations (1814)–(1817) now also contain transfer factors σ and σ' for the columns. The meaning of the factors σ will be evident from an inspection of Fig. 399–402 in conjunction with the equations. The moment corresponding to the "jump" or discontinuity at the column—i.e., in this case the difference (M_4-M_5)—will be distributed over the lower and upper column in proportion to the factors η [see Fig. 392 and the formulas (1807)].

The flexibilities b_n and b'_n for the columns are given by the formulas (1812) and are therefore known. When the elastic equations have been solved, all the factors ι and σ will also be known. In particular, it will then likewise be possible to write down the beam flexibilities as expressed by (1753), or the beam stiffnesses as expressed by (1764), namely:

$$\left.\begin{array}{ll} b_\mu = k_\mu(2-\iota_\mu) = 1/S_\mu & b_\nu = k_\nu(2-\iota_\nu) = 1/S_\nu \\ b'_\mu = k_\mu(2-\iota'_\mu) = 1/S'_\mu & b'_\nu = k_\nu(2-\iota'_\nu) = 1/S'_\nu. \end{array}\right\} \quad (1818)$$

In the case represented in Fig. 401 the moment $X_4 = 1$ is distributed over the three other members in proportion to their stiffnesses $S_n : S'_n : S$. In particular, the following condition must be satisfied:

$$M_5 = +1 \cdot \sigma_N = \frac{S_\nu}{(S_n + S'_n) + S_\nu} = \frac{1/b_\nu}{1/s_N + 1/b_\nu} = \frac{s_N}{s_N + b_\nu}. \quad (1819)$$

The total moment falling to the share of the two columns (upper and lower) is:

$$M_4 - M_5 = 1 - \sigma_N = \frac{S_n + S'_n}{(S_n + S'_n) + S_\nu} = \frac{1/s_N}{1/s_N + 1/b_\nu} = \frac{b_\nu}{s_N + b_\nu}, \quad (1820)$$

and for the columns individually (having due regard to the algebraic sign):

$$\left.\begin{array}{l} M_{Nn} = -\dfrac{S_n}{(S_n+S'_n)+S_\nu} = -\dfrac{1/b_n}{1/s_N+1/b_\nu} = -\dfrac{b_\nu}{b_n}\cdot\sigma_N \\[6pt] M'_{Nn} = +\dfrac{S'_n}{(S_n+S'_n)+S_\nu} = \dfrac{1/b'_n}{1/s_N+1/b_\nu} = \dfrac{b_\nu}{b'_n}\cdot\sigma_N. \end{array}\right\} \quad (1821)$$

Furthermore, with reference to Fig. 392 and formulas (1808), we have:

$$M_{Nn} = -(1-\sigma_N)\eta_n \quad \text{and} \quad M'_{Nn} = +(1-\sigma_N)\eta'_n. \quad (1822)$$

The values (1821) should be identical with the values (1822), which can readily be demonstrated with the aid of the relations (1819), (1820) and (1807). From equation (1819) we can also obtain the following relationship, which will be required later on:

$$\sigma_N \cdot b_\nu = (1-\sigma_N) s_N. \quad (1823)$$

For the case represented in Fig. 402 we similarly obtain the following expressions, embodying σ'_N and b'_μ:

$$\sigma'_N = \frac{s_N}{b'_\mu + s_N} \qquad \bigg| \qquad M_{Nn} = + \frac{b'_\mu}{b_n} \cdot \sigma'_N = + (1 - \sigma'_N)\, \eta_n \qquad \bigg| \qquad (1824)$$

$$\text{and}$$

$$1 - \sigma'_N = \frac{b'_\mu}{b'_\mu + s_N}\,. \qquad \bigg| \qquad M'_{Nn} = - \frac{b'_\mu}{b'_n} \cdot \sigma'_N = - (1 - \sigma'_N)\, \eta'_n\,. \qquad \bigg| \qquad (1825)$$

$$\sigma'_N \cdot b'_\mu = (1 - \sigma'_N)\, s_N. \tag{1826}$$

(γ) The external loads envisaged in Fig. 394, 395 and 396 produce in the s times statically indeterminate main system (Fig. 397) the moment diagrams shown in Fig. 405, 407a and 408a. In the statically determinate auxiliary system (Fig. 398) the same loads produce the moment diagrams shown in Fig. 405, 407b and 408b. As the column line in the main system is still statically indeterminate to the first degree, we shall also require the conditions represented in Fig. 407c and 407d (see below).

For each $(r-1)$ times statically indeterminate system we can write a virtual-work equation as follows:

$$X_i \cdot a_{ii} + a_{i0} = 0, \tag{1827}$$

whence we immediately obtain:

$$X_i = - \frac{a_{i0}}{a_{ii}} = - a_{i0} \cdot n_{ii} \quad \text{where} \quad n_{ii} = \frac{1}{a_{ii}}\,. \tag{1828}$$

In geometrical terms equation (1827) means that the relative rotation of the two sides of the cut, where the pin-joint is assumed to have been introduced, must be zero. In the present case the values a are the total rotations (multiplied by a factor $\mathfrak{T}_k$) of the sides of the cut (at the pin-joint) resulting from the "$X_i = 1$ condition" and from the so-called "M^0 condition". These values a can be directly obtained, as integral combinations of moment diagrams, from appropriate tables (e.g., in *Belastungsglieder*, page 258/259). When using the so-called reduction theorem for determining those values a, one of the diagrams to be combined may, in each case, be taken from the statically determinate auxiliary system. Owing to this helpful fact, all the combinations applicable in the case of the four members meeting at the joint under consideration will refer only to those four members.

For example, for $i = 4, 5$ the elastic equations are:

$$X_4 \cdot a_{44} + a_{40} = 0 \quad \text{and} \quad X_5 \cdot a_{55} + a_{50} = 0. \tag{1829}$$

a_{44} and a_{55} are obtained by combining the moment diagrams of Fig. 401 and 403, and of Fig. 402 and 404, having due regard to equation (1818):

$$\begin{aligned} a_{44} &= k_\mu (2 - \iota'_\mu) + \sigma_N \cdot k_\nu (2 - \iota_\nu) = b'_\mu + \sigma_N \cdot b_\nu \\ a_{55} &= \sigma'_N \cdot k_\mu (2 - \iota'_\mu) + k_\nu (2 - \iota_\nu) = \sigma'_N \cdot b'_\mu + b_\nu, \end{aligned} \qquad \bigg\} \qquad (1830)$$

or, making use of the relationships (1823) and (1826):

$$a_{44} = b'_\mu + (1 - \sigma_N)\, s_N; \qquad a_{55} = (1 - \sigma'_N)\, s_N + b_\nu. \tag{1831}$$

Note: Instead of the triangular diagram for span ν in Fig. 403, or for span μ in Fig. 404, it would also have been possible to take the triangular diagrams (shown by dotted lines) for the columns n or n'. The same values would have to be obtained for a_{44} and a_{55} as those indicated above.

The reciprocals of the values a_{ii} are designated as "principal influence coefficients":

$$n_{44} = \frac{1}{a_{44}} = \frac{1}{b'_\mu + (1-\sigma_N)\, s_N} \qquad n_{55} = \frac{1}{a_{55}} = \frac{1}{(1-\sigma'_N)\, s_N + b_\nu}. \qquad (1832)$$

By combining the diagrams of Fig. 401 and Fig. 405 we obtain a_{40}, and similarly by combining Fig. 402 and Fig. 405 we obtain a_{50}, for loading applied to the beam:

$$a_{40} = (-\mathfrak{L}_\mu \cdot \iota'_\mu + \mathfrak{R}_\mu \cdot 1)\, k_\mu + (\mathfrak{L}_\nu \cdot 1 - \mathfrak{R}_\nu \cdot \iota_\nu)\, \sigma_N\, k_\nu$$
$$a_{50} = (-\mathfrak{L}_\mu \cdot \iota'_\mu + \mathfrak{R}_\mu \cdot 1)\, \sigma'_N \cdot k_\mu + (\mathfrak{L}_\nu \cdot 1 - \mathfrak{R}_\nu \cdot \iota_\nu)\, k_\nu,$$

or in general, for $\boldsymbol{L} = \mathfrak{L} k$ and $\boldsymbol{R} = \mathfrak{R} k$:

$$\left.\begin{aligned} a_{40} &= -\boldsymbol{L}_\mu \cdot \iota'_\mu && + \boldsymbol{R}_\mu \cdot 1 && + \boldsymbol{L}_\nu \cdot \sigma_N && - \boldsymbol{R}_\nu \cdot \sigma_N\, \iota_\nu \\ a_{50} &= -\boldsymbol{L}_\mu \cdot \sigma'_N\, \iota'_\mu && + \boldsymbol{R}_\mu \cdot \sigma'_N && + \boldsymbol{L}_\nu \cdot 1 && - \boldsymbol{R}_\nu \cdot \iota_\nu. \end{aligned}\right\} \qquad (1833)$$

With reference to equation (1828) the expressions for X_i will now be:

$$\left.\begin{aligned} X_4 &= -\frac{a_{40}}{a_{44}} = -a_{40} \cdot n_{44} \\ X_5 &= -\frac{a_{50}}{a_{55}} = -a_{50} \cdot n_{55}. \end{aligned}\right\} \qquad (1834)$$

It will readily be seen that the results so far obtained with regard to the four members meeting at the joint N (Fig. 393) are immediately applicable

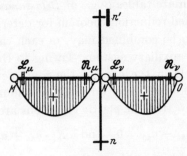

Fig. 405

M^0 diagrams for main system or auxiliary system due to loading on beam spans as represented in Fig. 394

also to the adjacent sets of members meeting at the joints M and O of the framework, subject to appropriate cyclic transposition of the subscripts. Thus, from equations (1832) we directly obtain:

$$n_{33} = \frac{1}{(1-\sigma'_M)\, s_M + b_\mu} \quad \text{and} \quad n_{66} = \frac{1}{b'_\nu + (1-\sigma_O)\, s_O} \qquad (1835)$$

Furthermore, from (1834):

$$X_3 = -a_{30} \cdot n_{33} \quad \text{and} \quad X_6 = -a_{60} \cdot n_{66}. \tag{1836}$$

The values a_{30} and a_{60} can be written down in analogy with (1833). The four values a_{i0}, which contain only the four load terms $\mathfrak{L}_\mu$, $\mathfrak{R}_{\mu'}$, $\mathfrak{L}_\nu$ and $\mathfrak{R}_\nu$ (as indicated in Fig. 405), can be summarized as follows:

$$\begin{array}{c|cccc}
 & \boldsymbol{L_\mu} & \boldsymbol{R_\mu} & \boldsymbol{L_\nu} & \boldsymbol{R_\nu} \\ \hline
a_{30} = & +1 & -\iota_\mu & -\iota_\mu \sigma_N & +\iota_\mu \sigma_N \iota_\nu \\
a_{40} = & -\iota'_\mu & +1 & +\sigma_N & -\sigma_N \iota_\nu \\
a_{50} = & -\sigma'_N \iota'_\mu & +\sigma'_N & +1 & -\iota_\nu \\
a_{60} = & +\iota'_\nu \sigma'_N \iota'_\mu & -\iota'_\nu \sigma'_N & -\iota'_\nu & +1.
\end{array} \tag{1837}$$

From this scheme (1837) we can, having regard to equations (1834) and (1836), obtain a corresponding scheme for the values X_i themselves, namely:

$$\begin{array}{c|cccc}
 & \boldsymbol{L_\mu} & \boldsymbol{R_\mu} & \boldsymbol{L_\nu} & \boldsymbol{R_\nu} \\ \hline
X_3 = & -n_{33} & +n_{34} & +n_{35} & -n_{36} \\
X_4 = & +n_{43} & -n_{44} & -n_{45} & +n_{46} \\
X_5 = & +n_{53} & -n_{54} & -n_{55} & +n_{56} \\
X_6 = & -n_{63} & +n_{64} & +n_{65} & -n_{66}.
\end{array} \tag{1838}$$

The relationship between the new n_{ik} matrix in (1838) and the matrix in (1837) can be represented symbolically by the following array:

$$\begin{array}{cccc}
\cdot \iota_\mu & \cdot \sigma_N & \cdot \iota_\nu & \\ \hline
(n_{33}) \rightarrow & n_{34} \rightarrow & n_{35} \rightarrow & n_{36} \\
n_{43} \leftarrow & (n_{44}) \rightarrow & n_{45} \rightarrow & n_{46} \\
n_{53} \leftarrow & n_{54} \leftarrow & (n_{55}) \rightarrow & n_{56} \\
n_{63} \leftarrow & n_{64} \leftarrow & n_{65} \leftarrow & (n_{66}) \\ \hline
\cdot \iota'_\mu & \cdot \sigma'_N & \cdot \iota'_\nu &
\end{array} \tag{1839}$$

All the elements n_{ik} are derivable from the elements n_{ii} of the principal diagonal (running from top left to bottom right, as indicated), in conjunction with the factors ι and σ on proceeding to the right and with the factors ι' and σ' on proceeding to the left.

The matrix (1839) of the "influence coefficients n_{ik}" is the so-called "conjugate matrix" of the coefficients a_{ik} of the set of trinomial equations to be solved. As the matrix of these last-mentioned coefficients is symmetric about its principal diagonal, i.e., $a_{ik} = a_{ki}$, the conjugate matrix associated with it must—for purely mathematical reasons—likewise be symmetric about its principal diagonal, i.e., the following identity condition must be satisfied:

$$n_{ik} = n_{ki}. \tag{1840}$$

Further details are given in Section C. For the purpose of the present discussion we may accept this property as an established fact. By way of demonstration we shall verify the validity of (1840) for one case; for instance, it must be shown that:

$$(n_{45} = n_{44} \cdot \sigma_N) = (n_{54} = n_{55} \cdot \sigma'_N), \tag{1841}$$

or according to equation (1832)

$$\frac{\sigma_N}{b'_\mu + (1 - \sigma_N)s_N} = \frac{\sigma'_N}{(1 - \sigma'_N)s_N + b_\nu}.$$

Multiplying the left member of the equation by σ'_N/σ'_N, and the right member by σ_N/σ_N, we obtain:

$$\frac{\sigma_N}{b'_\mu + (1 - \sigma_N)s_N} = \frac{\sigma'_N}{(1 - \sigma'_N)s_N + b_\nu}.$$

Replacing the underlined terms by the relations (1826) and (1823), it is finally found that both members of the equation reduce to:

$$n_{45} = n_{54} = \frac{\sigma_N \sigma'_N}{s_N(1 - \sigma_N \sigma'_N)}. \tag{1842}$$

Having regard to matrix symmetry ($n_{ik} = n_{ki}$), the array represented in (1839) can more conveniently be rearranged as follows:

$$
\begin{array}{c|cccc|c}
 & (n_{33}) & n_{34} & n_{35} & n_{36} & \\
\cdot\, \iota_\mu & \downarrow & \uparrow & \uparrow & \uparrow & \cdot\, \iota'_\mu \\
 & n_{43} & (n_{44}) & n_{45} & n_{46} & \\
\cdot\, \sigma_N & \downarrow & \downarrow & \uparrow & \uparrow & \cdot\, \sigma'_N \\
 & n_{53} & n_{54} & (n_{55}) & n_{56} & \\
\cdot\, \iota_\nu & \downarrow & \downarrow & \downarrow & \uparrow & \cdot\, \iota'_\nu \\
 & n_{63} & n_{64} & n_{65} & (n_{66}) & \\
\end{array}
\tag{1843}
$$

In this arrangement (1843) the carry-over of the moments for each individual load term $\mathfrak{L}$ or $\mathfrak{R}$ corresponds to the columns of the matrix.

In the main part of this volume the following variant notation has been adopted for formal reasons:

$$X_3 = M_{M\mu} \qquad X_4 = M_{N\mu} \qquad X_5 = M_{N\nu} \qquad X_6 = M_{O\nu}. \qquad (1844)$$

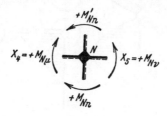

Fig. 406
Joint N shown detached from remainder of framework

For the joint N, which is assumed to be detached from the rest of the framework by the application of cuts to the members, the equilibrium condition yields the relation (see Fig. 406):

$$+ M_{Nn} + X_4 - M'_{Nn} - X_5 = 0 \qquad (1845)$$

or, writing it as a relationship between the differences of the beam moments and of the column moments:

$$M_{Nn} - M'_{Nn} = -(X_4 - X_5) = -D_N . \qquad (1846)$$

where $D_N = X_4 - X_5 = M_N - M_N$ is the "differential moment", i.e., the difference between the moments acting at the ends of the two beam spans adjacent to the joint under consideration. With reference to equations (1808) we can immediately write:

$$M_{Nn} = (-X_4 + X_5)\eta_n \qquad M'_{Nn} = (+X_4 - X_5)\eta'_n. \qquad (1847)$$

(δ) If loading acts on the columns only (as in Fig. 395), then the load terms a_{40} and a_{50} of equations (1829) can be obtained by combining Fig. 401, or Fig. 402, with Fig. 407b. At the same time we can introduce the load terms multiplied by k, namely:

$$L_\mu = \mathfrak{L}_\mu k_\mu \qquad R_\mu = \mathfrak{R}_\mu k_\mu$$
$$L_\nu = \mathfrak{L}_\nu k_\nu \qquad R_\nu = \mathfrak{R}_\nu k_\nu$$

and we thus obtain:

$$\left.\begin{aligned} a_{40} &= -(1-\sigma_N)\,\eta_n(\boldsymbol{R}_n - \boldsymbol{L}_n \cdot \varepsilon_n) + (1-\sigma'_N)\,\eta'_n(\boldsymbol{L}'_n - \boldsymbol{R}'_n \cdot \varepsilon'_n) \\ a_{50} &= +(1-\sigma_N)\,\eta_n(\boldsymbol{R}_n - \boldsymbol{L}_n \cdot \varepsilon_n) - (1-\sigma'_N)\,\eta'_n(\boldsymbol{L}'_n - \boldsymbol{R}'_n \cdot \varepsilon'_n). \end{aligned}\right\} \qquad (1848)$$

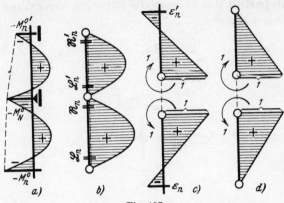

Fig. 407

M^0 diagrams due to loading on columns as envisaged in Fig. 395
(a) M^0 diagram for main system
(b) M^0 diagram for auxiliary system
(c) State $M_N^0 = 1$ for elastically restrained column line
(d) State $M_N^0 = 1$ for auxiliary system

With the abbreviated notation:

$$\mathfrak{B}_n = R_n - L_n \cdot \varepsilon_n \quad \text{and} \quad \mathfrak{B}_n' = L_n' - R_n' \cdot \varepsilon_n' \tag{1849}$$

and furthermore:

$$\mathfrak{B}_N = -\mathfrak{B}_n \cdot \eta_n + \mathfrak{B}_n' \cdot \eta_n' \tag{1850}$$

the above expressions can be simplified to:

$$a_{40} = +\mathfrak{B}_N(1-\sigma_N) \quad \text{and} \quad a_{50} = -\mathfrak{B}_N(1-\sigma_N'). \tag{1851}$$

In conjunction with equations (1834) we finally obtain for the values X_i:

$$\left.\begin{array}{l} X_4 = -\mathfrak{B}_N(1-\sigma_N)n_{44} = -\mathfrak{B}_N(+n_{44}-n_{45}) \\ X_5 = +\mathfrak{B}_N(1-\sigma_N')n_{55} = +\mathfrak{B}_N(-n_{54}+n_{55}). \end{array}\right\} \tag{1852}$$

On the basis of considerations similar to those adopted in Section γ for obtaining X_3 and X_6, the above set of equations (1852) can immediately be extended as follows:

$$\left.\begin{array}{ll} X_3 = +\mathfrak{B}_N \cdot w_{3N} \quad \text{where} & w_{3N} = +n_{34}-n_{35} \\ X_4 = -\mathfrak{B}_N \cdot w_{4N} & w_{4N} = +n_{44}-n_{45} \\ X_5 = +\mathfrak{B}_N \cdot w_{5N} & w_{5N} = -n_{54}+n_{55} \\ X_6 = -\mathfrak{B}_N \cdot w_{6N}, & w_{6N} = -n_{64}+n_{65}. \end{array}\right\} \tag{1853}$$

For the beam moments we shall again use the variant notation as indicated in (1844). With regard to the column moments it must be borne in mind that the influences resulting from $D_N = X_4 - X_5$, in accordance with equations (1847) and Fig. 392, must be superimposed upon the moments already acting in the main system as represented in Fig. 407a. The "support moments" of the column line considered as a two-span continuous girder with elastic conditions of restraint (see Fig. 407a) are calculated as follows:

Elastic equation:

$$M_N^0 \cdot a_{nn} + a_{n0} = 0 \qquad M_N^0 = -a_{n0}/a_{nn}.$$

For a_{nn} combine Fig. 407c with Fig. 407d

$$a_{nn} = k_n(2 - \varepsilon_n) + k_n'(2 - \varepsilon_n') = b_n + b_n';$$

For a_{n0} combine Fig. 407c with Fig. 407b

$$a_{n0} = -\mathfrak{L}_n k_n \varepsilon_n + \mathfrak{R}_n k_n + \mathfrak{L}_n' k_n' - \mathfrak{R}_n' k_n' \varepsilon_n'$$
$$= (\mathbf{R}_n - \mathbf{L}_n \cdot \varepsilon_n) + (\mathbf{L}_n' - \mathbf{R}_n' \cdot \varepsilon_n') = \mathfrak{B}_n + \mathfrak{B}_n'.$$

(1854)

Hence we obtain:

$$M_N^0 = -\frac{\mathfrak{B}_n + \mathfrak{B}_n'}{b_n + b_n'}. \qquad (1855)$$

and, furthermore, for the external moments:

$$M_n^0 = -(\mathfrak{L}_n + M_N^0)\varepsilon_n \qquad M_n^{0'} = -(\mathfrak{R}_n' + M_N^0)\varepsilon_n'. \qquad (1856)$$

The final column moments will now be:

$$M_{Nn} = M_N^0 - (X_4 - X_5)\eta_n \qquad M_{Nn}' = M_N^0 + (X_4 - X_5)\eta_n' \qquad (1857)$$
$$M_n = -(\mathfrak{L}_n + M_{Nn})\varepsilon_n \qquad M_n' = -(\mathfrak{R}_n' + M_{Nn}')\varepsilon_n'. \qquad (1858)$$

For the action of an external nodal moment D_N (rotational moment acting upon the joint N), as represented in Fig. 396, the load terms a_{40} and

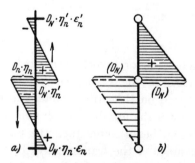

Fig. 408

M^0 diagrams due to D_N as envisaged in Fig. 396
(a) For main system
(b) For auxiliary system (either upper or lower triangle)

a_{50} of equation (1829) are obtained by combining Fig. 401, or Fig. 402, with Fig. 408b:
$$a_{40} = + D_N(1-\sigma_N)\,\eta'_n(2-\varepsilon'_n)\,k'_n = + D_N(1-\sigma_N)\,\eta'_n b'_n \\ a_{50} = - D_N(1-\sigma'_N)\,\eta'_n(2-\varepsilon'_n)\,k'_n = - D_N(1-\sigma'_N)\,\eta'_n b'_n. \quad (1859)$$

Instead of the positive upper triangle in Fig. 408b we could also have considered the negative lower triangle (shown as an intermittent line), in which case the following expressions would be obtained:
$$a_{40} = + D_N(1-\sigma_N)\,\eta_n b_n \\ a_{50} = - D_N(1-\sigma'_N)\,\eta_n b_n.$$

From equation (1807) it follows that:
$$(\eta_n b_n = \eta'_n b'_n) = s_N, \quad (1860)$$

so that we can write:
$$a_{40} = + D_N s_N (1-\sigma_N) \\ a_{50} = - D_N s_N (1-\sigma'_N). \quad (1861)$$

If we furthermore write $D_N s_N = \mathfrak{B}_N$ then the above values a will be similar in form to the values a in (1851). Hence the relations (1852) and (1853) are also directly applicable to the case where the moment D_N is acting upon the joint.

The expressions for the column moments will, however, be as follows:
$$M_{Nn} = -(D_N + X_4 - X_5)\eta_n \qquad M'_{Nn} = +(D_N + X_4 - X_5)\eta'_n; \quad (1862)$$
$$M_n = - M_{Nn} \cdot \varepsilon_n \qquad\qquad M'_n = - M'_{Nn} \cdot \varepsilon'_n. \quad (1863)$$

(ϵ) In treating the problem of the four-member joint N (Fig. 393) *according to Method II*, a stiffness coefficient K is assigned to each horizontal member and a stiffness coefficient S is assigned to each column, namely:
$$K_\mu = \frac{J_\mu}{J_k}\cdot\frac{l_k}{l_\mu} \quad K_\nu = \frac{J_\nu}{J_k}\cdot\frac{l_k}{l_\nu}; \qquad K_n = \frac{J_n}{J_k}\cdot\frac{l_k}{h_n} \quad K'_n = \frac{J'_n}{J_k}\cdot\frac{l_k}{h'_n}; \quad (1864)$$
$$S_n = \frac{K_n}{2-\varepsilon_n} \qquad S'_n = \frac{K'_n}{2-\varepsilon'_n}. \quad (1865)$$

The analysis of a single-story or two-story multi-bay framework that is geometrically indeterminate to the sth degree (or "s times geometrically indeterminate") can be based on that of a geometrically determinate main system. The number s denotes the number of columns (or column lines) and therefore also the number of beam joints (i.e., the joints, or nodes, in the continuous horizontal member of the framework). In this main system all the beam joints, which in reality are elastically rotatable (elastic rotational restraint), are assumed to be rigidly fixed so as to be incapable of rotation: in Fig. 409 this is symbolised by the black squares at the joints concerned.

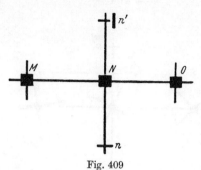

Fig. 409
Geometrically determinate main system with rigidly fixed joints represented symbolically by black squares

The procedure of establishing the actual elastic equations with the aid of self-equilibrating states of strain $Y_N = 1$ for the geometrically determinate main system will be discussed in Section C. We shall here merely consider the effects of the known values $Y_N = \varphi_N \cdot \mathfrak{T}_k/3$, i.e., of the angles of rotation of the joints multiplied by a factor $2EJ_k/l_k$ [cf. equations (1706)]. For this purpose we can make use of the self-equilibrating states of strain $Y_N = 1$ relating to the $(s-1)$ times geometrically indeterminate system. Each of these states of strain affects the entire multi-bay framework. The three which are pertinent to the joint and members under consideration are represented diagrammatically in Fig. 410–412: for the sake of simplicity only the rotated joints are shown and not the deflected shapes of the members connecting them (cf. Fig. 389b).

In the case of Fig. 410 carry-over of rotation occurs only from left to right; in the case of Fig. 411 it occurs in both directions; and in the case of Fig. 412 it occurs only from right to left. Having regard to the general equation (1800) we can write down the following relations:

for Fig. 410:

$$\begin{aligned}
{}_\top Y_M &= +1 & &= +1 \\
\left| Y_N \right. &= -Y_M \cdot j_\mu &&= -j_\mu \\
{}_\downarrow Y_O &= -Y_N \cdot j_\nu &&= +j_\mu \cdot j_\nu;
\end{aligned}$$

for Fig. 412:

$$\begin{aligned}
{}^\uparrow Y_M &= -Y_N \cdot j'_\mu &&= +j'_\nu \cdot j'_\mu \\
\left| Y_N \right. &= -Y_O \cdot j'_\nu &&= -j'_\nu \\
{}_- Y_O &= +1 &&= +1.
\end{aligned}$$

(1866) and (1868)

for Fig. 411:

$$\begin{aligned}
{}^\uparrow Y_M &= -Y_N \cdot j'_\mu = -j'_\mu \\
{}_\dashv Y_N &= +1 \quad\quad = +1 \\
{}_\downarrow Y_O &= -Y_N \cdot j_\nu = -j_\nu;
\end{aligned}$$

(1867)

— 404 —

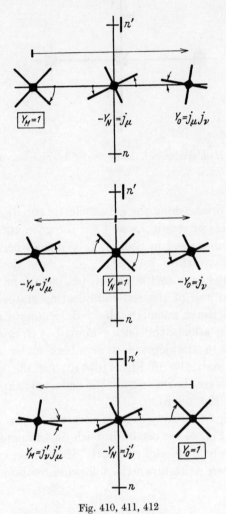

Fig. 410, 411, 412

Self-equilibrating states of strain $Y_N = 1$ for the $(s-1)$ times geometrically indeterminate system

For the columns the stiffnesses S_n and S'_n can be calculated from the formulas (1865) and can therefore be considered to be known quantities. When the elastic equations have been solved, all the values j and j' are known. Hence, with the aid of the relations (1771), the beam stiffnesses as envisaged in equations (1764) can be written down, as follows:

$$\left. \begin{array}{ll} 3S_\mu = K_\mu(2 - j_\mu) & 3S_\nu = K_\nu(2 - j_\nu) \\ 3S'_\mu = K_\mu(2 - j'_\mu) & 3S'_\nu = K_\nu(2 - j'_\nu). \end{array} \right\} \quad (1869)$$

Note: To avoid the introduction of fractions into these expressions, the factor 3 has been applied to the stiffness values.

The joint stiffness (multiplied by a factor 3) of the joint N will—by virtue of equation (1792)—therefore be:

$$3S_N = K_\mu(2 - j'_\mu) + 3S_n + 3S'_n + K_\nu(2 - j_\nu). \tag{1870}$$

From equations (1796) it follows that the moment distribution factors are:

$$\mu'_\mu = \frac{S'_\mu}{S_N} \quad \mu_n = \frac{S_n}{S_N} \quad \mu'_n = \frac{S'_n}{S_N} \quad \mu_\nu = \frac{S_\nu}{S_N}\,; \\ (\mu_\mu + \mu_n + \mu'_n + \mu_\nu = 1). \tag{1871}$$

Each external rotational moment acting upon the joint N of the s times geometrically indeterminate system under consideration will be distributed over the four members meeting at that joint in proportion to the factors μ in (1871).

The external loads as envisaged in Fig. 394 and 395 must not produce any rotation of the joints in the geometrically determinate main system (Fig. 409). This "$Y_N = 0$ condition" can evidently be satisfied by assuming all four members meeting at each beam joint to have full end-fixity, and, on this assumption, it is permissible to combine the four individual fixed-end moments into one resultant "joint fixing moment" for the joint as a whole.

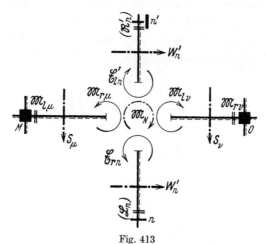

Fig. 413

"State of fixity" of the four-member joint N in the geometrically determinate main system; in this diagram the joint is assumed to be "cut open".

With reference to Fig. 413 this fixing moment is therefore:

$$\mathfrak{M}_N = -\mathfrak{M}_{r\mu} + \mathfrak{M}_{l\nu} - \mathfrak{E}_{rn} + \mathfrak{E}'_{ln}. \tag{1872}$$

All moments having a clockwise direction of rotation are reckoned as positive (see the Section "Sign Conventions" in the Introduction). For the beam spans the relevant fixed-end moments are those indicated in the formulas (1703); for the columns the restraining moments are those indicated in formulas (1730) and (1721), having due regard to the footnote to formula (1733).

We thus obtain the following expressions for the fixed-end and restraining moments:

$$\mathfrak{M}_{r\mu} = +\frac{1}{3}\mathfrak{L}_\mu - \frac{2}{3}\mathfrak{R}_\mu \qquad \mathfrak{M}_{l\nu} = -\frac{2}{3}\mathfrak{L}_\nu + \frac{1}{3}\mathfrak{R}_\nu;$$
$$\mathfrak{C}_{rn} = -\frac{\mathfrak{R}_n - \mathfrak{L}_n \cdot \varepsilon_n}{2 - \varepsilon_n} \qquad \mathfrak{C}'_{ln} = -\frac{\mathfrak{L}'_n - \mathfrak{R}'_n \cdot \varepsilon'_n}{2 - \varepsilon'_n}.$$
(1873)

The determining equation for the rotation of the joint must, in its original form, be as follows for the $(s\text{-}1)$ times geometrically indeterminate system:

$$\varphi_N \cdot S_N + \mathfrak{M}_N = D_N. \tag{1874}$$

This equation states that the algebraic sum of the two moments $\varphi_N \cdot S_N$ and $\mathfrak{M}_N$ acting upon the joint N must be zero or be equal to the external rotational moment D_N (if any) acting upon that joint. *All three joint moments occurring in equation (1874) are considered to have a clockwise direction of rotation.* The factor φ_N in the first term of (1874) is the angle of rotation of the joint multiplied by a factor $\mathfrak{T}_k$, while the factor S_N in that same term is the joint stiffness multiplied by $1/\mathfrak{T}_k$ as envisaged in equation (1792).

It is, however, more convenient to use the angles of rotation Y_N multiplied by a factor $\mathfrak{T}_k/3$ as in equations (1706), i.e., in general:

$$\varphi = \mathfrak{T}_k \cdot \varphi^w = 3Y, \tag{1875}$$

$$Y = \frac{\varphi}{3} = \frac{\mathfrak{T}_k}{3} \cdot \varphi^w = \frac{2EJ_k}{l_k} \cdot \varphi^w. \tag{1876}$$

Equation (1874) thus becomes:

$$Y_N \cdot 3S_N = -\mathfrak{M}_N + D_N = \mathfrak{K}_N. \tag{1877}$$

In this expression the negative joint restraining moment and the external rotational moment have been combined into the "joint load term" $\mathfrak{K}_N$, so that we can now write:

$$\boxed{\mathfrak{K}_N = +\mathfrak{M}_{r\mu} - \mathfrak{M}_{l\nu} + \mathfrak{C}_{rn} - \mathfrak{C}'_{ln} + D_N.} \tag{1878}$$

Finally, for equation (1877) we obtain:

$$Y_N = \frac{\mathfrak{K}_N}{3S_N}. \tag{1879}$$

The final end moments of the members at joint N, due to loading as indicated in Fig. 394, 395 and 396, are obtained from the first equation (1731) or the first equation (1725), according as the end of the member concerned is a right-hand or a left-hand end. We thus obtain:

(1880)
$$\begin{aligned}
M_{N\mu} &= \mathfrak{M}_{r\mu} - 3S'_{\mu} \cdot \frac{\mathfrak{K}_N}{3S_N} = \mathfrak{M}_{r\mu} - \mu'_{\mu} \cdot \mathfrak{K}_N \\
M_{N\nu} &= \mathfrak{M}_{l\nu} + 3S_{\nu} \cdot \frac{\mathfrak{K}_N}{3S_N} = \mathfrak{M}_{l\nu} + \mu_{\nu} \cdot \mathfrak{K}_N \\
M_{Nn} &= \mathfrak{E}_{rn} - 3S_{n} \cdot \frac{\mathfrak{K}_N}{3S_N} = \mathfrak{E}_{rn} - \mu_{n} \cdot \mathfrak{K}_N \\
M'_{Nn} &= \mathfrak{E}'_{ln} + 3S'_{n} \cdot \frac{\mathfrak{K}_N}{3S_N} = \mathfrak{E}'_{ln} + \mu'_{n} \cdot \mathfrak{K}_N .
\end{aligned}$$

With regard to the two beam moments we have $\mathfrak{E}_{r\mu} = \mathfrak{M}_{r\mu}$ and $\mathfrak{E}_{l\nu} = \mathfrak{M}_{l\nu}$ by virtue of the full end-fixity of both ends of the beam spans in the geometrically determinate main system. In the second expression for each of the moments in (1880) the moment distribution factors μ according to (1871) have been introduced.

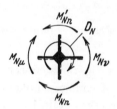

Fig. 414
Joint N in final condition, shown detached from remainder of framework

A check is provided by the requirement that the final end moments of the members, conceived as acting upon the joint and inclusive of the external rotational moment, must satisfy the equilibrium condition $\Sigma M = 0$, i.e., with reference to Fig. 414 we must have:

$$M_{N\mu} - M_{N\nu} + M_{Nn} - M'_{Nn} + D_N = 0. \tag{1881}$$

Substituting the values M as given by the formulas (1880) and rearranging, we obtain:

$$(\mathfrak{M}_{r\mu} - \mathfrak{M}_{l\nu} + \mathfrak{E}_{rn} - \mathfrak{E}'_{ln} + D_N) - \mathfrak{K}_N(\mu'_{\mu} + \mu_{\nu} + \mu_{n} + \mu'_{n}) = 0.$$

As the first expression in brackets is equal to $\mathfrak{K}_N$ by virtue of equation (1878) and as the second expression in brackets is equal to unity by virtue of equation (1871), it is seen that the checking condition (1881) is indeed satisfied.

For the sake of completeness the following should be noted with regard to the second expression for the member end moments in (1880). Each of these final moments comprises two terms. The first term is the fixed-end moment for the condition with fixed joints as envisaged in Fig. 413. The second term in these expressions may be regarded as the distributed proportions of a rotational moment $\mathfrak{K}_N$ acting upon the joint N of the s times geometrically indeterminate system (see Fig. 415). This rotational moment $\mathfrak{K}_N$ is sometimes designated as the "balancing moment".

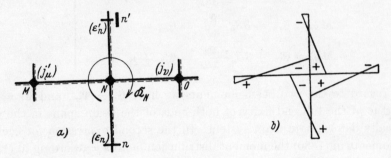

Fig. 415

"Balancing condition" of the joint N in the s times geometrically indeterminate system, i.e., in the given system

(a) Application of the balancing moment (b) Bending moment diagram belonging to it

(ζ) The results of the investigations concerning the four members intersecting at the joint N are, of course, similarly applicable to the adjacent joints M and O. Thus we directly obtain by virtue of formula (1879):

$$Y_M = \frac{\mathfrak{K}_M}{3 S_M} \quad \text{and} \quad Y_O = \frac{\mathfrak{K}_O}{3 S_O}, \qquad (1882)$$

where the values $\mathfrak{K}$ and S are analogous to those given in the expressions (1878) and (1870).

If the multi-bay framework as a whole is subjected only to the three load influences in the form of the "joint load terms" $\mathfrak{K}_M$, $\mathfrak{K}_N$, $\mathfrak{K}_O$ then the corresponding values Y and their carry-over effects, for the joint and four intersecting members under consideration, can evidently—on the basis of the sets of equations (1866), (1867) and (1868)—be summarized in the following array:

$$\left. \begin{array}{r|ccc} & \mathfrak{K}_M & \mathfrak{K}_N & \mathfrak{K}_O \\ \hline Y_M = & +\dfrac{1}{3 S_M} & -\dfrac{j_\mu}{3 S_M} & +\dfrac{j_\mu \cdot j_\nu}{3 S_M} \\ Y_N = & -\dfrac{j'_\mu}{3 S_N} & +\dfrac{1}{3 S_N} & -\dfrac{j_\nu}{3 S_N} \\ Y_O = & +\dfrac{j'_\nu \cdot j'_\mu}{3 S_O} & -\dfrac{j'_\nu}{3 S_O} & +\dfrac{1}{3 S_O} \end{array} \right\} \qquad (1883)$$

For the sake of simplicity, the "influence coefficients" in the matrix will be represented by the symbols u_{ik}:

$$\begin{array}{c|ccc}
 & \mathfrak{K}_M & \mathfrak{K}_N & \mathfrak{K}_O \\ \hline
Y_M = & +u_{mm} & -u_{mn} & +u_{mo} \\
Y_N = & -u_{nm} & +u_{nn} & -u_{no} \\
Y_O = & +u_{om} & -u_{on} & +u_{oo}.
\end{array} \qquad (1884)$$

The process of determining the elements u_{ik} from the elements u_{ii} of the principal diagonal can be symbolized by the following array:

$$\begin{array}{ccc}
\cdot j_\mu & \cdot j_\nu & \\ \hline
(u_{mm}) \rightarrow & u_{mn} \rightarrow & u_{mo} \\
& \searrow & \\
u_{nm} \leftarrow & (u_{nn}) \rightarrow & u_{no} \\
\\
u_{om} \leftarrow & u_{on} \leftarrow & (u_{oo}) \\ \hline
\cdot j'_\mu & \cdot j'_\nu &
\end{array} \qquad (1885)$$

The foregoing and further treatment of the problem is, from the mathematical point of view, essentially similar to the foregoing treatment applied to Method I, [cf. the matrices (1838), (1839) and (1843)]. In particular, for the matrix (1884) we again have the identity condition corresponding to equation (1840), namely:

$$u_{ik} = u_{ki}, \qquad (1886)$$

The array (1885) can therefore be more conveniently rearranged as follows:

$$\begin{array}{c|ccc|c}
\cdot j_\mu & (u_{mm}) & u_{mn} & u_{mo} & \cdot j'_\mu \\
 & \downarrow & \uparrow & \uparrow & \\
 & u_{nm} & (u_{nn}) & u_{no} & \\
\cdot j_\nu & \downarrow & \downarrow & \uparrow & \cdot j'_\nu \\
 & u_{om} & u_{on} & (u_{oo}) &
\end{array} \qquad (1887)$$

In the above form (1887) the carry-over of the rotation for each individual load term $\mathfrak{K}$ corresponds to the columns of the matrix.

With the aid of formulas (1705) we can now write for the final beam moments:

$$\begin{aligned}
M_{M\mu} &= \mathfrak{M}_{l\mu} + K_\mu(2Y_M + Y_N) & M_{N\mu} &= \mathfrak{M}_{r\mu} - K_\mu(Y_M + 2Y_N) \\
M_{N\nu} &= \mathfrak{M}_{l\nu} + K_\nu(2Y_N + Y_O) & M_{O\nu} &= \mathfrak{M}_{r\nu} - K_\nu(Y_N + 2Y_O).
\end{aligned} \qquad (1888)$$

The two different sets of expressions for the beam moments—according to (1888) and (1880) respectively—must, of course, yield the same final results.

The identities will immediately be evident if we substitute:

$$Y_M = -j'_\mu \cdot Y_N \quad \text{and} \quad Y_O = -j_\nu \cdot Y_N \qquad (1889)$$

into equations (1888), in conjunction with the stiffnesses according to equations (1869).

Having regard to equations (1731) and (1732) for the lower column and equations (1725) and (1726) for the upper column, the final column moments for the column line N are:

$$\left.\begin{array}{ll} M_{Nn} = \mathfrak{C}_{rn} - 3 S_n \cdot Y_N & M_n = -(M_{Nn} + \mathfrak{L}_n)\varepsilon_n; \\ M'_{Nn} = \mathfrak{C}'_{ln} + 3 S'_n \cdot Y_N & M'_n = -(M'_{Nn} + \mathfrak{R}_n)\varepsilon'_n. \end{array}\right\} \qquad (1890)$$

C. The Laterally Restrained Multi-Bay Rigid Framework or Continuous Beam on Elastically Restrained Supports

1. Analysis by the Method of Forces (Method I)

The analysis can most suitably be demonstrated for a "mixed" system as represented in Fig. 416. The designations adopted are explained in the Introduction. The analysis of the joint with four intersecting members, treated as a separate entity, as given in Section B.2.c, is relevant to the present discussion.

The four-bay frame in Fig. 416 is really $6 \times 3 - 2 \times 1 - 1 \times 2 = 14$ times statically indeterminate (statically indeterminate to the 14th degree).[1] However, as the elastic restraints of the columns are assumed to be known (the values ε being given), the degree of statical indeterminacy will be reduced to $r = 14 - 4 - 2 = 8$.

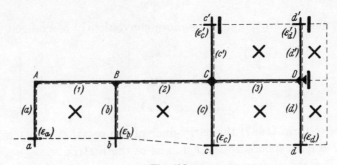

Fig. 416
Frame chosen as example

---

[1] The formula employed is $r = 3f - b$, where f is the number of closed "circuits", including those in which one member is constituted by the rigid foundation, while b is the number of stress resultants or support reactions which would have to be introduced at all the movable connections or supports in order to establish full structural continuity or fixity. In the case of Fig. 416 there are six such "circuits" (marked by crosses). At each column head one reaction is zero (vertical force) and at the support D two are zero (vertical force and moment).

— 411 —

As a suitable "main system" for establishing the redundants (statically indeterminate quantities) we shall adopt the structure represented in Fig. 417, which is still statically indeterminate to the 2nd degree. The structure represented in Fig. 418, which is fully statically determinate, will be adopted as the "auxiliary system". In Fig. 417 the six beam end moments X_i (which are conceived as moments acting at the ends of the lengths of beam detached from the columns by the introduction of cuts) constitute the statically indeterminate quantities.

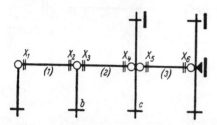

Fig. 417
Main system statically indeterminate to 2nd degree

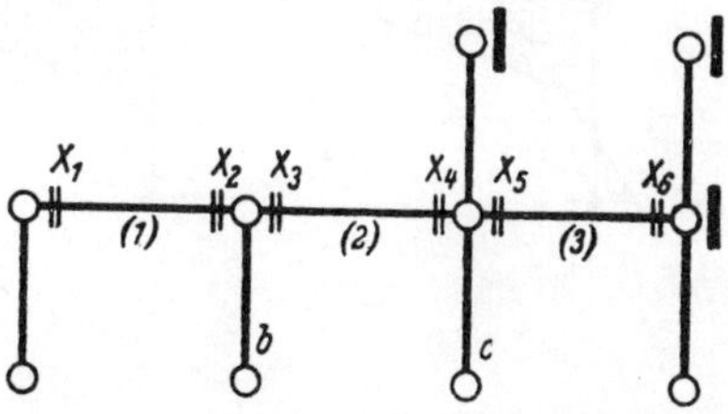

Fig. 418
Statically determinate auxiliary system

The six requisite elastic equations can be represented in the following form:

$$\begin{array}{c|cccccc|l}
 & X_1 & X_2 & X_3 & X_4 & X_5 & X_6 & \\
\hline
1) & a_{11} & a_{12} & a_{13} & a_{14} & a_{15} & a_{16} & = -a_{10} \\
2) & a_{21} & a_{22} & a_{23} & a_{24} & a_{25} & a_{26} & = -a_{20} \\
3) & a_{31} & a_{32} & a_{33} & a_{34} & a_{35} & a_{36} & = -a_{30} \\
4) & a_{41} & a_{42} & a_{43} & a_{44} & a_{45} & a_{46} & = -a_{40} \\
5) & a_{51} & a_{52} & a_{53} & a_{54} & a_{55} & a_{56} & = -a_{50} \\
6) & a_{61} & a_{62} & a_{63} & a_{64} & a_{65} & a_{66} & = -a_{60} \\
\end{array} \quad (1891)$$

where the coefficients a_{ik} are the coefficients δ_{ik} of the virtual-work equation multiplied by a factor $\mathfrak{T}_k$.[1]

As has already been stated in Section A, the elastic equations (1891) are geometrical compatibility conditions which here relate to the six junctions between the beam spans and the columns. The "self-equilibrating states of stress" are obtained for $X_i = 1$, i.e., when unit values of these moments X_i are acting upon the structure, and are represented in Fig. 419. In compliance with the law of reciprocal action, each of these moments $X_i = 1$ must be applied as a pair of equal and opposite moments acting at the two edges of the cut. At the tops of the single-story columns the ordinate 1 therefore

---

[1] See also *Belastungsglieder*, page 257.—For lack of space it has, unfortunately, not been possible to reproduce here the authors' treatise "Das Prinzip der virtuellen Arbeit und die Maxwell-Mohrsche Arbeitsgleichung" ("The Principle of Virtual Work and the Maxwell-Mohr Energy Equation"), which is in manuscript.

has the full magnitude (i.e., unity) (diagrams for X_1, X_2 and X_3 in Fig. 419), whereas in the case of a column line extending over a height of two storys the unit moment is split up into the portions η and η', as indicated in Fig. 392 (diagrams for X_4, X_5 and X_6 in Fig. 419).

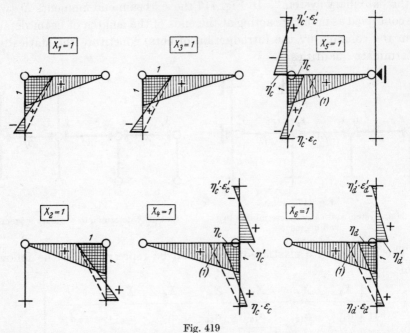

Fig. 419

Self-equilibrating states of stress $X_i = 1$ for main system, upper diagrams for odd values of i, lower diagrams for even values of i. (The same states of stress for the auxiliary system are shown by broken lines.)

It will readily be seen that the bending influence exercised by each of the six conditions $X_i = 1$ can, in each case, extend only over the two adjacent members (each column line being regarded as one member). It is merely for the sake of clarity that the various conditions have been shown separated in Fig. 419. The moment diagrams and their algebraic signs call for no special comment. In addition, the conditions $X_i = 1$ are shown for the statically determinate auxiliary system (Fig. 418) as broken lines in Fig. 419: in the case of the column lines the unit moment can be distributed over the upper and lower column in arbitrarily assumed proportions; in the present case a distribution based on $\eta = 1$ and $\eta' = 0$ has been adopted throughout.

The coefficients a_{ik} of the set of equations (1891) can be determined with the aid of the table on page 258 and 259 of *Belastungsglieder*. With regard to the several "moment diagram combinations" the so-called "reduction theorem" may advantageously be employed. This theorem states that, for each required combination, one of the moment diagrams can be taken from the statically determinate auxiliary system.[1] It is furthermore immediately evident from Fig. 419 that all the coefficients a_{ik} situated outside the two stepped lines in (1891) are zero, so that we are left with a set of trinomial equations. The result obtained by combining the moment diagrams is represented in (1892). The matrix of the coefficients in this array now consists of the principal diagonal and two equal adjoining diagonals, i.e., the matrix is symmetrical about its principal diagonal.

$$
\begin{array}{c|cccccc|cl}
 & X_1 & X_2 & X_3 & X_4 & X_5 & X_6 & & \\
\hline
1) & O_1 & k_1 & \cdot & \cdot & \cdot & \cdot & = -a_{10} & \\
2) & k_1 & O_2 & -b_b & \cdot & \cdot & \cdot & = -a_{20} & \\
3) & \cdot & -b_b & O_3 & k_2 & \cdot & \cdot & = -a_{30} & (1892) \\
4) & \cdot & \cdot & k_2 & O_4 & -s_C & \cdot & = -a_{40} & \\
5) & \cdot & \cdot & \cdot & -s_C & O_5 & k_3 & = -a_{50} & \\
6) & \cdot & \cdot & \cdot & \cdot & k_3 & O_6 & = -a_{60}. & \\
\end{array}
$$

The elements of the principal diagonal are as follows:

$$O_1 = b_a + 2k_1 \qquad O_3 = b_b + 2k_2 \qquad O_5 = s_C + 2k_3$$
$$O_2 = 2k_1 + b_b \qquad O_4 = 2k_2 + s_C \qquad O_6 = 2k_3 + s_D \qquad (1893)$$

where furthermore—as well as in the two adjoining diagonals of the matrix—we have:

$$b_n = k_n(2 - \varepsilon_n) \quad \text{where} \quad n = a, b, c, c', d, d' \qquad (1894)$$

and also:

$$s_C = \frac{b_c \cdot b_c'}{b_c + b_c'} \quad \text{and} \quad s_D = \frac{b_d \cdot b_d'}{b_d + b_d'}. \qquad (1895)$$

According to formulas (1812), (1813), (1753) and (1806) these are the flexibilities of the individual columns and the column lines.

The solution of the set of equations (1892) can be set out in the form of an array as represented in (1896), inasmuch as each unknown X_i must be dependent upon each load term a_{i0}. The influence coefficients n_{ik}—which have yet to be determined—constitute the so-called conjugate matrix with respect to the matrix of the coefficients a_{ik}. The actual process of solving the equations, i.e., determining the values $n_{ik} = n_{ki}$, is carried out by means of so-called backward reduction followed by forward recursion, or vice versa.

---

[1] By applying the reduction theorem, each value a_{ik} directly acquires its simplest form, whereas if this theorem were not employed, involved and awkward transformations would sometimes be necessary. The advantages of the abbreviated method of computation are especially evident in dealing with conditions of elastic restraint or with main systems possessing a high degree of statical indeterminacy (for instance, in the preliminary analysis of the four-member joint as dealt with in Section B). As has already been pointed out, a more exhaustive treatment of the applications of the Principle of Virtual Work (and the reduction theorem) would be outside the scope of the present volume.

	$+a_{10}$	$+a_{20}$	$+a_{30}$	$+a_{40}$	$+a_{50}$	$+a_{60}$	
$X_1 =$	$-n_{11}$	$+n_{12}$	$+n_{13}$	$-n_{14}$	$-n_{15}$	$+n_{16}$	
$X_2 =$	$+n_{21}$	$-n_{22}$	$-n_{23}$	$+n_{24}$	$+n_{25}$	$-n_{26}$	(1896)
$X_3 =$	$+n_{31}$	$-n_{32}$	$-n_{33}$	$+n_{34}$	$+n_{35}$	$-n_{36}$	
$X_4 =$	$-n_{41}$	$+n_{42}$	$+n_{43}$	$-n_{44}$	$-n_{45}$	$+n_{46}$	
$X_5 =$	$-n_{51}$	$+n_{52}$	$+n_{53}$	$-n_{54}$	$-n_{55}$	$+n_{56}$	
$X_6 =$	$+n_{61}$	$-n_{62}$	$-n_{63}$	$+n_{64}$	$+n_{65}$	$-n_{66}$.	

With regard to the algebraic signs and their sequence in (1896) the reader is referred back to the corresponding array (1838), which was obtained by a different method. The reduction and recursion procedures developed below are fully explained in *Durchlaufträger*, Vol. I, page 67 et seq. In the present case we have:

either
backward reduction: n:

$\iota_3 = \dfrac{k_3}{r_6}$ $r_6 = O_6$

$\sigma_C = \dfrac{s_C}{r_5}$ $r_5 = O_5 - k_3 \cdot \iota_3$

         $r_4 = O_4 - s_C \cdot \sigma_C$

$\iota_2 = \dfrac{k_2}{r_4}$

$\sigma_b = \dfrac{b_b}{r_3}$ $r_3 = O_3 - k_2 \cdot \iota_2$

$\iota_1 = \dfrac{k_1}{r_2}$ $r_2 = O_2 - b_b \cdot \sigma_b$

         $r_1 = O_1 - k_1 \cdot \iota_1$;

$\left(\sigma_a = \dfrac{b_a}{r_1}\right)$;

with
forward recursion:

$n_{11} = \dfrac{1}{r_1}$

$n_{22} = \dfrac{1}{r_2} + n_{11} \cdot \iota_1^2$

$n_{33} = \dfrac{1}{r_3} + n_{22} \cdot \sigma_b^2$ (1897)

$n_{44} = \dfrac{1}{r_4} + n_{33} \cdot \iota_2^2$ and

$n_{55} = \dfrac{1}{r_5} + n_{44} \cdot \sigma_C^2$ (1898)

$n_{66} = \dfrac{1}{r_6} + n_{55} \cdot \iota_3^2$.

or
forward reduction:

$\iota_1' = \dfrac{k_1}{v_1}$ $v_1 = O_1$

$\sigma_b' = \dfrac{b_b}{v_2}$ $v_2 = O_2 - k_1 \cdot \iota_1'$

$\iota_2' = \dfrac{k_2}{v_3}$ $v_3 = O_3 - b_b \cdot \sigma_b'$

         $v_4 = O_4 - k_2 \cdot \iota_2'$

$\sigma_C' = \dfrac{s_C}{v_4}$ $v_5 = O_5 - s_C \cdot \sigma_C'$

$\iota_3' = \dfrac{k_3}{v_5}$ $v_6 = O_6 - k_3 \cdot \iota_3'$;

$\left(\sigma_D' = \dfrac{s_D}{v_6}\right)$;

with
backward recursion:

$n_{66} = \dfrac{1}{v_6}$

$n_{55} = \dfrac{1}{v_5} + n_{66} \cdot \iota_3'^2$

$n_{44} = \dfrac{1}{v_4} + n_{55} \cdot \sigma_C'^2$ (1899)

$n_{33} = \dfrac{1}{v_3} + n_{44} \cdot \iota_2'^2$ and

$n_{22} = \dfrac{1}{v_2} + n_{33} \cdot \sigma_b'^2$ (1900)

$n_{11} = \dfrac{1}{v_1} + n_{22} \cdot \iota_1'^2$.

In analogy with (1839) and (1843) the array representing the procedure whereby the matrix elements n_{ik} can be determined will have the following form:

$$\begin{array}{c|cccccc|c}
 & (n_{11}) & n_{12} & n_{13} & n_{14} & n_{15} & n_{16} & \\
\cdot \iota_1 & \downarrow & \uparrow & \uparrow & \uparrow & \uparrow & \uparrow & \cdot \iota_1' \\
 & n_{21} & (n_{22}) & n_{23} & n_{24} & n_{25} & n_{26} & \\
\cdot \sigma_b & \downarrow & \downarrow & \uparrow & \uparrow & \uparrow & \uparrow & \cdot \sigma_b' \\
 & n_{31} & n_{32} & (n_{33}) & n_{34} & n_{35} & n_{36} & \\
\cdot \iota_2 & \downarrow & \downarrow & \downarrow & \uparrow & \uparrow & \uparrow & \cdot \iota_2' \\
 & n_{41} & n_{42} & n_{43} & (n_{44}) & n_{45} & n_{46} & \\
\cdot \sigma_C & \downarrow & \downarrow & \downarrow & \downarrow & \uparrow & \uparrow & \cdot \sigma_C' \\
 & n_{51} & n_{52} & n_{53} & n_{54} & (n_{55}) & n_{56} & \\
\cdot \iota_3 & \downarrow & \downarrow & \downarrow & \downarrow & \downarrow & \uparrow & \cdot \iota_3' \\
 & n_{61} & n_{62} & n_{63} & n_{64} & n_{65} & (n_{66}) & \\
\end{array} \qquad (1901)$$

The values σ_a and σ_D' indicated in brackets in the reduction sequences (1897) and (1899) are clearly the transfer factor for the end column A and the transfer factor for the end column line D for an external rotational moment acting there (e.g., cantilever extension of the horizontal member of the framework).

In the system consisting of four members intersecting at a joint, as investigated in Section B (see Fig. 393 and 397), the joints M, N, O may be designated as B, C, D, and the spans μ, ν by 2, 3. The joint in question can then be regarded as being identical with joint C in Fig. 416. For instance, equation (1841) will then be:

$$(n_{45} = n_{44} \cdot \sigma_C) = (n_{54} = n_{55} \cdot \sigma_C'). \qquad (1902)$$

The principal influence coefficient n_{44} from (1900), in conjunction with $\sigma_C' = s_C/v_4$ from (1899) and then with $n_{55} \cdot \sigma_C' = n_{44} \cdot \sigma_C$ from (1902), becomes:

$$n_{44} = \left(\frac{1}{v_4} + n_{55} \cdot \sigma_C' \cdot \frac{s_C}{v_4}\right) = \frac{1}{v_4} + n_{44} \cdot \sigma_C \cdot \frac{s_C}{v_4},$$

or, omitting the part in parentheses and rearranging:

$$n_{44}\left(\frac{v_4}{v_4} - \frac{\sigma_C s_C}{v_4}\right) = \frac{1}{v_4},$$

whence we finally obtain:

$$n_{44} = \frac{1}{v_4 - s_C \sigma_C} = \frac{1}{b_2' + s_C(1 - \sigma_C)} \cdot \qquad (1903)$$

Similarly, n_{55} from (1898), in conjunction with $\sigma_C = s_C/r_5$ from (1897) and then with $n_{44} \cdot \sigma_C = n_{55} \cdot \sigma_C'$ from (1902), becomes:

$$n_{55} = \frac{1}{r_5} + n_{55} \cdot \sigma_C' \cdot \frac{s_C}{r_5},$$

whence finally:

$$n_{55} = \frac{1}{r_5 - s_C \cdot \sigma_C'} = \frac{1}{s_C(1 - \sigma_C') + b_3} \cdot \qquad (1904)$$

The beam flexibilities $b_2' = k_2(2 - \iota_2')$ and $b_3 = k_3(2 - \iota_3)$ occurring in (1903) and (1904) here correspond to the values b_μ' and b_ν of equation (1818).

The expressions (1903) and (1904) are in agreement with the expressions (1832) that were obtained by a different method.

Finally, with $-s_C \sigma_C = r_4 - O_4$ and with $-s_C \cdot \sigma'_C = v_5 - O_5$ we can respectively obtain:

$$n_{44} = \frac{1}{r_4 + v_4 - O_4} \qquad n_{55} = \frac{1}{r_5 + v_5 - O_5}. \tag{1905}$$

By virtue of the formulas (1903), (1904) or (1905) the principal influence coefficients can be computed directly and in any desired order, without having recourse to the recursion sequences (1898) or (1900)—provided only that the two reductions (1897) and (1899) have first been carried out.

For the purpose of determining the load terms a_{i0} of the set of equations (1892) or of the array (1896), the so-called "M^0 conditions for the main system" are represented in Fig. 420–422. For convenience these are shown separately for beam, column and joint loading. The same conditions for the auxiliary system are shown by broken lines in Fig. 421 and 422. In principle, all the values a_{i0} can now be determined by combination either of the moment diagrams for the main system (as represented in Fig. 419) with the moment diagrams for the auxiliary system (as represented in Fig. 420–422) or, alternatively, by combination of the moment diagrams for the auxiliary system (as represented in Fig. 419) with the moment diagrams for the main system (as represented in Fig. 420–422). It is worth noting that Fig. 421 and 422 are related to Fig. 407 and 408 for the general case of four intersecting members, so that further explanatory comment is unnecessary.

For arbitrary vertical beam loading we have:

$$a_{10} = \mathfrak{L}_1 k_1 = \boldsymbol{L}_1, \qquad a_{20} = \mathfrak{R}_1 k_1 = \boldsymbol{R}_1 \text{ etc., up to } a_{60} = \mathfrak{R}_3 k_3 = \boldsymbol{R}_3. \tag{1906}$$

In this connection the reader is referred to the final solution arrays (12) for Frame Shape 1 and (1227) for Frame Shape 37.

For arbitrary horizontal column loading and for external rotational moments acting upon joints we obtain the following expressions:

$$\left.\begin{aligned}
a_{10} &= -\mathfrak{L}_a \varepsilon_a k_a + \mathfrak{R}_a k_a - D_A(2-\varepsilon_a) k_a = -\mathfrak{B}_A, \\
\text{where} \quad \mathfrak{B}_A &= D_A \cdot b_a + L_a \cdot \varepsilon_a - R_a. \\
a_{20} &= -a_{30} = L_b \cdot \varepsilon_b - R_b + D_B \cdot b_b = +\mathfrak{B}_B. \\
a_{40} &= -a_{50} = (L_c \cdot \varepsilon_c - R_c)\eta_c + (L'_c - R'_c \cdot \varepsilon'_c)\eta'_c + D_C \cdot s_C \\
&= \mathfrak{B}_C = D_C \cdot s_C - \mathfrak{B}_c \cdot \eta_c + \mathfrak{B}'_c \cdot \eta'_c, \\
\text{where} \quad \mathfrak{B}_c &= (R_c - L_c \cdot \varepsilon_c)\eta_c \quad \text{and} \quad \mathfrak{B}'_c = (L'_c - R'_c \cdot \varepsilon'_c)\eta'_c. \\
\text{similarly} \quad a_{60} &= \mathfrak{B}_D = D_D \cdot s_D - \mathfrak{B}_d \cdot \eta_d + \mathfrak{B}'_d \cdot \eta'_d, \\
\text{where} \quad \mathfrak{B}_d &= (R_d - L_d \cdot \varepsilon_d)\eta_d \quad \text{and} \quad \mathfrak{B}'_d = (L'_d - R'_d \cdot \varepsilon'_d)\eta'_d.
\end{aligned}\right\} \tag{1907}$$

Since a_{20} and a_{30}, as also a_{40} and a_{50}, differ only in that they have opposite signs, each pair of related n_{ik} columns can be combined into one w_{ib} and w_{iC} column. In this connection the reader is referred to the final solution arrays (17) for Frame Shape 1 and (1236) for Frame Shape 37: in particular, the columns A and B of (17), and the columns C and D of (1236).

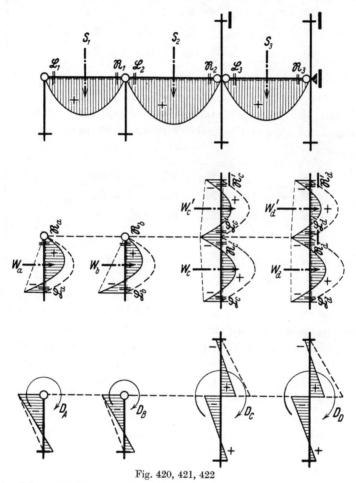

Fig. 420, 421, 422

"M^0 conditions" for vertical beam loading (top), horizontal column loading (center), and external rotational moment at beam joints (bottom), relating in all cases to the main system. (The same conditions for the auxiliary system are shown dotted.)

2. Analysis by the Deformation Method (Method II)

The analysis will again be discussed with reference to Fig. 416. This multi-bay framework is 4 times geometrically indeterminate in consequence of the four elastically rotatable joints it contains (namely, the joints of the horizontal member). The geometrical unknowns Y_N are the rotations φ_N of the joints multiplied by $(\mathfrak{T}_k/3 = 2EJ_k/l_k)$.

The four requisite equations for the determination of Y_N can at once be written down as follows:

$$\begin{array}{c|cccc|l}
 & Y_A & Y_B & Y_C & Y_D & \\
\hline
a) & S_{aa} & S_{ab} & \cdot & \cdot & = \mathfrak{K}_A \\
b) & S_{ba} & S_{bb} & S_{bc} & \cdot & = \mathfrak{K}_B \\
c) & \cdot & S_{cb} & S_{cc} & S_{cd} & = \mathfrak{K}_C \\
d) & \cdot & \cdot & \mathcal{S}_{dc} & S_{dd} & = \mathfrak{K}_D
\end{array} \qquad (1908)$$

With reference to the preliminary analysis of an arbitrary joint N with four intersecting members as envisaged in Section B, page 403 et seq., the symbols $\mathfrak{K}_N$ represent the joint load terms (= balancing moments = negative restraining moments) as envisaged in formulas (1878) and (1877). We obtain in particular:

$$\begin{aligned}
\mathfrak{K}_A &= \quad\quad\quad - \mathfrak{M}_{l1} + \mathfrak{E}_{ra} \quad\quad\quad + D_A \\
\mathfrak{K}_B &= + \mathfrak{M}_{r1} - \mathfrak{M}_{l2} + \mathfrak{E}_{rb} \quad\quad\quad + D_B \\
\mathfrak{K}_C &= + \mathfrak{M}_{r2} - \mathfrak{M}_{l3} + \mathfrak{E}_{rc} - \mathfrak{E}'_{lc} + D_C \\
\mathfrak{K}_D &= + \mathfrak{M}_{r3} \quad\quad\quad + \mathfrak{E}_{rd} - \mathfrak{E}'_{ld} + D_D .
\end{aligned} \qquad (1909)$$

The moments acting upon the joints of the framework under consideration are summarized in Fig. 423.

Furthermore the values S_{ii} of the principal diagonal of the matrix of coefficients of the set of equations (1908) are the joint stiffnesses (multiplied by 3) in the geometrically determinate main system, i.e., the fictitious joint moments S_{ii} in each case produce a unit rotation—multiplied by $(3/\mathfrak{T}_k (= l_k/2EJ_k)$—of the joint upon which the moment concerned is acting. Finally, the values $S_{ik} = S_{ki}$ are the requisite restraining moments at the relevant adjacent joints due to the action of those fictitious joint moments. The situation is conveniently summarized in Fig. 424 and 425.

For the sake of completeness it must be emphasized that the four equations (1908) are conditions of statical equilibrium. For instance, equation c) of (1908) is as follows:

$$S_{cb} \cdot Y_B + S_{cc} \cdot Y_C + S_{cd} \cdot Y_D = \mathfrak{K}_C .$$

In this equation all the moments are envisaged as having clockwise rotation. The left member of the equation consists of the joint moment $S_{cc} \cdot Y_C$ (see Fig. 424) and the two restraining moments $S_{cb} \cdot Y_B$ and $S_{cd} \cdot Y_D$ (see Fig. 425). The sum of these three moment terms must be equal to the joint moment $\mathfrak{K}_C$ constituting the right member of the equation.

The expressions for S_{ii} may be obtained with the aid of the general formula (1870), having due regard to the fact that certain terms must be omitted and that all j'_μ and j_ν must be zero. We thus obtain:

$$\left.\begin{aligned} S_{aa} &= \phantom{2K_1+{}}3S_a \phantom{{}+3S'_c} + 2K_1 = K_A \\ S_{bb} &= 2K_1 + 3S_b \phantom{{}+3S'_c} + 2K_2 = K_B \\ S_{cc} &= 2K_2 + 3S_c + 3S'_c + 2K_3 = K_C \\ S_{dd} &= 2K_3 + 3S_d + 3S'_d \phantom{{}+2K_3} = K_D. \end{aligned}\right\} \quad (1910)$$

The portions due to the beam stiffnesses are also derivable from the general formulas (1764) in the following general form, taking $\iota = \iota' = \tfrac{1}{2}$ for full fixity at the opposite end of the member under consideration:

$$3S_\nu = \frac{3K_\nu}{3/2} = 2K_\nu. \quad (1911)$$

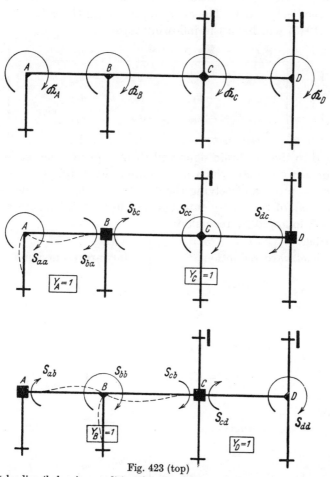

Fig. 423 (top)
Joint loading (balancing condition) for the geometrically indeterminate system
Fig. 424 (centre) and Fig. 425 (bottom)
Self-equilibrating states of strain $Y_A = 1$, $Y_B = 1$, $Y_C = 1$ and $Y_D = 1$

Having regard to the moment carry-over determined by $\iota = \iota' = \tfrac{1}{2}$, the fictitious restraining moments at the opposite ends of the members (beam

spans) will be:
$$(S_{ik} = S_{ki}) = \frac{1}{2} \cdot 2K_\nu = K_\nu. \tag{1912}$$

Hence the array (1908) can be rewritten as follows:

$$\begin{array}{c|cccc|l}
 & Y_A & Y_B & Y_C & Y_D & \\
\hline
a) & K_A & K_1 & \cdot & \cdot & = \mathfrak{K}_A \\
b) & K_1 & K_B & K_2 & \cdot & = \mathfrak{K}_B \\
c) & \cdot & K_2 & K_C & K_3 & = \mathfrak{K}_C \\
d) & \cdot & \cdot & K_3 & K_D & = \mathfrak{K}_D.
\end{array} \tag{1913}$$

The further treatment of the problem is essentially similar to that of Method I (see Section C.1). The array representing the solution of the set of equations (1913) will have the following form:

$$\begin{array}{c|cccc}
 & \mathfrak{K}_A & \mathfrak{K}_B & \mathfrak{K}_C & \mathfrak{K}_D \\
\hline
Y_A = & +u_{aa} & -u_{ab} & +u_{ac} & -u_{ad} \\
Y_B = & -u_{ba} & +u_{bb} & -u_{bc} & +u_{bd} \\
Y_C = & +u_{ca} & -u_{cb} & +u_{cc} & -u_{cd} \\
Y_D = & -u_{da} & +u_{db} & -u_{dc} & +u_{dd}.
\end{array} \tag{1914}$$

With regard to the algebraic signs and their sequence the reader is referred back to the array (1883), which was obtained by a different method.

The actual process of solving the equations (1913), i.e., determining the values $u_{ik} = u_{ki}$ of the "conjugate matrix" in (1914), is again accomplished by reduction followed by recursion:

either
backward reduction

$$j_3 = \frac{K_3}{r_d} \quad r_d = K_D$$
$$\quad r_c = K_C - K_3 \cdot j_3$$
$$j_2 = \frac{K_2}{r_c}$$
$$\quad r_b = K_B - K_2 \cdot j_2$$
$$j_1 = \frac{K_1}{r_b}$$
$$\quad r_a = K_A - K_1 \cdot j_1;$$

with
forward recursion

$$u_{aa} = \frac{1}{r_a}$$
$$u_{bb} = \frac{1}{r_b} + u_{aa} \cdot j_1^2$$
$$u_{cc} = \frac{1}{r_c} + u_{bb} \cdot j_2^2$$
$$u_{dd} = \frac{1}{r_d} + u_{cc} \cdot j_3^2.$$

(1915) and (1916)

or
forward reduction

$$j_1' = \frac{K_1}{v_a} \quad v_a = K_A$$
$$\quad v_b = K_B - K_1 \cdot j_1'$$
$$j_2' = \frac{K_2}{v_b}$$
$$\quad v_c = K_C - K_2 \cdot j_2'$$
$$j_3' = \frac{K_3}{v_c}$$
$$\quad v_d = K_D - K_3 \cdot j_3';$$

with
backward recursion

$$u_{dd} = \frac{1}{v_d}$$
$$u_{cc} = \frac{1}{v_c} + u_{dd} \cdot j_3'^2$$
$$u_{bb} = \frac{1}{v_b} + u_{cc} \cdot j_2'^2$$
$$u_{aa} = \frac{1}{v_a} + u_{bb} \cdot j_1'^2.$$

(1917) and (1918)

In this connection reference may be made to the formulas (23)–(26) for Frame Shape 1 and the formulas (1246)–(1249) for Frame Shape 37, which have exactly the same form as the above formulas (1915)–(1918). The differences between single-story and two-story frames are manifested only in the "joint coefficients" (22), (1245) and (1910) respectively.

It is unnecessary here to write down the computation array associated with the symmetrical influence coefficient matrix for the values u_{ik}, since it is exactly as indicated in (28) and (1251).

The arrays for the auxiliary moments Y_N—i.e., for the rotations at the joints expressed dimensionally as moments—are likewise identical in the case of Frame Shape 1, Frame Shape 37 and the "mixed" framework represented in Fig. 416: see the arrays (31), (1254) and (1914).

The direct forms of the principal influence coefficients—e.g., see (27) and (1250)—can be obtained as follows. For example, from (1914) we obtain for $\mathfrak{K}_C = 1$ in conjunction with $\mathfrak{K}_A = \mathfrak{K}_B = \mathfrak{K}_D = 0$:

$$(Y_A = +u_{ac}), \qquad Y_B = -u_{bc}, \qquad Y_C = +u_{cc}, \qquad Y_D = -u_{dc}. \tag{1919}$$

From equation c) of (1913) we furthermore obtain:

$$K_2 \cdot Y_B + K_C \cdot Y_C + K_3 \cdot Y_D = 1. \tag{1920}$$

Substituting the particular values of Y, as expressed in (1919), into equation (1920), yields:

$$-K_2 \cdot u_{bc} + K_C \cdot u_{cc} - K_3 \cdot u_{dc} = 1. \tag{1921}$$

Finally, according to (28) or (1251):

$$u_{bc} = u_{cc} \cdot j_2' \quad \text{and} \quad u_{dc} = u_{cc} \cdot j_3. \tag{1922}$$

Substituting these values into equation (1921), we obtain:

$$u_{cc}(K_C - K_2 \cdot j_2' - K_3 \cdot j_3) = u_{cc}(r_c + v_c - K_C) = 1, \tag{1923}$$

whence ultimately:

$$u_{cc} = \frac{1}{r_c + v_c - K_C}. \tag{1924}$$

The denominator of u_{cc} or the first expression in parentheses in the equation (1923) can, in conjunction with K_C from (1910), furthermore be rearranged to:

$$3S_C = K_2(2 - j_2') + 3S_c + 3S_c' + K_3(2 - j_3) \tag{1925}$$

this being the joint stiffness (multiplied by 3) of the four times geometrically indeterminate system under consideration; it is, in fact, identical with the expression (1870) (if the appropriate subscripts are introduced), which was obtained by a different method.

With regard to the determination of the final end moments of the members, the reader is referred to equations (1888) and (1890) in Section B.

3. Validity of the Formulas for Frames with Oblique Members

(a) All the fifty different laterally restrained (i.e., no "side-sway") Frame Shapes for which formulas are given in the main section of this book have horizontal beams and vertical columns. It is evident that the formulas —those derived by Method I as well as those derived by Method II—retain their validity if one or more members of the framework are oblique. The only essential requirement for the validity of these formulas is that the

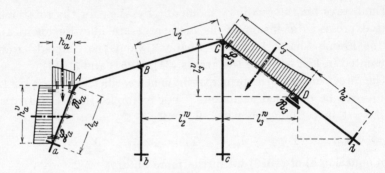

Fig. 426

Example of a multi-bay laterally restrained framed structure with sloping members and oblique loads—analysable according to "Frame Shape 1"

geometrical shape of the multi-bay framework is such that, taking due account of the supports, no translatory displacements of joints can occur. For example, the formulas given for Frame Shape 1 are also directly applicable to the framework represented in Fig. 426.

The beam span lengths l_v and the column lengths h_n, as envisaged in formulas (1) and (20), are always understood to be the actual lengths of the members, and it is these lengths that must be used for calculating the flexibility coefficients k_v, k_n or, alternatively, the stiffness coefficients K_v, K_n. In other words, any particular framework member has only one coefficient k (or, alternatively, K) which always relates to its actual length.

The direction of any given load acting upon a member may in itself be quite arbitrary. It should be noted, however, that the load terms $\mathfrak{L}$ and $\mathfrak{R}$, or $\mathfrak{M}_l$ and $\mathfrak{M}_r$, are always related to the orthogonal projection of the actual length of the member (i.e., the projection on to a line at right angles to the direction of the load). Furthermore a given oblique load can immediately be resolved into a horizontal and a vertical component. The same considerations apply to the k-fold load terms L and R, because k is merely a factor associated with the member under consideration and is independent of the direction of the load acting upon that member. These matters are dealt with in greater detail in the volume *Belastungsglieder*, page 27 and 28, in conjunction with pages 17–20. For example, the equations (69) given in

that book are immediately seen also to be valid in the form:
$$L = L_w + L_v \quad \text{and} \quad R = R_w + R_v \tag{1926}$$
and also:
$$\mathfrak{M}_l = \mathfrak{M}_{lw} + \mathfrak{M}_{lv} \quad \text{and} \quad \mathfrak{M}_r = \mathfrak{M}_{rw} + \mathfrak{M}_{rv}. \tag{1927}$$

In cases where oblique members or oblique loads occur, however, it is necessary to pay closer attention to the behaviour of the axial forces. A helpful consideration is that external point loads of arbitrary direction, acting upon joints of the framework, do not produce any bending moments in the laterally restrained framework, but are merely transmitted to the supports through the medium of axial forces in the members. Further information on the subject is given in *Durchlaufträger*, Vol. II, Section K, page 329–338 (in particular, Par. 2b, page 334–338).

(b) The structure represented in Fig. 427 (or any structure of similar type) can directly be analysed in accordance with "Frame Shape 37", Method II.[1] Each of the extra members marked e, f, g, h within a circle has

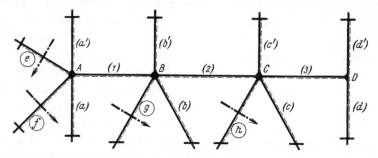

Fig. 427

Example of multi-bay laterally restrained framed structure with additional members as compared with "Frame Shape 37"

its appropriate stiffness coefficient K as envisaged in the formulas (1243) and its appropriate stiffness S as envisaged in the formulas (1244). The expressions for the joint coefficients (1245) must now be modified as follows:

$$\left. \begin{aligned} K_A &= 3(S_a + S'_a) + 2K_1 + 3(S_e + S_f) \\ K_B &= 2K_1 + 3(S_b + S'_b) + 2K_2 + 3S_g \\ K_C &= 2K_2 + 3(S_c + S'_c) + 2K_3 + 3S_h \\ K_D &= 2K_3 + 3(S_d + S'_d). \end{aligned} \right\} \tag{1928}$$

The further calculation, from formula (1246) to (1251), is similar to that indicated on pages 268/269.

---

[1] In principle the calculation could alternatively be carried out by Method I. In that case, however, instead of the column line flexibilities s_N, appropriately expanded expressions of the form (1804) would have to be introduced, and the column distribution factors η would have to be represented in accordance with equation (1805). Furthermore—though only for loading condition 2—similar appropriate expressions would have to be found for the auxiliary moments M_N^0.

If the extra members are subjected to arbitrary loads, the joint load terms must likewise be appropriately modified. The additional terms in the present case would be:

$$\mathfrak{R}_A = +\mathfrak{C}_{re} + \mathfrak{C}_{rf} \qquad \mathfrak{R}_B = +\mathfrak{C}_{rg} \qquad \mathfrak{R}_C = +\mathfrak{C}_{rh}. \tag{1929}$$

The formulas (1254) and (1255) again remain unchanged, while the column top moments (1256) are increased by:

$$\left.\begin{array}{ll} M_{Ae} = \mathfrak{C}_{re} - Y_A \cdot 3S_e & M_{Bg} = \mathfrak{C}_{rg} - Y_B \cdot 3S_g; \\ M_{Af} = \mathfrak{C}_{rf} - Y_A \cdot 3S_f; & M_{Ch} = \mathfrak{C}_{rh} - Y_C \cdot 3S_h. \end{array}\right\} \tag{1930}$$

The check relationships (1260) must, finally, be correspondingly modified by adding the moments $+ M_{Ae} + M_{Af}, + M_{Bg}$ and $+ M_{Ch}$.

4. Internal Loads

(a) With regard to the basic formulas for temperature variations and displacements of the supports, the reader is referred to Loading Conditions 134 and 135 in the auxiliary book *Belastungsglieder*. Furthermore the derivations given in *Durchlaufträger*, Vol. II, Section M, page 355–367, can usefully be consulted.

In particular, the case of "non-uniform temperature variation", as envisaged in Loading Condition 134 of *Belastungsglieder*, calls for no further comment, since the load terms $\mathfrak{L} = \mathfrak{R}$ and $\mathfrak{M}_l = \mathfrak{M}_r$ due to Δt of the individual members will, indeed, cause deflections of members but not displacements of joints. As regards the other "internal" loading conditions, which do cause joint displacements, it will be appropriate to give a short discussion. For this purpose reference will, by way of example, be made to Frame Shape 37.

(b) Referring to Fig. 428 it will be assumed that the beam BC (length l_2) undergoes a uniform rise in temperature of t_2 degrees. Now if the whole multi-bay frame is conceived as a statically determinate system with hinged joints (as, for instance, in Fig. 418 according to Method I), the beam in question will be able to elongate freely to the left by an amount

$$\Delta l_2 = \varepsilon_t \cdot t_2 \cdot l_2 \tag{1931}$$

The geometrically visible result of this will be that the four columns a, a', b, b' will undergo rotations, namely:

$$\psi_a = -\frac{\Delta l_2}{h_a} \qquad \psi_a' = +\frac{\Delta l_2}{h_a'} \qquad \psi_b = -\frac{\Delta l_2}{h_b} \qquad \psi_b' = +\frac{\Delta l_2}{h_b'}. \tag{1932}$$

According to *Belastungsglieder*, page 226 and 227, the rotations (1932) will in turn give rise to the k-fold load terms:

$$L_n = -R_n = \mathfrak{T}_k \cdot \psi_n, \quad \text{for} \quad n = a, a', b, b'. \tag{1933}$$

According to Method I, these must now be substituted into the column load terms (1234). Taking due account of signs, we finally obtain the present case:

$$\mathfrak{B}_n = + \mathfrak{T}_k \cdot \frac{\Delta l_2}{h_n} \cdot (1 + \varepsilon_n), \quad \text{for} \quad n = a, a', b, b'; \tag{1934}$$

i.e.: in this case all the four column load terms are positive. Furthermore, the two joint load terms according to formula (1235) are:

$$\mathfrak{B}_A = -\mathfrak{B}_a \eta_a + \mathfrak{B}'_a \eta'_a \qquad \mathfrak{B}_B = -\mathfrak{B}_b \eta_b + \mathfrak{B}'_b \eta'_b. \tag{1935}$$

All the other formulas (1235)–(1242) remain unchanged. Only with regard to the formulas (1240) and (1242) it must be noted that they merely contain the simple load terms. In general, we can therefore write:

$$\left[\left(\mathfrak{L}_n = \frac{L_n}{k_n}\right) \quad \text{or} \quad \left(\mathfrak{R}'_n = \frac{R'_n}{k'_n}\right)\right] = \underbrace{-\frac{\mathfrak{T}_k}{k_n} \cdot \frac{\Delta l_2}{h_n} = -\mathfrak{T}_n \cdot \frac{\Delta l_2}{h_n}}_{\text{for} \quad n = a, a', b, b'.}, \tag{1936}$$

If Method II is employed, the $1/k_n$-fold load terms (1933), or the load terms (1936), must be substituted into the expressions for the fixed-end moments (1252). A comparison of the values $\mathfrak{E}$ (1252) with the values $\mathfrak{B}$ (1234) shows that we may simply write:

$$\mathfrak{E}_n = -\frac{\mathfrak{T}_k}{k_n} \cdot \frac{\Delta l_2}{h_n} \cdot \frac{1 + \varepsilon_n}{2 - \varepsilon_n}, \quad \text{for} \quad n = a, a', b, b'; \tag{1937}$$

or, since $1/k_n = K_n$ and $K_n/(2 - \varepsilon_n) = S_n$:

$$(\mathfrak{E}_{rn} \quad \text{or} \quad \mathfrak{E}'_{ln}) = \underbrace{-\mathfrak{T}_k \cdot \frac{\Delta l_2}{h_n} \cdot S_n (1 + \varepsilon_n),}_{\text{for} \quad n = a, a', b, b'.} \tag{1938}$$

The two joint load terms according to the formulas (1253) reduce to:

$$\mathfrak{R}_A = +\mathfrak{E}_{ra} - \mathfrak{E}'_{la} \qquad \mathfrak{R}_B = +\mathfrak{E}_{rb} - \mathfrak{E}'_{lb}. \tag{1939}$$

All the other formulas (1254)–(1260) remain unchanged. With regard to the formulas (1257) and (1259) the equation (1936) must again be taken into account.

(c) Let it furthermore be assumed that the column base c has subsequently subsided a distance δ_c, as represented in Fig. 428. In the statically determinate structure with hinged joints, as envisaged above, the points C and c'—i.e., the whole column line—will likewise be displaced downward by the same amount.

The geometrical effect of this is, however, manifested merely in the two rotations:

$$\psi_2 = \delta_c/l_2 \quad \text{and} \quad \psi_3 = -\delta_c/l_3 \tag{1940}$$

— 426 —

These in turn will give rise to the k-fold load terms:

$$L_2 = -R_2 = \mathfrak{T}_k \cdot \psi_2 = +\mathfrak{T}_k \cdot \delta_c/l_2$$
$$L_3 = -R_3 = \mathfrak{T}_k \cdot \psi_3 = -\mathfrak{T}_k \cdot \delta_c/l_3.$$
(1941)

in accordance with *Belastungsglieder*, page 226.

The values (1941) can now, according to Method I, immediately be substituted into the formulas (1227) and furthermore into the formulas (1228)–(1231).

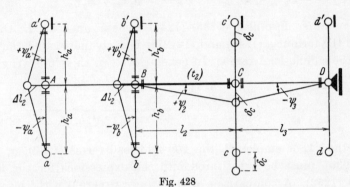

Fig. 428

Uniform temperature rise t_2 in beam span l_2, in conjunction with settlement δ_c of bearing c

According to Method II the load terms appear as fixed-end moments. Having regard to *Belastungsglieder*, page 227, we obtain:

$$\mathfrak{M}_{r2} = -\mathfrak{M}_{l2} = \mathfrak{T}_2 \cdot \psi_2 = +\mathfrak{T}_2 \cdot \delta_c/l_2$$
$$\mathfrak{M}_{r3} = -\mathfrak{M}_{l3} = \mathfrak{T}_3 \cdot \psi_3 = -\mathfrak{T}_3 \cdot \delta_c/l_3.$$
(1942)

With the values expressed in (1942) the three joint load terms as envisaged in the formulas (1253) become:

$$\mathfrak{R}_B = \qquad -\mathfrak{M}_{l2} = -\mathfrak{T}_2 \cdot \delta_c/l_2$$
$$\mathfrak{R}_C = \mathfrak{M}_{r2} - \mathfrak{M}_{l3} = +\mathfrak{T}_2 \cdot \delta_c/l_2 + \mathfrak{T}_3 \cdot \delta_c/l_3$$
$$\mathfrak{R}_D = \mathfrak{M}_{r3} \qquad = \qquad -\mathfrak{T}_3 \cdot \delta_c/l_3.$$
(1943)

The further analysis proceeds in accordance with the formulas (1254)–(1260), omitting all the load terms not appropriate to the case under consideration.

5. Influence Lines and Influence Line Equations

(a) In Section O, Sub-Sections 1, 2, 3 and 5 of *Durchlaufträger*, Vol. II, by the present authors, influence lines for continuous beams are dealt with in some detail, and their derivation and interpretation are indicated. The information given there is, in principle, also applicable to multi-bay rigid frames. To save space, the basic theory will therefore not be repeated here; instead, the reader is referred to the above-mentioned source for further information on the subject. The numerical examples given there—in Par. 4—are likewise in a form suitable for application to multi-bay frames.[1] For the present purpose it will be sufficient to give brief instructions for establishing influence line equations with reference to "Frame Shape 1", which has been selected as a typical example, with additional information relating to "Frame Shapes 12 and 13", i.e., without going into the underlying relationships.

In general, analytical influence line equations have to be established with regard to all the principal quantities for a unit load $P = 1$ moving along the entire length of the continuous horizontal member, or beam, of the rigid frame. In each case it is best to start from "Loading Condition 1: All the beam spans carrying arbitrary vertical loading" for each type of frame under consideration. If the unit load $P = 1$ can also travel along cantilevered extensions projecting from the two ends of the continuous horizontal member, then "Loading Condition 3: External rotational moments applied at all the joints of the horizontal member" should be adopted as an additional starting point. Of course, in this case only the two outermost joints of the horizontal member must be considered. [Finally, it is possible to produce influence lines for the effects of horizontal point loads applied to the columns of the rigid frame at any particular level. In this case it is appropriate to start from "Loading Condition 2: All the columns carrying arbitrary horizontal loading."]

On the basis of Loading Conditions 1 and 3 (according to Method I) the influence line equations for all the end moments X_i of the beam spans ("support moments") can be directly obtained. The "basic influence lines" can then be used for establishing the influence line equations for the "span moments", i.e., for the bending moment at any desired point of a beam span, as well as for any desired column moments, shear forces, and reactions at the supports. In accordance with standard practice, positive ordinates are plotted upwards from the axis of the horizontal member, and negative ordinates are plotted downwards, in all diagrams. [If influence lines have to be drawn for the columns, positive ordinates are plotted to the left, negative ordinates to the right.]

For the k-fold load terms L_ν and R_ν of the beams and for the external rotational moments D_A and D_N (where N in each case denotes the last beam joint) it is merely necessary to substitute the following values:

$$\boldsymbol{L_\nu} = l_\nu k_\nu \cdot \omega'_D \quad \text{and} \quad \boldsymbol{R_\nu} = l_\nu k_\nu \cdot \omega_D \qquad (1944)$$

and for the relevant end ordinates of the cantilevers:

$$\boldsymbol{D_A} = -1 \cdot l_A \quad \text{and} \quad \boldsymbol{D_N} = +1 \cdot l_N, \qquad (1945)$$

---

[1] See also the concise discussion of influence lines in Kleinlogel: *Rahmenformeln*, 12th and 13th ed., page 454 – 460, publ. Wilhelm Ernst & Sohn, Berlin.

where l_A and l_N are the lengths of the cantilevers. The symbols ω'_D and ω_D represent so-called omega functions; these are further explained in the auxiliary book *Belastungsglieder*, which also contains numerical tables for these functions. Furthermore:

$$\left.\begin{array}{ll} \omega'_D = \xi' - \xi'^3 \quad \text{and} \quad \omega_D = \xi - \xi^3 \\ \text{where} \qquad \xi' = x'/l \qquad \text{and} \qquad \xi = x/l, \\ \text{while always} \qquad x' + x = l \qquad \text{and} \qquad \xi' + \xi = 1. \end{array}\right\} \quad (1946)$$

[For the k_n-fold load terms L_n and R_n of the columns we may similarly write:

$$\boldsymbol{L}_n = h_n k_n \cdot \omega'_D \quad \text{and} \quad \boldsymbol{R}_n = h_n k_n \cdot \omega_D \quad (1947)$$

and furthermore for the load terms $\mathfrak{B}_N$ of the joints (for $N = A, B, C \ldots$):

$$\mathfrak{B}_N = -h_n k_n (\omega_D - \varepsilon_n \cdot \omega'_D) = -h_n k_n \cdot \omega_{Tn}.] \quad (1948)$$

Each influence line equation for any particular quantity consists of as many single equations as there are beam spans in the multi-bay rigid frame under consideration—plus two equations arising from cantilevers, if any. [Where necessary, each loaded column will require one additional equation.] The total influence line equation for a support moment X_i will then be composed of elements having the basic form:

$$\left.\begin{array}{c} \eta_\nu = \pm e'_{i\nu} \cdot \omega'_D \mp e_{i\nu} \cdot \omega_D \\ \text{with the coefficients:} \\ e'_{i\nu} = l_\nu k_\nu \cdot n_{il} \qquad e_{i\nu} = l_\nu k_\nu \cdot n_{ir}. \end{array}\right\} \quad (1949)$$

In these expressions the subscripts have the following meaning:

$\nu = 1, 2, 3\ldots$ span numbers
$i = 1, 2, 3\ldots$ sequence numbers of the moments X
$l = 1, 3, 5\ldots$ sequence of odd values i
$r = 2, 4, 6\ldots$ sequence of even values i

Furthermore, the end ordinates of the cantilevers have the form:

$$\eta_A = \pm l_A \cdot b_a \cdot n_{i1} \quad \text{and} \quad \eta_N = \pm l_N \cdot b_n \cdot n_{in}, \quad (1950)$$

where the subscript n in each case denotes the last number i.

Note: In the case of two-story multi-bay frames the flexibilities b_a and b_b of the individual columns in the expressions (1950) must be replaced by the "column line" flexibilities S_A and S_N, i.e., the flexibilities of the combined upper and lower columns.

[In the event of loads acting on columns the additional elements of the equations will correspondingly be:

$$\eta_n = \mp e_{in}(\omega_D - \varepsilon_n \cdot \omega'_D) \quad \text{where} \quad e_{in} = h_n k_n \cdot w_{in}. \quad (1951)$$

For end columns, w is replaced by n.]

(b) As has already been stated, "Frame Shape 1" has been selected to serve as an example. The process of establishing the influence line equations for the support moments X_i is carried out with the aid of the formulas (12) for the ordinates in the beam spans, and with the aid of the formulas (17) for the cantilever end ordinates—having due regard to the equations (1944), (1945), (1949), (1950).

Influence line equations for the support moment $X_1 = M_{A1}$
(M_{A1} line; see Fig. 429)

Spans: Cantilevers:

$$\left.\begin{aligned}
\eta_1 &= -e'_{11} \cdot \omega'_D + e_{11} \cdot \omega_D = -e'_{11} \cdot \omega'_{T1} & \eta_A &= -l_A b_a \cdot n_{11} \\
\eta_2 &= +e'_{12} \cdot \omega'_D - e_{12} \cdot \omega_D = +e'_{12} \cdot \omega'_{T2} & & \\
\eta_3 &= -e'_{13} \cdot \omega'_D + e_{13} \cdot \omega_D = -e'_{13} \cdot \omega'_{T3}; & \eta_D &= +l_D b_d \cdot n_{16};
\end{aligned}\right\} \quad (1952)$$

where: furthermore:

$$\begin{aligned}
e'_{11} &= l_1 k_1 \cdot n_{11} & e_{11} &= l_1 k_1 \cdot n_{12} & \omega'_{T1} &= \omega'_D - \iota_1 \cdot \omega_D \\
e'_{12} &= l_2 k_2 \cdot n_{13} & e_{12} &= l_2 k_2 \cdot n_{14} & \omega'_{T2} &= \omega'_D - \iota_2 \cdot \omega_D \\
e'_{13} &= l_3 k_3 \cdot n_{15} & e_{13} &= l_3 k_3 \cdot n_{16}; & \omega'_{T3} &= \omega'_D - \iota_3 \cdot \omega_D.
\end{aligned}$$

[Columns—having due regard to equations (1948) and (1951):

$$\left.\begin{aligned}
\eta_a &= -e_{1a} \cdot \omega_{Ta} &= -(h_a k_a \cdot n_{11}) \cdot (\omega_D - \varepsilon_a \cdot \omega'_D) \\
\eta_b &= -e_{1b} \cdot \omega_{Tb} &= -(h_b k_b \cdot w_{1b}) \cdot (\omega_D - \varepsilon_b \cdot \omega'_D) \\
\eta_c &= +e_{1c} \cdot \omega_{Tc} &= +(h_c k_c \cdot w_{1c}) \cdot (\omega_D - \varepsilon_c \cdot \omega'_D) \\
\eta_d &= -e_{1d} \cdot \omega_{Td} &= -(h_d k_d \cdot n_{16}) \cdot (\omega_D - \varepsilon_d \cdot \omega'_D).
\end{aligned}\right\} \quad (1953)$$

Note: In the further examples of influence lines—with the exception of the M_{Bb} line, the M_{Z1} line and the H line—the values η_n relating to the columns will be omitted.]

Influence line equations for the support moment $X_2 = M_{B1}$
(M_{B1} line; see Fig. 430)

Spans: Cantilevers:

$$\left.\begin{aligned}
\eta_1 &= +e'_{21} \cdot \omega'_D - e_{21} \cdot \omega_D = -e_{21} \cdot \omega_{T1} & \eta_A &= +l_A b_a \cdot n_{21} \\
\eta_2 &= -e'_{22} \cdot \omega'_D + e_{22} \cdot \omega_D = -e'_{22} \cdot \omega'_{T2} & & \\
\eta_3 &= +e'_{23} \cdot \omega'_D - e_{23} \cdot \omega_D = +e'_{23} \cdot \omega'_{T3}; & \eta_D &= -l_D b_d \cdot n_{26}.
\end{aligned}\right\} \quad (1954)$$

where: furthermore:

$$\begin{aligned}
e'_{21} &= l_1 k_1 \cdot n_{21} & e_{21} &= l_1 k_1 \cdot n_{22} & \omega_{T1} &= \omega_D - \iota'_1 \cdot \omega'_D \\
e'_{22} &= l_2 k_2 \cdot n_{23} & e_{22} &= l_2 k_2 \cdot n_{24} & \omega'_{T2} &= \omega'_D - \iota_2 \cdot \omega_D \\
e'_{23} &= l_3 k_3 \cdot n_{25} & e_{23} &= l_3 k_3 \cdot n_{26}; & \omega'_{T3} &= \omega'_D - \iota_3 \cdot \omega_D.
\end{aligned}$$

— 430 —

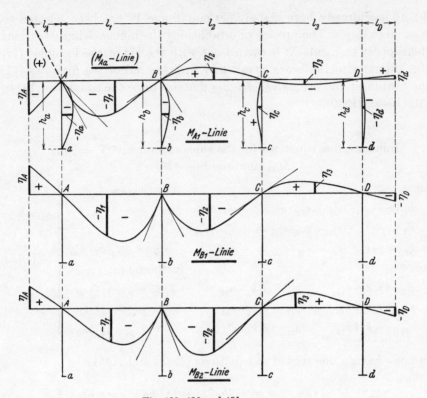

Fig. 429, 430 and 431
Influence lines for the support moments (beam moments): $X_1 = M_{A1}$ (top), $X_2 = M_{B1}$ (centre) and $X_3 = M_{B2}$ (bottom)

Influence line equations for the support moment $X_3 = M_{B2}$
(M_{B2} line; see Fig. 431)

$$\left.\begin{array}{ll}\text{Spans:} & \text{Cantilevers:} \\[4pt] \eta_1 = +e'_{31} \cdot \omega'_D - e_{31} \cdot \omega_D = -e_{31} \cdot \omega_{T1} & \eta_A = +l_A\, b_a \cdot n_{31} \\ \eta_2 = -e'_{32} \cdot \omega'_D + e_{32} \cdot \omega_D = -e'_{32} \cdot \omega'_{T2} & \\ \eta_3 = +e'_{33} \cdot \omega'_D - e_{33} \cdot \omega_D = +e'_{33} \cdot \omega'_{T3}; & \eta_D = -l_D\, b_d \cdot n_{36}; \\[6pt] \text{where:} & \text{furthermore:} \\[4pt] e'_{31} = l_1 k_1 \cdot n_{31} \quad e_{31} = l_1 k_1 \cdot n_{32} & \omega_{T1} = \omega_D - \iota'_1 \cdot \omega'_D \\ e'_{32} = l_2 k_2 \cdot n_{33} \quad e_{32} = l_2 k_2 \cdot n_{34} & \omega'_{T2} = \omega'_D - \iota_2 \cdot \omega_D \\ e'_{33} = l_3 k_3 \cdot n_{35} \quad e_{33} = l_3 k_3 \cdot n_{36}; & \omega'_{T3} = \omega'_D - \iota_3 \cdot \omega_D. \end{array}\right\} \quad (1955)$$

The influence line equations for $X_4 = M_{C2}$, $X_5 = M_{C3}$ and $X_6 = M_{D3}$ are established in similar fashion.

For establishing the influence line equation for the moment M_Z at an arbitrary point Z of a member (this point being defined by the abscissae z

and z') we must start from the basic equation:

$$M_Z = M_Z^0 + \frac{z'}{l} \cdot X_l + \frac{z}{l} \cdot X_r = M_Z^0 + \zeta' \cdot X_l + \zeta \cdot X_r, \qquad (1956)$$

where X_l in each case denotes a left-hand, and X_r a right-hand end moment of the member, while M_Z^0 is the bending moment at the point Z of the simply-supported beam. With regard to M_Z^0 we must distinguish between the case where the unit load $P = 1$ is located on the portion z or on the portion z' of the member in question. (See Fig. 129, page 409, in *Durchlaufträger*, Vol. II). Hence the influence line for the beam span in which the reference point under consideration is located will display two branches, and its equation will, in general, be:

$$\left. \begin{array}{l} \eta = +z' \cdot \xi \\ \eta' = +z \cdot \xi' \end{array} \right\} + \zeta' \cdot \eta_l + \zeta \cdot \eta_r. \qquad (1957)$$

For all the other spans the terms M_Z^0 are absent, and the equation simply becomes:

$$\eta = +\zeta' \cdot \eta_l + \zeta \cdot \eta_r. \qquad (1958)$$

In the equations (1957) and (1958) η_l and η_r denote the influence line equations for the support moment at the left-hand and at the right-hand end of the span respectively.

As an example with reference to "Frame Shape 1" the following equations will now be given.

Influence line equations for the span moment M_{Z1}
(M_{Z1} line; see Fig. 432)

$$\left. \begin{array}{l} \text{Spans:} \\ \begin{array}{l} \eta_1 = +z_1' \cdot \xi \\ \eta_1' = +z_1 \cdot \xi' \end{array} \Big\rangle - m_1' \cdot \omega_D' - m_1 \cdot \omega_D \quad \begin{array}{l} = +z_1' \cdot \xi \\ = +z_1 \cdot \xi' \end{array} \Big\rangle - \begin{bmatrix} m_1' \cdot \omega_{T1}' \\ m_1 \cdot \omega_{T1} \end{bmatrix}^* \\ \eta_2 = \quad \cdots \quad - m_2' \cdot \omega_D' + m_2 \cdot \omega_D \quad = \cdots \quad - m_2' \cdot \omega_{T2}' \\ \eta_3 = \quad \cdots \quad + m_3' \cdot \omega_D' - m_3 \cdot \omega_D \quad = \cdots \quad + m_3' \cdot \omega_{T3}'; \\ \text{Cantilevers:} \\ \qquad \eta_A = -\frac{l_A b_a}{l_1 k_1} \cdot m_1' \qquad \eta_D = -\frac{l_D b_d}{l_3 k_3} \cdot m_3; \\ \text{where:} \\ \quad m_1' = +\zeta_1' \cdot e_{11}' - \zeta_1 \cdot e_{21}' \qquad m_1 = -\zeta_1' \cdot e_{11} + \zeta_1 \cdot e_{21} \\ \quad m_2' = -\zeta_1' \cdot e_{12}' + \zeta_1 \cdot e_{22}' \qquad m_2 = -\zeta_1' \cdot e_{12} + \zeta_1 \cdot e_{22} \\ \quad m_3' = -\zeta_1' \cdot e_{13}' + \zeta_1 \cdot e_{23}' \qquad m_3 = -\zeta_1' \cdot e_{13} + \zeta_1 \cdot e_{23}; \\ \text{and furthermore:} \\ \begin{bmatrix} \omega_{T1}' = \omega_D' + \mu_1 \cdot \omega_D \\ \omega_{T1} = \omega_D + \mu_1' \cdot \omega_D' \end{bmatrix} \text{with } \begin{array}{l} \mu_1 = m_1/m_1' \\ \mu_1' = m_1'/m_1 \end{array} \Big]^* \quad \begin{array}{l} \omega_{T2}' = \omega_D' - \iota_2 \cdot \omega_D \\ \omega_{T3}' = \omega_D' - \iota_3 \cdot \omega_D. \end{array} \end{array} \right\} \quad (1959)$$

* Of the two forms enclosed in square brackets, either the upper or the lower one is valid depending on whether μ_1 or μ_1' is less than unity.

[Columns:
$$\eta_a = -m_a \cdot \omega_{Ta} \quad \text{where} \quad m_a = +\zeta_1' \cdot e_{1a} - \zeta_1 \cdot e_{2a}$$
$$\eta_b = +m_b \cdot \omega_{Tb} \qquad\qquad m_b = -\zeta_1' \cdot e_{1b} + \zeta_1 \cdot e_{2b}$$
$$\eta_c = -m_c \cdot \omega_{Tc} \qquad\qquad m_c = -\zeta_1' \cdot e_{1c} + \zeta_1 \cdot e_{2c}$$
$$\eta_d = +m_d \cdot \omega_{Td} \qquad\qquad m_d = -\zeta_1' \cdot e_{1d} + \zeta_1 \cdot e_{2d};$$
(1960)

where the symbols ω_{Tn} have the same significance as in the set of equations (1953).]

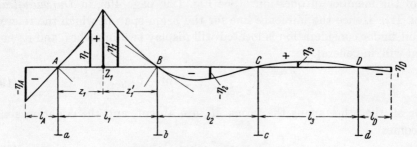

Fig. 432
Influence line for the span moment M_{z1}

With further reference to "Frame Shape 1", the column top moment M_{Aa}, in accordance with equation (13) or (18), is:

$$M_{Aa} = +X_1 \quad \text{or} \quad M_{Aa} = -D_A + X_1. \tag{1961}$$

This means that the M_{Aa} line for all three spans and for the right-hand cantilever is identical with the $(X_1 = M_{A1})$ line—and that only the end ordinate of the left-hand cantilever, with reference to the first equation (1945), changes into:

$$\eta_A = +l_A - l_A b_a \cdot n_{11}. \tag{1962}$$

See the influence line represented in Fig. 429, with the portion relating to the left-hand cantilever shown as a broken line.

For the column top moment M_{Bb} the expression according to equation (13) as well as (18) is:

$$M_{Bb} = -X_2 + X_3. \tag{1963}$$

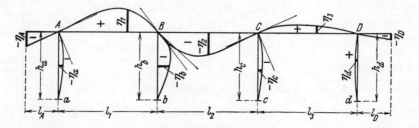

Fig. 433
Influence for the column head moment M_{Bb}

— 433 —

Hence the M_{Bb} line is the line obtained by subtracting the X_2 line from the X_3 line.

We can now write down the expressions for the

Influence line for the column top moment M_{Bb}
(M_{Bb} line; see Fig. 433)

Spans: Cantilevers:

$$\left.\begin{aligned}
\eta_1 &= -e'_{b1} \cdot \omega'_D + e_{b1} \cdot \omega_D = +e_{b1} \cdot \omega_{T1} \qquad \eta_A = -l_A\, b_a \cdot w_{1b} \\
\eta_2 &= -e'_{b2} \cdot \omega'_D + e_{b2} \cdot \omega_D = -e'_{b2} \cdot \omega'_{T2} \\
\eta_3 &= +e'_{b3} \cdot \omega'_D - e_{b3} \cdot \omega_D = +e'_{b3} \cdot \omega'_{T3} \qquad \eta_D = -l_D\, b_d \cdot w_{6b};
\end{aligned}\right\} \quad (1964)$$

where:

$$\begin{aligned}
e'_{b1} &= +e'_{21} - e'_{31} = l_1 k_1 \cdot w_{1b} & e_{b1} &= +e_{21} - e_{31} = l_1 k_1 \cdot w_{2b} \\
e'_{b2} &= -e'_{22} + e'_{32} = l_2 k_2 \cdot w_{3b} & e_{b2} &= -e_{22} + e_{32} = l_2 k_2 \cdot w_{4b} \\
e'_{b3} &= -e'_{23} + e'_{33} = l_3 k_3 \cdot w_{5b} & e_{b3} &= -e_{23} + e_{33} = l_3 k_3 \cdot w_{6b};
\end{aligned}$$

furthermore:

$$\omega_{T1} = \omega_D - \iota'_1 \cdot \omega'_D \qquad \omega'_{T2} = \omega'_D - \iota_2 \cdot \omega_D \qquad \omega'_{T3} = \omega'_D - \iota_3 \cdot \omega_D.$$

[Columns:
$$\left.\begin{aligned}
\eta_a &= -e_{ba} \cdot \omega_{Ta} \text{ where } & e_{ba} &= +e_{2a} - e_{3a} \\
\eta_b &= -e_{bb} \cdot \omega_{Tb} & e_{bb} &= +e_{2b} + e_{3b} \\
\eta_c &= -e_{bc} \cdot \omega_{Tc} & e_{bc} &= -e_{2c} + e_{3c} \\
\eta_d &= +e_{bd} \cdot \omega_{Td} & e_{bd} &= -e_{2d} + e_{3d};
\end{aligned}\right\} \quad (1965)$$

In these expressions the symbols ω_{Tn} have the same significance as in the set of equations (1953).]

The $(M_{Cc} = -X_4 + X_5)$ line can be obtained in similar fashion. For the M_{Dd} line we have the relationship $M_{Dd} = -X_6$ and $\eta_D = -l_D(1 - b_d n_{66})$.

The influence lines for the column base moments M_n are, according to equation (14), in each case obtained by multiplying the ordinates of the corresponding M_{Nn} line by $(-\epsilon_n)$. [If the influence line portions η_n for the columns are required, however, equation (19) should be taken into account, i.e., the term $-h_n \epsilon_n \cdot \omega'_D$ must in each case be added.]

For establishing the influence line equation for the shear force Q_Z at an arbitrary point Z of a span, the determining relationship is:

$$Q_Z = Q_Z^0 + (X_r - X_l)/l \tag{1966}$$

The two symbols X have the same significance as in equation (1956), while Q_Z^0 is the shear force at the point Z of the simply-supported beam. Here again we must distinguish between the case where the unit load $P = 1$ is located on the portion z or on the portion z' of the span. (See Fig. 128, page 408, of *Durchlaufträger*, Vol. II). Example:

Influence line equations for the shear force Q_{Z1}

(Q_{Z1} line; see Fig. 434)

Spans:

$$\left.\begin{array}{l}\begin{aligned}\eta_1 &= -\xi \\ \eta'_1 &= +\xi'\end{aligned}\Big\rangle + \vartheta'_{11}\cdot\omega'_D - \vartheta_{11}\cdot\omega_D = \Big\langle\begin{array}{l}-\xi \\ +\xi'\end{array}\Big\rangle + \left[\begin{array}{l}+\vartheta'_{11}\cdot\omega'_{T1} \\ -\vartheta_{11}\cdot\omega_{T1}\end{array}\right]^{\,1} \\ \eta_2 = \quad\cdot\cdot\quad -\vartheta'_{12}\cdot\omega'_D + \vartheta_{12}\cdot\omega_D = \quad\cdot\cdot\cdot\cdot\quad -\vartheta'_{12}\cdot\omega'_{T2} \\ \eta_3 = \quad\cdot\cdot\quad +\vartheta'_{13}\cdot\omega'_D - \vartheta_{13}\cdot\omega_D = \quad\cdot\cdot\cdot\cdot\quad +\vartheta'_{13}\cdot\omega'_{T3}; \\ \text{Cantilevers:} \qquad\qquad \text{where:} \\ \eta_A = +\frac{l_A\,b_a}{l_1\,k_1}\cdot\vartheta'_{11} \qquad \vartheta'_{11} = (e'_{11}+e'_{21})/l_1 \quad \vartheta_{11} = (e_{11}+e_{21})/l_1 \\ \qquad\qquad\qquad\qquad \vartheta'_{12} = (e'_{12}+e'_{22})/l_1 \quad \vartheta_{12} = (e_{12}+e_{22})/l_1 \\ \eta_D = -\frac{l_D\,b_d}{l_3\,k_3}\cdot\vartheta_{13}; \quad \vartheta'_{13} = (e'_{13}+e'_{23})/l_1 \quad \vartheta_{13} = (e_{13}+e_{23})/l_1; \\ \qquad\qquad\qquad \text{the values } e \text{ are as indicated in the} \\ \qquad\qquad\qquad \text{sets of equations (1952) and (154)} \\ \text{furthermore:} \\ \left[\begin{array}{l}\omega'_{T1} = \omega'_D - \tau_1\cdot\omega_D \\ \omega_{T1} = \omega_D - \tau'_1\cdot\omega'_D\end{array}\right] \text{where} \begin{array}{l}\tau_1 = \vartheta_{11}/\vartheta'_{11} \\ \tau'_1 = \vartheta'_{11}/\vartheta_{11}\end{array}\bigg]^{\,1} \quad \begin{array}{l}\omega'_{T2} = \omega'_D - \iota_2\cdot\omega_D \\ \omega'_{T3} = \omega'_D - \iota_3\cdot\omega_D.\end{array}\end{array}\right\} \quad (1967)$$

Note: With reference to Fig. 434: for Z_1 located directly to the right of joint A, i.e., for the Q_{l1} line, the influence line in span 1 is constituted by η'_1 only; for Z_1 located directly to the left of joint B, i.e., for the Q_{r1} line, the influence line in span 1 is constituted by η_1 only.

In the same way as the Q_{Z1} line we can also obtain the Q_{Z2} line, which is represented in Fig. 435. This diagram also shows the two limiting cases for the influence line in span 2: the Q_{l2} line constituted by η'_2 only, and the Q_{r2} line constituted by η_2 only.

The procedure for establishing the influence line equation for a total reaction at a support, or for an axial force in a column, is based on the general statical relationship:

$$V = V_\mu + V_\nu = -Q_{r\mu} + Q_{l\nu}, \tag{1968}$$

where μ and ν denote the sequence numbers of any two successive beam spans between which the support under consideration is situated. We thus obtain, for example, the V_b line represented in Fig. 436, conforming to the equation:

$$V_b = -Q_{r1} + Q_{l2}$$

---

[1] Of the two forms enclosed in square brackets, either the upper or the lower one is valid depending on whether τ_1 or τ'_1 is less than unity.

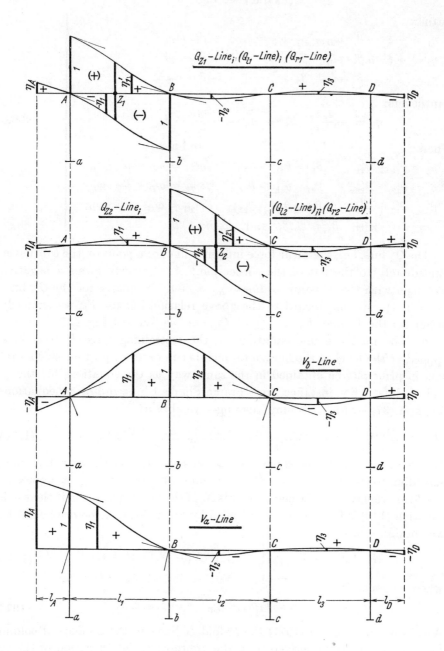

Fig. 434, 435, 436 and 437
Influence lines for the shear forces Q_{z1} and Q_{z2} and for the reactions V_b and V_a

by superposition of the negative Q_{r1} line (Fig. 434) and the Q_2 line, as follows:

Influence line equations for the support reaction V_b
(V_b line; see Fig. 436)

Spans:
$$\eta_1 = +\xi - \beta_1' \cdot \omega_D' + \beta_1 \cdot \omega_D = +\xi + \beta_1 \cdot \omega_{T1}$$
$$\eta_2 = +\xi' + \beta_2' \cdot \omega_D' - \beta_2 \cdot \omega_D = +\xi' + \beta_2' \cdot \omega_{T2}'$$
$$\eta_3 = \qquad -\beta_3' \cdot \omega_D' + \beta_3 \cdot \omega_D = \qquad -\beta_3' \cdot \omega_{T3}';$$

Cantilevers:
$$\eta_A = -\frac{l_A\, b_a}{l_1\, k_1} \cdot \beta_1' \qquad \eta_D = +\frac{l_D\, b_d}{l_3\, k_3} \cdot \beta_3; \tag{1969}$$

where: \hspace{4em} and also:
$$\beta_1' = \vartheta_{11}' + \vartheta_{21}' \qquad \beta_1 = \vartheta_{11} + \vartheta_{21} \qquad \omega_{T1} = \omega_D - \beta_{b1}' \cdot \omega_D'$$
$$\beta_2' = \vartheta_{12}' + \vartheta_{22}' \qquad \beta_2 = \vartheta_{12} + \vartheta_{22} \qquad \omega_{T2}' = \omega_D' - \beta_{b2} \cdot \omega_D$$
$$\beta_3' = \vartheta_{13}' + \vartheta_{23}' \qquad \beta_3 = \vartheta_{13} + \vartheta_{23}; \qquad \omega_{T3}' = \omega_D' - \iota_3 \cdot \omega_D;$$

with $\beta_{b1}' = \beta_1'/\beta_1$ and $\beta_{b2} = \beta_2/\beta_2'$.

The Q_A line, i.e., the shear force influence line for a point on the cantilever immediately to the left of the corner joint A, obviously forms a negative rectangle with the constant ordinate $\eta_A = -1$. (Similarly for the Q_D line: $\eta_D = +1$). Having regard to the above relationship for V_b, we similarly obtain for the V_a line: $V_a = -Q_A + Q_{l1}$, and we thus get Fig. 437.

(c) The influence line equation for the restraining force H (the force applied at the level of the horizontal member in order to prevent side-sway) can, in principle, be obtained in the same way as the equations for the Q_v and V_n lines. For instance, for "Frame Shape 1", according to equations (36) and (37)—without column loading—we obtain:

$$H = -\frac{M_{Aa}(1+\varepsilon_a)}{h_a} - \frac{M_{Bb}(1+\varepsilon_b)}{h_b} - \frac{M_{Cc}(1+\varepsilon_c)}{h_c} - \frac{M_{Dd}(1+\varepsilon_d)}{h_d}. \tag{1970}$$

That is to say, the H line is obtained by superposition of the four M_{Nn} lines, according to equation (1970). If the column top moments M_{Nn} as expressed in (13) are replaced by the end moments X_i of the beam spans, and if the whole equation (170) is furthermore multiplied by an arbitrary reference height h_k, we obtain:

$$(H \cdot h_k) = -X_1 \cdot \gamma_a(1+\varepsilon_a) + (X_2 - X_3)\gamma_b(1+\varepsilon_b) + \tag{1971}$$
$$+ (X_4 - X_5)\gamma_c(1+\varepsilon_c) + X_6 \cdot \gamma_d(1+\varepsilon_d),$$

where in general:

$$\gamma_n = h_k/h_n, \quad \text{for} \quad n = a, b, c, d. \tag{1972}$$

According to equation (1971) the h_k-fold H line—in the absence of column loading—can also be derived from the appropriate superposition of the six X_i lines.

If the H line is required also to include the columns, equations (38) and (39) would have to be taken as the starting point, i.e., equation (1970) would additionally have to include the terms:

$$\sum (\mathfrak{S}_{In} - \mathfrak{L}_n \cdot \varepsilon_n)/h_n, \quad \text{for} \quad n = a, b, c, d. \tag{1973}$$

In this connection it will be helpful to recall "Betti's Law", which states that every influence line of a statical quantity may be conceived as a fictitious deflection curve. All this has already been adequately explained in *Durchlaufträger*, Vol. II, Section O, Par. 5, page 458–464. In general, let y denote the ordinates of that fictitious deflection curve and let η be the ordinates of the influence lines; then the following general relationship will hold good:

$$\eta = -y, \tag{1974}$$

i.e., in order to use a deflection curve as an influence line it is merely necessary to change the algebraic sign.

Applying Betti's Law to "Frame Shape 1", we find that the influence line for H (the H line) is equal to the negative deflection curve produced by giving the continuous horizontal member of the frame a displacement of unit length in the positive direction of H, i.e., from right to left.

This horizontal unit displacement of the horizontal member acts as an "internal load" and as such affects all the columns, inasmuch as each column head undergoes the unit displacement applied to the horizontal member, with the result that each column undergoes a rotation ψ_n. In Fig. 438 displacements have been represented (to an exaggerated horizontal scale) for the statically determinate hinged system, while the dotted curves are meant to represent the deflection curve of the statically indeterminate frame. For the sake of completeness it should be noted that the displacement to the left is reckoned as negative in relation to the ends of the columns, i.e., we here have for the relative displacement of the ends of the members: $\delta = -1$.

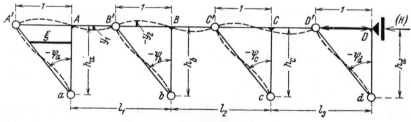

Fig. 438

Displacement $\delta = -1$ of the entire horizontal member in the statically determinate hinged system (the deflection curve for the statically indeterminate framework is shown dotted)

According to the auxiliary book *Belastungsglieder*, Loading Condition

135, the k_n-fold load terms with $\psi_n = -1/h_n$ and $\gamma_n = h_k/h_n$ become as follows:

$$L_n = -R_n = \mathfrak{T}_k \cdot \psi_n = -\frac{\mathfrak{T}_k}{h_n} = -\frac{\mathfrak{T}_k}{h_k} \cdot \gamma_n. \tag{1975}$$

Furthermore, according to equation (16), the column load terms become:

$$\mathfrak{B}_N = L_n \cdot \varepsilon_n - R_n = -\frac{\mathfrak{T}_k}{h_k} \cdot \gamma_n (1 + \varepsilon_n). \tag{1976}$$

Finally, the end moments X_i of the beam spans due to the displacement $\delta = -1$ can be obtained from the set of equations (17). Here the factor $\mathfrak{T}_k/h_k$, which is common to all the $\mathfrak{B}_N$, can be factored out. Setting, in general:

$$X_i = \frac{\mathfrak{T}_k}{h_k} \cdot x_i, \quad \text{and therefore} \quad x_i = \frac{X_i \cdot h_k}{\mathfrak{T}_k}, \tag{1977}$$

the expression for the "specific" moments x_i will be as follows:

	$\gamma_a(1+\varepsilon_a)$	$\gamma_b(1+\varepsilon_b)$	$\gamma_c(1+\varepsilon_c)$	$\gamma_d(1+\varepsilon_d)$
$x_1 =$	$-n_{11}$	$-w_{1b}$	$+w_{1c}$	$-n_{16}$
$x_2 =$	$+n_{21}$	$+w_{2b}$	$-w_{2c}$	$+n_{26}$
$x_3 =$	$+n_{31}$	$-w_{3b}$	$-w_{3c}$	$+n_{36}$
$x_4 =$	$-n_{41}$	$+w_{4b}$	$+w_{4c}$	$-n_{46}$
$x_5 =$	$-n_{51}$	$+w_{5b}$	$-w_{5c}$	$-n_{56}$
$x_6 =$	$+n_{61}$	$-w_{6b}$	$+w_{6c}$	$+n_{66}$.

(1978)

The set of expressions (1978) for x_i obviously corresponds to a displacement $\delta = -h_k/\mathfrak{T}_k$ of the horizontal member.

The column top moments associated with this displacement, obtained with the aid of equations (18), are:

$$x_{Aa} = +x_1 \quad x_{Bb} = -x_2 + x_3 \quad x_{Cc} = -x_4 + x_5 \quad x_{Dd} = -x_6 \tag{1979}$$

Having regard to

$$\mathfrak{L}_n = \mathfrak{T}_n \cdot \psi_n = -\frac{\mathfrak{T}_n}{h_k} \cdot \gamma_n = -\frac{\mathfrak{T}_k}{h_k} \cdot \frac{\gamma_n}{k_n}, \tag{1980}$$

the corresponding column base moments, obtained with the aid of equation (19), are:

$$x_n = (\gamma_n/k_n - x_{Nn})\varepsilon_n, \quad \text{for} \quad n = a, b, c, d. \tag{1981}$$

The relations (1977) are, of course, similarly valid for x_{Nn}, equations (1979), and x_n, equations (1981).

In Fig. 439 the diagram for the "specific" moments x has been represented, for which we must now determine the deflection curve. In this connection the reader is again referred to *Durchlaufträger*, Vol. II, Section N, Par. 4, page 396–400. For the present case, only the equation (534) given in that book has to be considered, omitting the $\mathfrak{M}^0$ term.

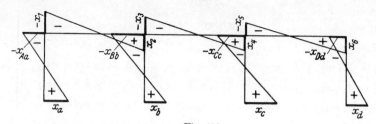

Fig. 439
$h_k/\mathfrak{T}_k$-fold moment diagram due to displacement $\delta = -1$

We thus generally have for each span (writing $M_L \triangleq X_l$ and $M_R \triangleq X_r$):

$$\mathfrak{T}_k \cdot y_\nu = l_\nu k_\nu (X_l \cdot \omega_D' + X_r \cdot \omega_D). \tag{1982}$$

In this expression X_l again always denotes the left-hand end moment and X_r the right-hand end moment of a beam span. If in equation (1982) the values X_i are replaced by the first expression (1977), the distortion factor $\mathfrak{T}_k$ will be eliminated from the whole equation, and we obtain for the beam spans:

$$y_\nu = \frac{l_\nu k_\nu}{h_k}(x_l \cdot \omega_D' + x_r \cdot \omega_D) = -\eta_\nu, \tag{1983}$$

for $\nu = 1, 2, 3; \quad l = 1, 3, 5; \quad r = 2, 4, 6.$

For the complete deflection curve of the columns we similarly obtain the corresponding general equation:

$$y_n = -\xi + \frac{h_n k_n}{h_k}(x_n \cdot \omega_D' + x_{Nn} \cdot \omega_D) = -\eta_n, \tag{1984}$$

for $n = a, b, c, d$ and $N = A, B, C, D.$

In the above expression the additional term $-\xi$ represents the displacement ordinate in the statically determinate condition of the structure (see Fig. 438).

The deflection curve equations (1983) and (1984) are, with due regard to the negative sign, identical with the influence line equations for the restraining force H, for each particular member concerned. By way of illustration, these equations are here given in full for "Frame Shape 1":

Influence line equations for the restraining force H
(H line; see Fig. 440)

Spans: Cantilevers:

$$\begin{aligned}
\eta_1 &= +f_1' \cdot \omega_D' - f_1 \cdot \omega_D \\
\eta_2 &= +f_2' \cdot \omega_D' - f_2 \cdot \omega_D \\
\eta_3 &= +f_3' \cdot \omega_D' - f_3 \cdot \omega_D;
\end{aligned} \qquad \begin{aligned}
\eta_A &= -\frac{l_A}{l_1}(2f_1' - f_1) \\
\eta_D &= +\frac{l_D}{l_3}(2f_3 - f_3');
\end{aligned} \tag{1985}$$

where:

$$f_1' = \frac{l_1 k_1}{h_k}(-x_1) \qquad f_2' = \frac{l_2 k_2}{h_k}(-x_3) \qquad f_3' = \frac{l_3 k_3}{h_k}(-x_5)$$

$$f_1 = \frac{l_1 k_1}{h_k} \cdot x_2 \qquad f_2 = \frac{l_2 k_2}{h_k} \cdot x_4 \qquad f_3 = \frac{l_3 k_3}{h_k} \cdot x_6.$$

— 440 —

Columns:

$$\eta_a = +\xi - f'_a \cdot \omega'_D + f_a \cdot \omega_D$$
$$\eta_b = +\xi - f'_b \cdot \omega'_D + f_b \cdot \omega_D$$
$$\eta_c = +\xi - f'_c \cdot \omega'_D + f_c \cdot \omega_D$$
$$\eta_d = +\xi - f'_d \cdot \omega'_D + f_d \cdot \omega_D$$

where:

$$f'_a = \gamma_a k_a \cdot x_a \qquad f_a = \gamma_a k_a(-x_{Aa})$$
$$f'_b = \gamma_b k_b \cdot x_b \qquad f_b = \gamma_b k_b(-x_{Bb})$$
$$f'_c = \gamma_c k_c \cdot x_c \qquad f_c = \gamma_c k_c(-x_{Cc})$$
$$f'_d = \gamma_d k_d \cdot x_d \qquad f_d = \gamma_d k_d(-x_{Dd}).$$

(1986)

Note: With reference to equations (1985) and (1986): In accordance with Fig. 439 the values x_l and x_{Nn} placed in parentheses are, as a rule, negative; they have, therefore, been provided with a negative sign, in order to obtain positive values for the coefficients f. All the values x in the above set of equations are obtained from the numerically evaluated equations (1978), (1979) and (1981). The η_v and η_n of (1985) and (1986) can, of course, be rearranged in the form ω'_T and ω_T respectively by placing the larger of the two values f outside brackets.

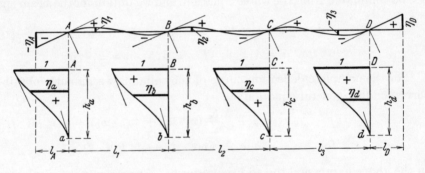

Fig. 440
Complete influence line for the restraining force H

(d) If the first beam span is elastically restrained at its left-hand end (or if the last beam span is elastically restrained at its right-hand end), the principles for establishing the influence line equations, as envisaged under points (b) and (c) of the present Section, will remain unchanged if the mode of representation as given on page 102 for "Frame Shape 13" is chosen as the initial form for the end moments of the beam spans. The only formal difference occurs in the function ω_{T1} in (1954) in that now the given value ε'_1 takes the place of ι'_1.

With regard to Loading Condition 1 as applied to, for instance, "Frame Shape 12", the coefficients e'_{v1} and e_{v1} of the equations η_1 for the support moments X_i do, admittedly, at first present themselves in a different form—but it can readily be shown that these coefficients are essentially similar to those relating to "Frame Shape 1" or "Frame Shape 13". In the formulas (425) the expressions corresponding to span 1 will now be:

$$\boldsymbol{L}_1 = l_1 k_1 \cdot \omega'_D \qquad \boldsymbol{R}_1 = l_1 k_1 \cdot \omega_D; \qquad \boldsymbol{\mathfrak{B}}_1 = l_1 k_1 (\omega_D - \varepsilon'_1 \cdot \omega'_D). \qquad (1987)$$

For $X_2 = M_{B1} = -\mathfrak{B}_1 \cdot n_{22}$ we thus obtain:

$$\eta_1 = -(l_1 k_1 \cdot n_{22}) \cdot (\omega_D - \varepsilon_1' \cdot \omega_D') = -e_{21} \cdot \omega_{T1}. \tag{1988}$$

Furthermore, for $M_{A1} = [X_1] = -(\mathfrak{A}_1 + M_{B1})\varepsilon_1'$:

$$\eta_1 = -l_1 \varepsilon_1' \cdot \omega_D' + e_{21} \varepsilon_1'(\omega_D - \varepsilon_1' \cdot \omega_D'), \tag{1989}$$

$$\eta_1 = -e_{11}' \cdot \omega_D' + e_{11} \cdot \omega_D = -e_{11}' \cdot \omega_{T1}',$$

where:

$$e_{11}' = l_1(1 + k_1 n_{22} \varepsilon_1') \varepsilon_1' \qquad e_{11} = l_1 k_1 n_{22} \varepsilon_1' \tag{1990}$$

and:

$$\omega_{T1}' = \omega_D' - \iota_1 \cdot \omega_D \quad \text{where} \quad \iota_1 = e_{11}/e_{11}'. \tag{1991}$$

The coefficient e_{21} in (1988) is the same as in (1954). The carry-over factor ϵ_1' for "Frame Shape 12 and 13" corresponds to ι_1' for "Frame Shape 1". Hence, having regard to the computation schemes (442) and (9):

$$(n_{12} = n_{21}) = n_{22} \cdot \varepsilon_1' \triangleq n_{22} \cdot \iota_1', \tag{1992}$$

so that the coefficients e_{21}' would also become equal.

The coefficient e_{11} in (1990) will, having regard to equation (1992), be equal to the same coefficient in (1952). The coefficients e'_{11} in (1990) and (1952) must comply with the following identity condition:

$$e_{11}' = [l_1(1 + k_1 \underline{n_{22} \varepsilon_1'}) \varepsilon_1' = l_1 k_1 n_{11}]! \tag{1993}$$

Complementary to (1992), we obtain from (442):

$$(n_{12} = n_{21}) = \underline{n_{22} \cdot \varepsilon_1'} = n_{11} \cdot \iota_1. \tag{1994}$$

The underlined product in equation (1993) can, with reference to (1994), be replaced by its equivalent $n_{11}\iota_1$. Solving for n_{11} we thus obtain:

$$n_{11} = \frac{1}{k_1(1/\varepsilon_1' - \iota_1)}. \tag{1995}$$

As this value n_{11} is the same as that in (439), it follows that the statement of identity, as envisaged above, is indeed valid.

(e) The formulas for the moments according to Method II are not so suitable as a starting point for deriving influence line equations, because all the moments—and, in particular, also the end moments of the beam spans—are composed of several terms. In this connection see, for example, the sets of equations (30)–(33) relating to "Frame Shape 1". The derivation of influence line equations according to Method II will therefore not be considered to come within the scope of this book. Suffice it to point out that the following functions would have to be employed for the fixed-end moments:

$$\left. \begin{array}{ll} \mathfrak{M}_l = -l \cdot \omega_{DT}' & \omega_{DT}' = \xi'^2 - \xi'^3 = \xi \xi'^2 = \xi' \omega_R \\ \mathfrak{M}_r = -l \cdot \omega_{DT} & \text{where} \quad \omega_{DT} = \xi^2 - \xi^3 = \xi^2 \xi' = \xi \omega_R; \end{array} \right\} \tag{1996}$$

and furthermore:

$$\omega'_{DT} = \frac{2\omega'_D - \omega_D}{3} = \omega'_D - \omega_R \qquad \omega_{DT} = \frac{2\omega_D - \omega'_D}{3} = \omega_D - \omega_R. \qquad (1997)$$

6. Numerical Example 1

(a) Preliminary note

In this example we shall consider a two-story three-span frame corresponding to "Frame Shape 37", as illustrated in Fig. 441, where l and h represent the lengths of members (beams and columns, respectively), J denotes moment of inertia, and the values ϵ characterise the semi-rigidly fixed ends of the columns. The calculations are carried out according to both Method I and Method II. In order to obtain an accurate comparison of the final results, all the computations have been performed rigorously, which necessitated the use of common fractions. All the numerical data recorded in Fig. 441 have intentionally been given somewhat haphazard values, though care has been taken to avoid making the arithmetic unduly complicated. The structure considered in this example must be conceived as bearing only an approximate resemblance to an actual structure which might be encountered in practice.

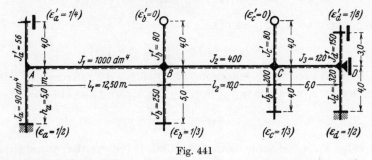

Fig. 441

System corresponding to "Frame Shape 37" chosen as numerical example

(b) Analysis according to Method I (Method of Forces)

See the formulas (1216) – (1242).

Flexibility coefficients of all the individual members, with $J_k = J_2 = 400$ ft^4 and $l_k = l_2 = 10.0$ ft, according to equation (1216):

$$k_1 = \frac{400}{1000} \cdot \frac{12.5}{10.0} = \frac{1}{2}, \quad k_2 = \frac{400}{400} \cdot \frac{10.0}{10.0} = 1, \quad k_3 = \frac{400}{120} \cdot \frac{6.0}{10.0} = 2;$$

$$k_a = \frac{400 \cdot 5.0}{90 \cdot 10.0} = \frac{20}{9}, \quad k_b = \frac{400 \cdot 5.0}{250 \cdot 10.0} = \frac{4}{5}, \quad k_c = \frac{400 \cdot 4.0}{200 \cdot 10.0} = \frac{4}{5}, \quad k_d = \frac{400 \cdot 4.0}{320 \cdot 10.0} = \frac{1}{2};$$

$$k'_a = \frac{400 \cdot 4.0}{56 \cdot 10.0} = \frac{20}{7}, \quad k'_b = k'_c = \frac{400 \cdot 4.0}{80 \cdot 10.0} = 2, \quad k'_d = \frac{400 \cdot 3.0}{150 \cdot 10.0} = \frac{4}{5}.$$

Flexibilities of the columns according to equations (1217):

$$b_a = \frac{20}{9} \cdot \frac{3}{2} = \frac{10}{3}, \quad b_b = b_c = \frac{4}{5} \cdot \frac{5}{3} = \frac{4}{3}, \quad b_d = \frac{1}{2} \cdot \frac{3}{2} = \frac{3}{4};$$

$$b'_a = \frac{20}{7} \cdot \frac{7}{4} = 5, \quad b'_b = b'_c = 2 \cdot 2 = 4, \quad b'_d = \frac{4}{5} \cdot \frac{15}{8} = \frac{3}{2}.$$

Column line flexibilities according to equations (1218):

$$s_A = \frac{1}{\frac{3}{10} + \frac{1}{5}} = 2, \quad s_B = s_C = \frac{1}{\frac{3}{4} + \frac{1}{4}} = 1, \quad s_D = \frac{1}{\frac{4}{3} + \frac{2}{3}} = \frac{1}{2}.$$

Column distribution factors according to equations (1219):

$$\eta_a = 2 \cdot \frac{3}{10} = \frac{3}{5} \qquad \eta_b = \eta_c = 1 \cdot \frac{3}{4} = \frac{3}{4} \qquad \eta_d = \frac{1}{2} \cdot \frac{4}{3} = \frac{2}{3};$$

$$\eta'_a = 2 \cdot \frac{1}{5} = \frac{2}{5} \qquad \eta'_b = \eta'_c = 1 \cdot \frac{1}{4} = \frac{1}{4} \qquad \eta'_d = \frac{1}{2} \cdot \frac{2}{3} = \frac{1}{3}.$$

Diagonal coefficients according to equations (1220):

$$O_1 = 2 + 1 = 3 \qquad O_3 = 1 + 2 = 3 \qquad O_5 = 1 + 4 = 5$$
$$O_2 = 1 + 1 = 2 \qquad O_4 = 2 + 1 = 3 \qquad O_6 = 4 + 1/2 = 9/2.$$

Note: For the "exact" numerical calculation that it is intended to carry out, it would appear to be expedient to determine the principal influence coefficients direct. For this purpose it is necessary to apply the two reduction sequences.

Backward reduction, according to (1221):

$$\iota_3 = 2 \cdot \frac{2}{9} = \frac{4}{9}$$
$$\sigma_C = 1 \cdot \frac{9}{37} = \frac{9}{37}$$
$$\iota_2 = 1 \cdot \frac{37}{102} = \frac{37}{102}$$
$$\sigma_B = 1 \cdot \frac{102}{269} = \frac{102}{269}$$
$$\iota_1 = \frac{1}{2} \cdot \frac{269}{436} = \frac{269}{872}$$

$$r_6 = \frac{9}{2}$$
$$r_5 = 5 - 2 \cdot \frac{4}{9} = \frac{37}{9}$$
$$r_4 = 3 - 1 \cdot \frac{9}{37} = \frac{102}{37}$$
$$r_3 = 3 - 1 \cdot \frac{37}{102} = \frac{269}{102}$$
$$r_2 = 2 - 1 \cdot \frac{102}{269} = \frac{436}{269}$$
$$r_1 = 3 - \frac{1}{2} \cdot \frac{269}{872} = \frac{4963}{1744}.$$

Forward reduction, according to (1222):

$$\iota'_1 = \frac{1}{2} \cdot \frac{1}{3} = \frac{1}{6}$$
$$\sigma'_B = 1 \cdot \frac{12}{23} = \frac{12}{23}$$
$$\iota'_2 = 1 \cdot \frac{23}{57} = \frac{23}{57}$$
$$\sigma'_C = 1 \cdot \frac{57}{148} = \frac{57}{148}$$
$$\iota'_3 = 2 \cdot \frac{148}{683} = \frac{296}{683}$$

$$v_1 = 3$$
$$v_2 = 2 - \frac{1}{2} \cdot \frac{1}{6} = \frac{23}{12}$$
$$v_3 = 3 - 1 \cdot \frac{12}{23} = \frac{57}{23}$$
$$v_4 = 3 - 1 \cdot \frac{23}{57} = \frac{148}{57}$$
$$v_5 = 5 - 1 \cdot \frac{57}{148} = \frac{683}{148}$$
$$v_6 = \frac{9}{2} - 2 \cdot \frac{296}{683} = \frac{4963}{1366}.$$

Principal influence coefficients direct, according to equations (1225):

$$n_{11} = \frac{1744}{4963}, \qquad n_{22} = \frac{1}{\dfrac{436}{269} + \dfrac{23}{12} - 2} = \frac{3228}{4963},$$

$$n_{33} = \frac{1}{\dfrac{269}{102} + \dfrac{57}{23} - 3} = \frac{2346}{4963}, \qquad n_{44} = \frac{1}{\dfrac{102}{37} + \dfrac{148}{57} - 3} = \frac{2109}{4963},$$

$$n_{55} = \frac{1}{\dfrac{37}{9} + \dfrac{683}{148} - 5} = \frac{1332}{4963}, \qquad n_{66} = \frac{1366}{4963}.$$

Matrix for the influence coefficients, according to the computation scheme (1226):

$$\begin{array}{c}
\left|\begin{array}{c} -\dfrac{269}{872} \\ \dfrac{102}{269} \\ \dfrac{37}{102} \\ \dfrac{9}{37} \\ \dfrac{4}{9} \\ \downarrow \end{array}\right| \cdot
\left|\begin{array}{cccccc}
\mathbf{\dfrac{1744}{4963}} & \dfrac{538}{4963} & \dfrac{204}{4963} & \dfrac{74}{4963} & \dfrac{18}{4963} & \dfrac{8}{4963} \\
\dfrac{538}{4963} & \mathbf{\dfrac{3228}{4963}} & \dfrac{1224}{4963} & \dfrac{444}{4963} & \dfrac{108}{4963} & \dfrac{48}{4963} \\
\dfrac{204}{4963} & \dfrac{1224}{4963} & \mathbf{\dfrac{2346}{4963}} & \dfrac{851}{4963} & \dfrac{207}{4963} & \dfrac{92}{4963} \\
\dfrac{74}{4963} & \dfrac{444}{4963} & \dfrac{851}{4963} & \mathbf{\dfrac{2109}{4963}} & \dfrac{513}{4963} & \dfrac{228}{4963} \\
\dfrac{18}{4963} & \dfrac{108}{4963} & \dfrac{207}{4963} & \dfrac{513}{4963} & \mathbf{\dfrac{1332}{4963}} & \dfrac{592}{4963} \\
\dfrac{8}{4963} & \dfrac{48}{4963} & \dfrac{92}{4963} & \dfrac{228}{4963} & \dfrac{592}{4963} & \mathbf{\dfrac{1366}{4963}}
\end{array}\right| \cdot
\left|\begin{array}{c} \dfrac{1}{6} \\ \dfrac{12}{23} \\ \dfrac{23}{57} \\ \dfrac{57}{148} \\ \dfrac{296}{683} \end{array}\right|
\end{array}$$

Loading Condition 1: All the beams carrying arbitrary vertical loading

Beam moments, according to (1227):

	$\mathfrak{L}_1/2$	$\mathfrak{R}_1/2$	$\mathfrak{L}_2$	$\mathfrak{R}_2$	$2\mathfrak{L}_3$	$2\mathfrak{R}_3$
$X_1 = M_{A1} =$	-1744	$+\ 538$	$+\ 204$	$-\ \ 74$	$-\ \ 18$	$+\ \ \ \ 8$
$X_2 = M_{B1} =$	$+\ 538$	-3228	-1224	$+\ 444$	$+\ 108$	$-\ \ 48$
$X_3 = M_{B2} =$	$+\ 204$	-1224	-2346	$+\ 851$	$+\ 207$	$-\ \ 92$
$X_4 = M_{C2} =$	$-\ \ 74$	$+\ 444$	$+\ 851$	-2109	$-\ 513$	$+\ 228$
$X_5 = M_{C3} =$	$-\ \ 18$	$+\ 108$	$+\ 207$	$-\ 513$	-1332	$+\ 592$
$X_6 = M_{D3} =$	$+\ \ \ \ 8$	$-\ \ 48$	$-\ \ 92$	$+\ 228$	$+\ 592$	-1366

(1998)

all values to be multiplied by $1/4963$

Column moments at the beam joints, according to equations (1228) and (1230):

$$\left.\begin{array}{ll}
M_{Aa} = + X_1 \cdot 3/5 & M'_{Aa} = - X_1 \cdot 2/5 \\
M_{Bb} = (-X_2 + X_3) \cdot 3/4 & M'_{Bb} = (+X_2 - X_3) \cdot 1/4 \\
M_{Cc} = (-X_4 + X_5) \cdot 3/4 & M'_{Cc} = (+X_4 - X_5) \cdot 1/4 \\
M_{Dd} = - X_6 \cdot 2/3; & M'_{Dd} = + X_6 \cdot 1/3.
\end{array}\right\} \quad (1999)$$

Column moments at the opposite ends: according to equations (1229) and (1231), with the values ϵ as indicated in Fig. 441.

Load Example 1: Vertical point load $P_1 = 4963 \cdot 5/6 \approx 4136$ (lb) at the left-hand fifth point of l_1 (see Fig. 442).

Load terms according to the auxiliary book *Belastungsglieder*, table on page 70, row for $\alpha = 1/5$:

$$\mathfrak{L}_1 = P_1 l_1 \cdot 36/125 = 3 \cdot \underline{4963} \qquad \mathfrak{R}_1 = P_1 l_1 \cdot 24/125 = 2 \cdot \underline{4963}.$$

Beam moments, according to the array (1998) (the underlined factors 4963 cancel out):

$$\begin{aligned}
X_1 &= -2616 + 538 = -2078 & X_4 &= -111 + 444 = +333 \\
X_2 &= +807 - 3228 = -2421 & X_5 &= -27 + 108 = +81 \\
X_3 &= +306 - 1224 = -918 & X_6 &= +12 - 48 = -36.
\end{aligned}$$

Column moments, as obtained with these values of X_i, from the expressions (1999):

$$\begin{aligned}
M_{Aa} &= -1246{,}8 & M_a &= +623{,}4 & M'_{Aa} &= +831{,}2 & M'_a &= -207{,}8 \\
M_{Bb} &= +1127{,}25 & M_b &= -375{,}75 & M'_{Bb} &= -375{,}75 & M'_b &= 0{,}0 \\
M_{Cc} &= -189{,}0 & M_c &= +63{,}0 & M'_{Cc} &= +63{,}0 & M'_c &= 0{,}0 \\
M_{Dd} &= +24{,}0, & M_d &= -12{,}0; & M'_{Dd} &= -12{,}0, & M'_d &= +1{,}5.
\end{aligned}$$

The bending moment diagrams have been drawn to scale in Fig. 442. It should be noted that $M^0_{P1} = P_1 l_1 \times 4/25 = 8271.6$. The "fixed points" I for the various load examples are indicated in Fig. 442, 443 and 444.

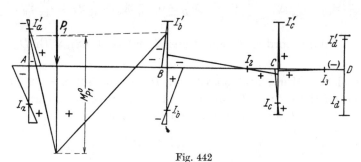

Fig. 442
Bending moment diagram for "Load Example 1"

Loading Condition 2: All the columns carrying arbitrary horizontal loading

Column load terms, according to equations (1234):

$$\begin{aligned}
\mathfrak{B}_a &= \mathfrak{R}_a \cdot 20/9 - \mathfrak{L}_a \cdot 10/9 & \mathfrak{B}'_a &= \mathfrak{L}'_a \cdot 20/7 - \mathfrak{R}'_a \cdot 5/7 \\
\mathfrak{B}_b &= \mathfrak{R}_b \cdot 4/5 - \mathfrak{L}_b \cdot 4/15 & \mathfrak{B}'_b &= \mathfrak{L}'_b \cdot 2 - 0 \\
\mathfrak{B}_c &= \mathfrak{R}_c \cdot 4/5 - \mathfrak{L}_c \cdot 4/15 & \mathfrak{B}'_c &= \mathfrak{L}'_c \cdot 2 - 0 \\
\mathfrak{B}_d &= \mathfrak{R}_d \cdot 1/2 - \mathfrak{L}_d \cdot 1/4; & \mathfrak{B}'_d &= \mathfrak{L}'_d \cdot 4/5 - \mathfrak{R}'_d \cdot 1/10.
\end{aligned}$$

Joint load terms, according to equation (1235):

$$\mathfrak{B}_A = -\frac{4}{3}\mathfrak{R}_a + \frac{2}{3}\mathfrak{L}_a + \frac{8}{7}\mathfrak{L}_a' - \frac{2}{7}\mathfrak{R}_a' \qquad \mathfrak{B}_B = -\frac{3}{5}\mathfrak{R}_b + \frac{1}{5}\mathfrak{L}_b + \frac{1}{2}\mathfrak{L}_b'$$

$$\mathfrak{B}_C = -\frac{3}{5}\mathfrak{R}_c + \frac{1}{5}\mathfrak{L}_c + \frac{1}{2}\mathfrak{L}_c' \qquad \mathfrak{B}_D = -\frac{1}{3}\mathfrak{R}_d + \frac{1}{6}\mathfrak{L}_d + \frac{4}{15}\mathfrak{L}_d' - \frac{1}{30}\mathfrak{R}_d'.$$

Beam moments, according to (1236), with values w according to (1237):

	$\mathfrak{B}_A$	$\mathfrak{B}_B$	$\mathfrak{B}_C$	$\mathfrak{B}_D$
$X_1 = M_{A1} =$	+ 1744	+ 334	− 56	+ 8
$X_2 = M_{B1} =$	− 538	− 2004	+ 336	− 48
$X_3 = M_{B2} =$	− 204	+ 1122	+ 644	− 92
$X_4 = M_{C2} =$	+ 74	− 407	− 1596	+ 228
$X_5 = M_{C3} =$	+ 18	− 99	+ 819	+ 592
$X_6 = M_{D3} =$	− 8	+ 44	− 364	− 1366

(2000)

all values to be multiplied by 1/4963

Support moments for the column line by itself, according to equation (1238):

$$b_a + b_a'' = 25/3, \qquad b_b + b_b' = b_c + b_c' = 16/3, \qquad b_d + b_d' = 9/4.$$

$$M_A^0 = -\frac{4}{15}\mathfrak{R}_a + \frac{2}{15}\mathfrak{L}_a - \frac{12}{35}\mathfrak{L}_a' + \frac{3}{35}\mathfrak{R}_a' \qquad M_B^0 = -\frac{3}{20}\mathfrak{R}_b + \frac{1}{20}\mathfrak{L}_b - \frac{3}{8}\mathfrak{L}_b'$$

$$M_C^0 = -\frac{3}{20}\mathfrak{R}_c + \frac{1}{20}\mathfrak{L}_c - \frac{3}{8}\mathfrak{L}_c' \qquad M_D^0 = -\frac{2}{9}\mathfrak{R}_d + \frac{1}{9}\mathfrak{L}_d - \frac{16}{45}\mathfrak{L}_d' + \frac{2}{45}\mathfrak{R}_d'.$$

Column moments: according to equations (1239) − (1242), with $D_N = 0$.

Load Example 2: Horizontal point load $P_c = 4963 \cdot 5/3 \approx 8272$ lb acting at the centre of the column h_c (see Fig. 443). Load terms according to the auxiliary book *Belastungsglieder*, Loading Condition 3, page 74:

$$\mathfrak{L}_c = \mathfrak{R}_c = \frac{3}{8} P_c h_c = \frac{5}{2} \cdot 4963; \qquad M_{Pc}^0 = \frac{P_c h_c}{4} = \frac{5}{3} \cdot 4963 = 8271{,}6 \cdots$$

Joint load term and M_C^0 moment:

$$\mathfrak{B}_C = -\frac{3-1}{5} \cdot \frac{5}{2} \cdot 4963 = -\underline{4963}; \qquad M_C^0 = -\frac{1}{4} \cdot 4963.$$

Beam moments, according to the array (2000) (the underlined factors 4963 cancel out):

$$X_1 = M_{A1} = + \ 56 \qquad X_3 = M_{B2} = - \ 644 \qquad X_5 = M_{C3} = - 819$$
$$X_2 = M_{B1} = - 336 \qquad X_4 = M_{C2} = + 1596 \qquad X_6 = M_{D3} = + 364.$$

Moments of the columns h_c and h'_c, according to equations (1239) – (1242):

$$M_{Cc} = -\frac{1}{4} \cdot 4963 - (1596 + 819) \cdot \frac{3}{4} = -3052,$$

$$M'_{Cc} = -\frac{1}{4} \cdot 4963 + (1596 + 819) \cdot \frac{1}{4} = -637;$$

$$M_c = -\left(\frac{5}{2} \cdot 4963 - 3052\right) \cdot \frac{1}{3} = -3118,5, \qquad M'_c = 0.$$

The other column moments, according to the same formulas:

$M_{Aa} = +33,6$	$M_a = -16,8$	$M'_{Aa} = -22,4$	$M'_a = +5,6$
$M_{Bb} = -231,0$	$M_b = +77,0$	$M'_{Bb} = +77,0$	$M'_b = 0,0$
$M_{Dd} = -242,6\ldots,$	$M_d = +121,3\ldots;$	$M'_{Dd} = +121,3\ldots$	$M'_d = -15,16\ldots$

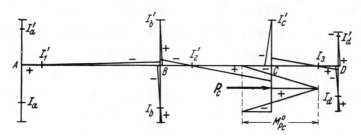

Fig. 443
Bending moment diagram for "Load Example 2"

Loading Condition 3: External rotational moments applied at all the joints of the horizontal member

Joint load terms, according to equation (1235), $\mathfrak{B}_N = D_N \cdot s_N$.
Beam moments, according to the formulas (1236), with the values w according to (1237), i.e., an array of numbers exactly similar to (2000):

	$2D_A$	D_B	D_C	$D_D/2$	
$X_1 = M_{A1} =$	$+1744$	$+334$	-56	$+8$	
$X_2 = M_{B1} =$	-538	-2004	$+336$	-48	
$X_3 = M_{B2} =$	-204	$+1122$	$+644$	-92	(2001)
$X_4 = M_{C2} =$	$+74$	-407	-1596	$+228$	
$X_5 = M_{C3} =$	$+18$	-99	$+819$	$+592$	
$X_6 = M_{D3} =$	-8	$+44$	-364	$-1366.$	

all values to be multiplied by 1/4963

Column moments at beam joint, according to equations (1239) and (1241):

$$\left.\begin{aligned}
M_{Aa} &= -(\boldsymbol{D_A} - X_1) \cdot 3/5 & M'_{Aa} &= +(\boldsymbol{D_A} - X_1) \cdot 2/5 \\
M_{Bb} &= -(\boldsymbol{D_B} + X_2 - X_3) \cdot 3/4 & M'_{Bb} &= +(\boldsymbol{D_B} + X_2 - X_3) \cdot 1/4 \\
M_{Cc} &= -(\boldsymbol{D_C} + X_4 - X_5) \cdot 3/4 & M'_{Cc} &= +(\boldsymbol{D_C} + X_4 - X_5) \cdot 1/4 \\
M_{Dd} &= -(\boldsymbol{D_D} + X_6) \cdot 2/3; & M'_{Dd} &= +(\boldsymbol{D_D} + X_6) \cdot 1/3.
\end{aligned}\right\} \quad (2002)$$

Column moments at the opposite ends: according to equations (1240) and (1242), with $\mathfrak{L}_n = \mathfrak{R}'_n = 0$.

Load Example 3: Rotational moment $D_B = +\ 4963$ (ft–lb) acting at joint B (see Fig. 444).

Beam moments, according to the array (2001) (the underlined factors 4963 cancel out):

$$X_1 = M_{A1} = +\ 334 \qquad X_3 = M_{B2} = +1122 \qquad X_5 = M_{C3} = -\ 99$$
$$X_2 = M_{B1} = -2004 \qquad X_4 = M_{C2} = -\ 407 \qquad X_6 = M_{D3} = +44.$$

Moments of the columns h_b and h'_b, according to equations (2002) etc.:

$$M_{Bb} = -(4963 - 2004 - 1122) \cdot 3/4 = -1377{,}75 \qquad M_b = +459{,}25$$
$$M'_{Bb} = +(4963 - 2004 - 1122) \cdot 1/4 = +\ 459{,}25 \qquad M'_b = 0{,}0.$$

The other column moments, according to equations (2002), (1240) and (1242):

$$M_{Aa} = +200{,}4 \qquad M_a = -100{,}2 \qquad M'_{Aa} = -133{,}6 \qquad M'_a = +33{,}4$$
$$M_{Cc} = +231{,}0 \qquad M_c = -\ 77{,}0 \qquad M'_{Cc} = -\ 77{,}0 \qquad M'_c = 0{,}0$$
$$M_{Dd} = -\ 29{,}3\ldots, \qquad M_d = +\ 14{,}6\ldots; \qquad M'_{Dd} = +\ 14{,}6\ldots \qquad M'_d = -\ 1{,}825.$$

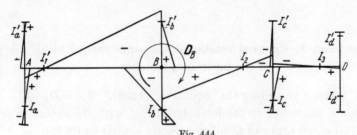

Fig. 444
Bending moment diagram for "Load Example 3"

(c) Analysis according to Method II (Deformation Method)
See the formulas (1243) – (1260):

Stiffness coefficients of all the individual members, with the same J_k and l_k as in Section (b), according to equations (1243) (in general: $K = 1/k$):

$$K_1 = 2$$
$$K_2 = 1 \qquad K_a = \frac{9}{20} \qquad K_b = K_c = \frac{5}{4} \qquad K_d = 2;$$
$$K_3 = 1/2; \qquad K'_a = \frac{7}{20} \qquad K'_b = K'_c = \frac{1}{2} \qquad K'_d = \frac{5}{4}.$$

Stiffnesses of the columns, according to equations (1244) (in general: $S = 1/b$):

$$S_a = \frac{3}{10} \qquad S_b = S_c = \frac{3}{4} \qquad S_d = \frac{4}{3}$$

$$S'_a = \frac{1}{5} \qquad S'_b = S'_c = \frac{1}{4} \qquad S'_d = \frac{2}{3}.$$

Joint coefficients, according to equations (1245):

$$K_A = \frac{9}{10} + \frac{3}{5} + 4 = \frac{11}{2} \qquad K_B = 4 + \frac{9}{4} + \frac{3}{4} + 2 = 9$$

$$K_C = 2 + \frac{9}{4} + \frac{3}{4} + 1 = 6 \qquad K_D = 1 + 4 + 2 = 7.$$

Note: The principal influence coefficients will again be determined direct, as under (b).

Backward reduction, according to (1246):

$$j_3 = \frac{1}{2} \cdot \frac{1}{7} = \frac{1}{14} \qquad r_d = 7$$

$$j_2 = 1 \cdot \frac{28}{167} = \frac{28}{167} \qquad r_c = 6 - \frac{1}{2} \cdot \frac{1}{14} = \frac{167}{28}$$

$$j_1 = 2 \cdot \frac{167}{1475} = \frac{334}{1475} \qquad r_b = 9 - 1 \cdot \frac{28}{167} = \frac{1475}{167}$$

$$r_a = \frac{11}{2} - 2 \cdot \frac{334}{1475} = \frac{14889}{2950}.$$

Forward reduction, according to (1247):

$$v_a = \frac{11}{2}$$

$$j'_1 = 2 \cdot \frac{2}{11} = \frac{4}{11} \qquad v_b = 9 - 2 \cdot \frac{4}{11} = \frac{91}{11}$$

$$j'_2 = 1 \cdot \frac{11}{91} = \frac{11}{91} \qquad v_c = 6 - 1 \cdot \frac{11}{91} = \frac{535}{91}$$

$$j'_3 = \frac{1}{2} \cdot \frac{91}{535} = \frac{91}{1070} \qquad v_d = 7 - \frac{1}{2} \cdot \frac{91}{1070} = \frac{14889}{2140}.$$

Principal influence coefficients direct, according to equations (1250):

$$u_{aa} = \frac{2950}{14889}, \qquad u_{bb} = \frac{1}{\frac{1475}{167} + \frac{91}{11} - 9} = \frac{1837}{14889},$$

$$u_{cc} = \frac{1}{\frac{167}{28} + \frac{535}{91} - 6} = \frac{2548}{14889}, \qquad u_{dd} = \frac{2140}{14889}.$$

— 450 —

Matrix for the influence coefficients, according to the computation array (1251); the common denominator is $14{,}889 = 3 \times 4963$:

$$\begin{array}{c} -\dfrac{334}{1475} \cdot \\ \dfrac{28}{167} \cdot \\ \dfrac{1}{14} \cdot \end{array} \left| \begin{array}{cccc} \dfrac{\mathbf{2950}}{14889} & \dfrac{668}{14889} & \dfrac{112}{14889} & \dfrac{8}{14889} \\ \dfrac{668}{14889} & \dfrac{\mathbf{1837}}{14889} & \dfrac{308}{14889} & \dfrac{22}{14889} \\ \dfrac{112}{14889} & \dfrac{308}{14889} & \dfrac{\mathbf{2548}}{14889} & \dfrac{182}{14889} \\ \dfrac{8}{14889} & \dfrac{22}{14889} & \dfrac{182}{14889} & \dfrac{\mathbf{2140}}{14889} \end{array} \right| \cdot \begin{array}{c} \dfrac{4}{11} \\ \dfrac{11}{91} \\ \dfrac{91}{1070} \end{array}$$

Loading Condition 4: Overall loading condition

Column restraining moments, according to equations (1252):

$$\left.\begin{array}{ll} \mathfrak{E}_{ra} = -\mathfrak{R}_a \cdot 2/3 + \mathfrak{L}_a \cdot 1/3 & \mathfrak{E}'_{la} = -\mathfrak{L}'_a \cdot 4/7 + \mathfrak{R}'_a \cdot 1/7 \\ \mathfrak{E}_{rb} = -\mathfrak{R}_b \cdot 3/5 + \mathfrak{L}_b \cdot 1/5 & \mathfrak{E}'_{lb} = -\mathfrak{L}'_b \cdot 1/2 \\ \mathfrak{E}_{rc} = -\mathfrak{R}_c \cdot 3/5 + \mathfrak{L}_c \cdot 1/5 & \mathfrak{E}'_{lc} = -\mathfrak{L}'_c \cdot 1/2 \\ \mathfrak{E}_{rd} = -\mathfrak{R}_d \cdot 2/3 + \mathfrak{L}_d \cdot 1/3; & \mathfrak{E}'_{ld} = -\mathfrak{L}'_d \cdot 8/15 + \mathfrak{R}'_d \cdot 1/15. \end{array}\right\} \quad (2003)$$

Joint load terms $\mathfrak{K}_N$: according to equations (1253):

Auxiliary moments, according to the formulas (1254):

$$\left.\begin{array}{c|cccc} & \mathfrak{K}_A & \mathfrak{K}_B & \mathfrak{K}_C & \mathfrak{K}_D \\ \hline Y_A = & +2950 & -668 & +112 & -8 \\ Y_B = & -668 & +1837 & -308 & +22 \\ Y_C = & +112 & -308 & +2548 & -182 \\ Y_D = & -8 & +22 & -182 & +2140 \end{array}\right\} \quad (2004)$$

all values to be multiplied by $1/(3 \times 4963)$

Beam moments, column moments and check relationships: according to equations (1255) − (1260).

The same *Load Examples* which have been considered with reference to Method I will now be considered with reference to Method II.

Load Example 1: Vertical concentrated load $P_1 = 4963 \times 5/6$ (lb) at left-hand fifth point of l_1 (see Fig. 442).

Fixing moments according to the auxiliary book *Belastungsglieder*, table on page 70, row for $\alpha = 1/5$:

$$\mathfrak{M}_{l1} = -P_1 l_1 \cdot 16/125 = -\frac{4}{3} \cdot 4963 \qquad \mathfrak{M}_{r1} = -P_1 l_1 \cdot 4/125 = -\frac{1}{3} \cdot 4963.$$

Auxiliary moments, according to the array (2004) (the underlined factors 4963 cancel out):

$$Y_A = +\frac{4}{3} \cdot \frac{2950}{3} + \frac{1}{3} \cdot \frac{668}{3} = +\frac{4156}{3}$$

$$Y_B = -\frac{4}{3} \cdot \frac{668}{3} - \frac{1}{3} \cdot \frac{1837}{3} = -\frac{1503}{3} = -501$$

$$Y_C = +\frac{4}{3} \cdot \frac{112}{3} + \frac{1}{3} \cdot \frac{308}{3} = +\frac{252}{3} = +84$$

$$Y_D = -\frac{4}{3} \cdot \frac{8}{3} - \frac{1}{3} \cdot \frac{22}{3} = -\frac{18}{3} = -6.$$

Beam moments, according to equations (1255):

$$M_{A1} = -\frac{4 \cdot 4963}{3} + \left(2 \cdot \frac{4156}{3} - \frac{1503}{3}\right) \cdot 2 = -2078$$

$$M_{B1} = -\frac{4963}{3} - \left(\frac{4156}{3} - 2 \cdot \frac{1503}{3}\right) \cdot 2 = -2421,$$

$M_{B2} = +(-2 \cdot 501 + 84) \cdot 1 = -918 \qquad M_{C3} = +(2 \cdot 84 - 6) \cdot 1/2 = +81$

$M_{C2} = -(-501 + 2 \cdot 84) \cdot 1 = +333, \qquad M_{D3} = -(84 - 2 \cdot 6) \cdot 1/2 = -36.$

Column moments at the beam joints, according to equations (1256) and (1258):

$M_{Aa} = -\frac{4156}{3} \cdot \frac{9}{10} = -1246{,}8 \qquad M'_{Aa} = +\frac{4156}{3} \cdot \frac{3}{5} = +831{,}2$

$M_{Bb} = +501 \cdot 9/4 \quad = +1127{,}25 \qquad M'_{Bb} = -501 \cdot 3/4 \quad = -375{,}75$

$M_{Cc} = -84 \cdot 9/4 \quad = -189{,}0 \qquad M'_{Cc} = +84 \cdot 3/4 \quad = +63{,}0$

$M_{Dd} = +6 \cdot 4 \quad = +24{,}0; \qquad M'_{Dd} = -6 \cdot 2 \quad = -12{,}0.$

Note: The end moments of the members as obtained by Method II are in exact agreement with those obtained by Method I.

Load Example 2: Horizontal point load $P_c = 4963 \times 5/3$ (lb) acting at the centre of the column h_c (see Fig. 443).

Load terms $\mathfrak{L}_c = \mathfrak{R}_c$ as under (b); restraining moments according to (2003); joint load term according to (1253):

$$\mathfrak{L}_c = \mathfrak{R}_c = \frac{5 \cdot 4963}{2}, \qquad \mathfrak{C}_{rc} = \mathfrak{K}_C = \frac{5 \cdot 4963}{2}\left(-\frac{3}{5} + \frac{1}{5}\right) = -\underline{4963}.$$

Auxiliary moments, according to (2004) (the underlined factors 4963 cancel out):

$$Y_A = -\frac{112}{3}, \qquad Y_B = +\frac{308}{3}, \qquad Y_C = -\frac{2548}{3}, \qquad Y_D = +\frac{182}{3}.$$

— 452 —

Beam moments, according to equations (1255):

$$M_{A1} = \left(-2 \cdot \frac{112}{3} + \frac{308}{3}\right) \cdot 2 = +56 \quad M_{B1} = \left(+\frac{112}{3} - 2 \cdot \frac{308}{3}\right) \cdot 2 = -336$$

$$M_{B2} = \left(+2 \cdot \frac{308}{3} - \frac{2548}{3}\right) \cdot 1 = -644 \quad M_{C2} = \left(-\frac{308}{3} + 2 \cdot \frac{2548}{3}\right) \cdot 1 = +1596$$

$$M_{C3} = \left(-2 \cdot \frac{2548}{3} + \frac{182}{3}\right) \cdot \frac{1}{2} = -819; \quad M_{D3} = \left(+\frac{2548}{3} - 2 \cdot \frac{182}{3}\right) \cdot \frac{1}{2} = -364.$$

Column moments at the beam joints, according to equations (1256) and (1258):

$$M_{Aa} = +\frac{112}{3} \cdot \frac{9}{10} = +33{,}6 \qquad M'_{Aa} = -\frac{112}{3} \cdot \frac{3}{5} = -22{,}4$$

$$M_{Bb} = -\frac{308}{3} \cdot \frac{9}{4} = -231{,}0 \qquad M'_{Bb} = +\frac{308}{3} \cdot \frac{3}{4} = +77{,}0$$

$$M_{Cc} = -4963 + \frac{2548}{3} \cdot \frac{9}{4} = -3052{,}0 \qquad M'_{Cc} = -\frac{2548}{3} \cdot \frac{3}{4} = -637{,}0$$

$$M_{Dd} = -\frac{182}{3} \cdot 4 = -242{,}6\ldots \qquad M'_{Dd} = +\frac{182}{3} \cdot 2 = +121{,}3\ldots$$

Note: The end moments of the members as obtained by Method II are in exact agreement with those obtained by Method I.

Load Example 3: Rotational moment $D_B = +4963$ (ft–lb) acting at joint B (see Fig. 444).

Joint load term, according to equations (1253):

$$\mathfrak{K}_B = +D_B = +\underline{4963}.$$

Auxiliary moments, according to (2004) (the underlined factors 4963 cancel out):

$$Y_A = -\frac{668}{3} \quad Y_B = +\frac{1837}{3} \quad Y_C = -\frac{308}{3} \quad Y_D = +\frac{22}{3}.$$

Beam moments, according to equations (1255):

$$M_{A1} = \left(-2 \cdot \frac{668}{3} + \frac{1837}{3}\right) \cdot 2 = +334 \quad M_{B1} = \left(+\frac{668}{3} - 2 \cdot \frac{1837}{3}\right) \cdot 2 = -2004$$

$$M_{B2} = \left(+2 \cdot \frac{1837}{3} - \frac{308}{3}\right) \cdot 1 = +1122 \quad M_{C2} = \left(-\frac{1837}{3} + 2 \cdot \frac{308}{3}\right) \cdot 1 = -407$$

$$M_{C3} = \left(-2 \cdot \frac{308}{3} + \frac{22}{3}\right) \cdot \frac{1}{2} = -99; \quad M_{D3} = \left(+\frac{308}{3} - 2 \cdot \frac{22}{3}\right) \cdot \frac{1}{2} = +44$$

Column moments at the beam joints, according to equations (1256) and (1258):

$$M_{Aa} = +\frac{668}{3} \cdot \frac{9}{10} = +200{,}4 \qquad M'_{Aa} = -\frac{668}{3} \cdot \frac{3}{5} = -133{,}6$$

$$M_{Bb} = -\frac{1837}{3} \cdot \frac{9}{4} = -1377{,}75 \qquad M'_{Bb} = +\frac{1837}{3} \cdot \frac{3}{4} = +459{,}25$$

$$M_{Cc} = +\frac{308}{3} \cdot \frac{9}{4} = +231{,}0 \qquad M'_{Cc} = -\frac{308}{3} \cdot \frac{3}{4} = -77{,}0$$

$$M_{Dd} = -\frac{22}{3} \cdot 4 = -29{,}3\ldots \qquad M'_{Dd} = +\frac{22}{3} \cdot 2 = +14{,}6\ldots$$

Check relationship at joint B, according to equations (1260):

$$+4963 - 1377{,}75 - 459{,}25 - 2004 - 1122 = 0.$$

Note: The end moments of the members as obtained by Method II are in exact agreement with those obtained by Method I.

Concluding remark: The results obtained by means of Method I for any particular loading must, of course, be in exact agreement with those obtained by means of Method II. A direct comparison of the final moments at the ends of the members is, however, possible only if in the case of Method I all the values $\mathfrak{B}$ and in the case of Method II all the values $\mathfrak{M}$, $\mathfrak{E}$ and $\mathfrak{K}$ are expressed by the "simple" load terms $\mathfrak{L}$ and $\mathfrak{R}$.

D. The Laterally Unrestrained Multi-Bay Framework

1. Analysis by Method I and Method II

(a) Single-story multi-bay frames

In the section "General Considerations" (Section A, page 362) some preliminary indications have been given as to the analysis of laterally unrestrained multi-bay rigid frames. It was pointed out that, in the first stage of the calculations, the structure must be treated as laterally restrained. This involves the application of a horizontal "restraining force" H to the continuous horizontal member. For each state of loading of the structure, this force H can be calculated, from considerations of statics, as a reaction at a support. The second stage of the calculations will then merely consist in determining the effect of a "correcting load" W acting at the level of the horizontal member, this load W being of equal magnitude but opposite in direction to the force H. On superimposing the two stages, the horizontal forces W and H cancel each other (this process is expressed by the equilibrium condition $\Sigma H = 0$ in the form $W - H = 0$), and the final moments and reactions for the laterally unrestrained framework are thus obtained.

In the further discussion of the above principles, "Frame Shape 1" and "Frame Shape 1v" will be taken as examples. In this connection the reader is furthermore referred to the analysis of joint displacements given in Section C.4.b and c and C.5.c.

The whole continuous horizontal member of the frame represented in Fig. 445 is assumed to have been given a displacement of arbitrary magnitude δ to the right. To start with, this displacement may be supposed to be applied to the so-called "kinematic" system, with hinges at all the joints. On completion of the displacement, the horizontal member is once again assumed to be laterally restrained: in Fig. 445 this condition is represented by the restraining bearing at D' (which is conceived as the bearing at D that has undergone a displacement δ along with the horizontal member).

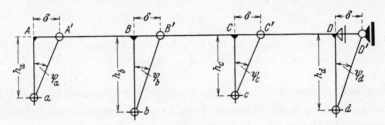

Fig. 445
Horizontal displacement of the entire horizontal member by an amount $\delta = h_k/\mathfrak{T}_k$ towards the right

The four rotations ψ_a, ψ_b, ψ_c and ψ_d are geometrical effects of the displacement. These rotations can be directly conceived as so-called "internal loading" of the, in itself, laterally restrained "Frame Shape 1" (cf. Section C.5.c mentioned above). In this connection "Loading Condition 2: All the columns carrying arbitrary horizontal loading" must obviously be considered.

For reasons of convenience we may write:

$$\delta = \frac{h_k}{\mathfrak{T}_k} = \frac{h_k \cdot l_k}{6EJ_k}, \quad \text{(units: 1/lb)} \tag{2005}$$

where h_k is an arbitrary reference column height and $\mathfrak{T}_k$ is the distortion factor of the system.

According to *Belastungsglieder*, Loading Condition 135, we have in general:

$$\mathfrak{L}_n = -\mathfrak{R}_n = \mathfrak{T}_n \cdot \psi_n = \mathfrak{T}_n \cdot \delta/h_n \tag{2006}$$

or, for the k-fold load terms:

$$L_n = -R_n = \mathfrak{T}_k \cdot \psi_n = \mathfrak{T}_k \cdot \delta/h_n. \tag{2007}$$

Substituting the value (2005) chosen for δ into equations (2006) and (2007), we obtain respectively:

$$\mathfrak{L}_n = -\mathfrak{R}_n = \frac{\mathfrak{T}_n}{h_n} \cdot \frac{h_k}{\mathfrak{T}_k} = K_n \cdot \gamma_n = \frac{\gamma_n}{k_n} \tag{2008}$$

and

$$L_n = -R_n = \frac{\mathfrak{T}_k}{h_n} \cdot \frac{h_k}{\mathfrak{T}_k} = \frac{h_k}{h_n} = \gamma_n. \tag{2009}$$

Note: The load terms therefore appear here as unknowns.

On treating the problem according to Method I, the column load terms according to equation (16) now become:

$$\mathfrak{B}_N = L_n \cdot \varepsilon_n - R_n = \gamma_n \cdot \varepsilon_n + \gamma_n = \gamma_n(1 + \varepsilon_n). \tag{2010}$$

The moments arising in consequence of the displacement $\delta = h_k/\mathfrak{T}_k$ of the horizontal member are termed "specific" moments and will be designated by the symbol m instead of M.

Beam moments according to formulas (17):

$$\left. \begin{array}{r|cccc} & \gamma_a(1+\varepsilon_a) & \gamma_b(1+\varepsilon_b) & \gamma_c(1+\varepsilon_c) & \gamma_d(1+\varepsilon_d) \\ \hline m_{A1} = & +n_{11} & +w_{1b} & -w_{1c} & +n_{16} \\ m_{B1} = & -n_{21} & -w_{2b} & +w_{2c} & -n_{26} \\ m_{B2} = & -n_{31} & +w_{3b} & +w_{3c} & -n_{36} \\ m_{C2} = & +n_{41} & -w_{4b} & -w_{4c} & +n_{46} \\ m_{C3} = & +n_{51} & -w_{5b} & +w_{5c} & +n_{56} \\ m_{D3} = & -n_{61} & +w_{6b} & -w_{6c} & -n_{66} \end{array} \right\} \tag{2011}$$

Note: The array (2011) must, of course, be in agreement with the array (1978), apart from the fact that all the algebraic signs are reversed. Hence $m_{A1} = -x_1$, $m_{B1} = -x_2$, $m_{B2} = -x_3$, etc. If all the moments $m_{N\nu}$ of the array (2011) were to be multiplied by $\mathfrak{T}_k/h_k$, they would correspond to a displacement of the horizontal member equal to $\delta = +1$.

Column moments at top, according to equations (18):

$$m_{Aa} = +m_{A1} \quad m_{Bb} = -m_{B1} + m_{B2} \quad m_{Cc} = -m_{C2} + m_{C3} \quad m_{Dd} = -m_{D3}. \tag{2012}$$

Column moments at base, according to equation (19) in conjunction with (2008):

$$m_n = -(\gamma_n/k_n + m_{Nn})\varepsilon_n, \quad \text{for} \quad n = a, b, c, d. \tag{2013}$$

If the problem is treated according to Method II, then the joint load terms as expressed in equations (30), in conjunction with the somewhat rearranged equation (29), will be:

$$\mathfrak{K}_N = \mathfrak{C}_{rn} = \frac{\mathfrak{L}_n \cdot \varepsilon_n - \mathfrak{R}_n}{2 - \varepsilon_n} = \frac{K_n}{2 - \varepsilon_n} \cdot \gamma_n(\varepsilon_n + 1)$$

and finally, with reference to equation (21):

$$\mathfrak{K}_N = \mathfrak{C}_{rn} = S_n \gamma_n (1 + \varepsilon_n). \tag{2014}$$

The specific moments, for $\delta = h_k/\mathfrak{T}_k$, are calculated from equations (31) – (34) as follows:

Auxiliary moments y (instead of Y):

$$\begin{array}{r|cccc}
 & S_a\gamma_a(1+\varepsilon_a) & S_b\gamma_b(1+\varepsilon_b) & S_c\gamma_c(1+\varepsilon_c) & S_d\gamma_d(1+\varepsilon_d) \\
\hline
y_A = & +u_{aa} & -u_{ab} & +u_{ac} & -u_{ad} \\
y_B = & -u_{ba} & +u_{bb} & -u_{bc} & +u_{bd} \\
y_C = & +u_{ca} & -u_{cb} & +u_{cc} & -u_{cd} \\
y_D = & -u_{da} & +u_{db} & -u_{dc} & +u_{dd}.
\end{array} \qquad (2015)$$

Beam span moments:

$$\begin{aligned}
m_{A1} &= +(2y_A + y_B)K_1 & m_{B1} &= -(y_A + 2y_B)K_1 \\
m_{B2} &= +(2y_B + y_C)K_2 & m_{C2} &= -(y_B + 2y_C)K_2 \\
m_{C3} &= +(2y_C + y_D)K_3 & m_{D3} &= -(y_C + 2y_D)K_3.
\end{aligned} \qquad (2016)$$

Column moments at top:

$$\begin{aligned}
m_{Aa} &= S_a\gamma_a(1+\varepsilon_a) - y_A \cdot 3S_a & m_{Cc} &= S_c\gamma_c(1+\varepsilon_c) - y_C \cdot 3S_c \\
m_{Bb} &= S_b\gamma_b(1+\varepsilon_b) - y_B \cdot 3S_b & m_{Dd} &= S_d\gamma_d(1+\varepsilon_d) - y_D \cdot 3S_d.
\end{aligned} \qquad (2017)$$

Column moments at base:

$$m_n = -(\gamma_n K_n + m_{Nn})\varepsilon_n, \quad \text{for} \quad n = a, b, c, d. \qquad (2018)$$

The basic form of the bending moment diagram due to a displacement $\delta = +h_k/\mathfrak{T}_k$ of the horizontal member is shown in Fig. 446.

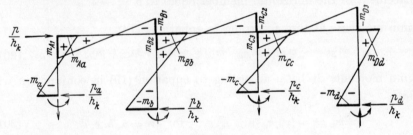

Fig. 446

Bending moment diagram (m) due to horizontal displacement $\delta = h_k/\mathfrak{T}_k$, equivalent to that due to a horizontal concentrated load $W = p/h_k$ applied at the level of the horizontal member

To the specific column moments as expressed by equations (2012)/(2013) according to Method I, or equations (2017)/(2018) according to Method II, correspond column shears of the magnitude:

$$(H_n = +Q_n) = \frac{m_{Nn} - m_n}{h_n} = \frac{(m_{Nn} - m_n)\gamma_n}{h_k} \qquad (2019)$$

or of the h_k-fold magnitude:

$$p_n = (H_n h_k = Q_n h_k) = (m_{Nn} - m_n)\gamma_n. \qquad (2020)$$

The sum of all the h_k-fold column shears is:

$$p = \sum p_n = p_a + p_b + p_c + p_d. \tag{2021}$$

Hence the moments m produced by a displacement $\delta = h_k/\mathfrak{T}_k$ of the horizontal member (see Fig. 446) may also be conceived as being produced by a horizontal concentrated load of magnitude p/h_k acting at the level of the horizontal member.

For an actual given horizontal concentrated load W acting on the laterally unrestrained rigid frame at the level of its horizontal member (for example, "Frame Shape 1v"), we can obviously obtain the corresponding bending moment diagram (M) from the specific bending moment diagram (m) simply by multiplying by the conversion factor:

$$U = \frac{W}{p/h_k} = \frac{W h_k}{p}. \tag{2022}$$

The final moment diagram (M) due to W—which for each member is affine with the moment diagram (m) in consequence of δ or p/h_k (see Fig. 446)—fundamentally has the form represented in Fig. 6. Having regard to the information given under the heading "Frame Shape 1v", including the "Supplementary Note", the above discussion calls for no further elaboration.

We shall now again write, in three alternative forms, the general superposition equation for laterally unrestrained multi-bay frames. All the moments occurring therein, and the restraining force, in each case relate to one particular loading condition and to one and the same point of the frame.

$$\boxed{M^v = M + M^{(W=H)} = M + m \cdot U = M + \left(\frac{m}{p}\right) \cdot (H\, h_k).} \tag{2023}$$

Where the symbols have the following general meanings:

M^v moments of the laterally unrestrained frame
M moments of the laterally restrained frame
$M^{(W=H)}$ moments due to a concentrated horizontal load W equal to the restraining force H
(m) specific moments according to equations (2011), (2012), (2013), or according to equations (2016), (2017), (2018), for a displacement of the horizontal member equal to $\delta = h_k/\mathfrak{T}_k$
U conversion factor according to equation (2022)

Furthermore, in the third form of equation (2023):

$\left(\dfrac{m}{p}\right)$ specific moments for a concentrated horizontal load $W = 1/h_k$ or $Wh_k = 1$,
$(H\, h_k)$ the h_k-fold restraining force (shear moment)

Equation (2023) is also valid for the superposition of forces (V_n, H_n, Q_n) if these are substituted in place of the moments (M, m).

(b) Two-story multi-bay frames

For the rigorous calculation of laterally unrestrained two-story multi-bay frames it is necessary to write down two displacement equations. As the two-story framed structures treated in this first volume do not actually possess an upper horizontal member, the condition as to simultaneous freedom of lateral displacement for the actual horizontal member A-N and the imaginary horizontal member a'-n' has been dispensed with. Only in the case of those frames having a single column line and in the case of the symmetrical frames having two column lines (Frame Shapes 45v, 48v, 49v and 50v) does the said displacement condition fully apply.

2. Validity restrictions for oblique members

In the case of the structure represented in Fig. 445, with vertical columns and horizontal beams, the entire three-span continuous horizontal member will undergo a horizontal displacement δ or, in other words, each of the four column tops will undergo the same displacement δ. The only geometrical manifestation will be that the columns undergo rotations ψ_n, which will in general differ in magnitude, since the height h_n is not necessarily the same for all the columns of the structure. On the other hand, no such rotations will occur in the case of the individual beam spans.

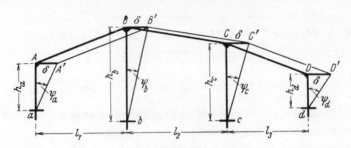

Fig. 447

Horizontal displacement δ of a multi-bay framework with a "broken" continuous connecting member and vertical columns

Fig. 447 represents a three-bay framed structure in which the columns are vertical but in which the various beam spans of the continuous connecting member are sloping. Now, if one or other of the four joints of this member is given a horizontal displacement δ (again assuming the structure to be a pin-jointed "kinematic" assembly), it will readily be seen that, for geometrical reasons, the three other joints must each undergo a horizontal displacement of the same magnitude δ, because the continuous connecting member over the columns moves as a single whole. Since the individual beam spans are merely displaced parallel to themselves (i.e., they undergo translatory displacement), they do not display rotations of the type ψ in this case either.

Hence the validity of the expressions derived in Section 1 for laterally unrestrained multi-bay frames can be summarized in the following general terms: The said derivation, and the computation schemes given for the laterally unrestrained frames (Frame Shapes denoted by subscript "v") in the main part of this book, are subject only to the requirement that all the columns of the structure must be parallel to one another. Neither the column tops nor the column bases need all be located at the same level, but may be located at arbitrary different levels in relation to one another.

In the case of two-story frames the above requirement must obviously apply to the columns of each story in itself, that is to say, the column lines need not be straight, but may be "broken" at the level of the continuous connecting member (which may itself have a "broken" shape), provided that all the columns of one and the same story are parallel.

3. Internal loads

The calculation procedure for temperature changes and displacements of bearings in the case of laterally unrestrained frames is, in principle, similar to that for external loads, i.e., the same three "steps" have to be carried out in the same sequence as, for instance, indicated for "Frame Shape 1v".

4. Influence lines and influence line equations

Just as moments (and forces) for the laterally unrestrained frame can be obtained by superposition in accordance with equation (2023), influence lines can similarly be obtained for a frame of this kind. The relevant superposition equation—likewise in three different forms—is as follows:

$$\eta^v = \eta + \eta^H \cdot M^{(W=1)} = \eta + \eta^H \cdot \left(m \cdot \frac{h_k}{p}\right) = \eta + (h_k \cdot \eta^H) \cdot \left(\frac{m}{p}\right). \qquad (2024)$$

Where the symbols have the following general meanings:

η^v influence line equation for any particular statical quantity in the laterally unrestrained frame

η influence line equation for the same statical quantity in the laterally restrained frame

η^H influence line equation for the restraining force H of the laterally restrained frame (H line)

$M^{(W=1)}$ moment (or force) at point for which influence line is required, due to a concentrated horizontal load $W = 1$ acting at level of horizontal member

Furthermore, in the second form of equation (2024):

$\left(m \cdot \dfrac{h_k}{p}\right)$ equivalent for $M^{(W=1)}$ with the factors

m specific moment as in equation (2023), at point for which influence line is required

$\left(\dfrac{h_k}{p}\right)$ conversion factor U according to equation (2022), but with $W = 1$

Finally, in the third form of equation (2024):

$(h_k \cdot \eta^H)$ the h_k-fold influence line

$\left(\dfrac{m}{p}\right)$ specific moment (or specific force) as in equation (2023), at point for which influence line is required

In terms of diagrams, the superposition equation (2024) can be explained as follows. The ordinate η represents each of the influence line diagrams Fig. 429 – 437, i.e., the M_{A1}, M_{B1}, M_{B2} line, the M_{Z1} line, the M_{Aa}, M_{Bb} line, the Q_{Z1}, Q_{Z2} line, and the V_a, V_b line. Each of these influence line diagrams must now be superimposed upon—that is to say: added to—the $(M^{(W=1)} = m.h_k/p)$-fold H line represented in Fig. 440. The algebraic sign of the relevant "factor" $(M^{(W=1)} = m.h_k/p)$ can be obtained from Fig. 446 for each of the statical quantities concerned.

Example: M_{A1}^v line

Moment factor of the H line according to formula (2024), represented by the symbol μ for convenience:

$$\mu = M_{A1}^{(W=1)} = m_{A1} \cdot \frac{h_k}{p}.$$

Spans:
$$\begin{aligned}\eta_1^v &= \eta_1 + \mu \cdot \eta_1^H = (-e'_{11} + \mu \cdot f'_1) \cdot \omega'_D + (e_{11} - \mu \cdot f_1) \cdot \omega_D \\ \eta_2^v &= \eta_2 + \mu \cdot \eta_2^H = (+e'_{12} + \mu \cdot f'_2) \cdot \omega'_D - (e_{12} + \mu \cdot f_2) \cdot \omega_D \\ \eta_3^v &= \eta_3 + \mu \cdot \eta_3^H = (-e'_{13} + \mu \cdot f'_3) \cdot \omega'_D + (e_{13} - \mu \cdot f_3) \cdot \omega_D;\end{aligned} \qquad (2025)$$

where the values η according to equations (1952) and (1985) have been substituted so as to obtain the representation involving omega functions.

Cantilevers: $\qquad \eta_A^v = \eta_A + \mu \cdot \eta_A^H \qquad \eta_D^v = \eta_D + \mu \cdot \eta_D^H.$

Columns (directly in the "omega" form):

$$\left.\begin{aligned}\eta_a^v &= +\mu \cdot \xi + (\varepsilon_a \cdot e_{1a} - \mu \cdot f'_a) \cdot \omega'_D + (\mu \cdot f_a - e_{1a}) \cdot \omega_D \\ \eta_b^v &= +\mu \cdot \xi + (\varepsilon_b \cdot e_{1b} - \mu \cdot f'_b) \cdot \omega'_D + (\mu \cdot f_b - e_{1b}) \cdot \omega_D \\ \eta_c^v &= +\mu \cdot \xi - (\varepsilon_c \cdot e_{1c} + \mu \cdot f'_c) \cdot \omega'_D + (\mu \cdot f_c + e_{1c}) \cdot \omega_D \\ \eta_d^v &= +\mu \cdot \xi + (\varepsilon_d \cdot e_{1d} - \mu \cdot f'_d) \cdot \omega'_D + (\mu \cdot f_d - e_{1d}) \cdot \omega_D;\end{aligned}\right\} \quad (2026)$$

where the values η according to equations (1953) and (1986) have been substituted so as to obtain the representation involving omega functions.

Example: V_b line

Force factor for the H line: $\quad \mu = V_b^{(W=1)} = v_b \cdot \dfrac{h_k}{p}$.

Spans:

$$\begin{aligned}
\eta_1^v &= \eta_1 + \mu \cdot \eta_1^H = +\xi \;\; + (-\beta_1' + \mu \cdot f_1') \cdot \omega_D' + (\beta_1 - \mu \cdot f_1) \cdot \omega_D \\
\eta_2^v &= \eta_2 + \mu \cdot \eta_2^H = +\xi' + (+\beta_2' + \mu \cdot f_2') \cdot \omega_D' - (\beta_2 + \mu \cdot f_2) \cdot \omega_D \qquad (2027)\\
\eta_3^v &= \eta_3 + \mu \cdot \eta_3^H = \quad\;\;\; (-\beta_3' + \mu \cdot f_3') \cdot \omega_D' + (\beta_3 - \mu \cdot f_3) \cdot \omega_D;
\end{aligned}$$

where the values η according to equations (1969) and (1985) have been substituted.

Cantilevers: $\quad \eta_A^v = \eta_A + \mu \cdot \eta_A^H \qquad \eta_D^v = \eta_D + \mu \cdot \eta_D^H.$

It is not considered necessary, for the purpose of the present discussion, to give diagrams of influence lines for laterally unrestrained framed structures. Whereas in the case of laterally restrained frames the influence lines can always be correctly represented in respect of shape and algebraic sign, this is not generally possible in the case of laterally unrestrained frames. It will readily be seen that, on superimposing the diagrams in Fig. 429 – 437 upon the diagram for the H line as represented in Fig. 440, changes of sign of the ordinates and the associated zero values of the ordinates are liable to occur within the spans. How and where this occurs will depend on the stiffness ratios of all the structural members relative to one another.

5. Numerical Example 2

(a) Preliminary note

In this example a single-story three-bay framework corresponding to "Frame Shape 1v" will be considered. The dimensions, moments of inertia and values ϵ for this structure are represented in Fig. 448. A comparison with the foregoing Example 1 (see Fig. 441) shows that in the present case the only difference is that the moments of inertia J_n of the columns have been given somewhat larger values, thus compensating, as it were, for the absence of the upper-story columns. The sole reason for this modification is to enable the calculations for the "first step" of the present example to be taken direct from Example 1, for the sake of convenience.

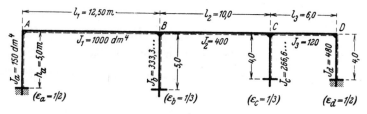

Fig. 448

System corresponding to "Frame Shape 1v" chosen as numerical example

(b) First step (laterally restrained frame)

See formulas (1)–(35). The flexibility coefficients of all the individual members, for $J_k = J_2 = 400$ ft^4 and $l_k = l_2 = 10.0$ ft, are according to equation (1):

$$k_1 = 1/2, \quad k_2 = 1, \quad k_3 = 2,$$

$$k_a = \frac{400}{150} \cdot \frac{5,0}{10,0} = \frac{4}{3}, \qquad k_b = \frac{400}{1000/3} \cdot \frac{5,0}{10,0} = \frac{3}{5},$$

$$k_c = \frac{400}{800/3} \cdot \frac{4,0}{10,0} = \frac{3}{5}, \qquad k_d = \frac{400}{480} \cdot \frac{4,0}{10,0} = \frac{1}{3}.$$

Flexibilities of the columns, according to equation (2):

$$b_a = \frac{4}{3} \cdot \frac{3}{2} = 2, \quad b_b = b_c = \frac{3}{5} \cdot \frac{5}{3} = 1, \quad b_d = \frac{1}{3} \cdot \frac{3}{2} = \frac{1}{2}.$$

Note: The values b_n for the single-story "Frame Shape 1" correspond to the values s_N of the two-story "Frame Shape 37". In the present case the values b_n of Example 2 are numerically equal to the values s_N of Example 1. Hence the further calculations as far as the coefficients n_{ik} must be the same. The matrix of influence coefficients of Example 1, as calculated according to the array (1226), can therefore be adopted unchanged for Example 2 [in this connection see also the array (9)]. The support moments X_i of the loading conditions 1, 2 and 3 likewise remain unchanged (for the same design loads!), whereas the column moments must of course have different numerical values. In the following, only the results for the three load examples considered in Example 1 will be given and the restraining forces calculated.

Load Example 1: Vertical point load $P_1 = 4963 \times 5/6$ at the left-hand fifth point of l_1.

Beam moments (as in Example 1):

$$X_1 = -2078 \quad X_3 = -918 \quad X_5 = +81$$
$$X_2 = -2421 \quad X_4 = +333 \quad X_6 = -36.$$

Column moments, according to equations (13) and (14):

$$M_{Aa} = -2078 \quad M_{Bb} = +1503 \quad M_{Cc} = -252 \quad M_{Dd} = +36;$$
$$\underline{M_a = +1039} \quad \underline{M_b = -501} \quad \underline{M_c = +84} \quad \underline{M_d = -18};$$
$$H_a h_a = -3117 \quad H_b h_b = +2004 \quad H_c h_c = -336 \quad H_d h_d = +54;$$
$$H_a = -623,4 \quad H_b = +400,8 \quad H_c = -84,0 \quad H_d = +13,5.$$

The column shears H_n are obtained from equation (36).

Restraining force at horizontal member, according to equation (37):

$$H = +623,4 - 400,8 + 84,0 - 13,5 = \underline{+293,1}. \qquad (2028)$$

Load Example 2: Horizontal point load $P_c = 4963 \times 5/3$ acting at the centre of the column h_c.

Load terms (as in Example 1):
$$\mathfrak{L}_c = \mathfrak{R}_c = 4963 \cdot 5/2; \qquad M^0_{Pc} = 4963 \cdot 5/3.$$

Joint load term, according to equation (16):
$$\mathfrak{B}_C = -\mathfrak{L}_c \cdot k_c (1-\varepsilon_c) = -\frac{4963 \cdot 5}{2} \cdot \frac{3}{5} \cdot \frac{2}{3} = -4963 \quad \text{(as before)}$$

Beam moments (as in Example 1):
$$X_1 = +56 \qquad X_3 = -644 \qquad X_5 = -819$$
$$X_2 = -336 \qquad X_4 = +1596 \qquad X_6 = +364.$$

Column moments, according to equations (18) and (19):

$M_{Aa} = +56 \qquad M_{Bb} = -308 \qquad M_{Cc} = -2415 \qquad M_{Dd} = -364;$
$M_a \;\;= -28 \qquad M_b \;\;= +102{,}6\ldots \qquad\qquad\qquad\qquad M_d \;\;= +182;$
$H_a h_a = +84 \qquad H_b h_b = -410{,}6\ldots \qquad\qquad\qquad\qquad H_d h_d = -546;$
$M_c = -(4963 \cdot 5/2 - 2415) \cdot 1/3 = -3330{,}83\ldots$

Shears at column tops, according to equations (36) and (38):
$Q_{ra} = H_a = +16{,}8 \qquad Q_{rb} = H_b = -82{,}13\ldots \qquad Q_{rd} = H_d = -136{,}5$
$Q_{rc} = -4963 \cdot 5/6 - 2415/4{,}0 + 3330{,}83\ldots/4{,}0 = -3906{,}875.$

Restraining force at horizontal member, according to equation (39):
$$H = -16{,}8 + 82{,}13\ldots + 3906{,}875 + 136{,}5 = \underline{+4108{,}7083}\ldots \quad (2029)$$

Load Example 3: Rotational moment $D_B = +4963$ acting at joint B:

Beam moments (as in Example 1):
$$X_1 = +334 \qquad X_3 = +1122 \qquad X_5 = -99$$
$$X_2 = -2004 \qquad X_4 = -407 \qquad X_6 = +44.$$

Column moments, according to equations (18) and (19):

$M_{Aa} = +334 \qquad M_{Bb} = -1837 \qquad M_{Cc} = +308 \qquad M_{Dd} = -44;$
$M_a \;\;= -167 \qquad M_b \;\;= +612{,}3\ldots \qquad M_c \;\;= -102{,}6\ldots \qquad M_d \;\;= +22;$
$H_a h_a = +501 \qquad H_b h_b = -2449{,}3\ldots \qquad H_c h_c = +410{,}6\ldots \qquad H_d h_d = -66;$
$H_a = +100{,}2 \qquad H_b = -489{,}86\ldots \qquad H_c = +102{,}6\ldots \qquad H_d = -16{,}5.$

Restraining force at horizontal member, according to equation (37):
$$H = -100{,}2 + 489{,}86\ldots - 102{,}6\ldots + 16{,}5 = \underline{+303{,}5}. \qquad (2030)$$

(c) Second step (effect of displacement of horizontal member)

See the formulas (40) – (45).

Column ratios, for $h_k = h_3 = h_4 = 4.0$ ft, according to equation (40):
$$\gamma_a = \gamma_b = 4{,}0/5{,}0 = 4/5 = 0{,}8, \qquad \gamma_c = \gamma_d = 1.$$

Specific load terms, according to equation (41):

$$\mathfrak{B}_A = \frac{4}{5} \cdot \frac{3}{2} = \frac{6}{5}, \quad \mathfrak{B}_B = \frac{4}{5} \cdot \frac{4}{3} = \frac{16}{15}, \quad \mathfrak{B}_C = 1 \cdot \frac{4}{3} = \frac{4}{3}, \quad \mathfrak{B}_D = 1 \cdot \frac{3}{2} = \frac{3}{2};$$

$$\mathfrak{L}_a = \frac{4}{5} \cdot \frac{3}{4} = \frac{3}{5}, \quad \mathfrak{L}_b = \frac{4}{5} \cdot \frac{5}{3} = \frac{4}{3}, \quad \mathfrak{L}_c = 1 \cdot \frac{5}{3} = \frac{5}{3}, \quad \mathfrak{L}_d = 1 \cdot 3 = 3.$$

Specific beam moments, according to array (17) and with reference to array (2011); influence coefficients as in array (2000) (the values $\mathfrak{B}_N$ are all reduced to the common denominator 15):

$\mathfrak{B}_N =$	$\frac{18}{15}$	$\frac{16}{15}$	$\frac{20}{15}$	$\frac{22{,}5}{15}$	
$m_{A1} =$	$+1744$	$+334$	-56	$+8$	(all values to be divided by 4963)
$m_{B1} =$	-538	-2004	$+336$	-48	
$m_{B2} =$	-204	$+1122$	$+644$	-92	
$m_{C2} =$	$+74$	-407	-1596	$+228$	
$m_{C3} =$	$+18$	-99	$+819$	$+592$	
$m_{D3} =$	-8	$+44$	-364	-1366	

On multiplying by the numerators of the values $\mathfrak{B}_N$:

$$\left.\begin{array}{l}
m_{A1} = +31392 \quad +5344 \quad -1120 \quad +180 = +35796 \\
m_{B1} = -9684 \quad -32064 \quad +6720 \quad -1080 = -36108 \\
m_{B2} = -3672 \quad +17952 \quad +12880 \quad -2070 = +25090 \\
m_{C2} = +1332 \quad -6512 \quad -31920 \quad +5130 = -31970 \\
m_{C3} = +324 \quad -1584 \quad +16380 \quad +13320 = +28440 \\
m_{D3} = -144 \quad +704 \quad -7280 \quad -30735 = -37455
\end{array}\right\} \text{(to be divided by } 15 \times 4963\text{)} \quad (2031)$$

Specific column moments, according to equations (18) and (19) and with reference to equations (2012) and (2013):

Column	a	b	c	d	
m_{Nn}	$+35796$	$+61198$	$+60410$	$+37455$	$: (15 \cdot 4963).$
$\mathfrak{L}_n$	44667	99260	124075	223335	
	80463	160458	184485	260790	(2032)
m_n	$-40231{,}5$	-53486	-61495	-130395	

The above calculation relates to Method I. Alternatively, the specific moments could have been calculated according to Method II as follows:

Stiffness coefficients of the horizontal members, according to equations (20), $K_\nu = 1/k_\nu$:

$$K_1 = 2 \quad K_2 = 1 \quad K_3 = 1/2.$$

Stiffnesses of the columns, according to equation (21), $S_n = 1/b_n$:

$$S_a = 1/2 \qquad S_b = S_c = 1 \qquad S_d = 2.$$

Specific joint load terms, according to equation (42):

$$\mathfrak{K}_A = \frac{1}{2} \cdot \frac{6}{5} = \frac{3}{5}, \quad \mathfrak{K}_B = 1 \cdot \frac{16}{15} = \frac{16}{15}, \quad \mathfrak{K}_C = 1 \cdot \frac{4}{3} = \frac{4}{3}, \quad \mathfrak{K}_D = 2 \cdot \frac{3}{2} = 3.$$

Specific auxiliary moments, according to array (31) and with reference to array (2015); influence coefficients as in array (2004) (the values $\mathfrak{K}_N$ are all reduced to the common denominator 15):

$\mathfrak{K}_N =$	$\frac{9}{15}$	$\frac{16}{15}$	$\frac{20}{15}$	$\frac{45}{15}$	
$y_A =$	$+2950$	-668	$+112$	-8	
$y_B =$	-668	$+1837$	-308	$+22$	(all values to be
$y_C =$	$+112$	-308	$+2548$	-182	divided by 3×4963)
$y_D =$	-8	$+22$	-182	$+2140$	

On multiplying by the numerators of the values $\mathfrak{K}_N$:

$$y_A = +26550 \quad -10688 \quad +2240 \quad -360 = 17742 = 5914 \cdot 3$$
$$y_B = -6012 \quad +29392 \quad -6160 \quad +990 = 18210 = 6070 \cdot 3$$
$$y_C = +1008 \quad -4928 \quad +50960 \quad -8190 = 38850 = 12950 \cdot 3$$
$$y_D = -72 \quad +352 \quad -3640 \quad +96300 = 92940 = 30980 \cdot 3$$

(all values to be divided by $15 \times 3 \times 4963$)

Specific beam moments, according to equations (32) and with reference to equations (2016):

$$m_{A1} = (11828 + 6070) \cdot 2 = 35796 \qquad -m_{B1} = (5914 + 12140) \cdot 2 = 36108$$
$$m_{B2} = (12140 + 12950) \cdot 1 = 25090 \qquad -m_{C2} = (6070 + 25900) \cdot 1 = 31970$$
$$m_{C3} = (25900 + 30980) : 2 = 28440 \qquad -m_{D3} = (12950 + 61960) : 2 = 37455.$$

(divide by 15×4963)   (divide by 15×4963)

Specific column top moments, according to equation (33) and with reference to equations (2017):

Column	a	b	c	d	
$\mathfrak{C}_{rn} = \mathfrak{K}_N$	$+44667$	$+79408$	$+99260$	$+223335$	$: (15 \cdot 4963).$
$-y_N \cdot 3 S_n$	-8871	-18210	-38850	-185880	
$m_{Nn} =$	$+35796$	$+61198$	$+60410$	$37455.$	

The calculation of the specific column base moments is the same by Method I and Method II: in this connection see equation (19) or (34) or, alternatively, (2013) or (2018).

The above calculations give exact numerical agreement of all the values m as determined by Method I and Method II.

Specific h_k-fold column shears, according to equation (43) or (2020):

Column	a	b	c	d	
m_{Nn}	35796,0	61198	60410	37455	$:(15\cdot 4963)$
$-m_n$	40231,5	53486	61495	130395	
	76027,5	114684	121905	167850	
$p_n =$	60822,0	91747,2	121905	167850	

Corresponding specific h_k-fold applied load, according to equation (44) or (2021):

$$p = \frac{\Sigma p_n}{15\cdot 4963} = \frac{442324{,}2}{15\cdot 4963} = \frac{29488{,}28}{4963} \approx 5{,}9416.$$

Conversion factor, according to equation (45) or (2022):

$$U = \frac{W\cdot h_k}{p} = \frac{W\cdot 4{,}0\cdot 4963}{29488{,}28} = \frac{4963}{7372{,}07}\cdot W. \qquad (2033)$$

(d) Third step (superposition)

For example, the second form of the general superposition equation (2023) is as follows:

$$M^v = M + m\cdot U. \qquad (2034)$$

For the three load examples chosen, the numerical values of M are given in Section (b) and those of m are given in Section (c). In the present particular case, in which the computations have been carried out with exact numerical values, all the values m have been written in the form $z/(15 \times 4963)$; hence in equation (2034) we have in general:

$$m\cdot U = \frac{z}{15\cdot 4963}\cdot \frac{4963\,W}{7372{,}07} = \frac{W\cdot z}{110581{,}05}. \qquad (2035)$$

For determining the moments M^v of the laterally unrestrained frame according to equation (2034) and having regard to equation (2035), rigorous exactness of the computations will be abandoned for reasons which will be readily understood. Rounding the values to whole units is sufficiently accurate for the purpose.

Load Example 1

According to (2028) we have: $W = H = 293{,}1$.

$$m\cdot U \approx z\cdot 293{,}1/110581 \approx 0{,}002651\cdot z.$$

Beam moments, according to formula (2034):

$$M^v_{A1} = -2078 + 0{,}002651 \cdot 35796 \approx -2078 + 95 \approx -1983$$
$$M^v_{B1} = -2421 - 0{,}002651 \cdot 36108 \approx -2421 - 96 \approx -2517;$$

similarly:

$$M^v_{B2} \approx -918 + 66 \approx -852 \qquad M^v_{C3} \approx +81 + 75 \approx +156$$
$$M^v_{C2} \approx +333 - 85 \approx +248 \qquad M^v_{D3} \approx -36 - 99 \approx -135.$$

Column moments, according to formula (2034):

Column	a	b	c	d
M_{Nn}	−2078	+1503	−252	+ 36
$m_{Nn} \cdot U$	+ 95	+ 162	+160	+ 99
$M^v_{Nn} \approx$	−1983	+1665	− 92	+135;
M_n	+1039	− 501	+ 84	− 18
$m_n \cdot U$	− 107	− 142	−163	−346
$M^v_n \approx$	+ 932	− 643	− 79	−364.

Check with $\sum H^v_n h_k = \sum (M^v_{Nn} - M^v_n)\gamma_n = 0$:

$(-1983 - 932)\,0{,}8 + (1665 + 643)\,0{,}8 + (-92 + 79) + (135 + 364) \approx 0$.

Load Example 2

According to (2029) we have: $W = H \approx 4108{,}7$.

$$m \cdot U \approx z \cdot 4108{,}7/110581 \approx 0{,}037156 \cdot z.$$

Beam moments, according to formula (2034):

$$M^v_{A1} \approx + 56 + 1330 \approx +1386 \qquad M^v_{B1} \approx - 336 - 1342 \approx -1678$$
$$M^v_{B2} \approx -644 + 932 \approx + 288 \qquad M^v_{C2} \approx +1596 - 1188 \approx + 408$$
$$M^v_{C3} \approx -819 + 1057 \approx + 238 \qquad M^v_{D3} \approx + 364 - 1392 \approx -1028.$$

Column moments, according to formula (2034):

Column	a	b	c	d
M_{Nn}	+ 56	− 308	−2415	− 364
$m_{Nn} \cdot U$	+1330	+2274	+2245	+1392
$M^v_{Nn} \approx$	+1386	+1966	− 170	+1028;
M_n	− 28	+ 103	−3331	+ 182
$m_n \cdot U$	−1495	−1988	−2285	−4845
$M^v_n \approx$	−1523	−1885	−5616	−4663.

Check with $\sum (M_{Nn}^v - M_n^v)\gamma_n - Ph_k/2 = 0$:

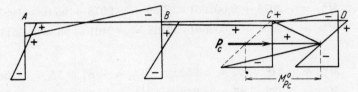

Fig. 449

Bending moment diagram for "Load Example 2" in respect of the laterally unrestrained multi-bay frame

Load Example 3

According to (2030) we have $W = H = 303{,}5$.
$$m \cdot U \approx 303{,}5/110581 \approx 0{,}002745 \cdot z.$$

Beam moments, according to formula (2034):

$M_{A1}^v \approx +\ 334 + 98 \approx +\ 432 \qquad M_{B1}^v \approx -2004 -\ 99 \approx -2103$

$M_{B2}^v \approx +1122 + 69 \approx +1191 \qquad M_{C2}^v \approx -\ 407 -\ 88 \approx -\ 495$

$M_{C3}^v \approx -\ 99 + 78 \approx -\ 21; \qquad M_{D3}^v \approx +\ 44 - 103 \approx -\ 59.$

Column moments, according to formula (2034):

Column	a	b	c	d
M_{Nn}	+334	−1837	+308	−44
$m_{Nn}\cdot U$	+98	+168	+166	+103
$M_{Nn}^v \approx$	+432	−1669	+474	+59;
M_n	−167	+612	−103	+22
$m_n \cdot U$	−110	−147	−169	−358
$M_n^v \approx$	−277	+465	−272	−336.

Check with $\sum H_n^v h_k = 0$:

$(432 + 277)\,0{,}8 - (1669 + 465)\,0{,}8 + (474 + 272) + (59 + 336) \approx 0.$

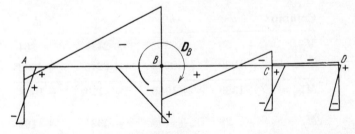

Fig. 450

Bending moment diagram for "Load Example 3" in respect of the laterally unrestrained multi-bay frame

(e) Final note

As the values b_n of the frame envisaged in Example 1 have been so chosen as to be in exact agreement with the values s_N of the frame envisaged in Example 2, the bending moment diagrams for the load examples 1, 2 and 3 of the laterally restrained and of the laterally unrestrained single-story frame can be directly compared with one another. The greatest difference—even involving some changes in algebraic sign—is found to occur between Fig. 443 and 449, i.e., for horizontal loading.